FORSYTH LIBRARY - FHSU
510H236
main Handbook of

2 1765 0004 3292 5

D1274307

HANDBOOK OF APPLICABLE MATHEMATICS

Volume III: Numerical Methods

HANDBOOK OF APPLICABLE MATHEMATICS

Chief Editor: Walter Ledermann

Editorial Board: Robert F. Churchhouse
Harvey Cohn
Peter Hilton
Emlyn Lloyd
Steven Vajda

Assistant Editor: Carol Jenkins

Volume I: ALGEBRA
Edited by Walter Ledermann, *University of Sussex*
and Steven Vajda, *University of Sussex*

Volume II: PROBABILITY
Emlyn Lloyd, *University of Lancaster*

Volume III: NUMERICAL METHODS
Edited by Robert F. Churchhouse, *University College Cardiff*

Volume IV: ANALYSIS
Edited by Walter Ledermann, *University of Sussex*
and Steven Vajda, *University of Sussex*

Volume V: GEOMETRY AND COMBINATORICS
Edited by Walter Ledermann, *University of Sussex*
and Steven Vajda, *University of Sussex*

Volume VI: STATISTICS
Edited by Emlyn Lloyd, *University of Lancaster*

HANDBOOK OF

APPLICABLE MATHEMATICS

Chief Editor: Walter Ledermann

Volume III: Numerical Methods

Edited by

Robert F. Churchhouse
University College Cardiff

A Wiley–Interscience Publication

JOHN WILEY & SONS

Chichester – New York – Brisbane – Toronto

Copyright © 1981 by John Wiley & Sons Ltd.

All rights reserved.

No part of this book may be reproduced by any means, nor transmitted, nor translated into a machine language without the written permission of the publisher.

British Library Cataloguing in Publication Data:

Handbook of applicable mathematics.
  Vol. 3: Numerical methods
  1. Mathematics
  I. Churchhouse, Robert F.
  510   QA36   79-42724
  ISBN 0 471 27947 1

Typeset by J. W. Arrowsmith Ltd, Bristol BS3 2NT, England and printed in the United States of America by Vail-Ballou Press, Inc., Binghamton, N.Y.

Contributing Authors

R. F. Churchhouse, University College, Cardiff, U.K.

A. M. Cohen, University of Wales Institute of Science & Technology, Cardiff, U.K.

D. E. Jones, University of Wales Institute of Science & Technology, Cardiff, U.K.

W. F. B. Jones, University College, Cardiff, U.K.

I. M. Khabaza, Queen Mary College, London, U.K.

M. H. Rogers, University of Bristol, Bristol, U.K.

P. R. Turner, University of Lancaster, Lancaster, U.K.

Contents

Introduction
to the
Handbook of Applicable Mathematics

Today, more than ever before, mathematics enters the lives of every one of us. Whereas, thirty years ago, it was supposed that mathematics was only needed by somebody planning to work in one of the 'hard' sciences (physics, chemistry), or to become an engineer, a professional statistician, an actuary or an accountant, it is recognized today that there are very few professions in which an understanding of mathematics is irrelevant. In the biological sciences, in the social sciences (especially economics, town planning, psychology), in medicine, mathematical methods of some sophistication are increasingly being used and practitioners in these fields are handicapped if their mathematical background does not include the requisite ideas and skills.

Yet it is a fact that there are many working in these professions who do find themselves at a disadvantage in trying to understand technical articles employing mathematical formulations, and who cannot perhaps fulfil their own potential as professionals, and advance in their professions at the rate that their talent would merit, for want of this basic understanding. Such people are rarely in a position to resume their formal education, and the study of some of the available textbooks may, at best, serve to give them some acquaintance with mathematical techniques, of a more or less formal nature, appropriate to current technology. Among such people, academic workers in disciplines which are coming increasingly to depend on mathematics constitute a very significant and important group.

Some years ago, the Editors of the present Handbook, all of them actively concerned with the teaching of mathematics with a view to its usefulness for today's and tomorrow's citizens, got together to discuss the problems faced by mature people already embarked on careers in professions which were taking on an increasingly mathematical aspect. To be sure, the discussion ranged more widely than that—the problem of 'mathematics avoidance' or 'mathematics anxiety', as it is often called today, is one of the most serious problems of modern civilization and affects, in principle, the entire community—but it was decided to concentrate on the problem as it affected professional effectiveness. There emerged from those discussions a novel format for presenting mathematics to this very specific audience. The intervening years have been spent in putting this novel conception into practice, and the result is the Handbook of Applicable Mathematics.

THE PLAN OF THE HANDBOOK

The 'Handbook' consists of two sets of books. On the one hand, there are (or will be!) a number of *guide books*, written by experts in various fields in which mathematics is used (e.g. medicine, sociology, management, economics). These guide books are by no means comprehensive treatises; each is intended to treat a small number of particular topics within the field, employing, where appropriate, mathematical formulations and mathematical reasoning. In fact, a typical guide book will consist of a discussion of a particular problem, or related set of problems, and will show how the use of mathematical models serves to solve the problem. Wherever any mathematics is used in a guide book, it is cross-referenced to an article (or articles) in the *core volumes*.

There are 6 core volumes devoted respectively to Algebra, Probability, Numerical Methods, Analysis, Geometry and Combinatorics and Statistics. These volumes are texts of mathematics—but they are no ordinary mathematical texts. They have been designed specifically for the needs of the professional adult (though we believe they should be suitable for any intelligent adult!) and they stand or fall by their success in explaining the nature and importance of key mathematical ideas to those who need to grasp and to use those ideas. Either through their reading of a guide book or through their own work or outside reading, professional adults will find themselves needing to understand a particular mathematical idea (e.g. linear programming, statistical robustness, vector product, probability density, round-off error); and they will then be able to turn to the appropriate article in the core volume in question and *find out just what they want to know*—this, at any rate, is our hope and our intention.

How then do the content and style of the core volumes differ from a standard mathematical text? First, the articles are designed to be read by somebody who has been referred to a particular mathematical topic and would prefer not to have to do a great deal of preparatory reading; thus each article is, to the greatest extent possible, self-contained (though, of course, there is considerable cross-referencing within the set of core volumes). Second, the articles are designed to be read by somebody who wants to get hold of the mathematical ideas and who does not want to be submerged in difficult details of mathematical proof. Each article is followed by a bibliography indicating where the unusually assiduous reader can acquire that sort of 'study in depth'. Third, the topics in the core volumes have been chosen for their relevance to a number of different fields of application, so that the treatment of those topics is not biased in favour of a particular application. Our thought is that the reader—unlike the typical college student!—will already be motivated, through some particular problem or the study of some particular new technique, to acquire the necessary mathematical knowledge. Fourth, this is a handbook, not an encyclopedia—if we do not think that a particular aspect of a mathematical topic is likely to be useful or interesting to the kind of reader we have in mind, we have omitted it. We have not set out to include everything known on a particular topic, and we are *not* catering for the professional mathematician! The Handbook has been written as a contribution to the practice of mathematics, not to the theory.

The reader will readily appreciate that such a novel departure from standard textbook writing—this is neither 'pure' mathematics nor 'applied' mathematics as traditionally interpreted—was not easily achieved. Even after the basic concept of the Handbook had been formulated by the Editors, and the complicated system of cross-referencing had been developed, there was a very serious problem of finding authors who would write the sort of material we wanted. This is by no means the way in which mathematicians and experts in mathematical applications are used to writing. Thus we do not apologize for the fact that the Handbook has lain so long in the womb; we were trying to do something new and we had to try, to the best of our ability, to get it right. We are sure we have not been uniformly successful; but we can at least comfort ourselves that the result would have been much worse, and far less suitable for those whose needs we are trying to meet, had we been more hasty and less conscientious.

It is, however, not only our task which has not been easy. Mathematics itself is not easy! The reader is not to suppose that, even with his or her strong motivation and the best endeavours of the editors and authors, the mathematical material contained in the core volumes can be grasped without considerable effort. Were mathematics an elementary affair, it would not provide the key to so many problems of science, technology and human affairs. It is universal, in the sense that significant mathematical ideas and mathematical results are relevant to very different 'concrete' applications—a single algorithm serves to enable the travelling salesman to design his itinerary, and the refrigerator manufacturing company to plan a sequence of modifications of a given model; and could conceivably enable an intelligence unit to improve its techniques for decoding the secret messages of a foreign power. Given this universality, mathematics cannot be trivial! And, if it is not trivial, then some parts of mathematics are bound to be substantially more difficult than others.

This difference in level of difficulty has been faced squarely in the Handbook. The reader should not be surprised that certain articles require a great deal of effort for their comprehension and may well involve much study of related material provided in other referenced articles in the core volumes—while other articles can be digested almost effortlessly. In any case, different readers will approach the Handbook from different levels of mathematical competence and we have been very much concerned to cater for all levels.

THE REFERENCING AND CROSS-REFERENCING SYSTEM

To use the Handbook effectively, the reader will need a clear understanding of our numbering and referencing system, so we will explain it here. Important items in the core volumes or the guidebooks—such as definitions of mathematical terms or statements of key results—are assigned sets of numbers according to the following scheme. There are six categories of such mathematical items, namely:

   (i) Definitions
  (ii) Theorems, Propositions, Lemmas and Corollaries
 (iii) Equations and other Displayed Formulae
  (iv) Examples
   (v) Figures
  (vi) Tables

Items in any one of these six categories carry a triple designation a.b.c. of arabic numerals, where 'a' gives the *chapter* number, 'b' the *section* number, and 'c' the number of the individual *item*. Thus items belong to a given category, for example, definitions are numbered in sequence within a section, but the numbering is independent as between categories. For example, in Section 5 of Chapter 3 (of a given volume), we may find a displayed formula labelled (5.3.7) and also Lemma 5.3.7 followed by Theorem 5.3.8. Even where sections are further divided into *subsections*, our numbering system is as described above, and takes no account of the particular subsection in which the item occurs.

As we have already indicated, a crucial feature of the Handbook is the comprehensive cross-referencing system which enables the reader of any part of any core volume or guide book to find his or her way quickly and easily to the place or places where a particular idea is introduced or discussed in detail. If, for example, reading the core volume on Statistics, the reader finds that the notion of a *matrix* is playing a vital role, and if the reader wishes to refresh his or her understanding of this concept, then it is important that an immediate reference be available to the place in the core volume on Algebra where the notion is first introduced and its basic properties and uses discussed.

Such ready access is achieved by the adoption of the following system. There are six core volumes, enumerated by the Roman numerals as follows:

   I Algebra
  II Probability
 III Numerical Methods
  IV Analysis
   V Geometry and Combinatorics
  VI Statistics

A reference to an item will appear in square brackets and will *typically* consist of a pair of entries [see A, B] where A is the volume number and B is the triple designating the item in that volume to which reference is being made. Thus '[see II, (3.4.5)]' refers to equation (3.4.5) of Volume II (Probability). There are, however, two exceptions to this rule. The first is simply a matter of economy!—if the reference is to an item in the same volume, the volume number designation (A, above) is suppressed; thus '[see Theorem 2.4.6]', appearing in Volume III, refers to Theorem 2.4.6 of Volume III. The second exception is more fundamental and, we contend, wholly natural. It may be that we feel the need to refer to a substantial discussion rather than to a single mathematical item (this could well have been the case in the reference to 'matrix', given as an example above). If we judge that such a comprehensive reference is appropriate, then the second entry B of the reference may carry only two numerals—or even, in an extreme case, only one. Thus the reference '[see I, 2.3]' refers to Section 3 of Chapter 2 of Volume I and recommends the reader to study that entire section to get a complete picture of the idea being presented.

Bibliographies are to be found at the end of each chapter of the core volumes and at the end of each guide book. References to these bibliographies appear in the text as '(Smith (1979))'.

It should perhaps be explained that, while the referencing *within* a chapter of a core volume or *within* a guide book is substantially the responsibility of the author of that part of the text, the cross-referencing has been the responsibility of the editors as a whole. Indeed, it is fair to say that it has been one of their heaviest and most exacting responsibilities. Any defects in putting the referencing principles into practice must be borne by the editors. The successes of the system must be attributed to the excellent and wholehearted work of our invaluable colleague, Carol Jenkins.

CHAPTER 1

Introduction to Numerical Methods

1.1. HISTORY

For as long as men have been able to count they have been faced with problems requiring a numerical answer. Some of these problems could be answered exactly: if a farmer sells 10 cows to another farmer at a price of 5 *whatsits* per cow the total sum changing hands is 50 *whatsits*, irrespective of what the value of a *whatsit* may be. Other problems however could only be answered approximately: the area of an irregularly shaped field, for example.

In the course of time mathematical tools such as geometry, algebra, trigonometry and the calculus were developed which enabled men to solve problems, which had baffled previous generations, by analytical means. Thus although the ancient Greeks knew that the area of a circle of radius R was πR^2 where π was a number approximately equal to $\frac{22}{7}$ they would have been astonished to know that, as anyone who has done enough calculus can easily prove,

$$\frac{\pi}{4} = 1 - \tfrac{1}{3} + \tfrac{1}{5} - \tfrac{1}{7} + \tfrac{1}{9} - \tfrac{1}{11} + \dots \tag{1.1.1}$$

a fact first discovered by Gregory in 1671. This formula, in theory, enables us to calculate the value of π as accurately as we wish but in practice it is not of much use, and other formulae are much to be preferred; but whichever formula we use we *must* carry out a numerical computation involving an appropriate number of terms of an infinite series for it has been proved that π cannot be exactly represented by any *finite* set of fractions combined together by any finite number of additions, subtractions, multiplications, divisions and root-extractions. We see therefore that if we were to use a series such as (1.1.1) to compute π we would inevitably be faced with a question of the type: how many terms of the series (1.1.1) do we need to take if we want our result to have a specified degree of accuracy?

Another problem that occupied mathematicians of ancient times was that of finding formulae enabling them to solve polynomials of various degrees such as

$$ax^2 + bx + c = 0 \tag{1.1.2}$$

1

$$ax^3 + bx^2 + cx + d = 0 \qquad\qquad (1.1.3)$$

$$ax^4 + bx^3 + cx^2 + dx + e = 0 \qquad\qquad (1.1.4)$$

$$ax^5 + bx^4 + cx^3 + dx^2 + ex + f = 0. \qquad\qquad (1.1.5)$$

The solution of the general quadratic (1.1.2) was known by the 10th Century; the solution of the general cubic (1.1.3) was found by Tartaglia in 1535, and the solution of the general quartic (1.1.4) was found by Ferrari in 1545, but in 1823 Abel proved that there could be no formula for solving the general quintic (1.1.5) and *a fortiori* no formula for solving polynomials of degree greater than 5. Since polynomials of degrees higher than 4 frequently arise how then are we to solve them? The answer is simple: we must use numerical methods.

1.2. NUMERICAL METHODS

By a *numerical method* we mean 'a set of rules for solving a problem, or problems of a particular type, involving only the operations of arithmetic'.

This definition is rather vague and, whilst it is pointless trying to refine it too much, a few additional remarks are required. Firstly: 'the operations of arithmetic' strictly refer to the 4 fundamental operations of addition, subtraction, multiplication and division but inclusion of root-extractions is allowed since these can in any case, be ultimately reduced to operations involving only $+$, $-$, $\times$ and $\div$. Secondly: the 'set of rules' might include some kind of test to see whether a certain condition has been satisfied, and if not, the set of rules are applied again; a method incorporating such a condition is called *iterative*. In theory, if the condition is never satisfied, the application of the numerical method might be never-ending; it is therefore essential that an iterative set contain some rule for deciding when the process is to be terminated even though the condition has not been satisfied.

Numerical methods have increased in importance following the widespread use of computers and new methods, particularly suitable for computers, have been introduced. Methods which offer advantages in accuracy but which were unpopular with human beings because of awkward coefficients or lengthy, tedious calculations present no problem to a computer and so have come back into favour. At the same time a method calling for human judgment might be very difficult to program satisfactorily on a computer, and so fall in popularity.

The development of numerical methods and the analysis of their convergence, accuracy and stability forms the subject of *numerical analysis*, a branch of mathematics which has assumed major importance with many active research workers using all available mathematical techniques. A full appreciation of *numerical analysis* requires considerable mathematical knowledge; an appreciation of *numerical methods* on the other hand is accessible to any scientist or engineer who is prepared to make some effort. In this book we describe the most commonly-used numerical methods for solving problems that are frequently encountered. All these methods have been analysed rigorously but to present the rigour would not only lengthen the book considerably but also demand a specialized knowledge that few non-mathematicians possess (or would wish to acquire). Some methods however will only produce the desired solution under certain conditions; it is important to know what these conditions are; if the conditions are easy to prove, we prove them; if not, we state them as a fact—the reader who is sufficiently interested can find the proofs elsewhere.

In the remainder of this chapter we introduce a few basic ideas and results which will be used as a foundation for the rest of the book. In a sense this is the only chapter in the book that it is essential for you to read and understand. Having understood this chapter you should be able to read certain other chapters without much difficulty. There is no need to read through the chapters in order; although they occasionally refer to one another they are, whenever possible, self-contained, apart from the ideas and results of this chapter.

1.3. ERRORS

We cannot use numerical methods and ignore the existence of *errors*. Errors come in a variety of forms, some are avoidable, some are not, and an appreciation of how errors arise and of how they affect the accuracy of a calculation is an essential preliminary to an understanding of numerical methods.

Suppose that we have two numbers, x and y and that x is the *true value* of some quantity and that y is an *approximation* to x; how do we measure *how good* an approximation y is to x? The simplest measure is obviously provided by taking the *difference* between x and y, and if this is sufficiently small (according to some criterion which we have in mind) we might say that 'y is a good approximation to x' or not. Using this as our basis we then define

$$\text{Error} = (\text{True Value}) - (\text{Approximate Value}).$$

We might equally well have defined Error as (Approximate Value) – (True Value), it makes no difference provided that we are consistent.

EXAMPLE 1.3.1. The value of π is $3\cdot14159265\ldots$. A very commonly used approximation to π is $\frac{22}{7}$; what is the error in this approximation?

Solution. We must convert $\frac{22}{7}$ to decimal form [see I, § 2.5] so that we can find the difference: now

$$\pi = 3\cdot14159265\ldots$$

and

$$\tfrac{22}{7} = \underline{3\cdot14285714}\ldots$$

and so

$$\text{Error} = \underline{-0\cdot00126449}\ldots.$$

Looking again at the numbers in this example we see that all three of them consist of nine digits followed by a line of dots, which indicate that the digits continue but that we are unable or unwilling, to continue writing them down. Such a situation arises whenever we have a number which cannot be represented exactly by a terminating decimal; this is a simple fact of arithmetic and is unavoidable. Whenever we use such numbers in a calculation we must decide how many of the digits we are going to take into account; we are thus led, inevitably, to the concept of *rounding*. Consider again the value of π and suppose that we wish to use it in a calculation; if we are prepared to write 2 digits after the decimal point (say) then we write

$$\pi \doteqdot 3 \cdot 14$$

(the symbol $\doteqdot$ means is approximately equal to) and the error is $+0 \cdot 00159265 \ldots$. If we are prepared to write 3 digits after the decimal point we *might* write

$$\pi \doteqdot 3 \cdot 141$$

the error being $+0 \cdot 00059265 \ldots$ but it is better to write instead

$$\pi \doteqdot 3 \cdot 142$$

since the error now is $-0 \cdot 00040734 \ldots$ and this is smaller, in absolute value, then $0 \cdot 00059265 \ldots$. It is clear that if we want to use only 3 digits after the decimal point then we should use the value $3 \cdot 142$ since the error is thereby minimized; we then say that the value of π *rounded to 3 decimal places* is $3 \cdot 142$.

This simple, but fundamentally important idea can now be defined formally.

DEFINITION 1.3.1.   Let x be any real number and let y be a real number having exactly k digits after its decimal point, then we say that y represents x *rounded to k decimal places* if and only if

$$|x - y| \leq \tfrac{1}{2} \times 10^{-k}.$$

A slight extension now enables us to measure how accurately one number approximates to another:

DEFINITION 1.3.2.   Let x be any real number and y an approximation to x then we say that y is *accurate to k decimal places* if

$$|x - y| \leq \tfrac{1}{2} \times 10^{-k}$$

but

$$|x - y| \geq \tfrac{1}{2} \times 10^{-(k+1)}.$$

EXAMPLE 1.3.2.   To how many decimal places is $\frac{22}{7}$ accurate as an approximation to π?

Solution.   We saw that

$$|\pi - \tfrac{22}{7}| = 0{\cdot}00126449\ldots$$

and

$$\tfrac{1}{2} \times 10^{-3} = 0{\cdot}0005 < 0{\cdot}00126449\ldots < 0{\cdot}005 = \tfrac{1}{2} \times 10^{-2}$$

—so the approximation is accurate to 2 decimal places.

Since the phrase 'decimal places' is frequently used it is usually abbreviated to 'd.p.', and this we shall do henceforward.

EXAMPLE 1.3.3.   Another approximation to π is $\tfrac{355}{113}$. To how many d.p. is it accurate?

Solution

$$\pi = \quad 3{\cdot}14159265\ldots$$
$$\tfrac{355}{113} = \quad 3{\cdot}14159292\ldots$$
$$\overline{\text{Error} = -0{\cdot}00000027}$$

and

$$\tfrac{1}{2} \times 10^{-7} < 0{\cdot}00000027 < \tfrac{1}{2} \times 10^{-6}.$$

The approximation is therefore accurate to 6 d.p.

Whenever we carry out a calculation involving numbers which we have rounded we necessarily introduce errors into the calculation and these are known as *rounding errors*. A rounding error may be positive or negative: when we round π to 2 d.p. we have a rounding error of $\pi - 3{\cdot}14 = +0{\cdot}00159\ldots$ —a positive rounding error; when we round π to 3 d.p. we have $\pi - 3{\cdot}142 = -0{\cdot}000407\ldots$ —a negative rounding error. In the course of a long calculation we may expect to introduce many rounding errors some positive, some negative and to combine these in complex ways. At the end of the calculation a *cumulative rounding error* will have been generated. We cannot know the magnitude of this error with any precision although we may be able to put a maximum bound on it, but in practice such a bound is likely to be wildly pessimistic since in theory we must assume that each individual rounding error is as large as possible and that all of them combine in the worst possible way. In reality of course the errors will be of different sizes, some positive, some negative and some cancellations of errors will tend to occur so that the cumulative error will be much smaller than might have been anticipated.

EXAMPLE 1.3.4.   Compute the sum $\sum_{n=11}^{19} (1/n)$ (i) to 6 d.p. and then (ii) rounding to 3 d.p. at every stage and compare the rounding error in (ii) with its theoretical maximum.

Solution.   Using a calculator we find, to 6 d.p.

$$\sum_{n=11}^{19} \frac{1}{n} = 0{\cdot}618771.$$

Rounding each reciprocal to 3 d.p. and adding we obtain the value $0·620$. The accumulated rounding error is therefore approximately $0·0012$ (in absolute value [see I (2.6.5)]).

Since we have rounded each reciprocal to 3 d.p. the individual rounding errors lie in the range $-\frac{1}{2} \times 10^{-3}$ to $+\frac{1}{2} \times 10^{-3}$ and so the maximum possible accumulated rounding error, since we are summing 9 terms, is $\frac{9}{2} \times 10^{-3} = 0·0045$, in absolute value. We therefore see that our observed rounding error is only about one quarter of the theoretical maximum.

Obviously if we add n numbers each of which has been rounded to k decimal places the maximum possible accumulated rounding error is, in absolute value

$$\frac{n}{2} \times 10^{-k}.$$

In practice it is likely to be much smaller than this and it can be proved that the *probable error* is proportional to $\sqrt{n} \times 10^{-k}$ rather than to $n \times 10^{-k}$.

Another type of error which is unavoidable in many numerical calculations is *truncation error*. We shall meet this frequently in this book and a simple example will serve to illustrate it at this stage. Suppose we wish to find the value of a certain function $f(x)$, for a specific value of x [see I, § 1.4.1], by using its Taylor Series expansion [see I, § 3.6]

$$f(x) = \sum_{n=0}^{\infty} a_n x^n.$$

Since we cannot, in general, add together an infinity of terms we must stop after a finite number, say N. This means that we are taking

$$f(x) \doteqdot \sum_{n=0}^{N-1} a_n x^n, \quad \text{and ignoring the rest,} \ \sum_{n=N}^{\infty} a_n x^n.$$

By 'chopping off' or 'truncating' the Taylor Series after N terms in this way we will have a *truncation error* equal to the sum of the terms ignored, i.e.

$$\text{Truncation Error} = \sum_{n=N}^{\infty} a_n x^n. \tag{1.3.1}$$

Note that this is quite different from rounding error, which arises from the terms we *include* in the sum, whereas truncation error arises from the terms which we *exclude*.

Truncation errors also arise when we use a numerical formula to estimate the value of an integral (see Chapter 7) or to compute a solution to a differential equation (Chapter 8).

Relative Error

The importance of an error can often be better appreciated if we compare the error to the quantity being approximated, i.e. we use the *relative error* which we define as the ratio Error/True Value. This is particularly useful when the True Value is either very small or very large.

There remain two other types of error. The first is *Empirical Error* which arises when data for a problem are acquired by some experimental means and so are of limited precision (which may be unknown); this is therefore caused by a limitation in instrumentation (including the eye) and so may be unavoidable. It is however important to bear it in mind since there is no point in carrying out calculations to, say, 6 d.p. when data involved in the problem are only known to 2 d.p.

The second type of error is *human error* (or *blunders*). When calculations are done by hand or with a desk machine such mistakes frequently occur. The most common types of errors, apart from straightforward incorrect addition, multiplication etc. are

(i) transposing 2 digits in copying a number e.g. writing 72615 instead of 76215;
(ii) repeating the wrong digit e.g. writing 34667 instead of 34467;
(iii) forgetting a negative sign in copying a value from a table or calculator.

Some numerical methods (e.g. Gaussian Elimination, see Chapter 3) include checks to guard against human errors, and these should be used whenever possible.

When a computer is being used such errors, or anything comparable, are unlikely to occur.

1.4. FINITE DIFFERENCES

The operation of calculating differences of various kinds, between values of a function at a set of points, is one of the most-used in numerical methods. The differencing operation is the equivalent for a function tabulated at discrete points of differentiation of a continuous function over an interval [see IV, § 3.1].

The three most commonly used types of differences are *forward*, *backward* and *central*.

Suppose that we have a function, $f(x)$, whose values have been tabulated at a set of points $x_1, x_2, \ldots, x_n$.

DEFINITION 1.4.1. The first forward difference of $f(x)$ at $x = x_i$ is

$$f(x_{i+1}) - f(x_i) \tag{1.4.1}$$

and is denoted by $\Delta f(x_i)$, or by Δf_i.

By applying this process repeatedly we generate the second forward difference, third forward difference etc. and so

DEFINITION 1.4.2.   The kth forward difference, $\Delta^k f(x_i)$ of $f(x)$ at x_i is defined for $k \geq 2$ by

$$\Delta^k f(x_i) = \Delta(\Delta^{k-1} f(x_i)) \qquad (1.4.2)$$

where $\Delta^1 = \Delta$.

EXAMPLE 1.4.1.   Find expressions for $\Delta^2 f(x_i)$, $\Delta^3 f(x_i)$.

Solution

$$
\begin{aligned}
\Delta^2 f(x_i) &= \Delta(\Delta f(x_i)) \\
&= \Delta(f(x_{i+1}) - f(x_i)) \\
&= (f(x_{i+2}) - f(x_{i+1})) - (f(x_{i+1}) - f(x_i)) \\
&= f(x_{i+2}) - 2f(x_{i+1}) + f(x_i) \\
\Delta^3 f(x_i) &= \Delta(\Delta^2 f(x_i)) \\
&= \Delta(f(x_{i+2}) - 2f(x_{i+1}) + f(x_i)) \\
&= f(x_{i+3}) - 3f(x_{i+2}) + 3f(x_{i+1}) - f(x_i).
\end{aligned}
$$

In a similar manner we define backward differences.

DEFINITION 1.4.3.   The first backward difference of $f(x)$ at $x = x_i$ is

$$f(x_i) - f(x_{i-1}) \qquad (1.4.3)$$

and is denoted by $\nabla f(x_i)$ or by ∇f_i.

DEFINITION 1.4.4.   The kth backward difference $\nabla^k f(x)$ at $x = x_i$ is defined by $k \geq 2$ by

$$\nabla^k f(x_i) = \nabla(\nabla^{k-1} f(x_i)) \qquad (1.4.4)$$

where $\nabla^1 = \nabla$.

Forward and backward differences introduce asymmetry and occasionally we need formulae which are symmetrical about the point at which the differences are being taken; such formulae may be based on central differences.

DEFINITION 1.4.5.   The first central difference of $f(x)$ at x_i is

$$f(x_{i+\frac{1}{2}}) - f(x_{i-\frac{1}{2}}) \qquad (1.4.5)$$

and is denoted by $\delta f(x_i)$ or by δf_i.

DEFINITION 1.4.6.   The kth central difference $\delta^k f(x)$ at $x = x_i$ is defined for $k \geq 2$ by

$$\delta^k f(x_i) = \delta(\delta^{k-1} f(x_i)) \qquad (1.4.6)$$

where $\delta^1 = \delta$.

Note that if the function values are only known at the points $x_1, x_2, \ldots, x_n$ then $f(x_{i+\frac{1}{2}})$ is indeterminate so that $\delta f(x_i)$ is not computable, however, we *can* compute $\delta^2 f(x_i)$ for

$$\delta^2 f(x_i) = \delta(f(x_{i+\frac{1}{2}}) - f(x_{i-\frac{1}{2}}))$$
$$= f(x_{i+1}) - 2f(x_i) + f(x_{i-1}) \qquad (1.4.7)$$

and, more generally, we can compute $\delta^{2k} f(x_i)$ for all positive integers k.

Finite differences as defined above are mainly of use when the set of points $\{x_i\}$ are regularly spaced, so that $x_{i+1} - x_i = h$ (say) and therefore $x_m = x_1 + (m-1)h$. Under these circumstances h is called *the interval size*, or *step size*. Henceforward when we use Δ, ∇ or δ we shall assume that the points x_i are regularly spaced. Formulae using finite differences based upon equal intervals are heavily used in numerical methods involving interpolation (Chapter 2), numerical integration (Chapter 7) and differential equations (Chapters 8, 9). The reason for this is not hard to see for if h is the interval size

$$\Delta f(x) = f(x+h) - f(x) \doteqdot h f'(x)$$

when h is sufficiently small since [see IV, Def. 3.1.1]

$$\lim_{h \to 0} \frac{f(x+h) - f(x)}{h} = f'(x).$$

Thus we can write

$$f'(x) \doteqdot \frac{1}{h} \Delta f(x),$$

which can be used for approximate differentiation and integration; and $f(x+h) = f(x) + \Delta f(x)$, which is the starting point for interpolation. These simple results can be developed, as we shall see in later chapters, to produce an extensive range of very useful formulae.

Finite differences can also be used to detect, and sometimes to correct, errors in tables of values of functions; before considering how this can be done let us construct an example of a *difference table* of a simple polynomial.

EXAMPLE 1.4.2. Construct a forward-difference table of $f(x) = x^3$ starting at $x = 1$ and taking $h = 1$.

Solution

$x = 1$	2	3	4	5	6
$f(x) = 1$	8	27	64	125	216
$\Delta f =$	7	19	37	61	91
$\Delta^2 f =$	12	18	24	30	
$\Delta^3 f =$		6	6	6	
$\Delta^4 f =$			0	0.	

Observing that $\Delta^3 f$ appears to be equal to 6 everywhere we can attempt to predict the value of 7^3 as follows: since $\Delta^3 f \equiv 6$ the next number in the $\Delta^2 f$ line must be 36, the next number in the Δf line must therefore be $91 + 36 = 127$ and hence 7^3 ought to be $216 + 127 = 343$, which it is. Let us therefore accept that when $f(x) = x^3$, $\Delta^3 f \equiv 6$; we shall see shortly that this is so; meanwhile the next example illustrates the use of the difference table in detecting errors.

EXAMPLE 1.4.3.   The numbers below are taken from a table of cubes; they contain an error; find it. Can you correct the erroneous term?

$f(x) = 9261$ 10648   12167   13834   15625   17576   19683   21952.

Solution.   We form the difference table

$\Delta f = 1387$ 1519   1667   1791   1951   2107   2269

$\Delta^2 f =$ 132   148   124   160   156   162

$\Delta^3 f =$ 16 -24 36 -4 6.

We know that the line $\Delta^3 f$ ought to consist of 6's; since it doesn't we have an indication of an error. The number -4 in the $\Delta^3 f$ ought to be 6; if it were, the number 160 in the $\Delta^2 f$ line would be 150, and the number 1791 in the Δf line would then be 1801 which implies that the number 13834 in the original line of values of $f(x)$ should be 13824, and this is correct since $(24)^3 = 13824$. Making this correction and recalculating the difference table we now find $\Delta^3 f \equiv 6$.

In this case of course we were told that the values given were from a table of x^3 and so detection and correction of the error was easy; but what happens when we don't know the analytical form of the function which has been tabulated? Can we still detect an error? The answer is 'if the function is sufficiently smooth we can detect an error', where by a 'smooth' function we mean one whose derivatives change slowly.

The result obtained above that when $f(x) = x^3$, $\Delta^3 f \equiv 6$ is a special case of a theorem which we quote without proof: 'If $f(x)$ is a polynomial of degree k [see I, § 14.1] then $\Delta^k f$ is a constant and $\Delta^{k+1} f \equiv 0$'.

The converse of this theorem is also true viz:

'If $f(x)$ is a function such that $\Delta^{k+1} f \equiv 0$ then $f(x)$ can be exactly represented by a polynomial of degree k or less.'

This result will be used in fitting polynomials to data for interpolation and other purposes.

These two theorems tell us that if $f(x)$ is not a polynomial then *no line* of the difference table will ever be entirely zero. This then leads us to ask what the difference table of a function which is not a polynomial might be. A simple example of such a function is $\sqrt{x}$ [see I, § 3.3].

EXAMPLE 1.4.4.   Form 4 lines of a difference table of $\sqrt{x}$ to 4 d.p. for $x = 1(1)9$. Note that here we have introduced the following notation: $x = 1(1)9$ means 'x starts with the value 1 and increases in steps of 1, stopping when $x = 9$'.

Solution

$f(x) = 1\cdot0000$ $1\cdot4142$ $1\cdot7321$ $2\cdot0000$ $2\cdot2361$ $2\cdot4495$ $2\cdot6458$ $2\cdot8284$ $3\cdot0000$

$\Delta f =$ $0\cdot4142$ $0\cdot3179$ $0\cdot2679$ $0\cdot2361$ $0\cdot2134$ $0\cdot1963$ $0\cdot1826$ $0\cdot1716$

$\Delta^2 f =$ $-0\cdot0963$ $-0\cdot0500$ $-0\cdot0318$ $-0\cdot0227$ $-0\cdot0171$ $-0\cdot0137$ $-0\cdot0110$

$\Delta^3 f =$ $+0\cdot0463$ $+0\cdot0182$ $+0\cdot0091$ $+0\cdot0056$ $+0\cdot0034$ $+0\cdot0027$

$\Delta^4 f =$ $-0\cdot0281$ $-0\cdot0091$ $-0\cdot0035$ $-0\cdot0022$ $-0\cdot0007$.

We see that some kind of pattern seems to exist viz:

(i) the rows consist of all positive or all negative numbers;
(ii) positive and negative rows alternate;
(iii) the numbers in each row decrease steadily in absolute value.

It is quite easy to prove that these 3 observations are sound and the rows of the difference table for $\sqrt{x}$ for $x = n(1)m$, where n, m are any positive integers, will possess these properties. So, although the difference table for $\sqrt{x}$ will never contain a line of zeros it does have definite characteristics which might be used to detect an error, as the next example shows.

EXAMPLE 1.4.5. The numbers below are taken from a table of square roots calculated at equidistant points but contain an error. Find it by using differences

$7\cdot2801$ $7\cdot3485$ $7\cdot4162$ $7\cdot4843$ $7\cdot5498$ $7\cdot6158$ $7\cdot6811$.

Solution. We form the forward difference table

$f(x) = 7\cdot2801$ $7\cdot3485$ $7\cdot4162$ $7\cdot4843$ $7\cdot5498$ $7\cdot6158$ $7\cdot6811$

$\Delta f =$ $0\cdot0684$ $0\cdot0677$ $0\cdot0681$ $0\cdot0655$ $0\cdot0660$ $0\cdot0653$

$\Delta^2 f =$ $-0\cdot0007$ $+0\cdot0004$ $-0\cdot0026$ $+0\cdot0005$ $-0\cdot0007$

The presence of an error is already clear; the numbers in Δf are not decreasing steadily; the numbers in $\Delta^2 f$ are not all of the same sign. Since a single error in $f(x)$ causes two consecutive errors in Δf, three consecutive errors in $\Delta^2 f$ etc we see that the erroneous terms can all be enclosed by the two lines shown in the table. These lines from two sides of an isoceles triangle [see V, § 1.1.3] with the erroneous term $(7\cdot4843)$ at the apex. Such a triangle is called the *error triangle* or *error fan*.

We now observe that the two terms in $\Delta^2 f$ which lie outside the error triangle are both equal to $-0\cdot0007$. We might therefore conjecture that the other three terms ought to be about $-0\cdot0007$ (since the numbers in the line are supposed to decrease when $f(x) = \sqrt{x}$). This conjecture is not quite correct since rounding errors can affect the last digit in $\Delta^2 f$ so that a term which appears to be $-0\cdot0007$ might be $-0\cdot0008$ or $-0\cdot0006$. However it is sufficiently close to indicate that we are in error by an amount approximately equal to $0\cdot0004 - (-0\cdot0007)$ i.e. $0\cdot0011$ so that the number $7\cdot4843$ should perhaps be about $7\cdot4843 - 0\cdot0011 = 7\cdot4832$. (In fact the correct value is $7\cdot4833 = \sqrt{56}$ to 4 d.p.) With this correction the $\Delta^2 f$ becomes almost constant, the slight variations being caused by rounding errors.

We have now seen how to *detect* and even try to *correct* a single error in a table of values of a function when we know something about the pattern of values which occur in the lines of the difference table. Can we however hope to be so successful when we know nothing in advance of the patterns to be expected? The next example illustrates how we must proceed when we have no idea what patterns to expect.

EXAMPLE 1.4.6.   The function below has been tabulated at equidistant points; it contains an error. Find the error and estimate its size.

$$4204 \quad 4463 \quad 4885 \quad 5640 \quad 6179 \quad 7034 \quad 8018 \quad 9124.$$

Solution

$$
\begin{array}{lcccccccc}
f = & 4204 & 4463 & 4885 & 5640 & 6179 & 7034 & 8018 & 9124 \\
\Delta f = & & 259 & 422 & 755 & 539 & 855 & 984 & 1106 \\
\Delta^2 f = & & & 163 & 333 & -216 & 316 & 129 & 122.
\end{array}
$$

On the basis of the numbers in the Δf line we conjecture that Δf ought to be increasing, there is only one exception (539) and this is presumed to be one of the terms in error. Since a single error in the $f(x)$ line causes two adjacent errors in Δf we must decide which of 755 or 855 is also wrong. The term 333 in the $\Delta^2 f$ line looks suspiciously large and this points to 755 in Δf as being wrong. We therefore draw the error fan as shown, implying that the term 5640 in the line of values of $f(x)$ is incorrect.

The terms outside the fan in the $\Delta^2 f$ line now look to be decreasing fairly slowly. The change from 163 to 129 takes 4 steps, an average of 8·5 per step; the next step takes us from 129 to 122, a drop of 7. We therefore conjecture that the $\Delta^3 f$ line (which we haven't put in the table for lack of evidence) might be something like

$$-10 \quad -9 \quad -8 \quad -7 \quad -7.$$

On this hypothesis the reconstructed difference table is as shown below:

$$
\begin{array}{lcccccccc}
f = & 4204 & 4463 & 4885 & 5460 & 6179 & 7034 & 8018 & 9124 \\
\Delta f = & & 259 & 422 & 595 & 719 & 855 & 984 & 1106 \\
\Delta^2 f = & & & 163 & 153 & 144 & 136 & 129 & 122 \\
\Delta^3 f = & & & & -10 & -9 & -8 & -7 & -7.
\end{array}
$$

This looks fairly consistent and is typical of a smooth function and encourages us to suggest that the incorrect term was 5640 and that its correct value should be about 5460.

(In fact this reconstruction is entirely correct; the table consists of the values of

$$f(x) = 10^4(\sqrt{x^3 + 5} - \log_e x) - 20{,}000$$

rounded to the nearest integer for $x = 1 \cdot 3(0 \cdot 1)2 \cdot 0$ and the value when $x = 1 \cdot 6$ is 5460.)

The remarkably good result obtained in this example arose from the fact that $f'''(x)$ [see IV, § 3.4] changes only slowly over the range of values of x covered by the table; had this not been the case our attempt to correct the erroneous term could not have been so successful. The example does however show how powerful the technique of forward differencing can be and it should be added that in practice we would not be limited to a table consisting of a few values of $f(x)$, a limitation imposed by the page size of this book, but would probably have a very large number of values available which would provide a great deal more evidence, so enabling us to pick out the patterns with much more confidence.

1.5.   DIVIDED DIFFERENCES

So far we have assumed that the functions we have been studying have all been tabulated at equidistant points. Whilst this is usually the case if the values are taken from a table there are situations where the function values are available at only a limited set of points and these are not equidistant. Such a situation may arise, for example, if data are obtained experimentally. Under such circumstances there is not much point in simply taking differences between the function values for adjacent differences would no longer be comparable, and the important property that 'the $(k + 1)$st forward difference of a polynomial of degree k is identically zero' would no longer be true.

Fortunately a simple generalization of forward differences designed to cover the case of non-equidistant points not only restores the 'polynomial property' just mentioned but also provides us with a concept of wide applicability and great power. This generalization is provided by *divided differences*.

DEFINITION 1.5.1.   If a function $f(x)$ takes values $f(x_i)$ at a set of points $\{x_i\}$ then *the first divided difference between x_i and x_j is defined to be*

$$[x_i x_j] = \frac{f(x_j) - f(x_i)}{x_j - x_i}. \tag{1.5.1}$$

There are several points to note in this definition. Firstly: we not only make no assumption about whether the points $\{x_i\}$ are equally spaced we do not even assume that they are in order i.e. we are *not* assuming that

$$x_1 < x_2 < x_3 \ldots < x_n.$$

Secondly: we do not restrict the definition to points which are adjacent to one another, the definition applies to *any* 2 (distinct) points in the set.

Thirdly: note that $[x_i x_j] = [x_j x_i]$.

Fourthly: from the definition we see that $[x_i x_j]$ is numerically equal to the tangent of the angle which the line joining the points $(x_i, f(x_i))$ and $(x_j, f(x_j))$ makes with the positive direction of the x-axis [see V (1.2.9)].

We now complete the definition of divided differences by defining those of higher order.

DEFINITION 1.5.2.   For $k \geq 2$ the kth divided difference $[x_1 x_2 \ldots x_{k+1}]$ between the points $x_1, x_2, \ldots, x_{k+1}$ is defined to be

$$[x_1 x_2 \ldots x_{k+1}] = \frac{[x_2 x_3 \ldots x_{k+1}] - [x_1 x_2 \ldots x_k]}{x_{k+1} - x_1}. \qquad (1.5.2)$$

For the sake of simplifying the notation in the definition we have listed the points as $x_1, x_2, \ldots, x_{k+1}$ but it is important to note that these can be *any* points taken in *any* order.

EXAMPLE 1.5.1.   If $f(x) = x^2$ find the values of $[1, 7, 4]$ and $[2, 1, 5]$.

Solution

(i) $\qquad\qquad [1, 7, 4] = \dfrac{[7, 4] - [1, 7]}{4 - 1} = \dfrac{[7, 4] - [1, 7]}{3},$

where

$$[7, 4] = \frac{49 - 16}{3} = 11$$

and

$$[1, 7] = \frac{49 - 1}{6} = 8$$

so that

$$[1, 7, 4] = \frac{11 - 8}{3} = 1.$$

(ii) $\qquad\qquad [2, 1, 5] = \dfrac{[1, 5] - [2, 1]}{5 - 2} = \dfrac{[1, 5] - [2, 1]}{3}$

where

$$[1, 5] = \frac{25 - 1}{4} = 6$$

and

$$[2, 1] = \frac{1 - 4}{-1} = 3$$

so that

$$[2, 1, 5] = \frac{6-3}{3} = 1.$$

It is not a coincidence that the values of $[1, 7, 4]$ and $[2, 1, 5]$ turn out to be the same in this case. Had we chosen any triad of points a, b, c we would find that when $f(x) = x^2$ $[a, b, c] = 1$ as we now show; for

$$[a, b, c] = \frac{[b, c] - [a, b]}{c - a}.$$

Now

$$[b, c] = \frac{c^2 - b^2}{c - b} = c + b$$

and

$$[a, b] = \frac{b^2 - a^2}{b - a} = b + a$$

so that

$$[a, b, c] = \frac{(c + b) - (b + a)}{c - a} = 1.$$

This result is itself a special case ($k = 2$) of the following theorem.

THEOREM 1.5.1. *If $f(x) = x^k$ and $\{x_i\}$ is any set of points then the kth divided difference between any $(k + 1)$ of the points $\{x_i\}$ is identically equal to 1 and any $(k + 1)$st divided difference is zero.*

It follows at once from the theorem that if $f(x) = a_n x^n + a_{n-1} x^{n-1} + \ldots a_0$ is a polynomial of the nth degree then the nth divided difference at any set of points will be identically equal to a_n and the $(n + 1)$st divided difference will be identically zero. With this result as a basis the problems of error detection, interpolation and curve-fitting all become soluble even when the points are not equidistant. The problems of interpolation and of curve-fitting are covered in Chapters 2 and 6. We have already met the error detection problem for equidistant points in this chapter, for non-equidistant points we need only use divided differences in the difference table instead of forward differences, the rest of the analysis for the detection of an error is unchanged, as the next example illustrates.

EXAMPLE 1.5.2. The following data were measured experimentally; use the method of divided differences to check for the possible presence of an error:

$x = 4 \cdot 75$	$5 \cdot 0$	$5 \cdot 5$	6	7	9	10	13	17
$y = 1 \cdot 179$	$1 \cdot 486$	$2 \cdot 094$	$2 \cdot 695$	$3 \cdot 890$	$6 \cdot 196$	$7 \cdot 335$	$10 \cdot 696$	$15 \cdot 092.$

Solution. Since the x-values are not regularly spaced we must use divided differences. The first three lines are (working to 3 d.p.)

1·179 1·486 2·094 2·695 3·890 6·196 7·335 10·696 15·092

  1·228 1·216 1·202 1·195 1·153 1·139 1·120   1·099

    −0·016 −0·014 −0·005 −0·014 −0·005 −0·005 −0·002.

A single error in $f(x)$ would produce 2 adjacent errors in the second of these lines and 3 adjacent errors in the third line. The numbers in the third line seem to contain a ripple which destroys a monotonic decreasing pattern [see IV, Def. 2.7.1]. If the 3 erroneous terms in this line are (−0·005, −0·014, −0·005) the original error is traced to $f(7) = 3·890$.

Correction of the error when we are using divided differences is slightly more complicated than when we are using forward differences but it may be carried out as follows. Suppose that the correct value of $f(7)$ is $3·890 + h$ then the two incorrect terms in the second line above should be $(1·195 + h)$ and $(1·153 − \frac{1}{2}h)$, and the three incorrect terms in the third line should be

$$(-0·005 + \tfrac{2}{3}h)(-0·014 - \tfrac{1}{2}h)(-0·005 - h/6)$$

and if the terms are to be monotonic decreasing in absolute value we must have

$$-0·005 + \tfrac{2}{3}h < -0·014 - \tfrac{1}{2}h$$

giving $h < -0·008$ (approximately) and also

$$-0·014 - \tfrac{1}{2}h < -0·005 - h/6$$

giving $h > -0·014$ (approximately).

Taking the average of these two extreme values we conjecture that $h \doteqdot -0·011$ so that $f(7)$ should have the value 3·879. (In fact $f(x) = 3x^{1/4} + x - 8$ and $f(7) = 3·880$ to 3 d.p.)

This technique is one that can be employed as a means of rejecting 'wild' readings of experimental data. Another much-used technique in such cases is that of 'smoothing' by least squares (see Chapter 6).

1.6. TAYLOR'S THEOREM

It is proved in books on Pure Mathematics [see IV, § 3.6] that if $f(x)$ is a function which is continuous and at least n times differentiable over the interval $(x, x + h)$ [see IV, Def. 2.10.1] then

$$f(x+h) = f(x) + \frac{h}{1!}f'(x) + \frac{h^2}{2!}f''(x) + \ldots + \frac{h^{n-1}}{(n-1)!}f^{n-1}(x) + R_n$$

where

$$R_n = \frac{h^n}{n!}f^n(x + \theta h)$$

where θ is some number in (0, 1).

This result (Taylor's Theorem) is one of the most useful in the whole of mathematics and particularly so in Numerical Analysis. By means of Taylor's Theorem many formulae for interpolation, numerical integration, solution of non-linear equations, and solution of differential equations can be obtained and their accuracy, validity and stability analysed.

EXAMPLE 1.6.1.   Verify Taylor's Theorem by finding the value of θ when $f(x) = x^{1/2}$, $x = 2$, $h = 0 \cdot 5$, and $n = 2$.

Solution.   When $n = 2$ Taylor's Theorem says

$$f(x + h) = f(x) + \frac{h}{1!} f'(x) + \frac{h^2}{2!} f''(x + h\theta)$$

and so, when $f(x) = x^{1/2}$

$$(x + h)^{1/2} = x^{1/2} + \tfrac{1}{2} h x^{-1/2} - \tfrac{1}{8} h^2 (x + h\theta)^{-3/2}.$$

Putting $x = 2$, $h = 0 \cdot 5$ gives

$$(2 \cdot 5)^{1/2} = 2^{1/2} + \tfrac{1}{4} (2)^{-1/2} - \tfrac{1}{32} (2 + 0 \cdot 5\theta)^{-3/2}$$

and computing to 4 d.p. this reduces to

$$(2 + 0 \cdot 5\theta)^{-3/2} = 32(-1 \cdot 5811 + 1 \cdot 4142 + 0 \cdot 1768) = 0 \cdot 3168$$

and so $(2 + 0 \cdot 5\theta) \doteqdot (0 \cdot 3168)^{-2/3} \doteqdot 2 \cdot 1518$ giving $\theta \doteqdot 0 \cdot 3036$. Thus we have verified that a number θ lying in $(0, 1)$ *does* exist.

1.7.   ILL-CONDITIONED PROBLEMS

The value of θ obtained in this last example is not very important since all we wished to verify was that $0 < \theta < 1$; this was fortunate since calculations of θ to any precision is not easy. Look again at the equation which defines θ, written in the form

$$(2 + 0 \cdot 5\theta)^{-3/2} = 32(2^{1/2} + \tfrac{1}{4} 2^{-1/2} - (2 \cdot 5)^{1/2}). \tag{1.7.1}$$

Each of the numbers on the right is computed to 4 d.p. so the total rounding error could be as big as

$$32(\tfrac{1}{2} \times 10^{-4} + \tfrac{1}{4} \times \tfrac{1}{2} \times 10^{-4} + \tfrac{1}{2} \times 10^{-4}) = 36 \times 10^{-4} = 0 \cdot 0036.$$

In addition the number in the brackets on the r.h.s. of (1.7.1) is very small, about $0 \cdot 00985$, to 5 d.p. and when we multiply by 32 it becomes $0 \cdot 3152$, but because of rounding errors we can only guarantee that the value lies in the range $(0 \cdot 3152 - 0 \cdot 0036)$ to $(0 \cdot 3152 + 0 \cdot 0036)$ i.e. in $(0 \cdot 3116, 0 \cdot 3188)$.

If we now compute the values for θ from these two extreme cases, working to 4 d.p. we have

$$(2 + 0 \cdot 5\theta_1) = (0 \cdot 3116)^{-2/3} \quad \text{and} \quad (2 + 0 \cdot 5\theta_2) = (0 \cdot 3188)^{-2/3}$$

and we find that $\theta_1 \doteqdot 0 \cdot 3514$ and $\theta_2 \doteqdot 0 \cdot 2857$, i.e. the value for θ is not even known to an accuracy of 1 d.p!

A problem such as this in which small changes in the value of one or more of the variables produces large changes in the answer is called *ill-conditioned*. Such ill-conditioning frequently occurs when, as in the example above, we have to find the difference between two nearly equal numbers so that rounding errors in the numbers suddenly become *relatively* of great significance. Ill-conditioned problems can be very difficult to solve satisfactorily and sometimes a numerical method may have to be modified considerably to avoid the possibility of ill conditioning. Such situations arise occasionally in problems involving matrix inversion (Chapter 4) and in Least Squares fitting of polynomials to data (Chapter 6).

R.F.C.

CHAPTER 2

Computation and Interpolation of Functions

2.1. INTRODUCTION

Certain functions are familiar to everyone who has studied algebra and trigonometry at school, viz. square roots, logarithms, sine, cosine and tangent. When we need to use such functions we usually look up their values in a table. Such tables typically will give the value of these various functions to 4 or more places of decimals for a regularly spaced range of values of the variable. If we want to know the value of a function at a point other than those listed in the table we must somehow compute it for ourselves and this will involve at least some *computation* and possibly *interpolation* or *extrapolation*. If we wish to know the value of a function to greater accuracy than that given by the tables we must both compute the function and convince ourselves that the desired accuracy really is being achieved, and in order to do this we may have to exercise considerable skill in deciding which of several methods is to be used for the computation.

2.2. METHODS OF COMPUTATION

A function may be defined in various ways such as:

(i) as a power series, e.g.

$$\sin(x) = x - \frac{x^3}{3!} + \frac{x^5}{5!} - \frac{x^7}{7!} + \dots ;$$

(ii) as an integral, e.g.

$$\log_e x = \int_1^x \frac{dt}{t};$$

(iii) as the solution of an equation, e.g.

$$f(x) = y \quad \text{where } y^3 - xy^2 \sin x - 2 \log x = 0;$$

19

(iv) as the solution of a differential equation, e.g.

$$f(x) = y \quad \text{where} \quad \frac{x^2 d^2 y}{dx^2} + \frac{xdy}{dx} + (x^3 - y^3) = 0.$$

Methods for the solution of problems of types (iii) and (iv) are dealt with in Chapters 5 and 8 respectively; methods for the evaluation of integrals are covered in Chapter 7; computation of functions defined by power series forms part of this chapter. Thus, having noted the way in which our function is defined, or the way in which we propose to compute it, we should turn to the appropriate chapter of the book.

Even for a simple function there may be several ways by which its value might be computed. Consider for example the problem of finding the value of $\sqrt{5}$ to 4 d.p.; the method we use will depend on whether we have tables available, or a calculator, or have to work by hand.

Method 1 (No tables or calculator available).   We use a hand calculation based on the method taught at school shown below:

	5·00000000	2·23606
	4·00	
42	100	
	84	
443	1600	
	1329	
4466	27100	
	26796	
447206	3040000	
	2683236	

which gives $\sqrt{5} = 2 \cdot 2361$ to 4 d.p.

Method 2 (No tables or calculator).   Use the binomial theorem [see I (3.10.1)] by writing

$$5 = 4(1 + 0 \cdot 25)$$

and so

$$5^{1/2} = 2(1 + 0 \cdot 25)^{1/2}$$

$$\doteq 2\left(1 + \frac{0 \cdot 25}{2} - \frac{(0 \cdot 25)^2}{8} + \frac{(0 \cdot 25)^3}{16} - \frac{5(0 \cdot 25)^4}{128} + \frac{7(0 \cdot 25)^5}{256}\right)$$

$$= 2 \cdot 2361 \text{ to 4 d.p.}$$

In this case it is not a trivial matter to know how many terms of the binomial expansion we should take—we return to this problem (estimation of truncation error) in Section 2.2.1.

Method 3 (Tables available)

(i) Look up the entry opposite '5' in the table of square roots; this gives 2·2361 to 4 d.p.
(ii) If a table of square roots is not available use a table of logarithms:

$$\sqrt{5} = \text{antilog} \left(\tfrac{1}{2} \log 5\right) = \text{antilog}_{10} (0\cdot34949) = 2\cdot2361.$$

Method 4 (Calculator, without square root facility, available). Use the Newton–Raphson iterative procedure [see § 5.4.1]; i.e. take $x_1 = 2$ as starting value and then calculate $x_2, x_3, \ldots$ iteratively from

$$x_{n+1} = \frac{1}{2}\left(x_n + \frac{5}{x_n}\right)$$

until two consecutive approximations agree to 4 d.p. Thus

$$x_1 = 2\cdot0000, \qquad x_2 = 2\cdot2500, \qquad x_3 = 2\cdot2361, \qquad x_4 = 2\cdot2361;$$

we conclude that $\sqrt{5} = 2\cdot2361$ to 4 d.p.

These four methods are fundamentally different. The first is a method of successive approximations in which we work out one more digit in the decimal expansion of $\sqrt{5}$ at each step. The second uses that number of terms of an infinite series sufficient to guarantee that the sum of the terms ignored (i.e. the truncation error) will not affect the fourth decimal place of the result. The third first uses the results of computations previously performed by someone else (by means unknown to us but in which we have confidence) and then, since the tables do not give an antilog corresponding to 0·34949, is completed by some simple interpolation by us. The fourth method is totally different in every way from the others; in the first place it is not obvious that it will converge [see I, § 1.2] to $\sqrt{5}$ (though this is not difficult to prove), nor is it easy to see how many steps (iterations) the process will take, but the most extraordinary feature is that the process is, to some extent, self-correcting—if we make a mistake we might still eventually obtain the right answer. This last feature is, of course, even less obvious than the other two; we shall not prove it but illustrate with an example.

EXAMPLE 2.2.1. Make deliberate errors in the calculation of $\sqrt{5}$ by the Newton–Raphson method (i) by starting at $x_1 = 25$, (ii) by starting at $x_1 = 1$ and then taking 1 from the value of x_3.

Solution. (i) Using the formula above we find $x_1 = 25$, $x_2 = 12\cdot6$, $x_3 = 6\cdot4984$, $x_4 = 3\cdot6339$, $x_5 = 2\cdot5049$, $x_6 = 2\cdot2505$, $x_7 = 2\cdot2361$—so convergence has occurred, taking 6 iterations instead of 3.

(ii) $x_1 = 1$, $x_2 = 3$, $x_3 = 2 \cdot 333$—at this point we make the mistake, as requested, and put $x_3 = 1 \cdot 3333$, then $x_4 = 2 \cdot 5417$, $x_5 = 2 \cdot 2544$, $x_6 = 2 \cdot 2361$—and we
see that convergence has again occurred, despite our error.

A method of computation which not only converges but apparently does so
very rapidly and which is self-correcting against error is clearly of immense
value. An important class of methods which possess the property of being
self-correcting against errors which are not too large are those of the type
where successive approximations x_1, x_2, x_3, ... to the required value are
computed by means of a function evaluation of the type

$$x_{n+1} = F(x_n).$$

Such methods are called *Iterative Methods*; they are dealt with in Chapter 5.
The Newton–Raphson is clearly of this type. Their advantages are

    (i) any desired precision is achievable;
    (ii) they are self-correcting against errors which are not too large.

They have one serious disadvantage: they may fail to converge; but if they *do*
converge they are usually very satisfactory and well suited to use on computers.
The question of convergence of such methods is discussed in Chapter 5.

2.2.1.   Direct Methods of Computation

By *a direct method of computation* of a function $f(x)$ we mean one in which no
iterative process is involved; the value of x is inserted into some formula and
after a fixed number of arithmetic operations the value is obtained. Method 1
and Method 2 are clearly of this type; Method 4 is not.

If the function to be computed is either well-known or is a combination of
well-known functions a power series expansion, valid for some range of values
of x, is probably available. Such power series expansions are obtained by means
of Taylor's theorem [see IV, (2.10.1) and § 3.6]. Among the best known are

$$\sin x = x - \frac{x^3}{3!} + \frac{x^5}{5!} - \frac{x^7}{7!} + \frac{x^9}{9!} - \dots \qquad \text{(valid for all } x)$$

$$\cos x = 1 - \frac{x^2}{2!} + \frac{x^4}{4!} - \frac{x^6}{6!} + \frac{x^8}{8!} - \dots \qquad \text{(valid for all } x)$$

$$\log_e (1+x) = x - \frac{x^2}{2} + \frac{x^3}{3} - \frac{x^4}{4} + \frac{x^5}{5} - \dots \qquad \text{(valid for } |x| < 1)$$

$$e^x = 1 + \frac{x}{1!} + \frac{x^2}{2!} + \frac{x^3}{3!} + \frac{x^4}{4!} + \dots \qquad \text{(valid for all } x)$$

to which might be added all polynomials and expansions by means of the
binomial theorem [see I (3.10.1)]

$$(1+x)^k = 1 + \binom{k}{1} x + \binom{k}{2} x^2 + \binom{k}{3} x^3 + \dots \qquad \text{(valid for } |x| < 1, \text{ see IV, § 1.10).}$$

Computation of functions defined by power series is straightforward provided we can answer two questions affirmatively:

(1) Is the power series valid for the particular value (or range of values) of the variable which I am proposing to use?
(2) Do I know how many terms of the power series I must take in order to ensure that the result will be accurate to the desired number of decimal places?

We can easily answer the first of these questions if the power series is either one of the standard functions or a combination of several such functions. The *region of convergence* [see IV, Theorem 1.10.1] of such series are well-known and all we need to do is to ensure that the values for which we are computing the function lie within the regions of convergence of *each* of the standard series involved.

If the power series is not a combination of standard functions its region of convergence may possibly be established by applying a variety of tests [see IV, § 1.10], but if the coefficients of the power series do not follow some kind of pattern it may be difficult, or impossible, to establish the region of convergence and we must then turn to other methods of computation.

The second question: estimation of truncation error is one of the most important in numerical work; it is also one which frequently causes beginners the most trouble and yet it need not do so. The concept of truncation error was introduced in Chapter One but we repeat it here, to emphasize its importance and its relevance to the problem we are considering. Suppose we have a function, $f(x)$, defined by a power series

$$f(x) = a_0 + a_1 x + a_2 x^2 + \ldots = \sum_{n=0}^{\infty} a_n x^n \qquad (2.2.1)$$

and that we wish to compute $f(x)$ at some particular value $x = x_0$ (say). Since we cannot add an infinite number of terms together we must approximate $f(x_0)$ by taking the first N terms (say) of the power series, viz.

$$f(x_0) \doteqdot \sum_{n=0}^{N-1} a_n x_0^n.$$

The remaining terms of the power series, viz. $\sum_{n=N}^{\infty} a_n x_0^n$ form the truncation error.

If we wish to compute $f(x_0)$ correct to k decimal places it is obviously essential that we choose N so that

$$|\text{Truncation Error}| < \tfrac{1}{2} \times 10^{-k}$$

i.e. so that

$$\left| \sum_{n=N}^{\infty} a_n x_0^n \right| < \tfrac{1}{2} \times 10^{-k}.$$

Unless the coefficients a_n are of a particularly simple kind we cannot hope to find an analytic expression for the right-hand side of (2.2.1) and will have to be satisfied with an approximation in the form of an inequality which *overestimates* the truncation error, i.e. we try to find a simple function $g(x_0, N)$ such that

$$\left| \sum_{r=N}^{\infty} a_n x_0^n \right| < |g(x_0, N)| < \tfrac{1}{2} \times 10^{-k}.$$

There is no general rule for finding such a function $g(x_0, N)$ but some examples illustrate the sort of approach we might take.

EXAMPLE 2.2.2.   If $-1 \le x \le 1$ *how many* terms of the series for e^x [see IV (2.11.1)]

$$e^x = \sum_{n=0}^{\infty} \frac{x^n}{n!}$$

should we take in order to guarantee that the results will be correct to 6 d.p?

Solution.   Suppose we take N terms of the series then the truncation error is given by:

$$\text{T.E.} = \sum_{n=N}^{\infty} \frac{x^n}{n!}$$

The coefficients are all positive so the maximum possible truncation error for $-1 \le x \le 1$ occurs at $x = 1$ and so

$$|\text{T.E.}| \le \sum_{n=N}^{\infty} \frac{1}{n!} = \frac{1}{N!} + \frac{1}{(N+1)!} + \frac{1}{(N+2)!} + \dots$$

$$= \frac{1}{N!}\left(1 + \frac{1}{(N+1)} + \frac{1}{(N+1)(N+2)} + \dots\right).$$

Since N is at least 1 the expression in the brackets is at most

$$1 + \frac{1}{2} + \frac{1}{(2)(3)} + \frac{1}{(2)(3)(4)} + \dots$$

which is itself less than

$$1 + \frac{1}{2} + \frac{1}{2^2} + \frac{1}{2^3} + \dots = 2.$$

It follows that

$$|\text{T.E.}| < \frac{2}{N!}.$$

Hence if we choose N so that

$$\frac{2}{N!} < \frac{1}{2} \times 10^{-6}$$

our results will certainly be correct to 6 d.p. Rearranging we see that we should choose N so that

$$N! > 4 \times 10^6$$

which is true when $N = 11$, but not when $N = 10$.

We therefore conclude that 11 terms will suffice.

The example illustrates the cardinal principle: *always overestimate the truncation error.* Thus we first took $x = 1$ because this produces the largest error, then we took $N = 1$ in the expression in the brackets since this again is the worst possible case, finally we replaced the series

$$1 + \frac{1}{2} + \frac{1}{(2)(3)} + \frac{1}{(2)(3)(4)} + \ldots$$

(whose sum, incidentally, is $e - 1 = 1 \cdot 71828 \ldots$) by the larger series whose sum was 2. Since $N = 11$ is sufficient even under these estimates for our purposes it follows *a fortiori* that $N = 11$ will certainly be good enough for 6 d.p. accuracy in practice, with a bit to spare. In fact if we take $N = 10$ the maximum truncation error is

$$\frac{1}{10!} + \frac{1}{11!} + \frac{1}{12!} + \ldots \doteqdot \frac{1}{3 \cdot 3} \times 10^{-6}$$

and we see, by carrying out the calculations more accurately, that 10 terms will suffice, so in using 11 terms we are erring on the safe side. Only in exceptional circumstances however will we be able to estimate the truncation error with such precision and it is, in any case, unnecessary—calculation of a few extra terms is a small price to pay in order to guarantee precision.

Next we consider a more difficult example which involves rounding errors as well as truncation errors.

EXAMPLE 2.2.3. Estimate the number of terms of the series

$$\sum_{n=1}^{\infty} \frac{x^n}{n^3}$$

that should be taken if we wish to compute its value correct to 8 d.p. when $-1 \le x \le 1$, and determine how accurately each term should be computed to prevent rounding errors affecting the result.

Solution. Suppose we take $(N - 1)$ terms then the truncation error (T.E.) is

$$\sum_{n=N}^{\infty} \frac{x^n}{n^3} = \frac{x^N}{N^3} + \frac{x^{N+1}}{(N+1)^3} + \ldots.$$

All the coefficients are positive and so |T.E.| is greatest when $x = 1$ so that

$$\max_{-1 \leq x \leq 1} |T.E.| = \frac{1}{N^3} + \frac{1}{(N+1)^3} + \frac{1}{(N+2)^3} + \cdots \qquad (2.2.2)$$

and for 8 d.p. accuracy we must have

$$|T.E.| \leq \tfrac{1}{2} \times 10^{-8}.$$

To estimate the sum of the series on the right-hand side of (2.2.2) we use a result from pure mathematics which tells us that if $f(n)$ is a decreasing positive valued function of n then [see Theorem 1.8.6]

$$\int_N^\infty f(x)\,dx < \sum_{n=N}^\infty f(n) < \int_{(N-1)}^\infty f(x)\,dx.$$

The function $f(n) = 1/n^3$ is decreasing and positive so that we can apply this theorem which gives

$$\int_N^\infty \frac{dx}{x^3}\,dx < \sum_{n=N}^\infty \frac{1}{n^3} < \int_{N-1}^\infty \frac{dx}{x^3}$$

i.e.

$$\frac{1}{2N^2} < \sum_{n=N}^\infty \frac{1}{n^3} < \frac{1}{2(N-1)^2}.$$

Hence we can be sure that

$$\max_{-1 \leq x \leq 1} |T.E.| < \frac{1}{2(N-1)^2}$$

and so if we choose N such that

$$\frac{1}{2(N-1)^2} \leq \tfrac{1}{2} \times 10^{-8}$$

precision to 8 d.p. will be guaranteed so far as truncation error is concerned i.e. we should choose N to be at least 10,001.

So far we have taken no account of the effect of *rounding errors* on the accuracy of the result. In this example we must expect the rounding errors to be quite significant for we are adding together over 10,000 terms. If we compute each separate term correct to k places of decimals the absolute rounding error for each term is at most $\tfrac{1}{2} \times 10^{-k}$ [see § 1.3] and so in adding 10,000 such terms the maximum possible accumulated rounding error is $\tfrac{1}{2} \times 10^{-(k-4)}$ and since this must not exceed $\tfrac{1}{2} \times 10^{-8}$ we see that we must choose $k = 12$. This is, in fact, no restriction for in order to compute $1/n^3$ sensibly when $n \doteq 10,000$ we *must* compute to 12 d.p. at least.

We conclude that 10,000 terms computed to 12 d.p. will suffice for 8 d.p. accuracy.

In this example our estimate for the rounding error was very pessimistic; when we add 10,000 terms the accumulated rounding error is likely to be nearer 100 times $\frac{1}{2} \times 10^{-k}$ rather than 10,000 times, but again, as in the case of truncation error we must take the worst possible case and overestimate whenever there is any doubt in order to guarantee the precision.

2.3. INTERPOLATION

If we have available the value of a function at a set of points $x_1, x_2, \ldots, x_n$ where (say) $x_1 < x_2 < x_3 \ldots < x_n$ and wish to estimate the value of the function at some other point y where $x_1 < y < x_n$ we have the problem of *interpolation*. If y lies outside the range of points, so that either $y < x_1$ or $y > x_n$ we have the intrinsically more difficult problem of *extrapolation*, which we discuss in section 2.4.

In this section we shall make no assumptions concerning the points $x_1, x_2, x_3, \ldots$ and shall develop methods of interpolation which are of general application. Frequently however the points at which function values are available are equally spaced and then other formulae may be preferable; we shall derive these formulae in sections 2.3.1 and 2.3.2.

Interpolation is familiar to anyone who has used mathematical tables, as a simple example illustrates. Suppose we have a table of 4-figure logarithms for all the integers from 1 to 99 and wish to estimate the value of log 59·7, the 'obvious' estimate, used by schoolchildren is:

$$\log 59 \cdot 7 \doteqdot \log 59 + 0 \cdot 7 \, (\log 60 - \log 59)$$

$$= 1 \cdot 7709 + 0 \cdot 7 (1 \cdot 7782 - 1 \cdot 7709)$$

$$= 1 \cdot 7709 + 0 \cdot 7 (0 \cdot 0073)$$

$$= 1 \cdot 7709 + 0 \cdot 0051$$

$$= 1 \cdot 7760$$

and the result is, in fact, correct to 4 d.p.

This 'obvious' method has a simple geometrical interpretation. Suppose that we have a function $f(x)$ with known values at x_1, x_2 and wish to find its value at a point x_0 where $x_1 < x_0 < x_2$, then we construct the figure below by drawing the curve $y = f(x)$ [see V, Chapter 3] and mark on it the points $P_1(x_1, f(x_1))$ and $P_2(x_2, f(x_2))$; our objective is to estimate the value of $f(x_0)$ and this is equivalent to trying to locate the point $P_0(x_0, f(x_0))$ on the curve given the positions of P_1 and P_2.

We now draw the straight line joining P_1 and P_2; this meets the line $x = x_0$ in a point Q and we take Q as our approximation to P_0. The value of y at Q is easily seen to be

$$f(x_1) + \left(\frac{x_0 - x_1}{x_2 - x_1}\right)(f(x_2) - f(x_1)) \tag{2.3.1}$$

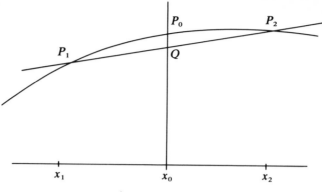

Figure 2.3.1

and since

$$\frac{f(x_2) - f(x_1)}{(x_2 - x_1)}$$

is $[x_1, x_2]$, the first divided difference of $f(x)$ at x_1, x_2 [see § 1.5], we see that the 'obvious' method of interpolation is equivalent to using the straight line interpolation formula

$$f(x_0) \doteqdot f(x_1) + (x_0 - x_1)[x_1 x_2].\qquad (2.3.2)$$

Written in this form the formula can be readily applied to the interpolation of additional values in a table of values of a function. The points (x_1, x_2) used in the interpolation should be chosen as close as possible to x_0, as the next example illustrates.

EXAMPLE 2.3.1.   From the section of a table of square roots of the integers estimate the value of $\sqrt{33\cdot4}$ using formula (2.3.2)

   (i) taking $x_1 = 33$, $x_2 = 34$;
   (ii) taking $x_1 = 32$, $x_2 = 35$;
   (iii) taking $x_1 = 25$, $x_2 = 36$.

Solution

$$x = \quad 30 \qquad 31 \qquad 32 \qquad 33 \qquad 34 \qquad 35$$
$$\sqrt{x} = 5\cdot477 \quad 5\cdot568 \quad 5\cdot657 \quad 5\cdot745 \quad 5\cdot831 \quad 5\cdot916$$

1st D.D. = $0\cdot091$ $0\cdot089$ $0\cdot088$ $0\cdot086$ $0\cdot085$

(i) $\sqrt{33\cdot4} \doteqdot 5\cdot745 + 0\cdot4(0\cdot086) = \underline{5\cdot779}$;
(ii) In this case we need to work out $[x_1 x_2]$, it is

$$\frac{5\cdot916 - 5\cdot657}{35 - 32} = (0\cdot086)$$

and our formula then gives

$$\sqrt{33 \cdot 4} \doteqdot 5 \cdot 657 + 1 \cdot 4(0 \cdot 086) = \underline{5 \cdot 777};$$

(iii) $\sqrt{33 \cdot 4} \doteqdot \sqrt{25} + \dfrac{33 \cdot 4 - 25}{36 - 25}(\sqrt{36} - \sqrt{25})$

$$= 5 + 0 \cdot 764 = \underline{5 \cdot 764}.$$

The correct value, to 3 d.p., is $5 \cdot 779$, in agreement with (i) and we observe that as the interval between x_1 and x_2 increases so the accuracy of the result decreases. If we look again at Figure (2.3.1) the reason for the declining accuracy becomes apparent: as P_1 and P_2 move further apart so the straight line $P_1 P_2$ becomes a less and less satisfactory approximation to the curve.

Formula (2.3.2) is the simplest of a class of formulae for interpolation, for instead of using a straight line joining 2 points as an approximation to the curve we can use a parabola [see V, § 1.3.3] joining 3 points, a cubic joining 4 points etc. etc. The interpolating parabola, based upon 3 points $P_1(x_1, f(x_1))$, $P_2(x_2, f(x_2))$, $P_3(x_3, f(x_3))$ is:

$$f(x) \doteqdot f(x_1) + (x - x_1)[x_1 x_2] + (x - x_1)(x - x_2)[x_1 x_2 x_3] \qquad (2.3.3)$$

where $[x_1 x_2 x_3]$ is the second divided difference [see § 1.5]. The values x_1, x_2, x_3 need not be in increasing (or decreasing) order.

EXAMPLE 2.3.2.   Use the interpolating parabola to estimate the value of $\sqrt{33 \cdot 4}$ taking $x_1 = 36$, $x_2 = 25$, $x_3 = 49$.

Solution.   We first form the divided difference table based upon the three points

$$x = 36 \qquad 25 \qquad 49$$

$$f(x) = \;\; 6 \qquad\;\; 5 \qquad\;\; 7$$

$$\text{1st D.D.} = \dfrac{-1}{-11} \qquad \dfrac{2}{24}$$

$$\text{2nd D.D.} = \dfrac{-1}{(11)(12)(13)}.$$

The interpolating parabola then gives

$$\sqrt{33 \cdot 4} \doteqdot 6 + (-2 \cdot 6)(\tfrac{1}{11}) + (-2 \cdot 6)(8 \cdot 4)(\tfrac{-1}{11})(\tfrac{1}{12})(\tfrac{1}{13})$$

$$= 6 - 0 \cdot 23636 + 0 \cdot 01272$$

$$= \underline{5 \cdot 7764} \text{ to 4 d.p.}$$

Considering the large interval sizes this result is better than we might have expected, the accuracy being 2 d.p.

There are two important points to be made relating to this example. The first is that since the values of $f(x)$ are equally spaced (whereas those of x are not) we have a problem of the type known as *inverse interpolation* and it would be better to deal with it as such by a method which we shall discuss in section 2.3.3. The second point is that our result is reasonably accurate because we have confined ourselves to values of x in an interval where the gradient of the curve [see IV, § 3.1.1] is not too steep; the gradient is $\frac{1}{2}x^{-1/2}$ and so it is essential to avoid the use of small values of x since no polynomial can satisfactorily represent a function in the region of a singularity of the function or of its derivative [see IV, § 9.8].

The reason for choosing the points in the order 36, 25, 49 in this example was to illustrate the principle that in using interpolation formulae we should always make use of function values at points closest to the point of interpolation. Our first choice therefore was that perfect square closest to 33·4, i.e. 36, then the next closest, 25, then 49. We could have made use of more points and fitted an interpolating polynomial of higher degree if we wished provided we do not use values of x too close to zero, for the reason given above. It is not difficult to prove that if we use the values of $f(x)$ at k points x_1, $x_2, \ldots, x_k$ to fit an interpolating polynomial of degree $(k-1)$ then the interpolated values of $f(x)$ is given by

$$f(x) \doteqdot f(x_1) + (x - x_1)[x_1 x_2] + (x - x_1)(x - x_2)[x_1 x_2 x_3] + \ldots$$
$$+ (x - x_1)(x - x_2) \ldots (x - x_{k-1})[x_1 x_2 \ldots x_k] \qquad (2.3.4)$$

where $[x_1 x_2 \ldots x_k]$ is the $(k-1)$st divided difference of the points $x_1, x_2, \ldots, x_k$ [see Def. 1.5.2].

This formula is known as *Newton's Divided Difference Formula*. Since the value of $[x_1 x_2 \ldots x_k]$ is independent of the order in which we take the points x_1, $x_2, \ldots x_k$ formula (2.3.4) gives the same result for any re-arrangement of the order of these points and this has an important practical consequence, viz: it is easy to make use of additional function values if we wish to increase the accuracy of the approximation. So, in the last example, we would have obtained the same result no matter in what order we used the 3 points but taking them in the order we did is natural for the reason given above: closeness to the point of interpolation; if we now decided to use a fourth point and fit a cubic the most natural choice would be to take $x_4 = 16$. In the next example we do this and perform the calculation with greater precision.

EXAMPLE 2.3.3.   Fit an interpolating cubic based upon 36, 25, 49, 16 to estimate $\sqrt{33\cdot4}$.

Solution.   It is only necessary to extend the divided difference table of the previous example by the addition of the fourth point (16) at the end of the first line and by one corresponding new difference on each subsequent line.

$$x = 36 \quad 25 \quad 49 \quad 16$$

$$f(x) = 6 \quad 5 \quad 7 \quad 4$$

$$\text{1st D.D.} = \frac{1}{11} \quad \frac{1}{12} \quad \frac{1}{11}$$

$$\text{2nd D.D.} = \frac{-1}{1716} \quad \frac{-1}{1188}$$

$$\text{3rd D.D.} = \frac{1}{77220}.$$

Formula (2.3.4) with $k = 4$ then gives

$$\sqrt{33\cdot4} \doteqdot 6 + \frac{(-2\cdot6)}{11} - \frac{(-2\cdot6)(8\cdot4)}{1716} + \frac{(-2\cdot6)(8\cdot4)(-15\cdot6)}{77220}$$

$$\doteqdot 5\cdot7808 \text{ to 4 d.p.}$$

The correct result to 4 d.p. is $5\cdot7792$ so the error is $0\cdot0016$, compared to $-0\cdot0029$ when we used only 3 points.

We now observe that this result could have been obtained more easily from the result of Example (2.3.2) since we have only introduced one extra function value $(x_4 = 16, f(x_4) = 4)$ so that

$$\text{(New approximation to } \sqrt{33\cdot4})$$

$$= \text{(Old approximation to } \sqrt{33\cdot4})$$

$$+ (33\cdot4 - 36)(33\cdot4 - 25)(33\cdot4 - 49)[36, 25, 49, 16]$$

and since, from the divided difference table,

$$[36, 25, 49, 16] = \frac{1}{77,220}$$

we have: $\sqrt{33\cdot4} \doteqdot 5\cdot7764 + 0\cdot0044 = 5\cdot7808$ as we obtained before.

Looked at like this the way in which we should use formula (2.3.4) for interpolation becomes apparent:

(i) From the data and the difference table decide what accuracy we might hope to achieve; let this be m decimal places;

(ii) Let $x_1, x_2, \ldots, x_n$ be the points at which function values are available and let x be the point at which we wish to interpolate;

(iii) Let y_1 be that point of the set $\{x_1, x_2, \ldots, x_n\}$ which is closest to x, let y_2 be the point which is second closest, and so on;

(iv) Use formula (2.3.4) with y_1 in place of x_1, y_2 in place of x_2 etc, stopping as soon as a term occurs which is less than $\frac{1}{2} \times 10^{-m}$ in absolute value [see § 1.3].

We cannot of course guarantee that our result will be correct to m d.p. for we are using a polynomial to approximate to a function which is probably not a polynomial and may be of unknown analytical form. We can however claim that we have made a reasonable estimate and we might be able to make some comment about the possible accuracy as the next example indicates.

EXAMPLE 2.3.4.   From the table of values of a function, $f(x)$, given below estimate the value of $f(4\cdot463)$.

$$x = 4\cdot0 \qquad 4\cdot1 \qquad 4\cdot2 \qquad 4\cdot3 \qquad 4\cdot4$$
$$f(x) = 0\cdot05882 \quad 0\cdot05615 \quad 0\cdot05365 \quad 0\cdot05131 \quad 0\cdot04912$$

$$x = 4\cdot5 \qquad 4\cdot6 \qquad 4\cdot7 \qquad 4\cdot8 \qquad 4\cdot9$$
$$f(x) = 0\cdot04706 \quad 0\cdot04513 \quad 0\cdot04331 \quad 0\cdot04160 \quad 0\cdot03998.$$

Solution.   Since the values are given to 5 d.p. there is no point in seeking accuracy greater than 5 d.p., but we cannot hope to achieve this because when we difference two terms we have a potential rounding error of 10^{-5} and then on dividing by the distance between them ($0\cdot1$ if they are adjacent) this potential error becomes 10^{-4}. When we introduce second differences the potential error becomes

$$\frac{2 \times 10^{-4}}{2 \times 10^{-1}} = 10^{-3}$$

and so on. Evidently we should not use too many terms in the formula since each new term reduces the possible accuracy by one decimal place! On the other hand unless we use *some* divided differences we cannot carry out the interpolation at all; it follows that we must proceed cautiously.

The first lines of the D.D. table are shown. It is clear that there is no point in using third divided differences since rounding errors in the original data could cause these to be in error by 100%. We therefore confine ourselves to first and second divided differences.

x	$f(x)$	1st D.D.	2nd D.D.
4·0	0·05882		
4·1	0·05615	−0·0267	+0·0085
4·2	0·05365	−0·0250	+0·0080
4·3	0·05131	−0·0234	+0·0075
4·4	0·04912	−0·0219	+0·0065
4·5	0·04706	−0·0206	+0·0065
4·6	0·04513	−0·0193	+0·0055
4·7	0·04331	−0·0182	+0·0055
4·8	0·04160	−0·0171	+0·0045
4·9	0·03998	−0·0162	

With $x = 4\cdot463$ we take $y_1 = 4\cdot5$, $y_2 = 4\cdot4$, $y_3 = 4\cdot6$ and formula (2.3.4) then gives

$$f(4\cdot463) \doteqdot f(4\cdot5) + (4\cdot463 - 4\cdot5)[4\cdot5, 4\cdot4]$$

$$+ (4\cdot463 - 4\cdot5)(4\cdot463 - 4\cdot4)[4\cdot5, 4\cdot4, 4\cdot6]$$

$$= 0\cdot04706 + (-0\cdot037)(-0\cdot0206) + (-0\cdot037)(+0\cdot063)(0\cdot0065)$$

$$= 0\cdot04706 + 0\cdot00076 - 0\cdot00002 = \underline{0\cdot04780}.$$

On the basis of the argument above we would not have any justification for quoting this result to more than 3 d.p. $(0\cdot048)$; had we stopped after the first D.D. however we might have quoted 4 d.p. $(0\cdot0478)$—and this value is the same when we used the second D.D. Consequently we tentatively suggest $0\cdot0478$ as a 4 d.p. approximation to $f(4\cdot463)$.

It happens, in this case, that the function $f(x)$ is known viz

$$f(x) = \frac{1}{1 + x^2}$$

and so

$$f(4\cdot463) = 0\cdot047804874 \text{ to 9 d.p.}$$

and we see that our 4 d.p. estimate is correct, and, although we had no right to expect it the result is correct to 5 d.p.

In this example the data points were equally spaced and alternative methods are available for dealing with such cases as we now see.

2.3.1.   Interpolation Based on Equidistant Points

When the values of a function are available at equally spaced points a formula for interpolation due to Everett is commonly used, particularly if the points at which interpolation is required do not lie too near the beginning or end of the table. Everett's formula uses central differences of even order [see § 1.4].

Suppose that we know the value of a function at a set of equidistant points $x_1, x_2, \ldots, x_n, \ldots$, the corresponding function values being $f(x_1), f(x_2), \ldots, f(x_n) \ldots$ and that we wish to estimate its value at a point x which lies between x_n and x_{n+1}. Since the points x_i are equidistant we have

$$x_{n+1} - x_n = x_n - x_{n-1} = \ldots = x_2 - x_1 = h \text{ (say)}$$

and so we write

$$x = x_n + ph \tag{2.3.5}$$

where $0 < p < 1$.

We also write $f_n = f(x_n)$ so that

$$\delta f_n = \delta f(x_n) = f(x_{n+\frac{1}{2}}) - f(x_{n-\frac{1}{2}}) \tag{2.3.6}$$

and, as shown at (1.4.7)

$$\delta^2 f_n = f_{n+1} - 2f_n + f_{n-1}. \tag{2.3.7}$$

Finally, it is convenient to write

$$q = (1-p) \qquad (2.3.8)$$

so that $x = x_{n+1} - (1-p)h$. Then Everett's interpolation formula may be written.

$$f(x) = f(x_n + ph) = qf_n + \frac{(q+1)q(q-1)}{3!}\delta^2 f_n$$

$$+ \frac{(q+2)(q+1)q(q-1)(q-2)}{5!}\delta^4 f_n + \ldots + pf_{n+1} + \frac{(p+1)p(p-1)}{3!}\delta^2 f_{n+1}$$

$$+ \frac{(p+2)(p+1)p(p-1)(p-2)}{5!}\delta^4 f_{n+1} + \ldots \qquad (2.3.9)$$

or, more succinctly

$$f(x) = f(x_n + ph) = qf_n + \binom{q+1}{3}\delta^2 f_n + \binom{q+2}{5}\delta^4 f_n + \ldots$$

$$+ pf_{n+1} + \binom{p+1}{3}\delta^2 f_{n+1} + \binom{p+2}{5}\delta^4 f_{n+1} + \ldots \qquad (2.3.10)$$

[see I (3.8.1)].

EXAMPLE 2.3.5.   Use Everett interpolation to estimate the value of $f(3 \cdot 627)$ from the values given below

$$x = 3 \cdot 45 \quad\quad 3 \cdot 5 \quad\quad 3 \cdot 55 \quad\quad 3 \cdot 60 \quad\quad 3 \cdot 65 \quad\quad 3 \cdot 70 \quad\quad 3 \cdot 7$$
$$f(x) = 1 \cdot 11282 \ \ 1 \cdot 11927 \ \ 1 \cdot 12559 \ \ 1 \cdot 13178 \ \ 1 \cdot 13786 \ \ 1 \cdot 14382 \ \ 1 \cdot 14968.$$

Solution.   The value of h, from (2.3.5), is $0 \cdot 05$ and we take $x_n = 3 \cdot 60$ so that the values of p and q are $0 \cdot 54$ and $0 \cdot 46$ respectively.

We form the central difference table of even powers of δ using (2.3.7) repeatedly first on the values of f, then on the values of $\delta^2 f$ and so on; the resulting table is shown below

x	$f(x)$	$\delta^2 f$	$\delta^4 f$
3·45	1·11282		
3·50	1·11927	−0·00013	
3·55	1·12559	−0·00013	+0·00002
3·60	1·13178	−0·00011	−0·00003
3·65	1·13786	−0·00012	+0·00003
3·70	1·14382	−0·00010	
3·75	1·14968		

The small fluctuating values of $\delta^4 f$ indicate the presence of rounding error and suggest that we should ignore them, using only the first two terms on each line of (2.3.9) viz

$$f(3{\cdot}627) \doteqdot (0{\cdot}46)(1{\cdot}13178) + \frac{(1{\cdot}46)(0{\cdot}46)(-0{\cdot}54)}{6}(-0{\cdot}00011)$$

$$+ (0{\cdot}54)(1{\cdot}13786) + \frac{(1{\cdot}54)(0{\cdot}54)(-0{\cdot}46)}{6}(-0{\cdot}00012)$$

$$= 0{\cdot}52063 + 0{\cdot}61445 \quad \text{(to 5 d.p.)}$$

$$= 1{\cdot}13508.$$

Had we used the values of $\delta^4 f$ the result would have been the same for the additional terms are too small to affect the result to 5 d.p. The value obtained is correct to 5 d.p. as we now verify by direct calculation since the function tabulated is $\sqrt{\log_e x}$.

EXAMPLE 2.3.6. Given that the function in the previous example is $\sqrt{\log_e x}$, compute the value of $f(3{\cdot}627)$ directly using $f(3{\cdot}60)$ and the Taylor Series for $\log_e (1+x)$ [see IV, Example 3.6.3].

Solution

$$\log_e (3{\cdot}627) = \log_e (3{\cdot}60) + \log_e (1{\cdot}0075).$$

Now

$$\log_e (1+x) = x - \frac{x^2}{2} + \frac{x^3}{3} - \frac{x^4}{4} + \ldots \qquad (2.3.11)$$

and we may find $\log_e (1{\cdot}0075)$ from (2.3.11), but how many terms should we use? The values in the table are given to 5 d.p. and if we put $x = 0{\cdot}0075$ in (2.3.11) and use k terms the truncation error will be less than

$$\frac{(0{\cdot}0075)^{k+1}}{(k+1)} \qquad (2.3.12)$$

in absolute value (since the terms of (2.3.11) oscillate in sign when x is positive), hence for 5 d.p. accuracy we must choose k so that

$$\frac{(0{\cdot}0075)^{k+1}}{k+1} < \tfrac{1}{2} \times 10^{-5} \qquad (2.3.13)$$

the smallest solution of which is $k = 2$. We therefore take

$$\log_e (1{\cdot}0075) \doteqdot 0{\cdot}0075 - \frac{(0{\cdot}0075)^2}{2} = 0{\cdot}00747 \quad \text{(to 5 d.p.)}$$

and so

$$\log_e (3{\cdot}627) \doteqdot (1{\cdot}13178)^2 + 0{\cdot}00747 = 1{\cdot}28840.$$

Finally therefore

$$f(3 \cdot 627) = \sqrt{\log_e (3 \cdot 627)} \div \sqrt{1 \cdot 28840} = 1 \cdot 13508 \quad \text{to 5 d.p.}$$

in agreement with the result obtained before.

Everett's formula uses information from the tabular points closest to and symmetrically about the point of interpolation. The number of tabular points used is 2 if no differences are used, 4 if δ^2 differences are used, 6 if δ^4 differences are used, and so on. In Example 2.3.4 we used 3 tabular points to estimate a value for $f(4 \cdot 463)$ but since $4 \cdot 463$ is close to the centre of the table and the tabular points are equally spaced Everett's formula may be used and if we take differences up to δ^2 we will be using 4 tabular points and would expect a better result than we obtained in Example 2.3.4, and in fact we find

$$f(4 \cdot 463) \doteqdot (0 \cdot 37)(0 \cdot 04912) + \frac{(1 \cdot 37)(0 \cdot 37)(-0 \cdot 63)}{6}(0 \cdot 00013)$$

$$+ (0 \cdot 63)(0 \cdot 04706) + \frac{(1 \cdot 63)(0 \cdot 63)(-0 \cdot 37)}{6}(0 \cdot 00013)$$

$$\doteqdot 0 \cdot 047807$$

so that the error is less than $0 \cdot 000003$.

2.3.2.　Subtabulation

Sometimes we have available the values of a function $f(x)$ tabulated at a certain interval, h, and we wish to construct a table of values based upon a smaller interval such as $\frac{1}{2}h$ or $\frac{1}{10}h$. Clearly we are faced with a special form of the problem of interpolation and it is generally referred to as the problem of *subtabulation*.

There are two methods by which subtabulation might be achieved:

(i) by direct calculation, if $f(x)$ is a known function which is readily computable;

(ii) by interpolation if $f(x)$ is unknown, or is known but difficult to compute directly.

Only the second method concerns us here, it has the considerable advantage that it is very easy to program on a computer and the program can be applied to any function whose tabulated values can be provided. If we use Everett's formula the coefficients in (2.3.9) can be worked out in advance since the required values of p are known. Thus in the simplest case where $p = \frac{1}{2}$, so that we are doubling the number of points of tabulation, we have from (2.3.9)

$$f(x_n + \tfrac{1}{2}h) = \tfrac{1}{2}f_n - \tfrac{1}{16}\delta^2 f_n + \tfrac{3}{256}\delta^4 f_n - \tfrac{5}{2048}\delta^6 f_n + \ldots$$

$$+ \tfrac{1}{2}f_{n+1} - \tfrac{1}{16}\delta^2 f_{n+1} + \tfrac{3}{256}\delta^4 f_{n+1} - \tfrac{5}{2048}\delta^6 f_{n+1} + \ldots . \qquad (2.3.14)$$

Formulae based on Everett interpolation, such as (2.3.14) are very useful for subtabulation except at points close to either end of the table where values of δ^2, δ^4 etc may not be available. For subtabulation near these extremal points Newton's interpolation formula (2.3.4) is available.

Let the function values be known at the equidistant points $x_1 < x_2 < \ldots < x_n$ so that $x_{i+1} - x_i = h$ and suppose that we wish to subtabulate. For points in the interval (x_1, x_2) or in the interval (x_{n-1}, x_n) Everett's formula cannot be used so we use (2.3.4) but take advantage of the fact that the natural choice of points on which to base the interpolation is in the order $x_1, x_2, x_3, \ldots$ for the former interval and $x_n, x_{n-1}, x_{n-2} \ldots$ for the latter, and furthermore the points are equally spaced.

If we apply (2.3.4) to interpolate a value at $x = x_1 + ph$ where $0 < p < 1$ then x lies in (x_1, x_2) and (2.3.4) becomes

$$f(x_1 + ph) \doteqdot f(x_1) + ph[x_1 x_2] + p(p-1)h^2[x_1 x_2 x_3] + \ldots . \qquad (2.3.15)$$

Now

$$[x_1 x_2] = \frac{f(x_2) - f(x_1)}{x_2 - x_1} = \frac{f(x_2) - f(x_1)}{h} = \frac{1}{h}\Delta f_1$$

and so

$$[x_1 x_2 x_3] = \frac{[x_2 x_3] - [x_1 x_2]}{x_3 - x_1} = \frac{1}{2h^2}(\Delta f_2 - \Delta f_1) = \frac{\Delta^2 f_1}{2h^2}$$

and proceeding in this way we find

$$[x_1 x_2 \ldots x_{k+1}] = \frac{\Delta^k f_1}{h^k k!}. \qquad (2.3.16)$$

Substitution of (2.3.16) in (2.3.15) gives *Newton's Forward Difference Formula*:

$$f(x_1 + ph) \doteqdot f_1 + p\,\Delta f_1 + \frac{p(p-1)}{2!}\Delta^2 f_1 + \frac{p(p-1)(p-2)}{3!}\Delta^3 f_1 + \ldots . \qquad (2.3.17)$$

Similarly for points in the interval (x_{n-1}, x_n) we may use *Newton's Backward Difference Formula*:

$$f(x_n - ph) \doteqdot f_n - p\,\nabla f_n + \frac{p(p-1)}{2!}\nabla^2 f_n - \frac{p(p-1)(p-2)}{3!}\nabla^3 f_n + \ldots . \qquad (2.3.18)$$

EXAMPLE 2.3.7.   Subtabulate the following table, changing the interval from 0·1 to 0·05

$x = 1\cdot0$	1·1	1·2	1·3	1·4
$f(x) = 0\cdot62996$	0·60980	0·59118	0·57392	0·55786

$x = 1\cdot5$	1·6	1·7	1·8	1·9
$f(x) = 0\cdot54288$	0·52887	0·51573	0·50338	0·49174.

Solution.   The central difference table of orders 2 and 4 is shown below

x	$f(x)$	δ^2	δ^4
1·0	0·62996		
1·1	0·60980	0·00154	
1·2	0·59118	0·00136	0·00002
1·3	0·57392	0·00120	0·00004
1·4	0·55786	0·00108	0·00001
1·5	0·54288	0·00097	0·00001
1·6	0·52887	0·00087	0·00002
1·7	0·51573	0·00079	0·00000
1·8	0·50338	0·00071	
1·9	0·49174		

Since $p = \frac{1}{2}$ we may use (2.3.14) and since $\frac{3}{256}\delta^4 < 10^{-6}$ in all cases we need only take the first two pairs of terms viz

$$f(x_n + \tfrac{1}{2}h) \doteq \tfrac{1}{2}(f_n + f_{n+1}) - \tfrac{1}{16}\delta^2(f_n + f_{n+1}). \tag{2.3.19}$$

Formula (2.3.19) can only be applied from $x_n = 1\cdot1$ to $x_n = 1\cdot7$ inclusive; outside this range at least one of the required δ^2 values is not available. Using (2.3.19) within this range we obtain

$$
\begin{array}{llllll}
x = 1\cdot15 & 1\cdot25 & 1\cdot35 & 1\cdot45 & 1\cdot55 \\
f(x) = 0\cdot60031 & 0\cdot58239 & 0\cdot56575 & 0\cdot55024 & 0\cdot53576 \\
\\
x = 1\cdot65 & 1\cdot75 \\
f(x) = 0\cdot52220 & 0\cdot50946.
\end{array}
$$

For the two remaining values, at $x = 1\cdot05$ and $x = 1\cdot85$ we use (2.3.17) and (2.3.18) respectively. For the former we require some forward differences, the appropriate section of the table being

x	$f(x)$	Δ	Δ^2	Δ^3	Δ^4
1·0	0·62996				
		−0·02016			
1·1	0·60980		0·00154		
		−0·01862		−0·00018	
1·2	0·59118		0·00136		+0·00002
		−0·01726		−0·00016	
1·3	0·51392		0·00120		
		−0·01606			
1·4	0·55786				

From (2.3.17) with $p = \frac{1}{2}$

$$f(1\cdot05) \doteqdot 0\cdot62996 - \tfrac{1}{2}(0\cdot02016) - \tfrac{1}{8}(0\cdot00154) - \tfrac{1}{16}(0\cdot00018) - \tfrac{5}{128}(0\cdot00002)$$

i.e. $f(1\cdot05) \doteqdot 0\cdot61968.$

Finally using (2.3.18) with $x_n = 1\cdot9$, $p = \frac{1}{2}$, $h = 0\cdot1$ and the appropriate section of the table of backward differences, viz.

x	$f(x)$	∇	∇^2	∇^3	∇^4
1·5	0·54288				
		−0·01401			
1·6	0·52887		0·00087		
				−0·00008	
		−0·01314			0·00000
1·7	0·51573		0·00079		
				−0·00008	
		−0·01235			
			0·00071		
1·8	0·50338				
		−0·01164			
1·9	0·49174				

we find

$$f(1\cdot85) \doteqdot 0\cdot49174 + \tfrac{1}{2}(0\cdot01164) - \tfrac{1}{8}(0\cdot00071) + \tfrac{1}{16}(0\cdot00008)$$

i.e. $f(1\cdot85) \doteqdot 0\cdot49747.$

The merits of these relatively simple formulae are apparent from the fact that all the values interpolated in this example are correct to 5 d.p. as may easily be verified since $f(x) = (1+x)^{-2/3}$.

2.3.3.   Inverse Interpolation

The problem of inverse interpolation is that of finding a point at which a function takes a specific value. If the function is defined analytically the problem can be dealt with by one of the methods for solving non-linear equations (Chapter 5) but if the function is defined only by means of a table of values techniques based upon interpolation formulae must be used. We suppose then that we have a set of points $x_1, x_2, \ldots, x_n$ and known function values $f(x_1), f(x_2), \ldots, f(x_n)$ at these points and we wish to find a point x at which $f(x)$ takes a specified value z (say).

We shall shortly assume that the points $x_1, x_2, \ldots, x_n$ are equally spaced (as is the case with most tables) for if not the problem must be treated as one of ordinary interpolation with unequal interval sizes, using Newton's divided differences formulae; we saw a special case of this in Examples 2.3.2 and 2.3.3 but we now illustrate the use of the method in a more general case.

EXAMPLE 2.3.8.   Given the following data estimate the value of x which makes $f(x) = 3.5$.

$$x = 4\cdot236 \quad 5\cdot018 \quad 6\cdot104 \quad 6\cdot851 \quad 7\cdot304 \quad 7\cdot698$$
$$f(x) = 2\cdot756 \quad 3\cdot043 \quad 3\cdot427 \quad 3\cdot782 \quad 3\cdot834 \quad 3\cdot963.$$

Solution.   Since the problem is one of inverse interpolation we form a divided difference table of x with respect to $f(x)$—the inverse of our usual procedure. The table, up to third divided differences is shown below

$f(x)$	x	1st D.D.	2nd D.D.	3rd D.D.
2·756	4·236			
		2·725		
3·043	5·018		0·154	
		2·828		0·004
3·427	6·104		0·158	
		2·929		−0·042
3·682	6·851		0·125	
		2·980		0·257
3·834	7·304		0·263	
		3·054		
3·963	7·698			

The changes in the values of the third divided differences make them of doubtful value; even the second divided differences are not too satisfactory and it is clear that we cannot hope to obtain a very accurate result.

If we take $x = 3\cdot5$, $x_1 = 3\cdot427$, $x_2 = 3\cdot682$ and use only first divided differences then, putting $x = f^{-1}(y)$ $(=f^{-1}f(x))$, (2.3.4) gives

$$f^{-1}(3\cdot5) \doteqdot f^{-1}(3\cdot427) + (3\cdot5 - 3\cdot427)2\cdot929$$
$$= 6\cdot104 + (0\cdot073)(2\cdot929) = \underline{6\cdot318}.$$

If we take $x_2 = 3\cdot834$ and introduce the second divided difference (2.3.4) brings in the extra term

$$+(3\cdot5 - 3\cdot427)(3\cdot5 - 3\cdot834)0\cdot125 = \underline{-0\cdot003}$$

which gives the estimate

$$f^{-1}(3\cdot5) \doteqdot \underline{6\cdot315}.$$

It would be unwise to proceed any further and we leave our estimate as 6·315. In this example the tabulated function was $f(x) = (x^2 + 3)^{1/3}$ so that the value of x which makes $f(x) = 3\cdot5$ is $6\cdot314665\ldots$ i.e. 6·315 to 3 d.p. and we see that our estimate is correct to 3 d.p. We have however been lucky, on the basis of the difference table we had no reason to expect such accuracy. A similar calculation for the case $x = 3\cdot85$ yields only 2 d.p. accuracy and it is not difficult to construct

examples where the value obtained by this method is a very poor approximation to the true value. In practice of course we will not know the true value and
if we are simply given a set of values of a function at non-equidistant points we
will have no alternative to using a method essentially the same as that used
above. The object of these remarks therefore is not to suggest using an
alternative method, for there is none, but to note that the estimated accuracy of
the result should be based upon the range and size of the numbers in the table of
divided differences rather than on hope. If we look again at Example 2.3.8 and
recognise that our result was correct to 3 d.p. only because of a rather doubtful
use of the second divided difference we must admit that a result correct to 2 d.p.
was the best that we could reasonably expect.

The more typical case of inverse interpolation arises when the x values are
equally spaced thus providing a feature which we should exploit. A method
which has proved to be satisfactory for such cases is one based upon the
iterative use of Everett's formula (2.3.9) which we first illustrate with a simple
example.

EXAMPLE 2.3.9.   From the table below estimate the value of x for which
$f(x) = 0.5$

x	$f(x)$	δ^2	δ^4
0·2	0·1987		
0·3	0·2955	−0·0029	
0·4	0·3894	−0·0039	+0·0001
0·5	0·4794	−0·0048	+0·0001
0·6	0·5646	−0·0056	0·0000
0·7	0·6442	−0·0064	
0·8	0·7174		

Solution.   The required value evidently lies between 0·5 and 0·6, let it therefore be $x = 0.5 + ph$ (where $h = 0.1$). From Everett's formula (2.3.9) we have

$$0.5 = f(0.5 + ph) = qf(0.5) + \frac{(q+1)q(q-1)}{6}\delta^2 f(0.5) + \ldots$$

$$+ pf(0.6) + \frac{(p+1)p(p-1)}{6}\delta^2 f(0.6) + \ldots \qquad (2.3.20)$$

As a first approximation we ignore all but the linear terms on the right-hand
side of (2.3.20) which gives

$$0.5 \doteqdot q(0.4794) + p(0.5646)$$

and since $q = 1 - p$ this tells us that $p \doteqdot 0.2418$, so that $q \doteqdot 0.7582$.

We now substitute these approximations for p, q in the *non-linear* terms in (2.3.20) so producing, if we ignore terms in δ^4 and higher

$$0 \cdot 5 \doteqdot q(0 \cdot 4794) + \frac{(1 \cdot 7582)(0 \cdot 7582)(-0 \cdot 2418)}{6}(-0 \cdot 0048)$$

$$+ p(0 \cdot 5646) + \frac{(1 \cdot 2418)(0 \cdot 2418)(-0 \cdot 7582)}{6}(-0 \cdot 0056)$$

and solving this linear equation for p produces a revised estimate $p \doteqdot 0 \cdot 2363$, so that $q \doteqdot 0 \cdot 7637$.

In this case the contributions to (2.3.20) made by the terms involving δ^4 are insignificant to 4 d.p. so we ignore them but attempt to improve our estimate for p by repeating the procedure above using the most recent estimates for p, q in the non-linear terms, viz.

$$0 \cdot 5 \doteqdot q(0 \cdot 4794) + \frac{(1 \cdot 7637)(0 \cdot 7637)(-0 \cdot 2363)}{6}(-0 \cdot 0048)$$

$$+ p(0 \cdot 5646) + \frac{(1 \cdot 2363)(0 \cdot 2363)(-0 \cdot 7837)}{6}(-0 \cdot 0056)$$

which produces a new approximation $p \doteqdot 0 \cdot 2364$. Further changes will now be insignificant so we accept the value $x = 0 \cdot 5 + 0 \cdot 2364h = 0 \cdot 5236$ to 4 d.p. as the solution.

The tabulated function in this problem was $\sin x$, where x is measured in radians. The exact solutions is therefore $\pi/6 = 0 \cdot 5236$ to 4 d.p.; so our solution is accurate to 4 d.p.

The method employed in Example 2.3.9 is now described more formally. (1) Locate an interval $[x_0, x_1]$ (where $x_1 = x_0 + h$) containing the solution; let the solution be $x = x_0 + ph$. (2) Compute a first approximation, p_1, to the value of p by solving the linear equation obtained by ignoring all but the linear terms of (2.3.9) viz:

$$f(x) = f(x_0 + ph) = (1 - p)f(x_0) + pf(x_1).$$

(3) Substitute p_1 and $q_1 (= 1 - p_1)$ in all the non-linear terms it is intended to use in (2.3.9) and so obtain a new linear equation for p which provides a revised estimate, p_2, for p. (4) Repeat (3) at each stage using the latest estimate for p to deduce a better estimate, so obtaining a sequence of approximations p_1, p_2, p_3, ... for p continuing until the process converges i.e. until consecutive values agree to the appropriate number of decimal places. If the final value for p is p_n (say) the solution is $x = x_0 + p_n h$.

If the analytical form of the tabulated function, $f(x)$ is known the problem of inverse interpolation can be solved by analytical methods which are covered in Chapter 5 but if $f(x)$ is of unknown analytic form we must rely on a method such as the one just described. Even when $f(x)$ is known this method is occasionally to be preferred to the analytical methods, particularly if $f(x)$ is a complicated function. In the next example the function involved is well-known and the problem could be solved by other means but the method of inverse interpolation yields a reasonably accurate result with little effort.

EXAMPLE 2.3.10.   From the table of the normal distribution function, $f(x)$, estimate the value of x which makes $f(x) = 0.75$.

Note:

$$f(x) = \frac{1}{\sqrt{2\pi}} \int_{-\infty}^{x} e^{-\frac{1}{2}t^2} \, dt \qquad [\text{see II, § 11.4.1}]$$

x	$f(x)$	δ^2	δ^4
0·4	0·6554		
0·5	0·6915	−0·0019	
0·6	0·7257	−0·0019	−0·0003
0·7	0·7580	−0·0022	+0·0002
0·8	0·7881	−0·0023	−0·0000
0·9	0·8159	−0·0024	
1·0	0·8413		

Solution.   The solution lies in $\langle 0.6, 0.7 \rangle$, let it be $x = 0.6 + ph$, where $h = 0.1$. The linear approximation to (2.3.9) gives (writing p_1 in place of p)

$$0.75 \doteqdot (1 - p_1)(0.7257) + p_1(0.7580)$$

which gives

$$p_1 \doteqdot \frac{0.0243}{0.0323} = \underline{0.7523}, \quad \text{hence } q_1 \doteqdot \underline{0.2477}.$$

Since δ^4 is small in absolute value and seems to be affected by rounding errors we will only use terms up to δ^2 in (2.3.9) which becomes, in its iterative form

$$0.75 \doteqdot (1 - p_2)(0.7257) + \frac{(q_1 + 1)(q_1)(q_1 - 1)}{6}(-0.0019)$$

$$+ p_2(0.7580) + \frac{(p_1 + 1)(p_1)(p_1 - 1)}{6}(-0.0022)$$

which gives, as our second estimate

$$p_2 = 0.7463, \quad \text{so that } q_2 = 0.2537.$$

The operation is now repeated with (p_3, q_3) in place of (p_2, q_2) and (p_2, q_2) in place of (p_1, q_1) the result is

$$p_3 = 0.7462, \qquad q_3 = \underline{0.2538}.$$

Further iterations produce no change in these values to 4 d.p. so we conclude
that

$$x = 0.6 + (0.7462)(0.1) = \underline{0.6746} \quad \text{to 4 d.p.}$$

The result is correct to 3 d.p., the true value being 0.6745.

2.3.4. Lagrange's Interpolation Formula

Newton's divided difference formula (2.3.4) can be used to find the polynomial of degree $(k-1)$ which takes the values $f(x_i)$ at the points $x_i(i =
1, \ldots, k)$. An alternative expression for this polynomial which is of consider-
able theoretical importance is due to Lagrange and is obtained as follows.
The function

$$\Pi_m(x) = \prod_{\substack{n=1 \\ n \neq m}}^{k} \frac{(x - x_n)}{(x_m - x_n)} \tag{2.3.21}$$

is a polynomial of degree $(k-1)$ which is zero at $x = x_1, x_2, \ldots, x_{m-1}$,
$x_{m+1}, \ldots, x_k$ and equal to 1 at $x = x_m$. It therefore follows that

$$P_k(x) = \sum_{m=1}^{k} f(x_m) \Pi_m(x) \tag{2.3.22}$$

is the polynomial of degree at most $(k-1)$ which takes the values $f(x_i)$ at the
points $x_i(i = 1, \ldots, k)$. Formulae (2.3.22) and (2.3.4) provide different
expressions for the same polynomial. Either may be used for interpolation at
points x other than the given points x_i; when (2.3.22) is used in this way it is
called *Lagrange's Interpolation Formula* and it can be proved that if $f(x)$ is at
least k-times differentiable then the error in using (2.3.22) is

$$f(x) - P_k(x) = \frac{f^{(k)}(\alpha)}{k!} \prod_{n=1}^{k} (x - x_n) \tag{2.3.23}$$

for some α in $\langle x_1, x_k \rangle$ (see e.g. Atkinson, pp. 110–111). The importance of
Lagrange's formula lies mainly in the relative simplicity of the expression
(2.3.23) for the error estimate. In practice Newton's formula is more convenient to use since additional points $(x_{k+1}, f(x_{k+1}))$ etc. can be introduced
more easily in (2.3.4) than in (2.3.22).

2.3.5. Other Methods of Interpolation

There are a number of other interpolation formulae which are similar in
appearance to Everett's formula (2.3.9) and which are sometimes preferred in
particular cases. Some of these formulae, including those due to Gauss, Bessel,
Stirling and Steffensen will be found in Fröberg, pp. 174–180.
When $f(x)$ is available in analytic form, so that computation of $f(x)$ and its
derivatives is possible, it is sometimes desirable to approximate $f(x)$ by a

polynomial which takes the same values and has the same derivatives as $f(x)$ at certain points. The polynomial may then be used for interpolation. Such methods of interpolation are covered in Chapter 6, section 5.

2.4. EXTRAPOLATION

When the value of x, of the point at which we wish to estimate the value of a function, lies outside the range of the set of points $\{x_1, x_2, \ldots, x_n\}$ at which the value of the function is known we have the problem of extrapolation.

Suppose, for simplicity, that

$$x_1 < x_2 < x_3 \ldots < x_n < x$$

then we use formula (2.3.4) or, if the points are equidistant (2.3.18), we will have to choose $y_1 = x_n$, $y_2 = x_{n-1}$, etc. The fact that all these points lie on the same side of x introduces a lack of symmetry which is unfortunate but unavoidable; in consequence we cannot expect our results to be as satisfactory as might be obtained by interpolation using the central difference formula (2.3.9). Let us see what happens if we re-work Example 2.3.4 using only points to one side of the point in which we are interested (4·463).

EXAMPLE 2.4.1.   Use the data of Example 2.3.4 to estimate a value for $f(4\cdot463)$ by extrapolation (i) to the left of 4·463, (ii) to the right of 4·463.

Solution.

(i) We take $y_1 = 4\cdot4$, $y_2 = 4\cdot3$, $y_3 = 4\cdot2$ and so obtain the estimate:

$$f(4\cdot463) \doteqdot f(4\cdot4) + (0\cdot063)[4\cdot4, 4\cdot3] + (0\cdot063)(0\cdot163)[4\cdot4, 4\cdot3, 4\cdot2]$$

$$= 0\cdot04912 - 0\cdot00138 + 0\cdot0007 = \underline{0\cdot04781}.$$

(ii) We take $y_1 = 4\cdot5$, $y_2 = 4\cdot6$, $y_3 = 4\cdot7$ and obtain

$$f(4\cdot463) \doteqdot f(4\cdot5) + (-0\cdot037)[4\cdot5, 4\cdot6] + (-0\cdot037)(-0\cdot137)[4\cdot5, 4\cdot6, 4\cdot7]$$

$$= 0\cdot04706 + 0\cdot00071 + 0\cdot00003 = \underline{0\cdot04780}.$$

The second result is correct to 5 d.p. and the first to 4 d.p. and might be considered satisfactory in the context of extrapolation. We note however that in the example above the point of extrapolation was only just outside the range of the set $x_1, x_2 \ldots x_n$ and we would expect that as we attempt to extrapolate at more and more distant points the results will become less and less accurate.

This intuitive feeling can easily be justified mathematically for as x moves further and further away from x_n so the factor

$$(x - x_n)(x - x_{n-1}) \ldots (x - x_{n-k-1}) \tag{2.4.1}$$

which multiplies the kth divided difference $[x_n x_{n-1} \ldots x_{n-k}]$ becomes larger and larger so that any rounding errors will become increasingly magnified.

Thus in Example 2.4.1 with $k = 2$, $x = 4\cdot463$, $x_n = 4\cdot4$, $x_{n-1} = 4\cdot3$ the value of (2.4.1) was $0\cdot010269$ and with a potential rounding error of 10^{-3} in the second divided difference the potential error in the product is *in this case* only about 10^{-5}. If however we take say, $x = 3\cdot0$, $x_n = 4\cdot0$, $x_{n-1} = 4\cdot1$ in (2.4.1) the potential rounding error arising from its product with the second divided difference is about 10^{-3} and in fact we would have

$$f(3\cdot0) \doteq f(4\cdot0) + (3\cdot0 - 4\cdot0)[4\cdot0, 4\cdot1] + (3\cdot0 - 4\cdot0)(3\cdot0 - 4\cdot1)[4\cdot0, 4\cdot1, 4\cdot2]$$

which gives

$$f(3\cdot0) \doteq 0\cdot05882 + 0\cdot02670 + 0\cdot00935 = 0\cdot9487$$

which is a very poor result since the true value is $0\cdot1$. In order to obtain a better result we would have had to use higher order divided differences but since these are subject to serious rounding errors it would be essential to have the values in the original table to more places of decimals. If the original function is tabulated to 9 d.p. for example and we use divided differences up to the sixth order a value of $0\cdot09995$ is obtained which demonstrates that accuracy is achievable, though at some cost, if the original data are known with sufficient precision—which will, unfortunately, rarely be the case.

2.4.1.   Special Forms of Extrapolation

There are three special forms of extrapolation which are widely used when circumstances are appropriate: Aitken's δ^2 extrapolation, Richardson's extrapolation and Euler's transformation.

Aitken's Extrapolation may be applied to any case where we have a set of values $x_1, x_2, x_3, \ldots$ approaching a limit x in such a way that successive errors $(x - x_n)$, $(x - x_{n+1})$ are in approximately constant (geometric) ratio to each other i.e.

$$(x - x_{n+1}) \doteq r(x - x_n) \tag{2.4.2}$$

where $|r| < 1$. In this case we therefore have

$$\frac{(x - x_{n+1})}{(x - x_n)} \doteq \frac{(x - x_n)}{(x - x_{n-1})} \tag{2.4.3}$$

and it follows that from (2.4.3) that

$$x \doteq x_{n+1} - \frac{(x_{n+1} - x_n)^2}{x_{n+1} - 2x_n + x_{n-1}} \tag{2.4.4}$$

or, using operator notation [see § 1.4]

$$x \doteq x_{n+1} - \frac{(\Delta x_n)^2}{\delta^2 x_n}$$

(from which the process derives part of its name). From (2.4.4) a new series of approximations x_n' are computed and these will converge more quickly to x.

For certain types of problems it is known on theoretical grounds that (2.4.4) holds, for example in a number of methods for solving non-linear equations (see Chapter 5) where the convergence to the solution is linear (see § 5.4.3).

EXAMPLE 2.4.2.   Apply Aitken's δ^2 extrapolation to the sequence of values

$$1 \cdot 14286 \quad 1 \cdot 20968 \quad 1 \cdot 23884 \quad 1 \cdot 25116 \quad 1 \cdot 25630$$

$$1 \cdot 25843 \quad 1 \cdot 25931 \quad 1 \cdot 25967$$

which is converging linearly to $2^{1/3}$.

Solution.   We form the Δ, δ^2 values and compute the extrapolated values (x_{n+1}) which are given on the right. Note that the alignment of the values in each line has been arranged to make the use of (2.4.4) as simple as possible. Calculations are performed to 5 d.p.

x_{n+1}	Δx_n	$\delta^2 x_n$	$(\Delta x_n)^2/\delta^2 x_n$	x'_{n+1}
1·14286				
1·20968	0·06682			
1·23884	0·02916	−0·03766	−0·02258	1·26142
1·25116	0·01232	−0·01684	−0·00901	1·26017
1·25630	0·00514	−0·00718	−0·00368	1·25998
1·25843	0·00213	−0·00301	−0·00151	1·25994
1·25931	0·00088	−0·00125	−0·00062	1·25993
1·25967	0·00036	−0·00052	−0·00025	1·25992

Convergence of the extrapolated values to 5 d.p. accuracy has now occurred ($2^{1/3} = 1 \cdot 25992105$ to 8 d.p.) which represents a considerable acceleration of convergence compared to the original sequence (see Example 5.3.2).

Aitken's process should not be applied when the convergence is not linear for then (2.4.2) does not hold and so (2.4.4) is invalid and no acceleration can be expected.

Richardson's Extrapolation is best explained in the context of numerical integration where the following situation arises: we are trying to estimate the value of an integral using a parameter, h. The exact value of the integral (I) and the approximate value $I(h_1)$ obtained when the parameter has the value h_1 are related by an expression of the type

$$I = I(h_1) + Ah_1^n \tag{2.4.5}$$

where n is a *known* integer.

If we now repeat the calculation using the parameter value $\frac{1}{2}h_1$ in place of h_1 we will have

$$I \doteqdot I(\tfrac{1}{2}h_1) + A\left(\frac{h_1}{2}\right)^n \tag{2.4.6}$$

and we can eliminate the terms involving A from (2.4.5) and (2.4.6) to give a new estimate for I, viz.

$$I \doteq \frac{2^n I(\frac{1}{2}h_1) - I(h_1)}{2^n - 1}. \tag{2.4.7}$$

Specific applications of (2.4.7) will be found in the section on Romberg integration in Chapter 6, section 2. Richardson's extrapolation is sometimes referred to as 'the deferred approach to the limit'.

Euler's Transformation is applicable when we are summing a series of terms of alternating signs.

Let

$$s = a_0 - a_1 + a_2 - a_3 + \dots \tag{2.4.8}$$

be a convergent series of terms of alternating signs so that we may take all $a_n > 0$. We write down the numbers $a_0, a_1, a_2 \dots$ and form a table of forward differences $\Delta, \Delta^2, \Delta^2 \dots$ for as many orders as significant [see § 1.4]. Then Euler's transformation is

$$s = \frac{1}{2} \sum_{n=0}^{\infty} (-\frac{1}{2})^n \Delta^n a_0 \tag{2.4.9}$$

and frequently accelerates convergence to a remarkable extent, as the example below shows.

EXAMPLE 2.4.3.   Use Euler's transformation to find the sum to 9 d.p. of the series

$$s = 1 - \frac{1}{4} + \frac{1}{7} - \frac{1}{10} + \dots = \sum_{n=0}^{\infty} \frac{(-1)^n}{3n+1}.$$

Solution.   When the terms of the series at first decrease rapidly and then more slowly it is usually worth summing the first few terms directly and then applying Euler's transformation to the remainder. This is so in the present case so we write (summing 10 terms directly)

$$s = (1 - \frac{1}{4} + \frac{1}{7} - \dots - \frac{1}{28}) + \sum_{n=10}^{\infty} \frac{(-1)^n}{3n+1} = s_1 + s_2 \quad \text{(say)}.$$

Then $s_1 = 0.818742968$ to 9 d.p. and we apply Euler's Transformation to s_2. The difference table is shown below but to save space all numbers in the table have been multiplied by 10^9.

Applying Euler's transformation to s_2 gives

$$s_2 = \frac{10^{-9}}{2} (32258065 + \frac{2846300}{2} + \frac{461562}{4} + \frac{103851}{8} + \frac{28981}{16} + \frac{9450}{32} + \frac{3473}{64}$$

$$+ \frac{1409}{128} + \frac{631}{256}) = 0.016905880.$$

Our estimate for s is therefore $s_1 + s_2 = \underline{0.835648848}$.

n	a_n	Δ	Δ^2	Δ^3	Δ^4	Δ^5	Δ^6
10	32258065						
11	29411765	−2846300					
12	27027027	−2384738	461562				
13	25000000	−2027027	357711	−103851			
14	23255814	−1744186	282841	−74870	28981		
15	21739130	−1516684	227502	−55339	19531	9450	
16	20408163	−1330967	185717	−41785	13554	5977	3473
17	19230769	−1177394	153573	−32144	9641	3913	2064
18	18181818	−1048951	128443	−25130	7014	2627	1286

It can be proved by analytical methods that

$$s = \frac{1}{3}\left(\log 2 + \frac{\pi}{\sqrt{3}}\right) = 0\cdot835648848 \quad \text{to 9 d.p.}$$

from which we see that our numerical evaluation is accurate to 9 d.p. Had we not used Euler's transformation but attempted to sum the series directly we would have had to take into account all terms a_n satisfying

$$|a_n| < \frac{10^{-9}}{2} \tag{2.4.10}$$

in order to achieve a comparable result i.e. more than 660,000,000 terms; instead of the 19 we used with Euler's method.

R.F.C.

REFERENCES

Atkinson, K. E. (1978). *An Introduction to Numerical Analysis*, Wiley.
Fröberg, C. E. (1970). *Introduction to Numerical Analysis*, Addison-Wesley.

PROGRAMS

In the Appendix, programs will be found for interpolation using (1) divided differences, (2) Everett's method.

CHAPTER 3

The Solution of Systems of Linear Equations

3.1. INTRODUCTION

The problem of solving a pair of simultaneous linear equations in two unknowns, usually denoted by x and y, has been one of the corner-stones of the algebra course for countless numbers of school-children for several generations. Usually only one method of solving this problem is taught: the method of eliminating one of the variables. As an alternative it is sometimes pointed out that a solution can also be found by representing the 2 equations by straight lines on graph paper, the point of intersection of the lines providing the required solution. Whilst this is an interesting method it unfortunately suffers from two disadvantages:

(i) the accuracy of the solution is limited by the scale of the graph paper;
(ii) the method cannot readily be generalized to the case of n equations in n unknowns.

The elimination method suffers from neither of these defects and so is rightly regarded as more important.

Schoolchildren are frequently asked to solve 2 equations in 2 unknowns, occasionally 3 equations in 3 unknowns and only very rarely 4 equations in 4 unknowns. One very good reason for this is rapid increase in the amount of effort required to solve n equations in n unknowns as n increases. It can easily be proved that the number of multiplications required when we use the elimination method to solve n equations is proportional to n^3. In addition, unless we use checking procedures, there is a non-zero chance of our making an undetected error every time we carry out an arithmetic operation $(+, -, \times \text{ or } \div)$ and so as n increases the probability that we will arrive at a correct solution of the equations rapidly approaches zero. The probability of our making an error during any one arithmetic operation depends of course upon whether we are working by hand, or with a desk machine or using a computer. When we are using a computer the probability of computer error is nowadays very small but there is a larger probability that the data will not be correctly put into the computer or that the program for solving the equations contains errors.

We now work through a simple example, solving a pair of equations by the elimination method and noting how many arithmetic operations we perform in the course of the solution.

EXAMPLE 3.1.1.   Solve $4x - 5y = 7$, $3x + 2y = 4$. Give the answers to 3 places of decimals.

Solution.   We write the equations underneath one another

$$4x - 5y = 7$$
$$3x + 2y = 4$$

and decide to eliminate x. We therefore multiply the first equation throughout by 3

$$12x - 15y = 21$$

and so have carried out 3 multiplications. We multiply the second equations throughout by 4

$$12x + 8y = 16$$

and so increase the number of multiplications so far to 6. Subtracting the second of these new equations from the first we obtain

$$-23y = 5$$

which yields, on division, $y = -5/23$ ($=0 \cdot 217$ to 3 d.p.). We have found one of the variables at a cost of 6 multiplications, 2 subtractions and 1 division. If we now repeat the process but eliminate y instead of x we eventually find that $x = 34/23$ ($= 1 \cdot 478$ to 3 d.p.). Using this method we have therefore performed 12 multiplications, 4 subtractions or additions and 2 divisions, a total of 18 arithmetic operations. Alternatively, having found the value of y we might have substituted it in either of the equations and so found the value of x; this method which is known as back-substitution requires only 1 multiplication, 1 subtraction and 1 division giving a total of 7, 3 and 2 respectively i.e. 12 arithmetic operations in all. We shall see in section 3.6 that the number of arithmetic operations required in the case of 2 equations can be reduced to 11 $(4 + 4 + 3)$.

In the sections that follow we shall examine a number of methods of solving systems of linear equations, pointing out the advantages and disadvantages of each. Before we can do this however we must do some preliminary ground-work, defining terms and finding conditions for systems of linear equations to have a unique solution.

3.2. SYSTEMS OF LINEAR EQUATIONS

In Example 3.1.1 we were given 2 equations in 2 unknowns, x and y. Had we been given only one equation, say the first, $4x - 5y = 7$ we would have been unable to find a unique solution. For example $x = 3$, $y = 1$ satisfies this equation, and so also does $x = 8$, $y = 5$ and indeed we can confirm at once that if t is any number at all then $x = 3 + 5t$, $y = 1 + 4t$ satisfy the equation. We call this a *parametric* solution, the arbitrary number t being called a parameter. Thus there are an infinite number of solutions. It is only when we introduce the second equation: $3x + 2y = 4$ that we can fix the value of t, and so obtain a unique solution. For if our parametric solution of the first equation is to satisfy the second equation we must have

$$3(3 + 5t) + 2(1 + 4t) = 4$$

which gives $23t = -7$, or $t = \frac{-7}{23}$. Substitution then provides us with the unique solution

$$x = 3 - \tfrac{35}{23} = \tfrac{34}{23}, \qquad y = 1 - \tfrac{28}{23} = \tfrac{-5}{23}$$

which we obtained before.*

The geometrical interpretation makes it obvious that we need two equations in order to find a unique solution. It also indicates that a pair of equations will sometimes fail to provide a solution; for if we have a pair of parallel lines, there is no point of intersection and hence no solution. Such a case arises, for example, if our first equation is, as before

$$4x - 5y = 7$$

and our second is

$$4x - 5y = 8.$$

The elimination method leads to

$$0y = -1$$

which has no finite solution. The parametric methods leads to

$$7 = 8$$

a contradiction which implies that our 2 equations are inconsistent, and again we have no solution.

What happens if we have 3 equations in 2 unknowns? Since 2 equations in 2 unknowns in general have a unique solution, (x_0, y_0) say, there can only be a solution to the 3 equations if (x_0, y_0) also satisfies the third equation. Obviously this is not likely to happen except in very special circumstances, and we conclude that in general there will be no solution to 3 equations in 2 unknowns.

---

* If a system of equations in the n variables $x_1, x_2, \ldots, x_n$ is *identically* satisfied when we express $x_1, x_2, \ldots, x_n$ as functions of the m variables $u_1, u_2, \ldots, u_m$, viz. $x_i = f_i(u_1, u_2, \ldots, u_m)$, $(i = 1, 2, \ldots, n)$, we say that the n functions $f_i(u_1, u_2, \ldots, u_m)$ provide a *parametric solution* of the original system of equations.

Once again the geometrical interpretation makes this obvious. The 3 equations correspond to 3 straight lines; each pair of lines (in general) meet in a point. Therefore we expect to find that there is a triangle of points formed by the three points of intersection of the pairs of lines. Only in exceptional circumstances will the three points coincide in a single point which, lying on all 3 lines will satisfy all three equations.

These simple arguments point the way to a more general result. If two equations in two unknowns 'usually' have a unique solution whereas three equations in two unknowns are 'unlikely' to have a solution we might expect that three equations in three unknowns will 'usually' have a unique solution whereas four equations in three unknowns are 'unlikely' to have a solution. This is indeed the case. The geometrical interpretation can still be applied but since we are now dealing with the intersection of 3 or 4 planes in 3-dimensional space it is a little harder to visualize. In higher dimensions still geometrical imagination becomes inadequate and it is simpler to deal with the general case purely algebraically. This we now do.

Suppose that we are given mn numbers $a_{1,1}, a_{1,2}, a_{1,3}, \ldots, a_{1,n}$; $a_{2,1}, \ldots, a_{2,n}; \ldots; a_{m,1}, \ldots, a_{m,n}$ and m numbers $b_1, b_2, \ldots, b_m$, and m equations in n unknowns $x_1, x_2, \ldots, x_n$, viz.

$$\left. \begin{array}{c} a_{1,1}x_1 + a_{1,2}x_2 + \ldots + a_{1,n}x_n = b_1 \\ a_{2,1}x_1 + a_{2,2}x_2 + \ldots + a_{2,n}x_n = b_2 \\ \vdots \qquad \vdots \qquad\qquad \vdots \qquad \vdots \\ a_{m,1}x_1 + a_{m,2}x_2 + \ldots + a_{m,n}x_n = b_m \end{array} \right\}. \tag{3.2.1}$$

Our object is to find out whether the system (3.2.1) of m linear equations in the n unknowns $x_1, x_2, \ldots, x_n$ has a unique solution and if so, to find it [see also I, § 5.10].

The system (3.2.1) can be very conveniently represented in the notation of *matrices* and *vectors* [see I, § 5.7]. Let $\mathbf{A} = (a_{ij})$ denote the $m \times n$ matrix of the numbers $a_{1,1}, \ldots, a_{m,n}$ and let $\mathbf{b} = (b_1, b_2, \ldots, b_m)'$ denote the column vector of the numbers $b_1, b_2, \ldots, b_m$ and $\mathbf{x} = (x_1, x_2, \ldots, x_n)'$ denote the column vector of the unknown numbers $x_1, x_2, \ldots, x_n$. Then the system (3.2.1) can be represented in the form

$$\mathbf{A}\mathbf{x} = \mathbf{b}. \tag{3.2.2}$$

$\mathbf{A}$ is called 'the coefficient matrix'. For example the system of 3 equations in 2 unknowns

$$4x - 5y = 7$$
$$3x + 2y = 4$$
$$5x + 34y = 0$$

can be written in matrix form as

$$\begin{pmatrix} 4 & -5 \\ 3 & 2 \\ 5 & 34 \end{pmatrix} \begin{pmatrix} x \\ y \end{pmatrix} = \begin{pmatrix} 7 \\ 4 \\ 0 \end{pmatrix}.$$

The conditions under which the system (3.2.2) has a unique solution are known and the most important cases are covered by the following theorem:

THEOREM 3.2.1
  (i) *If $m = n$ the system (3.2.2) has a unique solution if and only if the determinant* det **A** *of the matrix* **A** *is not equal to zero.*
  (ii) *In particular if $m = n$ and* **b** $= \mathbf{0}$ *the system (3.2.2) has the solution* **x** $= \mathbf{0}$. *If* det **A** $\neq 0$ *this is the only solution. If* det **A** $= 0$ *there will be other solutions.*
  (iii) *If $m < n$ the system (3.2.2) will, in general, have an infinity of solutions.*
  (iv) *If $m > n$ the system (3.2.2) will, in general, have no solution.*

For a discussion of Theorem 3.2.1, see Volume I, section 5.10.
  It is possible to be more specific about what happens in case (ii) when det **A** $= 0$ but such cases are mainly of theoretical interest. There is however an important different type of problem associated with case (iv) when, given that $m > n$, i.e. that we have more equations than unknowns, we wish to find a solution vector **x** $= (x_1, x_2, \ldots, x_m)'$ which *in some sense* 'best satisfies' the n equations. We shall return to this problem later (Chapter 6 of this volume).
  Since systems where $m = n$, are the most important in practice we shall now concentrate on methods for obtaining the unique solution, assuming it exists, in this case.

3.3. DIRECT METHODS OF SOLUTION

All the methods for obtaining the solution of a system of n linear equations in n unknowns can be classified as *direct* or *indirect*. The *direct* methods, of which the elimination method mentioned at the beginning is the best-known, are certain to find the solution, if one exists. The solution is found as exactly as rounding errors will allow with no intermediate approximations to the answer. Unfortunately such methods may not be suitable for the solution of large systems of equations on a computer. The *indirect* methods are all based on the idea of finding an approximate solution to the equations and gradually improving the approximation until some desired level of accuracy is achieved. Such methods, of which the Gauss–Seidel is probably the best example, have important advantages for large systems of equations of certain kinds. On the other hand such methods may fail to converge. All of this will be covered later in this chapter. We shall now consider the best-known direct method, which is:

3.3.1. Gaussian Elimination

In its essential details the method of Gaussian elimination is the generalization of the method used to solve the system of 2 equations in 2 unknowns at the beginning of the chapter, but there are a number of refinements, viz.

(i) the normalization of the pivot;
(ii) the use of a check-sum column when solving equations by hand or with a desk machine;
(iii) permutation of the rows or columns (or both) of the coefficient matrix.

These refinements are simply illustrated by examples [see also I, Example 5.8.1]. In the following example we solve a system of 3 equations using normalization of the pivot and a check-sum column.

EXAMPLE 3.3.1. Solve

$$4x - 2y + 3z = 9$$

$$2x - y + z = 3$$

$$x + 3y - z = 4$$

by the method of Gaussian elimination. Work to 2 places of decimals and use a check-sum column.

Solution.   We write the equations in the form

x	y	z	r.h.s.	c.s.
4	−2	3	9	14
2	−1	1	3	5
1	3	−1	4	7

Table 3.3.1

in which the first 3 columns show the coefficients of x, y and z in each equation. The fourth column shows the value of the right-hand side (r.h.s.) of each equation and the fifth column shows the check-sum (c.s.) formed by adding together all the numbers in each row. At every stage of the subsequent calculations the entry on any row of the check-sum column should be equal to the sum of all the other numbers in the same row. If this check fails at any stage an error has been made. The use of such a check-sum column is strongly recommended when equations are being solved by hand; it is not necessary when equations are being solved by computer, although even in this case it can provide a check on rounding errors.

The method of Gaussian elimination now proceeds by using the first equation to eliminate one of the variables from the other equations. Under normal circumstances we eliminate the left-most remaining variable; i.e. we first eliminate x, and then y and so on. Before carrying out the elimination however it is desirable to 'normalize the pivot' which means in this case that we make the coefficient of x in the first equation equal to 1. This is done by dividing all the numbers in the first row of 3.3.1 by 4. Table 3.3.1 is thus replaced by

x	y	z	r.h.s.	c.s.
1	−0·50	0·75	2·25	3·50
2	−1	1	3	5
1	3	−1	4	7

Table 3.3.2

Note that the check-sum entry for the first row still holds.

In order to eliminate x from the second equation we must multiply all the elements in the first row by 2 (= the coefficient of x in the second equation) and subtract them from the corresponding elements in the second row. Similarly we eliminate x from the third equation, but this time the multiplier is 1 (= the coefficient of x in the third equation). The advantage of normalizing the pivot is now clear: no divisions are required in order to carry out the elimination process.

When x has been eliminated in this way our equations have been reduced to the form

x	y	z	r.h.s.	c.s.
1	−0·50	0·75	2·25	3·50
		−0·50	−1·50	−2·00
	3·50	−1·75	1·75	3·50

Table 3.3.3

We now observe that we have not only eliminated x from the second equation; we have also eliminated y. This of course is purely fortuitous but it presents us with a problem for under normal circumstances we would have used the new second equation to eliminate y from the new third equation. Since the coefficient of y in the second equation is now zero we cannot do this. The problem is easily resolved however by interchanging the second and third equations, viz.

x	y	z	r.h.s.	c.s.
1	−0·50	0·75	2·25	3·50
	3·50	−1·75	1·75	3·50
	−0·50	−1·50	−2·00	

Table 3.3.4

The new pivot is the ringed term (3·50) and we normalize as before, producing

x	y	z	r.h.s.	c.s.
1	−0·50	0·75	2·25	3·50
	1	−0·50	0·50	1·00
		−0·50	−1·50	−2·00

Table 3.3.5

No elimination is necessary and we complete this stage by noting that the final pivot is the coefficient of z in the third equation (−0·50) so that on normalizing we have

x	y	z	r.h.s.	c.s.
1	−0·50	0·75	2·25	3·50
	1	−0·50	0·50	1·00
		1	3·00	4·00

Table 3.3.6

The solution is now found very easily. The third equation gives $z = 3$; back-substitution for z in the second equation tells us that

$$y - 0·50(3) = 0·50$$

i.e. $y = 2$. Finally the first equation then gives

$$x - 0·50(2) + 0·75(3) = 2·25$$

so that $x = 1$.

The solution of the original equations is therefore given by $(x, y, z) = (1, 2, 3)$.

We have worked through this example in detail in order to illustrate the method of Gaussian elimination, the use of check sums, normalization and permutation of rows. This last point was brought forcibly to our attention because the coefficient of y in Table 3.3.3 turned out to be zero. Had we been solving these equations on a computer we would have produced a machine fault (attempting to divide by zero) at this stage unless we had remembered to build into the program a test which ensured that the chosen pivot is not zero.

Interchanges of the rows is also desirable when the pivot is small. The reason for this is that when we divide the other elements of the row by a small number we increase the rounding errors. In order to minimize this effect we should at each stage interchange pairs of rows so that we use as the pivot the term which carries the largest coefficient (in absolute terms, it doesn't matter whether it is positive or negative).

3.3.2. Checking the Accuracy of the Solution

Having obtained the solution of the equations we should check on the accuracy of the solution. The most obvious way of doing this is to substitute the values of the variables into the equations and compare the numerical values so found with the values of the right-hand sides. Alternatively we can form 'the sum equation' by adding together all the left-hand sides of the equations and compare with the sum of the right-hand sides. Thus, in Example 3.3.1 the sum equation is

$$7x + 3z = 16$$

and since $x = 1$, $z = 3$ this is satisfied with zero error. In this case of course the sum check is hardly necessary since the equations have integral coefficients and integer solutions so that rounding errors could hardly have occurred. In the more general case when the equations do not have integer coefficients or solutions the use of the sum equation can provide a valuable check on the accuracy of the solution. The next example illustrates both this point and the need for row interchanges in order to avoid the use of small pivots.

EXAMPLE 3.3.2. Solve

$$1 \cdot 356x_1 + 2 \cdot 672x_2 - 1 \cdot 017x_3 = 6 \cdot 166$$
$$0 \cdot 845x_1 + 1 \cdot 667x_2 + 2 \cdot 214x_3 = 20 \cdot 158$$
$$0 \cdot 627x_1 + 3 \cdot 416x_2 - 1 \cdot 758x_3 = 2 \cdot 918$$

by Gaussian elimination, working to 3 places of decimals. Check the accuracy of the result by using the sum equation.

Solution. We proceed as we did in Example 3.3.1

x_1	x_2	x_3	r.h.s.	c.s.
(1·356)	2·672	−1·017	6·166	9·177
0·845	1·667	2·214	20·158	24·884
0·627	3·416	−1·758	2·918	5·203

Using the ringed term as the pivot and eliminating x_1 from the second and third equations produces

x_1	x_2	x_3	r.h.s.	c.s.
1	1·971	−0·750	4·547	6·768
	(0·002)	2·848	16·316	19·165
	2·180	−1·288	0·067	0·959

We choose the ringed term as pivot, even though it is very small. The consequences will become apparent shortly. Normalizing and eliminating x_2

from the third equation we have

x_1	x_2	x_3	r.h.s.	c.s.
1	1·971	−0·750	4·547	6·768
	1	1424·000	8158·000	9582·500
		−3105·608	−17784·373	−20888·891

(Note that the check-sum column is now in error by about one part in 20000 in the second and third rows).

The solution, as obtained from this triangular set of equations is $x_3 = 5\cdot727$, $x_2 = 2\cdot752$, $x_1 = 3\cdot418$. The sum equation is:

$$2\cdot828x_1 + 7\cdot755x_2 - 0\cdot561x_3 = 29\cdot242.$$

On substituting the values of the variables the left-hand side has the value 27·795 so that the sum equation indicates very serious errors somewhere in the calculation. On the assumption that these may be due to the use of the small pivot we return to that stage of the solution and interchange the second and third equations:

x_1	x_2	x_3	r.h.s.	c.s.
1	1·971	−0·750	4·547	6·768
	2·180	−1·288	0·067	0·959
	0·002	2·848	16·316	19·165

Using the new pivot we eventually reach the triangular form

x_1	x_2	x_3	r.h.s.	c.s.
1	1·971	−0·750	4·547	6.768
	1	−0·591	0·031	0·440
		2·849	16·316	19·164

which leads to the solution

$$x_3 = 5\cdot727, \qquad x_2 = 3\cdot416, \qquad x_1 = 2\cdot109$$

and substitution in the sum equation gives the value 29·242, in agreement with the r.h.s. The result can therefore be regarded as acceptable. The 'solution' obtained when we used the small pivot can now be seen to be very badly wrong.

If we substitute the values of x_1, x_2, x_3 in the 3 original equations the values obtained are 6·163, 20·156 and 2·923 so that even now our solution is not entirely satisfactory since there are 'residual errors' of +0·003, +0·002 and

$-0\cdot005$. We are however quite close to the true solution, which is

$$x_1 = 2\cdot117, \qquad x_2 = 3\cdot413, \qquad x_3 = 5\cdot727.$$

It is difficult to obtain a more accurate solution unless the calculations are performed to greater precision (i.e. carrying more than 3 decimal places at the intermediate stages) and then rounded to 3 places at the end. In this particular case a solution correct to 3 decimal places is obtained if we work to 4 decimal places at all intermediate stages and then round to 3 d.p. at the end. The values so obtained, before rounding are $x_1 = 5\cdot7269$, $x_2 = 3\cdot4130$, $x_3 = 2\cdot1171$ which round to the true solution quoted above. The fact that retention of a fourth decimal place throughout the calculations in this case produces an answer correct to three decimal places is a special case of a more general result, the proof of which is along the same lines as the argument at the end of Example 2.2.3, that if all the elements of a system of n linear equations are correct to k places of decimals then in order to obtain a solution which is correct to k places of decimals when we use the method of Gaussian elimination we must perform all the intermediate calculations to $(k + l)$ decimal places where $l = \log_{10} n$. Thus for $n < 10$ we must carry one extra decimal place, for $11 < n < 100$ we must carry 2 extra places and so on. This estimate for l is pessimistic, we *might*, for example get away with carrying only 1 extra decimal place when $n = 20$, but this cannot be guaranteed.

3.3.3.   The Gauss–Jordan Method

This method, sometimes known more simply as Jordan's method, is a variant of the method of Gaussian elimination. The only difference is that at each stage of the elimination process the appropriate variable is eliminated not only in all the equations *following* the pivotal equation but also in all the equations *preceding* it. We thus do more work at the elimination stage but the final stage of the Gaussian elimination method (back-substitution) is now no longer necessary since our system of equations will have been reduced to pure diagonal form from which the solution follows at once. All the refinements of the method of Gaussian elimination (normalization of the pivot, use of a check-sum column, row or column interchange) are also applicable to the Gauss–Jordan method.

EXAMPLE 3.3.3.   Solve

$$x + y + z + w = -2$$
$$x + 2y + 3z + 4w = -10$$
$$2x + 3y + 4z + w = 4$$
$$3x + 4y + z + 2w = -10$$

by the Gauss–Jordan method. Work to 2 d.p. and use a check-sum column.

Solution.   Proceeding as before we obtain the array

x	y	z	w	r.h.s.	c.s.
1	1	1	1	−2	2
1	2	3	4	−10	0
2	3	4	1	4	14
3	4	1	2	−10	0

Using the first equation we eliminate x from the other 3:

x	y	z	w	r.h.s.	c.s.
1	1	1	1	−2	2
0	①	2	3	−8	−2
0	1	2	−1	8	10
0	1	−2	−1	−4	−6

If we were prepared to interchange columns in order to obtain the largest pivot we would now transpose the columns for y and w; since however our primary purpose is to illustrate the method we will not interchange columns in this example and we therefore use the ringed term as pivot to eliminate y from the first, third and fourth equations obtaining:

x	y	z	w	r.h.s.	c.s.
1	0	−1	−2	6	4
0	1	2	3	−8	−2
0	0	0	−4	16	12
0	0	−4	−4	4	−4

Since the term which we would have chosen as pivot has vanished some kind of row or column interchange is forced upon us. An interchange of the third and fourth rows is very satisfactory so we rearrange the table.

x	y	z	w	r.h.s.	c.s.
1	0	−1	−2	6	4
0	1	2	3	−8	−2
0	0	⟨−4⟩	−4	4	−4
0	0	0	−4	16	12

Using the new pivot we eliminate z from the first and second equations:

x	y	z	w	r.h.s.	c.s.
1	0	0	−1	5	5
0	1	0	1	−6	−4
0	0	1	1	−1	1
0	0	0	⟨−4⟩	16	12

The elimination stage is then completed by using the fourth equation to eliminate the terms in w from the first three equations:

x	y	z	w	r.h.s.	c.s.
1	0	0	0	1	2
0	1	0	0	−2	−1
0	0	1	0	3	4
0	0	0	1	−4	−3

The solution: $x = 1$, $y = -2$, $z = 3$, $w = -4$ now follows immediately.

It is not obvious whether Gaussian elimination of the Gauss–Jordan method is to be preferred for the solution of systems of linear equations but we shall see, in section 3.6, that the number of multiplications and divisions required by the Gauss–Jordan method is about 50% more than the number required in Gaussian elimination so that for a large number of equations Gaussian elimination involves significantly less work.

3.3.4. Systems of Equations with Several Right-Hand Sides

If we have a system of equations with fixed coefficients which have to be solved for several different right-hand sides it may be worthwhile adopting a different approach to the problem. We remarked in section 3.2 that a system of linear equations may be expressed in matrix form

$$\mathbf{A}\mathbf{x} = \mathbf{b}$$

where $\mathbf{A}$ is the matrix of coefficients, $\mathbf{b}$ is the column vector of the right-hand side and $\mathbf{x}$ the column vector of the unknowns: The *theoretical* solution of these equations is then [cf. I, Theorem 5.8.1]

$$\mathbf{x} = \mathbf{A}^{-1}\mathbf{b}$$

$\mathbf{A}^{-1}$ being the matrix inverse to $\mathbf{A}$ [see I, § 6.4]. In practice the computation of the inverse matrix requires more effort than the solution of the equations by Gaussian elimination so that it is not a method to be recommended for the solution of a single system of equations. If, however, we have to solve the *same* set of equations for *several* right-hand sides computation of $\mathbf{A}^{-1}$ becomes more attractive and if the number of right-hand sides exceeds the number of equations the use of $\mathbf{A}^{-1}$ is theoretically more efficient than the use of Gaussian elimination. We shall therefore now consider the problem of computing $\mathbf{A}^{-1}$.

Given an $n \times n$ matrix $\mathbf{A}$ the inverse matrix $\mathbf{A}^{-1}$ is defined by the equation

$$\mathbf{A}\mathbf{A}^{-1} = \mathbf{I}$$

where $\mathbf{I}$ is the $n \times n$ unit matrix which has a 1 in every position on the main diagonal and zeros everywhere else [see I, (6.2.7)]. $\mathbf{A}^{-1}$ can therefore be found if we solve the systems of equations

$$\mathbf{A}\mathbf{x}_i = \mathbf{b}_i \qquad (i = 1, 2, \ldots, n)$$

where $\mathbf{A}$ is fixed and $\mathbf{b}_i$ is the column vector which consists of zeros in every position except the ith where it has a 1. The n column vectors $\mathbf{x}_1, \mathbf{x}_2, \ldots, \mathbf{x}_n$ which are so found are then the n columns of the matrix $\mathbf{A}^{-1}$. The n sets of equations can be conveniently solved simultaneously by the method of Gaussian elimination as we now illustrate, taking as our equations the same system as we used in Example 3.3.1.

EXAMPLE 3.3.4.   The system of equations used in Example 3.3.1 is to be solved for several right-hand sides. Use the method of Gaussian elimination to find the inverse matrix of the coefficients and then use the inverse to compute again the solution for the particular right-hand side of Example 3.3.1.

Solution.   We solve the equations simultaneously for the 3 right-hand sides

$$\begin{matrix} 1 & 0 & & 0 \\ 0 & 1 & \text{and} & 0. \\ 0 & 0 & & 1 \end{matrix}$$

The use of the check-sum column is even more desirable. Our system with 3 right-hand sides is thus represented by the scheme

x	y	z	$\mathbf{b}_1$	$\mathbf{b}_2$	$\mathbf{b}_3$	c.s.
4	−2	3	1	0	0	6
2	−1	1	0	1	0	3
1	3	−1	0	0	1	4

Proceeding exactly as in Example 3.3.1 we arrive eventually at the triangular scheme

x	y	z	$\mathbf{b}_1$	$\mathbf{b}_2$	$\mathbf{b}_3$	c.s.
1	−0·50	0·75	0·25	0	0	1·50
	1	−0·50	−0·07	0	0·29	0·71
		1	1	−2	0	0

As before we have interchanged rows 2 and 3, to avoid the zero pivot, and we note the rounding error in the check-sum column in row 2 caused by working to 2 d.p.

The solutions for the 3 right-hand sides are found by back-substitution and turn out to be

$$(x_1, y_1, z_1) = (-0\cdot28, 0\cdot43, 1)$$

$$(x_2, y_2, z_2) = (1, -1\cdot00, -2)$$

$$(x_3, y_3, z_3) = (0\cdot14, 0\cdot29, 0).$$

The inverse matrix is therefore, to 2 d.p.

$$\mathbf{A}^{-1} = \begin{pmatrix} -0\cdot28 & 1\cdot00 & 0\cdot14 \\ 0\cdot43 & -1\cdot00 & 0\cdot29 \\ 1\cdot00 & -2\cdot00 & 0\cdot00 \end{pmatrix}.$$

The inverse matrix calculated *exactly* is in fact

$$\begin{pmatrix} -\frac{2}{7} & 1 & \frac{1}{7} \\ \frac{3}{7} & -1 & \frac{2}{7} \\ 1 & -2 & 0 \end{pmatrix}.$$

In view of the presence of fractions involving multiples of $\frac{1}{7}$ no decimal representation of the inverse can be free of rounding errors.

Using the inverse matrix the solution of the set of equations with the right-hand side of Example 3.3.1 is

$$\mathbf{x} = \mathbf{A}^{-1}\mathbf{b} \doteq \begin{pmatrix} -0\cdot28 & 1\cdot00 & 0\cdot14 \\ 0\cdot43 & -1\cdot00 & 0\cdot29 \\ 1\cdot00 & -2\cdot00 & 0\cdot00 \end{pmatrix} \begin{pmatrix} 9 \\ 3 \\ 4 \end{pmatrix} = \begin{pmatrix} 1\cdot04 \\ 2\cdot03 \\ 3\cdot00 \end{pmatrix}$$

giving $x = 1\cdot04$, $y = 2\cdot03$, $z = 3\cdot00$. Since the exact solution is $x = 1$, $y = 2$, $z = 3$ the effect of rounding errors is apparent.

Now that we have the inverse matrix (approximately) the solution for the set of equations of Example 3.3.1 with *any* right-hand side, **b**, can be found quickly by forming $\mathbf{A}^{-1}\mathbf{b}$. It is obvious that if we are going to use this method the inverse matrix should be found as accurately as possible since rounding errors in the inverse matrix will inevitably produce rounding errors in the solution. There are a number of other methods for computing the inverse of a matrix; these are described in the chapter on Matrix Computations (Chapter 4 of this volume).

3.4.   INDIRECT METHODS: THE GAUSS–SEIDEL METHOD

It was mentioned in section 3.3 that the methods for solving systems of linear equations fall into two classes: direct and indirect. We have seen two of the direct methods in sections 3.3.1 and 3.3.3; these have the advantage that they will always find the solution if one exists. The indirect methods on the other hand may fail to converge to the solution but as they offer significant advantages when the number of equations to be solved is large, or when the matrix of coefficients is sparse (i.e. most of the coefficients are zero) they are very often used in cases where convergence to the solution can be guaranteed.

The most commonly used method is the Gauss–Seidel method. Before explaining the method formally we shall illustrate it by a simple example.

EXAMPLE 3.4.1.   Solve the equations

$$2x + y = 3$$

$$x - 2y = 4$$

by the Gauss–Seidel method.

Solution.   We begin by re-writing the equations in the form

(i) $x = \frac{1}{2}(3 - y)$

(ii) $y = \frac{1}{2}(x - 4)$

and we choose arbitrary starting values for x, y; i.e. we guess a solution, but the guess needn't be very good; the values $x = 0$, $y = 0$ will do as well as any.

Taking $y = 0$ in (i) we deduce that $x = \frac{3}{2}$. Taking $x = \frac{3}{2}$ in (ii) we deduce that $y = -\frac{5}{4}$. This completes the first cycle. Starting from the assumed 'solution' $(x, y) = (0, 0)$ we have obtained a new 'solution' $(x, y) = (\frac{3}{2}, -\frac{5}{4})$. We now repeat the process. Taking the value of y to be $-\frac{5}{4}$ we use (i) and deduce that $x = \frac{17}{8}$ and using this value in (ii) leads us to $y = -\frac{15}{16}$. Note that at every stage we make use of the most recently obtained value for x in (ii), and the most recently obtained value for y in (i)—this is the distinguishing feature of the Gauss–Seidel method.

The process of computing new values for x and y from old values is called 'an iteration'. Our progress towards the solution can be shown as in the table below, where all values have been rounded to 3 d.p.

Iteration number	x	y
0	0·000	0·000
1	1·500	−1·250
2	2·125	−0·938
3	1·969	−1·016
4	2·008	−0·996
5	1·998	−1·001
6	2·001	−1·000
7	2·000	−1·000

and all the subsequent iterations produce the same result: $x = 2\cdot000$, $y = -1\cdot000$ which is indeed the solution; so convergence has occurred, to 3 d.p. accuracy, after 7 iterations.

To show that the Gauss–Seidel method may fail to converge we shall now attempt to solve the same pair of equations as in Example 3.4.1 but we shall interchange the order of the equations.

EXAMPLE 3.4.2.   Solve the equations

$$x - 2y = 4$$

$$2x + y = 3$$

by the Gauss–Seidel method.

Solution.   We rewrite the equations in the form

(i) $x = 4 + 2y$

(ii) $y = 3 - 2x$

and, as before, start with $(x, y) = (0, 0)$. The table of values for (x, y) obtained at the completion of the first 4 iterations is shown below.

Iteration number	x	y
0	0	0
1	4	−5
2	−6	15
3	34	−65
4	−126	255

It is clear that convergence towards the solution $(x, y) = (2, -1)$ is certainly not taking place. It might reasonably be imagined that if we had chosen different starting values for (x, y) instead of $(0, 0)$ we might have obtained convergence in Example 3.4.2; but this is not so. Except in the unlikely case that our chosen starting values for (x, y) are exactly right, i.e. $(x, y) = (2, -1)$ the successive iterations in Example 3.4.2 will always fail to converge to the solution. For example suppose we start with $(x, y) = (2 \cdot 001, -1 \cdot 001)$ then the first 4 iterations produce

Iteration number	x	y
0	2·001	−1·001
1	1·998	−0·996
2	2·008	−1·016
3	1·968	−0·936
4	2·128	−1·256

and it can be seen that the error in both x and y is *increasing* by a factor of 4 in magnitude (but oscillating in sign) at each iteration. On the other hand we note that in Example 3.4.1 the iterations converge to the solution and the errors in x and y *decrease* by a factor of 4 at each iteration. This is not fortuitous. We shall see the explanation later (3.4.2), but before discussing the conditions under which the Gauss–Seidel process will converge we need to define the process more formally.

3.4.1.   Formal Definition of the Gauss–Seidel Process

Suppose that we have a system of n equations in n unknowns $x_1, x_2, \ldots, x_n$

$$\begin{aligned}
a_{11}x_1 + a_{12}x_2 + \ldots + a_{1n}x_n &= b_1 \\
a_{21}x_1 + a_{22}x_2 + \ldots + a_{2n}x_n &= b_2 \\
\vdots \qquad \vdots \qquad\quad \vdots \quad\ \ \vdots \\
a_{n1}x_1 + a_{n2}x_2 + \ldots + a_{nn}x_n &= b_n.
\end{aligned} \qquad (3.4.1)$$

We also suppose that the diagonal coefficients $a_{11}, a_{22}, \ldots, a_{nn}$ are all non-zero; if any of them are zero we permute the variables and equations until the non-zero diagonal elements condition is satisfied. The equation can then be written in the form:

$$
\begin{aligned}
x_1 &= a_{11}^{-1}(b_1 - a_{12}x_2 - a_{13}x_3 - \ldots - a_{1n}x_n) \\
x_2 &= a_{22}^{-1}(b_2 - a_{21}x_1 - a_{23}x_3 - \ldots - a_{2n}x_n) \\
&\vdots \qquad \vdots \\
x_n &= a_{nn}^{-1}(b_n - a_{n1}x_1 - a_{n2}x_2 - \ldots - a_{nn-1}x_{n-1}).
\end{aligned}
\qquad (3.4.2)
$$

The Gauss–Seidel process can now be described in *algorithmic form* as follows:

Step 1. Choose arbitrary starting values for $x_1, x_2, \ldots, x_n$ ($x_1 = x_2 = \ldots = x_n = 0$ will do). Call these 'the current values' of $x_1, x_2, \ldots, x_n$.

Step 2. Using the current values of $x_2, \ldots, x_n$ use the first of the equations (3.4.2) to compute a revised value for x_1. Let this revised value for x_1 be the new current value for x_1.

Step 3. Using the current values of $x_1, x_3, x_4, \ldots, x_n$ use the second of the equations (3.4.2) to compute a revised value for x_2. Let this revised value for x_2 be the new current value for x_2.
$$\vdots$$

Step $(n+1)$. Using the current values of $x_1, x_2, \ldots, x_{n-1}$ use the nth of the equations (3.4.2) to compute a revised value for x_n. Let this revised value for x_n be the new current value for x_n.

Step $(n+2)$. Compare the current values of $x_1, x_2, \ldots, x_n$ with their values at the end of the previous iteration. If none of them have changed by more than a specified amount (e.g. if the two sets of values agree to k places of decimals) accept the current values as the solution and go to step $(n+4)$ otherwise return to step 2 unless the number of iterations exceeds some integer N in which case go to step $(n+3)$.

Step $(n+3)$. Put out a message that after N iterations the iterations do not seem to be converging and stop.

Step $(n+4)$. Put out the solution (and, if desired, the number of iterations required for convergence to k places of decimals).

This algorithmic form of the process is particularly suitable for programming on a computer but it also defines the method which must, in essence, be followed if the calculations are being done by hand or using a desk calculator. In the next example we apply the process as described in the algorithm to the solution of a system of 3 equations.

EXAMPLE 3.4.3.   Solve the system of equations

$$3x + y - z = 1$$

$$x - 3y + z = 2$$

$$x + y + 3z = 3$$

by the Gauss–Seidel process. Continue for 50 iterations or until two consecutive solutions agree to 5 places of decimals for each variable separately.

Solution.   We rewrite the equations as

(i) $x = \frac{1}{3}(1 - y + z)$

(ii) $y = \frac{1}{3}(-2 + x + z)$

(iii) $z = \frac{1}{3}(3 - x - y)$

and we start with $x = y = z = 0$. Following the algorithm we record the current values of the variables at the end of each iteration as shown in the table below. We work to 6 d.p. at every stage so that we can be sure that two sets of values agree to 5 d.p.

Iteration number	x	y	z
0	0	0	0
1	0·333333	−0·555556	1·074074
2	0·876543	−0·016461	0·713306
3	0·576589	−0·236702	0·886704
4	0·707802	−0·135165	0·809121
5	0·648095	−0·180928	0·844278
6	0·675069	−0·160218	0·828283
7	0·662834	−0·169628	0·835478
8	0·668369	−0·165384	0·832338
9	0·665907	−0·167252	0·833728
10	0·666993	−0·166426	0·833144
11	0·666523	−0·166778	0·833418
12	0·666732	−0·166617	0·833295
13	0·666637	−0·166689	0·833351
14	0·666680	−0·166656	0·833325
15	0·666660	−0·166672	0·833337
16	0·666670	−0·166664	0·833331

The iterations are obviously converging but we have not yet reached the required level of accuracy because the changes in the values of x, y, z from the fifteenth to the sixteenth iteration are

0·000010,   0·000008   and   −0·000006

respectively and since we want each to be accurate to 5 d.p. we must continue the iterations until the maximum change in x, y, z does not exceed 0·000005 in absolute value. The next iteration leads to

Iteration number	x	y	z
17	0·666665	−0·166668	0·833334

and this is *just* good enough, but for safety we perform one more iteration:

18	0·666667	−0·166666	0·833333

which confirms the result that to 5 d.p. the solution is

$$x = 0·66667, \qquad y = -0·16667, \qquad z = 0·83333.$$

The exact solutions are in fact

$$x = \tfrac{2}{3}, \qquad y = -\tfrac{1}{6}, \qquad z = \tfrac{5}{6}.$$

3.4.2.   Convergence of the Gauss–Seidel Method

We have seen, in Examples 3.4.1 and 3.4.2 that the Gauss–Seidel method sometimes converges and sometimes does not. Obviously there is little point in using the method to try to solve a system of equations if it is not going to converge so we are naturally led to ask the question: under what conditions will the Gauss–Seidel method converge?

This question can be answered completely in the sense that there is a theorem (see Isaacson and Keller, pp. 61–81) which tells us that if the matrix of coefficients

$$\mathbf{A} = \begin{pmatrix} a_{11} & a_{12} & \cdots & a_{1n} \\ a_{21} & a_{22} & \cdots & a_{2n} \\ \vdots & \vdots & & \vdots \\ a_{n1} & a_{n2} & \cdots & a_{nn} \end{pmatrix}$$

of the equations possesses a certain property then the Gauss–Seidel method will converge no matter what starting values we choose for the variables. Unfortunately although this result is very satisfactory from the theoretical point of view it is not of much use in practice since the test of whether the matrix **A** has the necessary property involves a prohibitive amount of calculation if the number of equations is even moderately large. For practical use some test which can be easily applied to the matrix **A** is desirable. Inevitably such a test must be over-strict since otherwise it must be equivalent to the test given in the theorem mentioned above. If the test is to be simple and at the same time guarantee convergence, it must be of the form:

If the matrix **A** has a certain property, P, then the Gauss–Seidel method *will* certainly converge. The Gauss–Seidel *may* also converge even if **A** does not have the property P, but we cannot guarantee that it will do so. Such a condition is said to be 'sufficient' for convergence, but 'not necessary'.

It is not *a priori* certain that a test of sufficient simplicity will exist but fortunately it does. The test is given in the following theorem (for a proof of which see Isaacson and Keller, loc. cit.):

THEOREM 3.4.1. *If* $\mathbf{A} = (a_{ij})$ *is the matrix of coefficients of a system of linear equations and if the coefficients of* $\mathbf{A}$ *have the property that for all* $i = 1, 2, \ldots, n$

$$2|a_{ii}| > \sum_{j=1}^{n} |a_{ij}| \qquad (3.4.3)$$

then the Gauss–Seidel method applied to the system will converge no matter what initial values are chosen for the variables.

A matrix $\mathbf{A}$ which has the property (3.4.3) is said to be *diagonally dominant*.

Let us apply the theorem to the matrix of coefficients in the three examples above.

In Example 3.4.1

$$\mathbf{A} = \begin{pmatrix} 2 & 1 \\ 1 & -2 \end{pmatrix}$$

and the condition (3.4.3) is satisfied since

$$2 \times 2 > 2 + 1 \quad \text{and} \quad 2 \times |-2| > |-2| + 1.$$

In Example 3.4.2

$$\mathbf{A} = \begin{pmatrix} 1 & -2 \\ 2 & 1 \end{pmatrix}$$

and the condition (3.4.3) is not satisfied since $2 \times 1 < 1 + |-2|$.

In Example 3.4.3

$$\mathbf{A} = \begin{pmatrix} 3 & 1 & -1 \\ 1 & -3 & 1 \\ 1 & 1 & 3 \end{pmatrix}$$

and the condition is satisfied since in each of the three rows

$$2 \times 3 > 3 + 1 + 1.$$

Thus we could have been certain that the Gauss–Seidel method would converge in Examples 3.4.1 and 3.4.3. We could not, on the basis of the theorem, have been *certain* that Example 3.4.2 would produce a failure of the Gauss–Seidel method but the theorem at least warns us that we are on dangerous ground.

3.4.3.   Accelerating the Convergence of the Gauss–Seidel Process

The theory of the convergence of the Gauss–Seidel process enables us to develop a technique for accelerating the convergence. The relevant parts of the theory in a simplified form are as follows.

Suppose that at the kth iteration the value obtained for the variable x is x_k. Let $\hat{x}$ be the true value of x and let the error in x_k be e_k then

$$x_k = \hat{x} + e_k.$$

The theory of the convergence of the Gauss–Seidel process then shows that at the $(k+1)$st iteration the value obtained for x will be x_{k+1} where

$$x_{k+1} \doteqdot \hat{x} + \lambda e_k$$

where λ is a constant of absolute value less than unity (otherwise convergence will not occur). Similarly after $(k+2)$ iterations we will have

$$x_{k+2} \doteqdot \hat{x} + \lambda^2 e_k.$$

From these three equations we can deduce approximate values for x and λ, viz.

$$\hat{x} \doteqdot \frac{x_{k+2} x_k - x_{k+1}^2}{x_{k+2} - 2x_{k+1} + x_k} \quad \text{and} \quad \lambda \doteqdot \frac{x_{k+2} - x_{k+1}}{x_{k+1} - x_k}$$

and if the values of the numerator and denominator are not too small the estimate of $\hat{x}$ so produced will be better than any of the estimates provided by x_k, x_{k+1} or x_{k+2}. The estimated value of λ is also useful in that it provides us with an estimate of the speed of convergence of the Gauss–Seidel process—the smaller the absolute value of λ the faster the convergence will be.

If the values of the numerator and denominator in the estimate for $\hat{x}$ are too small the value obtained may be less accurate than those given by x_k, x_{k+1} or x_{k+2} because rounding errors are now becoming significant.

These points are illustrated in the next example.

EXAMPLE 3.4.4.   Apply the technique described above to estimate values for x, y, z, λ in Example 3.4.3 taking (i) $k = 5$, (ii) $k = 8$.

Solution.   (i) Taking $k = 5$ we have as our estimates

$$\hat{x} \doteqdot \frac{(0\cdot662834)(0\cdot648095) - (0\cdot657069)^2}{(0\cdot662834) - 2(0\cdot675069) + (0\cdot648095)} = \frac{-0\cdot026139}{-0\cdot039209} = 0\cdot666659$$

$$\hat{y} \doteqdot \frac{(-0\cdot169628)(-0\cdot180928) - (-0\cdot160218)^2}{(-0\cdot169628) - 2(-0\cdot160218) + (-0\cdot180928)} = \frac{+0\cdot005020}{-0\cdot030120} = -0\cdot166667$$

$$\hat{z} \doteqdot \frac{(0\cdot835478)(0\cdot844278) - (0\cdot828283)^2}{(0\cdot835478) - 2(0\cdot828283) + (0\cdot844278)} = \frac{0\cdot019323}{0\cdot023190} = 0\cdot833247$$

$$\lambda \doteqdot \frac{(0\cdot662834) - (0\cdot675069)}{(0\cdot675069) - (0\cdot648095)} = \frac{-0\cdot012235}{0\cdot026974} = -0\cdot453585.$$

We see at once that the estimates for x, y, z are significantly better than the best values so far found in the 7 iterations. The value of λ is also close to the true value, which can be shown to be

$$\lambda = -\frac{5+\sqrt{52}}{27} \doteq -0\cdot452.$$

(ii) Taking $k = 8$ our estimates are:

$$\hat{x} \doteq \frac{(0\cdot666993)(0\cdot668369)-(0\cdot665907)^2}{(0\cdot666993)-2(0\cdot665907)+(0\cdot668369)} = \frac{0\cdot002365}{0\cdot003548} = 0\cdot666573$$

$$\hat{y} \doteq \frac{(-0\cdot166426)(-0\cdot165384)-(-0\cdot167252)^2}{(-0\cdot166426)-2(-0\cdot167252)+(-0\cdot0165384)} = \frac{-0\cdot000449}{+0\cdot002694} = -0\cdot166667$$

$$\hat{z} \doteq \frac{(0\cdot833144)(0\cdot832338)-(0\cdot833728)^2}{(0\cdot833144)-2(0\cdot833728)+(0\cdot832338)} = \frac{-0\cdot001645}{-0\cdot001974} = 0\cdot833333$$

$$\lambda \doteq \frac{(0\cdot666993)-(0\cdot665907)}{(0\cdot665907)-(0\cdot668369)} = \frac{+0\cdot001086}{-0\cdot002462} = -0\cdot441105$$

and we see that although the estimates for $\hat{y}$ and $\hat{z}$ are now correct to 6 d.p. the estimates for $\hat{x}$ and λ are worse than those obtained in case (i). The presence of small terms in the denominators of the expressions for $\hat{x}$, $\hat{y}$, $\hat{z}$ and λ makes the values obtained for these variables very unreliable. Nevertheless, providing that checks are made to safeguard against the use of denominators which are too small, this technique for accelerating the convergence (which is known as Aitken extrapolation, see § 2.4.1) can be very useful in substantially reducing the number of iterations needed. If the technique is used after several consecutive iterations it should be fairly obvious whether consistent values of the variables are being obtained. For further comments on the use of Aitken extrapolation and other extrapolation techniques in general see section 2.4.

3.4.4. Jacobi's Method

In this method, which is similar in principle to the Gauss–Seidel method, we do not replace the values of the variables one at a time as they are computed but the entire set are replaced at the end of each *complete* iteration. To illustrate the method we shall re-work Example 3.4.1 using the Jacobi method.

EXAMPLE 3.4.5. Solve the equations

$$2x + y = 3$$
$$x - 2y = 4$$

using the Jacobi method.

Solution.   As in Example 3.4.1 we re-write the equations in the form

$$\text{(i)} \qquad x = \tfrac{1}{2}(3 - y)$$

$$\text{(ii)} \qquad y = \tfrac{1}{2}(x - 4)$$

and start the iteration off by taking $x = 0$, $y = 0$. From (i) we obtain the revised value $x = \tfrac{3}{2}$ and from (ii) keeping x at its original value, 0, we obtain the revised value for y: -2. Thus at the end of the *first complete iteration* we have the revised pair of values $x = \tfrac{3}{2}$, $y = -2$.

We now move to the second iteration taking $x = \tfrac{3}{2}$, $y = -2$ as our starting values. From (i) we obtain the new value for x: $\tfrac{5}{2}$, and from (ii), keeping x at $\tfrac{3}{2}$ we obtain the new value for y: $-\tfrac{3}{4}$. Thus at the end of the *second complete iteration* we have $x = \tfrac{5}{2}$, $y = -\tfrac{3}{4}$. Proceeding in this way we obtain the following table of values (rounded to 3 d.p.) at the end of each complete iteration.

Iteration number	x	y
0	0·000	0·000
1	1·500	−2·000
2	2·500	−1·250
3	2·125	−0·750
4	1·875	−0·938
5	1·969	−1·063
6	2·032	−1·016
7	2·008	−0·984
8	1·992	−0·996

Comparison with the table in Example 3.4.5 shows that the Jacobi method is converging to the correct result ($x = 2$, $y = -1$) but is doing so at only about half the speed of the Gauss–Seidel process. The values of x obtained in the Gauss–Seidel process after 1, 2, 3, 4—iterations are identical to the values of x obtained in the Jacobi process after 1, 3, 5, 7—iterations. Similarly the values for y obtained after 1, 2, 3, 4, ... iterations of the Gauss–Seidel process are the same as those obtained in the Jacobi process after 2, 4, 6, 8, ... iterations.

Theoretical analysis shows that, except in certain unimportant cases, the Gauss–Seidel method and Jacobi methods either both converge or both diverge and when they converge the Gauss–Seidel method does so in about half the number of iterations required by the Jacobi method. The evidence provided by Example 3.4.5 is therefore entirely consistent with the theory. Since the Gauss–Seidel method is no more difficult to use, either with a desk machine or programmed on a computer, it is usually preferred in practice to Jacobi's method.

3.5.   ILL-CONDITIONED EQUATIONS

Suppose that we have a system of n equations

$$a_{11}x_1 + a_{12}x_2 + \ldots + a_{1n}x_n = b_1$$
$$a_{21}x_1 + a_{22}x_2 + \ldots + a_{2n}x_n = b_2$$
$$\vdots \qquad\qquad \vdots$$
$$a_{n1}x_1 + a_{n2}x_2 + \ldots + a_{nn}x_n = b_n$$

and that we have an approximate solution given by $x_1 = s_1, x_2 = s_2, \ldots, x_n = s_n$. We substitute these values in the equations and obtain, say

$$a_{11}s_1 + a_{12}s_2 + \ldots + a_{1n}s_n = \beta_1$$
$$a_{21}s_1 + a_{22}s_2 + \ldots + a_{2n}s_n = \beta_2$$
$$\vdots \qquad\qquad \vdots$$
$$a_{n1}s_1 + a_{n2}s_2 + \ldots + a_{nn}s_n = \beta_n$$

We now ask the question: if $|\beta_i - b_i| < \varepsilon$ for $i = 1, 2, \ldots, n$ what can we say about $|s_i - x_i|$?

In effect what we are asking is: if the β_i are respectively very close to the b_i are the s_i necessarily very close to the true solutions x_i? We shall see that the answer is 'NO'.

Consider the system of three equations:

$$5x_1 + 16x_2 + 10x_3 = 1 \cdot 1$$

$$6x_1 + 29x_2 + 18x_3 = -0 \cdot 9$$

$$2x_1 + 8x_2 + 5x_3 = -0 \cdot 1.$$

We try $(s_1, s_2, s_3) = (1, 1, -2)$ as a possible approximate solution and find that, in the notation above, $(\beta_1, \beta_2, \beta_3) = (1, -1, 0)$ so that we have

$$|\beta_1 - b_1| = |\beta_2 - b_2| = |\beta_3 - b_3| = 0 \cdot 1$$

and we might reasonably conclude that $(1, 1, -2)$ is very close to the solution vector $(x_1, x_2, x_3)'$. We would, however, be quite wrong for the exact solution is

$$(x_1, x_2, x_3) = (1 \cdot 3, 5 \cdot 1, -8 \cdot 7)$$

so that, in particular,

$$|s_3 - x_3| = 67|\beta_3 - b_3|. \tag{3.5.1}$$

We now seek the explanation of this phenomenon.

In matrix notation our system of equations is

$$\mathbf{Ax = b}$$

and the solution, assuming that $\mathbf{A}$ is non-singular is

$$\mathbf{x = A^{-1}b.}$$

Suppose we have an approximate solution vector **s**. Substituting **s** in place of **x** we produce $\boldsymbol{\beta}$, an approximation to **b** i.e.

$$\mathbf{As} = \boldsymbol{\beta}.$$

Since $\boldsymbol{\beta}$ is supposed to be fairly close to **b** we write $\boldsymbol{\beta} = \mathbf{b} + \boldsymbol{\varepsilon}$ where $\boldsymbol{\varepsilon}' = (\varepsilon_1, \varepsilon_2, \ldots, \varepsilon_n)$ is the error vector. Similarly, since **s** is an approximation to **x** we write $\mathbf{s} = \mathbf{x} + \boldsymbol{\sigma}$ and so we have

$$\mathbf{A}(\mathbf{x} + \boldsymbol{\sigma}) = \mathbf{b} + \boldsymbol{\varepsilon}$$

or, since $\mathbf{Ax} = \mathbf{b}$

$$\mathbf{A}\boldsymbol{\sigma} = \boldsymbol{\varepsilon}$$

so that

$$\boldsymbol{\sigma} = \mathbf{A}^{-1}\boldsymbol{\varepsilon}.$$

Thus the relationship between the errors on the right-hand side of the equations and the errors in the approximate values of the variables depends upon the inverse of the matrix of the coefficients. If this inverse matrix, $\mathbf{A}^{-1}$, contains large elements it is clear that $\boldsymbol{\sigma}$ might well contain some large elements even though $\boldsymbol{\varepsilon}$ does not. In view of this we analyse the system of three equations above, where

$$\mathbf{A} = \begin{pmatrix} 5 & 16 & 10 \\ 6 & 29 & 18 \\ 2 & 8 & 5 \end{pmatrix} \quad \text{and} \quad \boldsymbol{\varepsilon} = \begin{pmatrix} -0 \cdot 1 \\ -0 \cdot 1 \\ +0 \cdot 1 \end{pmatrix}.$$

We find that

$$\mathbf{A}^{-1} = \begin{pmatrix} 1 & 0 & -2 \\ 6 & 5 & -30 \\ -10 & -8 & 49 \end{pmatrix}$$

and so $\boldsymbol{\sigma}' = (-0 \cdot 3, -4 \cdot 1, 6 \cdot 7)$. The true solution is therefore given by i.e.

$$\mathbf{x} = \mathbf{s} - \boldsymbol{\sigma}$$

$$\mathbf{x}' = (1, 1, -2) - (-0 \cdot 3, -4 \cdot 1, +6 \cdot 7)$$

so that $\mathbf{x}' = (1 \cdot 3, 5 \cdot 1, -8 \cdot 7)$, as remarked above.

If, more generally, our error vector $\boldsymbol{\varepsilon}$ is $(\varepsilon_1, \varepsilon_2, \varepsilon_3)'$ then the errors in the solution vector are seen to be

$$(\varepsilon_1 - 2\varepsilon_3, 6\varepsilon + 5\varepsilon_2 - 30\varepsilon_3, -10\varepsilon_1 - 8\varepsilon_2 + 49\varepsilon_3)'$$

and the result (3.5.1) arises because, in the case considered, $\varepsilon_1 = \varepsilon_2 = -\varepsilon_3 (= -0 \cdot 1)$, so that the error in x_3 will be $-67\varepsilon_1$.

The phenomenon is thus seen to be caused by large elements in the inverse of the matrix of coefficients and is therefore an intrinsic property of this particular system of equations. It is a particular case of the more general phenomenon of *ill-conditioned equations*.

DEFINITION 3.5.1.   A system of equations is said to be *ill-conditioned* when small changes in the coefficients produce large changes in the solution.

Ill-conditioning also occurs in other problems in numerical analysis, for example the fitting of a polynomial approximation to a given function by means of least squares (see Chapter 6).

When a system of equations is ill-conditioned it may be very difficult to obtain a solution which is accurate to the desired number of places. In such cases when the equations are being solved on a computer it may be necessary to use double-length facilities in order to obtain results which are correct to single-length precision. Unfortunately it is not in general possible to tell that a system of equations is ill-conditioned by looking at the coefficients and we will only become aware that the equations are ill-conditioned when we try to solve them. One useful technique, which is easy enough to implement on a computer and which helps to reveal that the equations are ill-conditioned, is worth mentioning. Instead of just solving the equations

$$\mathbf{Ax = b}$$

for one right-hand side **b** solve them for two by simultaneously solving

$$\mathbf{Ax = b + \delta b}$$

where the elements of **δb** are all small. Then if the two sets of solutions are very different the equations are ill-conditioned. The last example illustrates this point, for when

$$\mathbf{b}' = (1, -1, 0), \qquad \mathbf{x}' = (1, 1, -2)$$

and when

$$(\mathbf{b} + \mathbf{\delta b})' = (1 \cdot 1, -0 \cdot 9, -0 \cdot 1), \qquad \mathbf{x}' = (1 \cdot 3, 5 \cdot 1, -8 \cdot 7).$$

3.6. COMPARISON OF THE GAUSS AND JORDAN METHODS

These two methods are based upon the same fundamental idea, elimination of the variables one by one, but it is not obvious whether it is more efficient to do more work at the elimination stage and avoid back-substitution (as in the Jordan method) or not. We shall now show that, as remarked in section 3.3.3, Jordan's method requires about 50% more multiplications than Gauss' method when used on the same set of equations.

3.6.1. Analysis of the Gauss Method

We shall assume that we have n equations and that the coefficients are such that the pivot is always satisfactory. Let the equations be

$$a_{11}x_1 + a_{12}x_2 + \ldots + a_{1n}x_n = b_1$$
$$a_{21}x_1 + a_{22}x_2 + \ldots + a_{2n}x_n = b_2$$
$$\vdots \qquad \vdots \qquad\qquad \vdots \qquad \vdots$$
$$a_{n1}x_1 + a_{n2}x_2 + \ldots + a_{nn}x_n = b_n.$$

We divide the first equation throughout by a_{11} (cost: n divisions) and then subtract appropriate multiples of it from the 2nd, 3rd, ..., nth equations. Each such equation requires n multiplications and n subtractions, and there are $(n-1)$ equations in all. The total cost of this first stage is therefore

$$n(n-1) \qquad \text{multiplications}$$
$$n(n-1) \qquad \text{subtractions}$$

and

$$n \qquad \text{divisions.}$$

At the end of this first stage x_1 has been eliminated from all equations except the first.

We now proceed, in a similar way, to eliminate x from the 3rd, 4th, ..., nth equations. The cost is seen to be

$$(n-1)(n-2) \qquad \text{multiplications}$$
$$(n-1)(n-2) \qquad \text{subtractions}$$
$$(n-1) \qquad \text{divisions.}$$

We repeat the process to eliminate $x_3, x_4, \ldots, x_{n-1}$. The cost of eliminating x_k from the $(k+1)$st and subsequent equations is:

$$(n-k+1)(n-k) \qquad \text{multiplications}$$
$$(n-k+1)(n-k) \qquad \text{subtractions}$$
$$(n-k+1) \qquad \text{divisions.}$$

The total cost of the elimination process in the Gauss method is therefore

$$\sum_{k=1}^{n-1} (n-k+1)(n-k) \qquad \text{multiplications}$$

$$\sum_{k=1}^{n-1} (n-k+1)(n-k) \qquad \text{subtractions}$$

and

$$\sum_{k=1}^{n-1} (n-k+1) \qquad \text{divisions}$$

and evaluation of these sums gives the total of

$$\tfrac{1}{3}(n-1)(n)(n+1) \qquad \text{multiplications}$$
$$\tfrac{1}{3}(n-1)(n)(n+1) \qquad \text{subtractions}$$
$$\tfrac{1}{2}(n-1)(n) \qquad \text{divisions.}$$

The Gauss method is now completed by the process of back substitution. At the point where we have found $x_n, x_{n-1}, \ldots, x_{n-k+1}$ we will find x_{n-k} from an equation of the type

$$\alpha_{n-k}x_{n-k} + \ldots + \alpha_n x_n = \beta_n$$

and x_{n-k} is therefore found at a cost of

$$k \qquad \text{multiplications}$$
$$k \qquad \text{subtractions}$$

and

$$1 \qquad \text{division.}$$

The total cost of the back-substitution phase is therefore

$$\sum_{k=0}^{n-1} k = \tfrac{1}{2}n(n-1) \qquad \text{multiplications}$$

$$\sum_{k=0}^{n-1} k = \tfrac{1}{2}n(n-1) \qquad \text{subtractions}$$

$$\sum_{k=0}^{n-1} 1 = n \qquad \text{divisions.}$$

Combining the two sets of results we see that the total cost of the Gauss method is

$$\tfrac{1}{6}(n-1)(n)(2n+5) \qquad \text{multiplications}$$
$$\tfrac{1}{6}(n-1)(n)(2n+5) \qquad \text{subtractions}$$
$$\tfrac{1}{2}n(n+1) \qquad \text{divisions.}$$

(Thus for example when $n = 2$ we require 4 multiplications, 4 subtractions and 3 divisions, as mentioned in section 3.1.)

3.6.2. Analysis of the Jordan Method

Having worked out the analysis for the Gauss method in detail we can abbreviate the analysis for the Jordan method. Suppose that we have eliminated $x_1, x_2, \ldots, x_{k-1}$ from all but the 1st, 2nd, $\ldots, (k-1)$st equations respectively then we proceed to eliminate x_k from all but the kth equation. To do this we must

(1) divide the coefficients of $x_{k+1}, x_{k+2}, \ldots, x_n$ in the kth equation and b_k by the coefficient of x_k;
(2) Subtract suitable multiples of the kth equation from the other $(n-1)$ equations to eliminate x_k from these equations.
Step (1) takes $(n-k+1)$ divisions
Step (2) takes $(n-1)(n-k+1)$ multiplications and $(n-1)(n-k+1)$ subtractions.

The total number of operations involved is therefore:

$$\sum_{k=1}^{n} (n-1)(n-k+1) = \tfrac{1}{2}(n-1)n(n+1) \qquad \text{multiplications}$$

$$\sum_{k=1}^{n} (n-1)(n-k+1) = \tfrac{1}{2}(n-1)n(n+1) \qquad \text{subtractions}$$

and

$$\sum_{k=1}^{n} (n-k+1) = \tfrac{1}{2}n(n+1) \qquad \text{divisions.}$$

Since there is no back-substitution phase this completes the analysis. From the two sets of formulae we see that for large n the number of multiplications in the Gauss and Jordan methods are approximately $\tfrac{1}{3}n^3$ and $\tfrac{1}{2}n^3$ respectively. The number of subtractions is also $\tfrac{1}{3}n^3$ and $\tfrac{1}{2}n^3$ approximately and the number of divisions is $\tfrac{1}{2}n(n+1)$ in both cases, i.e. approximately $\tfrac{1}{2}n^2$ for large n. It follows that for large n the Gauss method requires only about $\tfrac{2}{3}$ of the number of multiplications or subtractions that the Jordan method requires, which is equivalent to the assertion made at the beginning of this section.

The type of analysis carried out above is a simple example of a kind which has attracted a lot of interest in recent years. If several methods are proposed for the solution of a problem on a computer which should we use? Assuming that the methods are equally good from the point of view of accuracy of the solution the deciding factor may be the time required on the computer for each method to complete the solution. In the case of direct methods such as the Gauss or Jordan the analysis can be done very precisely but in the case of indirect methods the number of iterations required before convergence to the desired precision is achieved may vary from one method to another and the uncertainty in this may make detailed analysis pointless.

R.F.C.

REFERENCES

For further discussion of the topics covered in this chapter the following texts can be recommended

Isaacson, E. and Keller, H. (1976). *Analysis of Numerical Methods*, Wiley.

Tewarson, P. (1973). *Sparse Matrices*, Academic Press.

Varga, R. (1962). *Matrix Iterative Analysis*, Prentice-Hall.

For a detailed analysis of the effect of rounding errors on the accuracy of the solution in Gaussian Elimination and other processes see

Wilkinson, J. H. (1963). *Rounding Errors in Algebraic Processes*, H.M.S.O., London.

PROGRAMS

In the Appendix, programs will be found for the solution of linear equations by (1) the Gauss–Seidel method, (2) the Gauss–Jordan method.

CHAPTER 4

Matrix Computations

4.1. MOTIVATION

The numerical solution of a wide variety of problems in applied sciences, econometric models, statistics, etc. involve matrix computations. There are many methods, programs and packages for matrix computations, but there are essentially only two problems: the solution of a system of linear equations [see I, §§ 5.7–5.10]

$$\mathbf{A}\mathbf{x} = \mathbf{b}, \qquad (4.1.1)$$

and the solution of the eigenvalue problem [see I, Chapter 7]

$$\mathbf{A}\mathbf{x} = \alpha\mathbf{x}. \qquad (4.1.2)$$

The first problem turns up in regression analysis and civil engineering structures, to take just two examples from different areas. The second problem turns up in principal components analysis and vibrations, taking the same two areas.

Often the problem to be solved does not directly involve matrix computations, for example the numerical solution of the many types of differential equations [see IV, Chapters 7 and 8]. This is reduced to a matrix problem by approximating the required unknown function by a vector representing the value of the function at a discrete set of points. *This discretization process* is applicable to any operator L, which may involve differentiation or integration and which operates on functions $f, g, \ldots$ of one or more independent variables, provided that L is linear, that is

$$L(af + bg) = aLf + bLg, \qquad (4.1.3)$$

where a and b are scalars not involving the independent variables. If the equation to be solved is

$$Ly = f, \qquad (4.1.4)$$

where y is the required unknown function, then this can be approximated by the system of linear equations (4.1.1) where $\mathbf{A}, \mathbf{x}, \mathbf{b}$ are discrete approximations of L, y, f respectively. If the equation to be solved is not linear in the required function y, then it is expressed in the form

$$Ly = f + M(y),\qquad(4.1.5)$$

where L is linear but M is not. Under suitable conditions this can be solved iteratively in the form

$$Ly^{(n+1)} = f + M(y^{(n)}).\qquad(4.1.6)$$

We start with an approximation $y^{(0)}$ and derive, by repeated solutions, successively improved approximations $y^{(1)}$, $y^{(2)}$, ... to the required solution y.

   Many different methods for solving the two basic problems have been designed with a view to reducing the cost of carrying out numerical computations to a high degree of accuracy. The cost is measured in terms of the number of arithmetic operations required (time) and the size of the computer memory needed to store the various coefficients and variables (space). The order of magnitude of the number of operations is called the *complexity* of the computation. If the number of unknowns (dimension of x) is n, then the number of operations is usually of order n^3 and the required space is of order n^2. Often n is of the order of hundreds and sometimes thousands, and the requirement of space and time becomes prohibitive in spite of the rapid advances in computer technology. Another difficulty is the loss of accuracy caused by the build-up of rounding errors; the accuracy of computer arithmetic may be high but is limited; the loss of accuracy may be large after only a few operations and is not necessarily caused by a large number of operations. To deal with the various problems listed above, a large number of different matrix methods have been developed. Often the method exploits a particular property of the matrix such as symmetry, band structure, etc. to minimize time or space or the build-up of rounding errors and is not appropriate if the matrix does not possess that particular property.

   In most of what follows it will be assumed that the coefficients and variables are real unless otherwise stated, and indeed this is the case in most applications. However, most methods continue to apply when the coefficients are complex; the necessary changes in formulating the methods and programming them are usually minor, especially when a programming language with complex arithmetic facilities is used.

   This chapter will adopt the notation, definitions and results of Chapters 6 and 7 of Volume I on matrices and eigenvalues.

4.2. THE CHOLESKI METHOD

   In this section we explore methods to minimize storage, arithmetic and loss of accuracy in the solution of $\mathbf{A}\mathbf{x} = \mathbf{b}$ when the matrix $\mathbf{A}$ is symmetric and positive definite [see I (6.7.5) and § 9.2], i.e.

$$\mathbf{A}' = \mathbf{A},\qquad(4.2.1)$$

and

$$\mathbf{x}'\mathbf{A}\mathbf{x} > 0 \qquad\qquad (4.2.2)$$

provided that $\mathbf{x} \neq \mathbf{0}$. For example if

$$\mathbf{x} = \begin{pmatrix} x_1 \\ x_2 \end{pmatrix} \quad\text{and}\quad \mathbf{A} = \begin{pmatrix} a_{11} & a_{12} \\ a_{21} & a_{22} \end{pmatrix} \ (a_{12} = a_{21})$$

then the quadratic form (4.2.2) becomes [see I, § 9.1]

$$\begin{aligned} \mathbf{x}'\mathbf{A}\mathbf{x} &= a_{11}x_1^2 + (a_{12} + a_{21})x_1 x_2 + a_{22}x_2^2 \\ &= a_{11}x_1^2 + 2a_{12}x_1 x_2 + a_{22}x_2^2. \end{aligned} \qquad (4.2.3)$$

Positive definite symmetric matrices occur frequently in applications.
Consider the system of linear equations $\mathbf{A}\mathbf{x} = \mathbf{b}$:

$$\begin{aligned} x_1 + 2x_2 + 3x_3 &= b_1 \\ 2x_1 + 8x_2 + 22x_3 &= b_2 \\ 3x_1 + 22x_2 + 82x_3 &= b_3. \end{aligned} \qquad (4.2.4)$$

We observe that the matrix of coefficients

$$\mathbf{A} = \begin{pmatrix} 1 & 2 & 3 \\ 2 & 8 & 22 \\ 3 & 22 & 82 \end{pmatrix} \qquad (4.2.5)$$

is symmetric. We subtract multiples $(2, 3)$ of the first equation from the second and third equation respectively to eliminate x_1, and a multiple (4) of the resulting second equation from the reduced third equation to eliminate x_2. We end up with the triangular system of linear equations $\mathbf{U}\mathbf{x} = \mathbf{c}$:

$$\begin{aligned} x_1 + 2x_2 + 3x_3 &= c_1, \\ 4x_2 + 16_3 &= c_2, \\ 9x_3 &= c_3, \end{aligned} \qquad (4.2.6)$$

where

$$\mathbf{c} = \begin{pmatrix} c_1 \\ c_2 \\ c_3 \end{pmatrix} = \begin{pmatrix} b_1 \\ b_2 - 2b_1 \\ b_3 - 4b_2 + 5b_1 \end{pmatrix}, \qquad (4.2.7)$$

and

$$\mathbf{U} = \begin{pmatrix} 1 & 2 & 3 \\ 0 & 4 & 16 \\ 0 & 0 & 9 \end{pmatrix}. \qquad (4.2.8)$$

The multipliers used in the elimination, augmented by 1's along the diagonal, form a lower triangular matrix

$$\mathbf{L} = \begin{pmatrix} 1 & 0 & 0 \\ 2 & 1 & 0 \\ 3 & 4 & 1 \end{pmatrix} \tag{4.2.9}$$

where each multiplier is placed in the position of the coefficient it eliminates. (Note that the multipliers refer to the subtraction of one equation from another.) We can verify, and one can prove in general, that the following formula holds:

$$\mathbf{A} = \mathbf{LU}. \tag{4.2.10}$$

This factorisation is possible provided that $\mathbf{A}$ is non-singular [see I, Definition 6.4.2] and that there are no zero divisors in the various stages. Thus, the traditional method for the solution of a system of n linear equations is first to reduce the 'square' system $\mathbf{Ax} = \mathbf{b}$ to an upper triangular system $\mathbf{Ux} = \mathbf{c}$ by an elimination process, and then to solve the resulting triangular system by simple back substitution. This procedure is equivalent to the factorisation of $\mathbf{A}$ as a product $\mathbf{LU}$; given $\mathbf{L}$ and $\mathbf{U}$ we can derive $\mathbf{c}$ from $\mathbf{b}$ by a forward substitution process to solve

$$\mathbf{Lc} = \mathbf{b}, \tag{4.2.11}$$

then derive $\mathbf{x}$ from $\mathbf{c}$ by a back substitution process to solve

$$\mathbf{Ux} = \mathbf{c}. \tag{4.2.12}$$

This procedure is applicable in general, and so far we have not used any special property of $\mathbf{A}$, but unless $\mathbf{A}$ is symmetric and positive definite, the method needs considerable modification such as rearrangements of the rows or columns in order to avoid zero divisors and to control the build-up of rounding errors.

If we forego the condition that $\mathbf{L}$ has 1's along the diagonal, then this process of elimination or factorization is no longer unique and can be carried out in different ways by different scalings giving

$$\mathbf{L} = \begin{pmatrix} 1 & 0 & 0 \\ 2 & 1 & 0 \\ 3 & 4 & 1 \end{pmatrix}, \quad \mathbf{U} = \begin{pmatrix} 1 & 2 & 3 \\ 0 & 4 & 16 \\ 0 & 0 & 9 \end{pmatrix} \tag{4.2.13}$$

or

$$\mathbf{L} = \begin{pmatrix} 1 & 0 & 0 \\ 2 & 4 & 0 \\ 3 & 16 & 9 \end{pmatrix}, \quad \mathbf{U} = \begin{pmatrix} 1 & 2 & 3 \\ 0 & 1 & 4 \\ 0 & 0 & 1 \end{pmatrix} \tag{4.2.14}$$

or

$$L = \begin{pmatrix} 1 & 0 & 0 \\ 2 & 2 & 0 \\ 3 & 8 & 3 \end{pmatrix}, \qquad U = \begin{pmatrix} 1 & 2 & 3 \\ 0 & 2 & 8 \\ 0 & 0 & 3 \end{pmatrix} \tag{4.2.15}$$

etc. These different forms can be derived as follows. Let

$$D = \begin{pmatrix} 1 & 0 & 0 \\ 0 & 4 & 0 \\ 0 & 0 & 9 \end{pmatrix} \tag{4.2.16}$$

be the diagonal matrix with elements 1, 4, 9. We note that if M is any 3×3 matrix, then DM is derived from M by multiplying (scaling) its rows by the corresponding diagonal elements of D. Similarly MD is derived from M by multiplying its columns by the corresponding diagonal elements of D. Using this remark we can verify that in (4.2.13) we have that

$$U = DL'. \tag{4.2.17}$$

This simple relation between L and U holds in general, provided that A is symmetric, and (4.2.10) becomes:

$$A = LU = LDL'. \tag{4.2.18}$$

We can factorize D as the product of 2 diagonal matrices $D_1 D_2$ in infinitely many ways corresponding to the factorizations of the diagonal elements of D. Each factorization $D_1 D_2$ gives a factorization

$$A = LDL' = (LD_1)(D_2 L'), \tag{4.2.19}$$

where the first factor LD_1 is lower triangular, and the second factor $D_2 L'$ is upper triangular.

The factorization $D_1 = D_2 = D^{1/2}$ has diagonal elements equal to the positive square root of the corresponding elements of D. In our example:

$$D_1 = D_2 = \begin{pmatrix} 1 & 0 & 0 \\ 0 & 2 & 0 \\ 0 & 0 & 3 \end{pmatrix}. \tag{4.2.20}$$

From among all the possible factorizations, (4.2.20) leads to the smallest overall numbers in magnitude in the various stages of the elimination. This is a very useful property in controlling the build-up of rounding errors through successive stages of the computation. Let

$$\mathbf{U} = \mathbf{D}^{1/2}\mathbf{L}'. \tag{4.2.21}$$

It follows that $\mathbf{U}' = \mathbf{L}\mathbf{D}^{1/2}$, and that

$$\mathbf{A} = \mathbf{U}'\mathbf{U}. \tag{4.2.22}$$

In our example:

$$\begin{pmatrix} 1 & 2 & 3 \\ 2 & 8 & 22 \\ 3 & 22 & 82 \end{pmatrix} = \begin{pmatrix} 1 & 0 & 0 \\ 2 & 2 & 0 \\ 3 & 8 & 3 \end{pmatrix}\begin{pmatrix} 1 & 2 & 3 \\ 0 & 2 & 8 \\ 0 & 0 & 3 \end{pmatrix}. \tag{4.2.23}$$

This is called the *Choleski factorization*; it is possible only if the original matrix **A** is symmetric and positive definite. If the matrix **A** is symmetric but not positive definite, then some of the elements of **D** will be negative, leading to imaginary or complex arithmetic resulting from the square roots; a more elaborate algorithm can avoid this. Alternatively we can use complex arithmetic, which is available in most programming languages, but the property which controls the build-up of rounding errors is lost.

Because the Choleski factorization is a fundamental step in many matrix computations, an informal description of the algorithm for the factorization $\mathbf{A} = \mathbf{U}'\mathbf{U}$ is given below.

```
for r: = 1(1)n
    u_rr: = √a_rr
    for i: = r + 1(1)n
        for j: = i(1)n
            a_ij: = a_ij − u_ri u_rj
        repeat j
    repeat i
repeat r.
```

In this notation the useful Algol convention is adopted whereby a loop of the form $s(1)t$ is not executed if $s > t$. It is worth pointing out a number of convenient properties of this algorithm. It makes use of elements on and above the diagonal only so that it is only necessary to store half the matrix by employing various facilities such as jagged arrays to map the upper triangular matrix onto a one-dimensional array. The corresponding elements of **U** and **A** can occupy the same locations, a new element of **U** is computed when the corresponding element in A is no longer required and can therefore overwrite it. Finally we observe that the order in which the elements of **A** or **U** are accessed is by rows, which is a useful organisation if the matrix is too large for the computer primary memory and the matrix has to be fetched from secondary memory one row at a time. An alternative version of the same algorithm is given below where the computation and access are organised by columns rather than rows. It is not proposed to describe other algorithms in such detail; these two algorithms should give the reader an idea of what is involved, for detailed descriptions see Wilkinson and Reinsch (1971).

$$
\begin{aligned}
&\text{for } j := 1(1)n \\
&\quad \text{for } i := 1(1)j \\
&\qquad \text{for } r := 1(1)i - 1 \\
&\qquad\quad a_{ij} := a_{ij} - u_{ri}u_{rj} \\
&\qquad \text{repeat } r \\
&\qquad u_{ij} := \begin{cases} a_{ij}/u_{ii} & \text{if } i < j \\ \sqrt{a_{jj}} & \text{if } i = j \end{cases} \\
&\quad \text{repeat } i \\
&\text{repeat } j.
\end{aligned}
$$

We conclude with two simple examples of the Choleski factorisation.

EXAMPLE 4.2.1.   Let

$$
\mathbf{A} = \begin{pmatrix} a_1 & b_1 \\ b_1 & a_2 \end{pmatrix}, \quad \text{then} \quad \mathbf{U} = \begin{pmatrix} u_{11} & u_{12} \\ 0 & u_{22} \end{pmatrix}
$$

where

$$
u_{11} = \sqrt{a_1},
$$
$$
u_{12} = b_1/u_{11},
$$
$$
u_{22} = \sqrt{(a_2 - u_{12}^2)}.
$$

EXAMPLE 4.2.2.   Let

$$
\mathbf{A} = \begin{pmatrix} a_1 & b_1 & 0 & 0 \\ b_1 & a_2 & b_2 & 0 \\ 0 & b_2 & a_3 & b_3 \\ 0 & 0 & b_3 & a_4 \end{pmatrix}, \quad \text{then} \quad \mathbf{U} = \begin{pmatrix} u_{11} & u_{12} & 0 & 0 \\ 0 & u_{22} & u_{23} & 0 \\ 0 & 0 & u_{33} & u_{34} \\ 0 & 0 & 0 & u_{44} \end{pmatrix}
$$

where

$$
u_{11} = \sqrt{a_1}, \qquad u_{12} = b_1/u_{11},
$$
$$
u_{22} = \sqrt{(a_2 - u_{12}^2)}, \qquad u_{23} = b_2/u_{22},
$$
$$
u_{33} = \sqrt{(a_3 - u_{23}^2)}, \qquad u_{34} = b_3/u_{33},
$$
$$
u_{44} = \sqrt{(a_4 - u_{34}^2)}.
$$

It is easy to see how to extend this procedure to an $n \times n$ matrix.

Example 4.2.2 illustrates a useful property of the Choleski method: if the matrix $\mathbf{A}$ is a band matrix, that is if all the elements outside a central band along the diagonal are zero, then the same applies to the corresponding elements of $\mathbf{U}$; thus band structure is preserved. This permits considerable savings in computer memory and arithmetic.

4.3. PARTITIONED MATRICES

It was mentioned in the last section that it is convenient to arrange matrix computations and storage by rows (or columns) if the matrix is too large for the primary memory. Another reason for this is the possibility of carrying out vector operations with some degree of parallel processing in some of the more powerful computers. The same convenience in storage management and speed through parallel processing can be achieved by partitioning the matrix into rectangular or square blocks [see I, § 6.6].

Any algorithm for computations for a matrix with scalar elements can be applied with simple minor modifications to a matrix whose elements are blocks. To show this, the results of Example 4.2.2 will be applied to a matrix $\mathbf{A}$ whose elements are matrices $\mathbf{A}_i$, $\mathbf{B}_i$ instead of scalars a_i, b_i.

EXAMPLE 4.3.1.   Let

$$\mathbf{A} = \begin{pmatrix} \mathbf{A}_1 & \mathbf{B}_1 & \mathbf{0} & \mathbf{0} \\ \mathbf{B}'_1 & \mathbf{A}_2 & \mathbf{B}_2 & \mathbf{0} \\ \mathbf{0} & \mathbf{B}'_2 & \mathbf{A}_3 & \mathbf{B}_3 \\ \mathbf{0} & \mathbf{0} & \mathbf{B}'_3 & \mathbf{A}_4 \end{pmatrix} \qquad (4.3.1)$$

where $\mathbf{A}_1$ is $n_1 \times n_1$, $\mathbf{A}_2$ is $n_2 \times n_2$, etc. and $\mathbf{B}_1$ is $n_1 \times n_2$, $\mathbf{B}_2$ is $n_2 \times n_3$, etc.; similarly, each $\mathbf{0}$ is a zero block of the appropriate dimension. If, in addition, $\mathbf{A}$ is also symmetric and positive definite, it follows that the same is true for each of the diagonal blocks $\mathbf{A}_i$. To carry out the necessary modification we define the *choleski* function which takes a positive definite symmetric matrix as an argument, and its value is the upper triangular matrix $\mathbf{U}$ given by the Choleski factorization (4.2.22). Thus we have:

$$\mathbf{U} = \text{choleski } (\mathbf{A}) \quad \text{if } \mathbf{U}'\mathbf{U} = \mathbf{A}. \qquad (4.3.2)$$

We now rewrite the algorithm of Example 4.2.2 using block matrices instead of scalars:

$$\begin{aligned}
\mathbf{U}_{11} &= \text{choleski } (\mathbf{A}_1), & \mathbf{U}_{12} &= \mathbf{U}_{11}^{-1}\mathbf{B}_1, \\
\mathbf{U}_{22} &= \text{choleski } (\mathbf{A}_2 - \mathbf{U}'_{12}\mathbf{U}_{12}), & \mathbf{U}_{23} &= \mathbf{U}_{22}^{-1}\mathbf{B}_2, \\
\mathbf{U}_{33} &= \text{choleski } (\mathbf{A}_3 - \mathbf{U}'_{23}\mathbf{U}_{23}), & \mathbf{U}_{34} &= \mathbf{U}_{33}^{-1}\mathbf{B}_3, \\
\mathbf{U}_{44} &= \text{choleski } (\mathbf{A}_4 - \mathbf{U}'_{34}\mathbf{U}_{34}), & \mathbf{U}_{13} &= \mathbf{0}, \qquad \mathbf{U}_{14} = \mathbf{0}, \mathbf{U}_{24} = \mathbf{0}.
\end{aligned} \qquad (4.3.3)$$

Unlike scalar multiplication, matrix multiplication is not commutative and we cannot change the order of the factors in expressions such as $\mathbf{U}_{11}^{-1}\mathbf{B}_1$, $\mathbf{U}'_{12}\mathbf{U}_{12}$, etc. [see I, § 6.2(vi)]. We refer to section 4.5 for methods of computing matrices of the type $\mathbf{U}_{11}^{-1}\mathbf{B}_1$ involving matrix inversion.

In spite of these differences, the modification needed to adopt an algorithm to a matrix with block elements is relatively straightforward. In this case the square root is replaced by the choleski function (4.3.2) and division is replaced by left multiplication by the inverse. To see the importance of this method, suppose that $n_1 = n_2 = n_3 = n_4 = 50$; instead of operating on a 200×200 matrix we are handling at most two 50×50 matrices at any one time, which is a much more manageable size of required primary computer memory, and may be considerably faster if an array processor is available to perform parallel arithmetic. Furthermore, considerable savings are achieved by ignoring the zero blocks.

4.4.   LEAST SQUARES METHODS

In this section we are concerned with the solution of $\mathbf{Ax} = \mathbf{b}$ when the coefficients may not be exact and an exact solution may not be required, or it may not exist or it may not be unique. This usually occurs in statistical computations. Let

$m = $ number of equations $=$ number of rows of $\mathbf{A}$,
$n = $ number of unknowns $=$ number of columns of $\mathbf{A}$,
$k = $ rank $(\mathbf{A}) = $ number of linearly independent rows (or columns) of $\mathbf{A}$

[see I, § 5.6]. If we assume that the m equations are soluble, then

$$k = \text{rank } (\mathbf{A}) = \text{rank } (\mathbf{A}, \mathbf{b}). \tag{4.4.1}$$

In the previous sections we were concerned with the case

$$m = n = k. \tag{4.4.2}$$

In this section we will consider in turn the following cases:

(i) $m > n,\ k = n$
(ii) $m < n,\ k = m$
(iii) $k < m,\ k < n$.

But first we recall briefly some results from matrix algebra. From the algebra of transposition [see I (6.5.1)] we have that

$$(\mathbf{AB})' = \mathbf{B}'\mathbf{A}', \qquad (\mathbf{ABC})' = \mathbf{C}'\mathbf{B}'\mathbf{A}', \quad \text{etc.} \tag{4.4.3}$$

Let $\mathbf{x}$ and $\mathbf{y}$ be two column vectors of dimension n. Then $\mathbf{x}'$ and $\mathbf{y}'$ are row vectors, say

$$\mathbf{x}' = (x_1, x_2, \ldots, x_n)$$
$$\mathbf{y}' = (y_1, y_2, \ldots, y_n). \tag{4.4.4}$$

By the definition of scalar products [$\mathbf{I}$(6.1.7) and $\mathbf{I}$(6.5.4)] we have that

$$\mathbf{x}'\mathbf{y} = \mathbf{y}'\mathbf{x} = x_1 y_1 + x_2 y_2 + \ldots + x_n y_n, \tag{4.4.5}$$

and

$$\mathbf{x}'\mathbf{x} = x_1^2 + x_2^2 + \ldots + x_n^2. \tag{4.4.6}$$

The length of a vector $\mathbf{x}$, sometimes called the *Euclidean norm*, or just *norm*, is defined by

$$\|\mathbf{x}\| = (\mathbf{x}'\mathbf{x})^{1/2} \qquad (4.4.7)$$

[see (6.2.6)]. Finally we recall the definition of the quadratic form $\mathbf{x}'\mathbf{A}\mathbf{x}$, a special case of which, when $n = 2$, is given by (4.2.3) [see I (6.2.7)].

(i) Consider first the case $\mathbf{A}\mathbf{x} = \mathbf{b}$ where $m > n$, $k = n$. For example,

$$2x_1 + 3x_2 = b_1$$
$$5x_1 + 7x_2 = b_2 \qquad (4.4.8)$$
$$4x_1 + 9x_2 = b_3$$

where $m = 3$, $n = 2$. In general, there is no solution $\mathbf{x}$ which satisfies $\mathbf{A}\mathbf{x} = \mathbf{b}$. For any vector $\mathbf{x}$ proposed as a solution we define the *residual* vector

$$\mathbf{r} = \mathbf{b} - \mathbf{A}\mathbf{x}. \qquad (4.4.9)$$

We seek a 'solution' $\mathbf{x}$ which minimizes $\|\mathbf{r}\|$, i.e. which minimizes the sum of the squares of the residuals; hence the title 'least squares method'. The sum of the squares is given by the following expression:

$$\mathbf{r}'\mathbf{r} = (\mathbf{b} - \mathbf{A}\mathbf{x})'(\mathbf{b} - \mathbf{A}\mathbf{x}), \qquad (4.4.10)$$

which after expansion simplifies to

$$\mathbf{r}'\mathbf{r} = \mathbf{x}'\mathbf{B}\mathbf{x} - 2\mathbf{x}'\mathbf{c} + \mathbf{b}'\mathbf{b}, \qquad (4.4.11)$$

where

$$\mathbf{B} = \mathbf{A}'\mathbf{A}, \qquad \mathbf{c} = \mathbf{A}'\mathbf{b}. \qquad (4.4.12)$$

Since rank $(\mathbf{A}) = n$, it follows from (4.4.12) that $\mathbf{B}$ is non-singular, positive definite and symmetric and so is $\mathbf{B}^{-1}$. Let $\mathbf{y} = \mathbf{B}\mathbf{x} - \mathbf{c}$, then using (4.4.11) we have

$$\mathbf{y}'\mathbf{B}^{-1}\mathbf{y} = (\mathbf{x}'\mathbf{B} - \mathbf{c}')\mathbf{B}^{-1}(\mathbf{B}\mathbf{x} - \mathbf{c}) = \mathbf{r}'\mathbf{r} - \mathbf{b}'\mathbf{b} + \mathbf{c}'\mathbf{B}^{-1}\mathbf{c} \geq 0$$

hence

$$\mathbf{r}'\mathbf{r} \geq \mathbf{b}'\mathbf{b} - \mathbf{c}'\mathbf{B}^{-1}\mathbf{c}$$

and

$$\min \mathbf{r}'\mathbf{r} = \mathbf{b}'\mathbf{b} - \mathbf{c}'\mathbf{B}^{-1}\mathbf{c} = \mathbf{b}'\mathbf{b} - \mathbf{c}'\mathbf{x}$$

when $\mathbf{y} = \mathbf{0}$, that is, when

$$\mathbf{B}\mathbf{x} = \mathbf{c}. \qquad (4.4.13)$$

To sum up, first compute the upper triangular elements b_{ij} of $\mathbf{B}$ as the scalar products of columns i and j of $\mathbf{A}$ and the elements c_i of $\mathbf{c}$ as the scalar products of $\mathbf{b}$ and column i of $\mathbf{A}$, then use section 4.2 to compute $\mathbf{U}$. Next compute $\mathbf{d} = \mathbf{U}'^{-1}\mathbf{c}$ by back substitution, $\mathbf{x} = \mathbf{U}^{-1}\mathbf{d}$ by forward substitution, and finally $\min \mathbf{r}'\mathbf{r} = \mathbf{b}'\mathbf{b} - \mathbf{d}'\mathbf{d}$ if required.

(ii) Consider next the case where $\mathbf{A}\mathbf{x} = \mathbf{b}$ represents a system of m linear equation in n unknowns and where $k = m < n$. For example

$$2x_1 + 5x_2 + 4x_3 = b_1$$
$$3x_1 + 4x_2 + 2x_3 = b_2. \tag{4.4.14}$$

By renumbering the unknowns and interchanging columns if necessary we can partition $\mathbf{A}$ and $\mathbf{x}$ as

$$\mathbf{A} = (\mathbf{B}, \mathbf{C}),$$

$$\mathbf{x} = \begin{pmatrix} \mathbf{y} \\ \mathbf{z} \end{pmatrix} \tag{4.4.15}$$

so that $\mathbf{B}$ is $m \times m$ and non-singular, $\mathbf{C}$ is $m \times n - m$, $\mathbf{y}$ is of dimension m and $\mathbf{z}$ of dimension $n - m$. In the above example

$$\mathbf{B} = \begin{pmatrix} 2 & 5 \\ 3 & 4 \end{pmatrix}, \qquad \mathbf{C} = \begin{pmatrix} 4 \\ 2 \end{pmatrix},$$

$$\mathbf{y} = \begin{pmatrix} x_1 \\ x_2 \end{pmatrix}, \qquad \mathbf{z} = (x_3). \tag{4.4.16}$$

The system of equations $\mathbf{A}\mathbf{x} = \mathbf{b}$ becomes

$$\mathbf{B}\mathbf{y} + \mathbf{C}\mathbf{z} = \mathbf{b}. \tag{4.4.17}$$

Hence

$$\mathbf{B}\mathbf{y} = \mathbf{b} - \mathbf{C}\mathbf{z}$$
$$\mathbf{y} = \mathbf{B}^{-1}(\mathbf{b} - \mathbf{C}\mathbf{z}). \tag{4.4.18}$$

Thus there is an infinity of solutions corresponding to the arbitrary values of $\mathbf{z}$. The best solution in this case is taken to mean the solution $\mathbf{x}$ with the smallest norm $\|\mathbf{x}\|$, i.e. the solution which minimizes $\mathbf{x}'\mathbf{x}$ subject to the constraint $\mathbf{A}\mathbf{x} = \mathbf{b}$. We use the method of Lagrange multipliers for constrained optimization [see IV, § 15.1.4]. Let

$$\mathbf{D} = \mathbf{A}\mathbf{A}'. \tag{4.4.19}$$

It follows from its definition that $\mathbf{D}$ is $m \times m$ positive definite symmetric. We use the Choleski method to solve for $\mathbf{w}$ the system of linear equations

$$\mathbf{D}\mathbf{w} = \mathbf{b}. \tag{4.4.20}$$

Then the required solution $\mathbf{x}$ is given by the formula

$$\mathbf{x} = \mathbf{A}'\mathbf{w}. \tag{4.4.21}$$

Thus when $m > n$ we use $\mathbf{A}'\mathbf{A}$; when $m < n$ we use $\mathbf{A}\mathbf{A}'$.

EXAMPLE 4.4.1.   We apply this method to 'solve' the equation

$$ax_1 + bx_2 + cx_3 = d.$$

According to the above interpretation the required solution is the 'foot of the perpendicular' from the origin to the plane represented by the equation, hence the solution is [see V (2.2.9)]

$$\frac{x_1}{a} = \frac{x_2}{b} = \frac{x_3}{c} = \frac{d}{\sqrt{(a^2 + b^2 + c^2)}}.$$

(iii) Finally we consider the general case $\mathbf{Ax} = \mathbf{b}$ where $\mathbf{A}$ is $m \times n$ of rank k when $k < m$ and $k < n$. By rearranging the rows and renumbering the unknowns if necessary we can rewrite $\mathbf{Ax} = \mathbf{b}$ in the form:

$$\begin{pmatrix} \mathbf{B} & \mathbf{C} \\ \mathbf{D} & \mathbf{E} \end{pmatrix} \begin{pmatrix} \mathbf{y} \\ \mathbf{z} \end{pmatrix} = \mathbf{b} \tag{4.4.22}$$

where

$$\mathbf{A} = \begin{pmatrix} \mathbf{B} & \mathbf{C} \\ \mathbf{D} & \mathbf{E} \end{pmatrix}, \qquad \mathbf{x} = \begin{pmatrix} \mathbf{y} \\ \mathbf{z} \end{pmatrix},$$

and $\mathbf{y}$ is of dimension k, $\mathbf{z}$ of dimension $n\text{-}k$, $\mathbf{B}$ is $k \times k$ non-singular, $\mathbf{C}$ is $k \times n\text{-}k$, $\mathbf{D}$ is $m\text{-}k \times k$ and $\mathbf{E}$ is $m\text{-}k \times n\text{-}k$. Let

$$\mathbf{B} = \mathbf{LU}. \tag{4.4.23}$$

This factorization is possible since $\mathbf{B}$ is square and non-singular [see (4.2.10)]. Becuase $\mathbf{A}$ is also of rank k, it follows that it can be factorized in the form

$$\mathbf{A} = \begin{pmatrix} \mathbf{B} & \mathbf{C} \\ \mathbf{D} & \mathbf{E} \end{pmatrix} = \mathbf{PQ}, \quad \text{where } \mathbf{P} = \begin{pmatrix} \mathbf{L} \\ \mathbf{F} \end{pmatrix}, \qquad \mathbf{Q} = (\mathbf{U}, \mathbf{G}), \tag{4.4.24}$$

$\mathbf{L}$ and $\mathbf{U}$ are derived from (4.4.23), $\mathbf{F}$ is derived by back substitution from the relation $\mathbf{FU} = \mathbf{D}$, $\mathbf{G}$ is derived by forward substitution from the relation $\mathbf{LG} = \mathbf{C}$.

Hence $\mathbf{Ax} = \mathbf{b}$ is equivalent to $\mathbf{PQx} = \mathbf{b}$. This is solved by first solving for $\mathbf{v}$ the system of linear equations

$$\mathbf{Pv} = \mathbf{b} \tag{4.4.25}$$

using the methods of (i), then solving for $\mathbf{x}$ the system of linear equations using the methods of (ii).

For more details and the underlying theory of generalised inverse [see Noble (1969), pp. 142–145].

To sum up, the solution of $\mathbf{Ax} = \mathbf{b}$ is given by the following three cases:

(i) $\mathbf{A}$ is $m \times n$ of rank $n < m$: we solve $\mathbf{A'Ax} = \mathbf{b}$;

(ii) $\mathbf{A}$ is $m \times n$ of rank $m < n$: then $\mathbf{x} = \mathbf{A'w}$ where $\mathbf{w}$ is obtained by solving

$$\mathbf{AA'w} = \mathbf{b};$$

(iii) $\mathbf{A}$ is $m \times n$ of rank $k < m, n$. We factorise $\mathbf{A}$ in the form $\mathbf{PQ}$ then use (i) to solve $\mathbf{Pv} = \mathbf{b}$ and (ii) to solve $\mathbf{Qx} = \mathbf{v}$.

4.5. MATRIX INVERSION

For the definition and properties of the inverse matrix $\mathbf{A}^{-1}$ of an $n \times n$ non-singular matrix $\mathbf{A}$ see I, § 6.4, and for explicit methods for the computation of the inverse matrix see I, § 6.4(i)–(iv). The computation is expensive in terms of time and space, fortunately this is often unnecessary. Usually we need to compute an expression of the form $\mathbf{A}^{-1}\mathbf{B}$ where $\mathbf{B}$ is $n \times m$ and m is much smaller than n. To do this we first factorise $\mathbf{A}$ in the form $\mathbf{LU}$ by the usual elimination process [see (4.2.10)]. Then we compute

$$\mathbf{A}^{-1}\mathbf{B} = (\mathbf{LU})^{-1}\mathbf{B} = \mathbf{U}^{-1}\mathbf{L}^{-1}\mathbf{B}. \qquad (4.5.1)$$

Premultiplying by $\mathbf{L}^{-1}$ is equivalent to carrying out the forward substitution process (4.2.11) on each column of $\mathbf{B}$ using the lower diagonal elements of $\mathbf{L}$ as the corresponding multipliers. Finally, premultiplying by $\mathbf{U}^{-1}$ is equivalent to carrying out the back substitution process (4.2.12) on each of the resulting columns.

Instead of operating on the columns of $\mathbf{B}$ separately, we can carry out the substitution operations on the rows of $\mathbf{B}$. This is equivalent to operating on all the columns simultaneously; it involves only a slight rearrangement of the order of the program loops but may be more convenient if $\mathbf{B}$ is stored by rows.

Computer libraries usually provide a subroutine to compute $\mathbf{A}^{-1}\mathbf{B}$ given $\mathbf{A}$ and $\mathbf{B}$. The solution of $\mathbf{Ax} = \mathbf{b}$ is the special case when $\mathbf{B} = \mathbf{b}$. The computation of the inverse matrix $\mathbf{A}^{-1}$, if required, is the special case when $\mathbf{B}$ is the unit matrix $\mathbf{I}$ [see I (6.2.7)].

It is sometimes required to find out how the solution of $\mathbf{Ax} = \mathbf{b}$ is changed as a result of changing some of the elements of $\mathbf{A}$. This is expressed as the change in the elements of the inverse matrix $\mathbf{A}^{-1}$ given by the following formula [Householder (1957)]

$$\mathbf{A}^{-1} - (\mathbf{A} + \mathbf{XUY})^{-1} = \mathbf{A}^{-1}\mathbf{X}(\mathbf{U}^{-1} + \mathbf{YA}^{-1}\mathbf{X})^{-1}\mathbf{YA}^{-1}. \qquad (4.5.2)$$

The right hand side of (4.5.2) is the *perturbation* in the inverse matrix $\mathbf{A}^{-1}$ as a result of the perturbation $\mathbf{XUY}$ in the matrix $\mathbf{A}$. The perturbation in the solution is

$$\mathbf{A}^{-1}\mathbf{X}(\mathbf{U}^{-1} + \mathbf{YA}^{-1}\mathbf{X})^{-1}\mathbf{YA}^{-1}\mathbf{b}, \qquad (4.5.3)$$

where $\mathbf{A}$ and $\mathbf{U}$ are non-singular matrices of orders n and m respectively, m being usually much smaller than n; in particular, when $m = 1$, then $\mathbf{U}$ becomes a scalar, the matrix $\mathbf{X}$ becomes a column vector and the matrix $\mathbf{Y}$ becomes a row vector. The associative law for matrix multiplication [see I (6.3.3)] gives several equivalent ways of evaluating (4.5.3), but some ways involve much less arithmetic than others. When m is much smaller than n, the following procedure is most efficient: first compute $\mathbf{Z} = \mathbf{A}^{-1}\mathbf{X}$, and then evaluate the products in the order indicated by the brackets, thus

$$\mathbf{Z}(((\mathbf{U}^{-1}+\mathbf{YZ})^{-1}\mathbf{Y})(\mathbf{A}^{-1}\mathbf{b})). \tag{4.5.4}$$

It is sometimes expedient to build up the solution of a large system of linear equations from smaller systems, as in the method of tearing in electrical networks [see Kron (1963), p. 40] done by partitioning $\mathbf{A}$ and expressing $\mathbf{A}^{-1}$ in terms of the smaller partitions and their inverses [see Bodewig (1959) pp. 217–266]. If

$$\mathbf{A}=\begin{pmatrix}\mathbf{B} & \mathbf{C}\\ \mathbf{C} & \mathbf{E}\end{pmatrix}, \qquad \mathbf{A}^{-1}=\begin{pmatrix}\mathbf{B}_1 & \mathbf{C}_1\\ \mathbf{D}_1 & \mathbf{E}_1\end{pmatrix} \tag{4.5.5}$$

then

$$\begin{aligned}
\mathbf{E}_1 &= (\mathbf{E}-\mathbf{DB}^{-1}\mathbf{C})^{-1},\\
\mathbf{C}_1 &= -\mathbf{B}^{-1}\mathbf{CE}_1,\\
\mathbf{B}_1 &= \mathbf{B}^{-1}+\mathbf{B}^{-1}\mathbf{CE}_1\mathbf{DB}^{-1},\\
\mathbf{D}_1 &= -\mathbf{E}_1\mathbf{DB}^{-1}.
\end{aligned} \tag{4.5.6}$$

$\mathbf{B}$ is $n \times n$ and $\mathbf{E}$ is $m \times m$, m being usually much smaller than n. If $m = 1$ then $\mathbf{E}$ becomes a scalar, $\mathbf{C}$ a column vector and $\mathbf{D}$ a row vector.

4.6. EIGENVALUES AND EIGENVECTORS

We assume the definitions and properties of eigenvalues and eigenvectors described in I, Chapter 7. The following numerical examples review some of them.

EXAMPLE 4.6.1.   Let

$$\mathbf{A}=\begin{pmatrix}5 & 2\\ 2 & 2\end{pmatrix}, \tag{4.6.1}$$

$$\mathbf{u}_1=\begin{pmatrix}-1\\ 2\end{pmatrix}, \qquad \mathbf{u}_2=\begin{pmatrix}2\\ 1\end{pmatrix}. \tag{4.6.2}$$

Then we can easily verify that:

$$\mathbf{A}\mathbf{u}_1 = \alpha_1\mathbf{u}_1,$$

and   (4.6.3)

$$\mathbf{A}\mathbf{u}_2 = \alpha_2\mathbf{u}_2,$$

where $\alpha_1 = 1$, $\alpha_2 = 6$. We can also verify that $\alpha_1 + \alpha_2 = \operatorname{tr}\mathbf{A}$ and $\alpha_1\alpha_2 = \det\mathbf{A}$ [see I (7.3.9) and (7.3.10)]. Let

$$\mathbf{S}=(\mathbf{u}_1,\mathbf{u}_2)=\begin{pmatrix}-1 & 2\\ 2 & 1\end{pmatrix},$$

then

$$S^{-1} = \begin{pmatrix} -0 \cdot 2 & 0 \cdot 4 \\ 0 \cdot 4 & 0 \cdot 2 \end{pmatrix},$$

and

$$S^{-1}AS = \text{diag}(\alpha_1, \alpha_2), \tag{4.6.4}$$

[see I (7.4.4)].

   In the above illustration we have used a 2×2 matrix A. Analogous results
hold for $n \times n$ matrices.
   We observe that the matrix A in (4.6.1) is symmetric and the following
properties of A hold generally for real symmetric matrices A: the eigenvalues
of A are real, the eigenvectors are orthogonal [see I, Theorem 7.8.1],

$$\mathbf{u}_i' \mathbf{u}_j = 0, \quad \text{when } i \neq j. \tag{4.6.5}$$

Returning to (4.6.2) we may replace the eigenvectors $\mathbf{u}_1, \mathbf{u}_2$ by the normalized
eigenvectors

$$\mathbf{p}_1 = \begin{pmatrix} -1/\sqrt{5} \\ 2/\sqrt{5} \end{pmatrix}, \quad \mathbf{p}_2 = \begin{pmatrix} 2/\sqrt{5} \\ 1/\sqrt{5} \end{pmatrix} \tag{4.6.6}$$

see [I (7.8.6)]. Then the matrix

$$P = (\mathbf{p}_1, \mathbf{p}_2) \tag{4.6.7}$$

satisfies the following equations:

$$P'P = PP' = I,$$
$$P^{-1} = P', \tag{4.6.8}$$

and

$$P'AP = \text{diag}(\alpha_1 \alpha_2). \tag{4.6.9}$$

A matrix P which satisfies (4.6.8) is said to be *orthogonal*.

EXAMPLE 4.6.2.   Let

$$A = \begin{pmatrix} 1 & 0 & 0 \\ 0 & 3 & 0 \\ 0 & 0 & 5 \end{pmatrix}. \tag{4.6.10}$$

Then the eigenvalues and corresponding eigenvectors of A are as follows:

$$\alpha_1 = 1, \mathbf{u}_1 = (1, 0, 0)',$$
$$\alpha_2 = 3, \mathbf{u}_2 = (0, 1, 0)', \tag{4.6.11}$$
$$\alpha_3 = 5, \mathbf{u}_3 = (0, 0, 1)'.$$

In the remainder of this section we deal with real symmetric matrices. The following expression

$$\frac{\mathbf{x}'\mathbf{A}\mathbf{x}}{\mathbf{x}'\mathbf{x}} \qquad (4.6.12)$$

occurs frequently in applications; it is sometimes called the *Rayleigh–Ritz* (*R–R*) ratio. If $\mathbf{x}$ is an eigenvector, then the *R–R* ratio is equal to the corresponding eigenvalue. If $\mathbf{x}$ is normalized so that $\mathbf{x}'\mathbf{x} = 1$, then the *R–R* ratio is equal to the quadratic form $\mathbf{x}'\mathbf{A}\mathbf{x}$. Finally we observe that the *R–R* ratio does not change in value if $\mathbf{x}$ is replaced by $k\mathbf{x}$, where k is a non-zero scalar. Let

$$\alpha_1 \leq \alpha_2 \leq \ldots \leq \alpha_n \qquad (4.6.13)$$

be the eigenvalues of $\mathbf{A}$ arranged in ascending order. For an arbitrary vector $\mathbf{x}$ we have [see I, Theorem 7.9.2]

$$\alpha_1 \leq \frac{\mathbf{x}'\mathbf{A}\mathbf{x}'}{\mathbf{x}'\mathbf{x}} \leq \alpha_n. \qquad (4.6.14)$$

When $\mathbf{x}$ is equal to the ith column of the unit matrix $\mathbf{I}$, the *R–R* ratio is equal to the ith diagonal element a_{ii} of $\mathbf{A}$, hence:

$$\alpha_1 \leq a_{ii} \leq \alpha_n \qquad (4.6.15)$$

$$\alpha_1 \leq \min_i a_{ii}, \qquad \alpha_n \geq \max_i a_{ii}. \qquad (4.6.16)$$

EXAMPLE 4.6.3.   Let

$$\mathbf{A} = \begin{pmatrix} 1 & 0{\cdot}1 & -0{\cdot}1 \\ 0{\cdot}1 & 3 & 0{\cdot}2 \\ -0{\cdot}1 & 0{\cdot}2 & 5 \end{pmatrix}, \qquad (4.6.17)$$

then $\alpha_1 \leq 1$, $\alpha_3 \geq 5$.

If $\mathbf{x}$ is an approximation to an eigenvector $\mathbf{u}$ then the *R–R* ratio is an improved approximation to the corresponding eigenvalue α. More precisely, if

$$\mathbf{x} = \mathbf{u} + O(\varepsilon), \qquad (4.6.18)$$

[see § 2.3], then

$$\frac{\mathbf{x}'\mathbf{A}\mathbf{x}}{\mathbf{x}'\mathbf{x}} = \alpha + O(\varepsilon^2). \qquad (4.6.19)$$

This provides the following quadratically convergent iterative process to compute α and $\mathbf{u}$: starting with $\mathbf{x}$, a guess or approximation to $\mathbf{u}$. We first compute

$$\alpha = \frac{\mathbf{x}'\mathbf{A}\mathbf{x}}{\mathbf{x}'\mathbf{x}}. \qquad (4.6.20)$$

Next we compute $\mathbf{p}$ by solving the system of linear equations:

$$(\mathbf{A} - \alpha\mathbf{I})\mathbf{p} = \mathbf{x}. \qquad (4.6.21)$$

Finally we compute the next improved approximation **x** by normalizing **p**:

$$\mathbf{x} = \mathbf{p}/(\mathbf{p}'\mathbf{p})^{1/2}. \tag{4.6.22}$$

This is a rapidly convergent process. The computations (4.6.20)–(4.6.22) have to be repeated two or three times at most, as illustrated in the following example. The solution of the system of linear equations (4.6.21) is relatively expensive but it is carried out two or three times only. As the approximation α improves, the matrix $\mathbf{A} - \alpha\mathbf{I}$ of (4.6.21) approaches a singular matrix and the solution **p** becomes large; none the less this apparently risky procedure is one of the most common and most efficient ways of computing the eigenvector **p** given a good approximation α [cf. the method of inverse iteration in § 4.8]; the approximation **x** in (4.6.21) need not be good, indeed **x** could be any arbitrary vector but the convergence is safer if **x** is an approximation to **p**. Thus (4.6.21) is used to compute **p** regardless of the method used to compute α [see Wilkinson (1963), pp. 142–143].

EXAMPLE 4.6.4.   To compute the eigenvalues and eigenvectors of the matrix of Example 4.6.3:

$$\mathbf{A} = \begin{pmatrix} 1 & 0 \cdot 1 & -0 \cdot 1 \\ 0 \cdot 1 & 3 & 0 \cdot 2 \\ -0 \cdot 1 & 0 \cdot 2 & 5 \end{pmatrix}.$$

Because the off-diagonal elements are relatively small, the eigenvalues and eigenvectors of the matrix (4.6.10) in Example 4.6.2 give a reasonable approximation to those of **A**. Starting with the approximation $\mathbf{x} = (1, 0, 0)'$, given by (4.6.11), the iterative application of (4.6.20), (4.6.21) and (4.6.22) gives the following Table 4.6.1:

	First approximation	Second approximation	Third approximation
x	1 0 0	0·99823 −0·05267 0·02759	0·99825 −0·05236 0·02750
α	1	0·99199	0·99199
p	−124·0625 6·5625 −3·4375	582·504 −30·5556 16·0445	

Table 4.6.1

We note that $\alpha = \alpha_1 = 0 \cdot 99199$ satisfies the inequality of Example 4.6.3. Similar computations can be used to compute the other eigenvalues and eigenvectors of **A** starting with the approximations given by (4.6.11).

4.7. NORMS FOR VECTORS AND MATRICES

We review briefly the definitions and basic properties of norms for vectors and matrices, for more details see section 6.2 in this volume. Let $\mathbf{x} = (x_1, x_2, \ldots, x_n)'$. Then

$$\|\mathbf{x}\|_p = \left(\sum_{i=1}^{n} |x_i|^p \right)^{1/p} \tag{4.7.1}$$

$$\|x\|_1 = |x_1| + |x_2| + \ldots + |x_n| \tag{4.7.2}$$

$$\|\mathbf{x}\|_\infty = \max_i |x_i| \tag{4.7.3}$$

$$\|\mathbf{x}\|_2 = (\mathbf{x}'\mathbf{x})^{1/2}. \tag{4.7.4}$$

If $\mathbf{x}$ is a complex vector, we have to replace $\mathbf{x}'$ by $\mathbf{x}^* = (\bar{\mathbf{x}}')$ in (4.7.4). Norms satisfy the following basic properties:

$$\|\mathbf{x}\| > 0 \quad \text{unless } \mathbf{x} = \mathbf{0} \tag{4.7.5}$$

$$\|k\mathbf{x}\| = |k| \|x\| \tag{4.7.6}$$

$$\|\mathbf{x}\| - \|\mathbf{y}\| \leq \|\mathbf{x} + \mathbf{y}\| \leq \|\mathbf{x}\| + \|\mathbf{y}\|. \tag{4.7.7}$$

The suffix p in (4.7.1) is often omitted because the results are usually true for any non-negative integer p; the 2-norm is the p-norm most frequently used, though the 1-norm and ∞-norm are easier to compute.

Before introducing matrix norms we introduce the following useful notation for row and column sums. Let $\mathbf{A} = (a_{ij})$. Then put

$$R_i = \sum_{j=1}^{n} |a_{ij}| \tag{4.7.8}$$

$$R'_i = R_i - |a_{ii}| \tag{4.7.9}$$

$$C_j = \sum_{i=1}^{n} |a_{ij}| \tag{4.7.10}$$

$$C'_j = C_j - |a_{jj}|. \tag{4.7.11}$$

Thus R_i and C_j are row and column sums, R'_i, C'_j are the corresponding off-diagonal sums. The norm of a matrix is defined in terms of the norm of vectors.

$$\|\mathbf{A}\|_p = \text{l.u.b.}_{\mathbf{x}} \frac{\|\mathbf{A}\mathbf{x}\|_p}{\|\mathbf{x}\|_p} \tag{4.7.12}$$

where l.u.b. stands for least upper bound [see § 2.3]. It is found that

$$\|\mathbf{A}\|_\infty = \max_i R_i \tag{4.7.13}$$

$$\|\mathbf{A}\|_1 = \max_j C_j \tag{4.7.14}$$

$$\|\mathbf{A}\|_2 = \max \alpha (\mathbf{A}'\mathbf{A})^{1/2}, \qquad (4.7.15)$$

where generally $\alpha(\mathbf{M})$ denotes an eigenvalue of $\mathbf{M}$. We note that $\mathbf{A}'\mathbf{A}$ is symmetric non-negative, hence its eigenvalues are real and non-negative [see I, Theorem 7.9.2] and hence the square root operation does not involve complex numbers. As in the case of vectors, $\mathbf{A}'\mathbf{A}$ has to be replaced by $\mathbf{A}^*\mathbf{A}$, when the coefficients a_{ij} are complex.

Matrix norms satisfy the following basic properties

$$\|\mathbf{A}\| > 0, \quad \text{unless } \mathbf{A} = \mathbf{0} \qquad (4.7.16)$$

$$\|k\mathbf{A}\| = |k|\,\|\mathbf{A}\| \qquad (4.7.17)$$

$$\|\mathbf{A}\| - \|\mathbf{B}\| \le \|\mathbf{A} + \mathbf{B}\| \le \|\mathbf{A}\| + \|\mathbf{B}\| \qquad (4.7.18)$$

$$\|\mathbf{AB}\| \le \|\mathbf{A}\|\,\|\mathbf{B}\|. \qquad (4.7.19)$$

EXAMPLE 4.7.1. Let

$$\mathbf{A} = \begin{pmatrix} 3 & 0\cdot 1 & 0\cdot 2 \\ -0\cdot 1 & 1 & 0\cdot 1 \\ 0\cdot 3 & 0\cdot 1 & 5 \end{pmatrix}. \qquad (4.7.20)$$

Then

$$R_1 = 3\cdot 3, \quad R_1' = 0\cdot 3, \quad R_2 = 1\cdot 2, \quad R_2' = 0\cdot 2, \quad R_3 = 5.4, \quad R_3' = 0\cdot 4,$$

$$C_1 = 3\cdot 4, \quad C_1' = 0\cdot 4, \quad C_2 = 1\cdot 2, \quad C_2' = 0\cdot 2, \quad C_3 = 5\cdot 3, \quad C_3' = 0\cdot 3,$$

$$\|\mathbf{A}\|_\infty = 5\cdot 4, \quad \|\mathbf{A}\|_1 = 5\cdot 3.$$

The 2-norm of a matrix $\mathbf{A}$ is sometimes called the *spectral radius* of $\mathbf{A}$ and is denoted by $\rho(\mathbf{A})$, thus

$$\rho(\mathbf{A}) = \max (\alpha(\mathbf{A}'\mathbf{A}))^{1/2}. \qquad (4.7.21)$$

[see I, Proposition 7.13.5]. We note that if $\mathbf{A}$ is symmetric then

$$\|\mathbf{A}\|_2 = \rho(\mathbf{A}) = \max |\alpha(\mathbf{A})|. \qquad (4.7.22)$$

Norms provide useful upper bounds for eigenvalues. let $\mathbf{Ax} = \alpha\mathbf{x}$. By applying (4.7.6) and (4.7.19) we get

$$\|\alpha\mathbf{x}\| = |\alpha|\,\|\mathbf{x}\| = \|\mathbf{Ax}\| \le \|\mathbf{A}\|\,\|\mathbf{x}\|$$

hence $|\alpha| \le \|\mathbf{A}\|$, i.e.

$$\alpha(\mathbf{A}) \le \|\mathbf{A}\|. \qquad (4.7.23)$$

EXAMPLE 4.7.2. Apply (4.7.23) to the matrix (4.7.20). Using the ∞-norm we get $\alpha(\mathbf{A}) \le 5\cdot 4$; using the 1-norm we get the sharper bound $\alpha(\mathbf{A}) \le 5\cdot 3$. Using the 2-norm we get $\alpha(\mathbf{A}) \le \rho(\mathbf{A})$, but this is not useful, since $\rho(\mathbf{A})$ is usually as difficult to compute as $\alpha(\mathbf{A})$.

EXAMPLE 4.7.3.   Let **D** be a diagonal matrix

$$\mathbf{D} = \text{diag}\,(a_1, a_2, \ldots, a_n). \tag{4.7.24}$$

Then

$$\|\mathbf{D}\|_1 = \|\mathbf{D}\|_2 = \|\mathbf{D}\|_\infty = \max_i |a_i|. \tag{4.7.25}$$

If **D** is the unit matrix **I** then we have:

$$\|\mathbf{I}\|_1 = \|\mathbf{I}\|_2 = \|\mathbf{I}\|_\infty = 1. \tag{4.7.26}$$

This gives a convenient estimate for the norm of the inverse of a non-singular matrix **A**. From the relation $\mathbf{I} = \mathbf{A}\mathbf{A}^{-1}$ we get by (4.7.19):

$$\|\mathbf{I}\| = 1 = \|\mathbf{A}\mathbf{A}^{-1}\| \le \|\mathbf{A}\|\,\|\mathbf{A}^{-1}\|.$$

Hence

$$\|\mathbf{A}^{-1}\| \ge 1/\|\mathbf{A}\|. \tag{4.7.27}$$

Norms provide useful conditions for a matrix to be non-singular [see I, Definition 6.4.2]. The matrix $\mathbf{A} = (a_{ij})$ is said to be *diagonally dominant* if

$$|a_{ii}| > R'_i \quad \text{for all } i, \tag{4.7.28}$$

where R'_i is defined by (4.7.9). Let

$$\mathbf{A} = \mathbf{D} + \mathbf{N}, \tag{4.7.29}$$

where **D** is the diagonal part and **N** the non-diagonal part of **A**. It follows from (4.7.28) that if **A** is diagonally dominant, then

$$\|\mathbf{D}^{-1}\mathbf{N}\|_\infty < 1. \tag{4.7.30}$$

If **A** is singular, then [see I, Proposition 6.4.1] there exists an **x** such that $\mathbf{A}\mathbf{x} = (\mathbf{D} + \mathbf{N})\mathbf{x} = 0$. Hence $\mathbf{D}\mathbf{x} = -\mathbf{N}\mathbf{x}$ and:

$$\mathbf{x} = -\mathbf{D}^{-1}\mathbf{N}\mathbf{x}.$$

Hence

$$\|\mathbf{x}\| = \|\mathbf{D}^{-1}\mathbf{N}\mathbf{x}\|.$$

Applying (4.7.19), we get:

$$\|\mathbf{x}\| \le \|\mathbf{D}^{-1}\mathbf{N}\|\,\|\mathbf{x}\|.$$

Hence

$$1 \le \|\mathbf{D}^{-1}\mathbf{N}\|.$$

But this contradicts (4.7.30), hence we have the following

THEOREM 4.7.1.   **A** *is non-singular if it is diagonally dominant* [see I, Proposition 7.13.1].

The same result holds when column sums are used instead of row sums. This theorem leads to two useful results for locating the eigenvalues α of a matrix $\mathbf{A}$ [see I, Propositions 7.13.2 and 7.13.3].

Let S_i be the circular disc in the complex plane [see I, § 2.7.2] of α with centre a_{ii} and radius R'_i: it is the set of all points α in the complex domain which satisfy the inequality

$$|a_{ii} - \alpha| \leq R'_i. \tag{4.7.31}$$

The discs $S_1, S_2, \ldots, S_n$ are sometimes called *Gershgorin* discs. If α is a point in the complex domain which is outside the union of the Gershgorin discs then $\mathbf{A} - \alpha\mathbf{I}$ is diagonally dominant. On the other hand, if α is an eigenvalue of $\mathbf{A}$ then $\mathbf{A} - \alpha\mathbf{I}$ is singular [see I (7.2.2)]. Hence we have the following

THEOREM 4.7.2.   *The eigenvalues α of $\mathbf{A}$ lie inside the Gershgorin discs.*

Let $T_1, T_2, \ldots, T_m (m \leq n)$ be the non-overlapping Gershgorin domains defined by the Gershgorin discs. For example if S_1 and S_2 overlap with each other but not with the other discs then $T_1 = S_1 \cup S_2$. By a continuity argument we can derive the following stronger theorem.

THEOREM 4.7.3.   *The number of eigenvalues inside T_i is the same as the number of overlapping discs which make up T_i.*

Similar results are obtained by using the column sums C'_i instead of the row sums R'_i.

EXAMPLE 4.7.4.   We apply these results to the matrix (4.7.20). S_1 has centre 3 and radius 0·3; S_2 has centre 1 and radius 0·2; S_3 has centre 5 and radius 0·3 (using row sums for the first two and column sum for the third to make the discs as small as possible). Since the discs are non-overlapping, each contains exactly one eigenvalue. The matrix is unsymmetric but real, hence any complex eigenvalues would occur in conjugate pairs [see I, Proposition 7.3.1]. This would contradict the previous statement. Hence the eigenvalues are real, the discs are reduced to intervals on the real axis, hence:

$$1 - 0·2 \leq \alpha_1 \leq 1 + 0·2,$$

$$3 - 0·3 \leq \alpha_2 \leq 3 + 0·3,$$

$$5 - 0·3 \leq \alpha_3 \leq 5 + 0·3.$$

EXAMPLE 4.7.6.   We illustrate various results by applying them to the matrix

$$\mathbf{A} = \begin{pmatrix} 2 & 0 & -0·2 \\ 0 & -1 & 0·1 \\ -0·2 & 0·1 & 5 \end{pmatrix}. \tag{4.7.31}$$

The matrix is symmetric, hence the eigenvalues α_1, α_2, α_3 are real. Let $\alpha_1 \le \alpha_2 \le \alpha_3$. Since $\|\mathbf{A}\|_\infty = 5\cdot3$, it follows that $|\alpha_i| \le 5\cdot3$, hence:

$$-5\cdot3 \le \alpha_1 \le \alpha_2 \le \alpha_3 \le 5\cdot3.$$

Applying theorems 4.7.2 and 4.7.3 we get:

$$-1\cdot1 \le \alpha_1 \le -0\cdot9,$$
$$1\cdot8 \le \alpha_2 \le 2\cdot2, \tag{4.7.32}$$
$$4\cdot7 \le \alpha_3 \le 5\cdot3.$$

Applying (4.6.16) we get:

$$\alpha_1 \le -1, \qquad \alpha_3 \ge 5.$$

Hence

$$-1\cdot1 \le \alpha_1 \le -1,$$
$$4\cdot7 \le \alpha_3 \le 5.$$

Let $\mathbf{S} = \text{diag}\,(k, 1, 1)$, then $\mathbf{S}^{-1} = \text{diag}\,(1/k, 1, 1)$ and

$$\mathbf{B} = \mathbf{S}^{-1}\mathbf{A}\mathbf{S} = \begin{pmatrix} 2 & 0 & -0\cdot2k \\ 0 & -1 & 0\cdot1 \\ -0\cdot2/k & 0\cdot1 & 5 \end{pmatrix}. \tag{4.7.33}$$

The matrices $\mathbf{A}$ and $\mathbf{B}$ have the same eigenvalues [see I, Theorem 7.4.1]. The Gershgorin disc S_1 for $\mathbf{B}$ has centre 2 and radius $0\cdot2k$; S_3 has centre 5 and radius $0\cdot1 + 0\cdot2/k$. By making k sufficiently small we can get sharper bounds for α_2 provided that S_1 and S_3 do not overlap. The smallest k is obtained when the two discs touch each other, that is when:

$$2 + 0\cdot2k = 5 - 0\cdot1 - 0\cdot2/k.$$

This gives $k = 0\cdot07$, hence $2 - 0\cdot2k \le \alpha_2 \le 2 + 0\cdot2k$, that is:

$$1\cdot986 \le \alpha_2 \le 2\cdot014,$$

which is much sharper than (4.7.32). Applying the same method to α_1 and α_3 we get:

$$-1\cdot0017 \le \alpha_1 \le -1,$$
$$5 \le \alpha_3 \le 5\cdot0157.$$

The actual values are:

$$\alpha_1 = -1\cdot00167,$$
$$\alpha_2 = 1\cdot98674,$$
$$\alpha_3 = 5\cdot01493,$$

and we check that $\alpha_1 + \alpha_2 + \alpha_3 = 6 = \text{tr}\,\mathbf{A} = 2 - 1 + 5$.

Next we use norms to compute estimates of errors or perturbations in the solution of the system of linear equations $\mathbf{A}\mathbf{x} = \mathbf{b}$. These perturbations in the solution are due to rounding errors in computer arithmetic or to inaccuracies in the coefficients of the equations. The expression

$$c(\mathbf{A}) = \|\mathbf{A}^{-1}\| \|\mathbf{A}\| \qquad (4.7.34)$$

occurs frequently in error estimates; it is called the *condition number*. From (4.7.27) we have that

$$c(\mathbf{A}) \geq 1. \qquad (4.7.35)$$

We use 2-norms (4.7.15) to derive some useful properties of the condition number. Since $\|\mathbf{A}\|_2 = \max (\alpha(\mathbf{A}'\mathbf{A}))^{1/2}$ [see (4.7.15)] it follows that:

$$\|\mathbf{A}^{-1}\| = 1/\min (\alpha(\mathbf{A}'\mathbf{A}))^{1/2},$$

and

$$c(\mathbf{A}) = \frac{\max (\alpha(\mathbf{A}'\mathbf{A}))^{1/2}}{\min (\alpha(\mathbf{A}'\mathbf{A}))^{1/2}}. \qquad (4.7.36)$$

If $\mathbf{A}$ is symmetric, this expression simplifies to

$$c(\mathbf{A}) = \frac{\max |\alpha(\mathbf{A})|}{\min |\alpha(\mathbf{A})|}. \qquad (4.7.37)$$

The result (4.7.35) follows also from the expressions (4.7.36) and (4.7.37).

Next we consider a matrix of the form $\mathbf{I} + \mathbf{E}$ where the coefficients of $\mathbf{E}$ are usually small. If $\mathbf{I} + \mathbf{E}$ singular, then there exists a non-zero $\mathbf{x}$ such that

$$(\mathbf{I} + \mathbf{E})\mathbf{x} = \mathbf{0},$$

[see I, Proposition 6.4.1]. Applying norms and (4.7.19) we get

$$\mathbf{x} = -\mathbf{E}\mathbf{x},$$

$$\|\mathbf{x}\| = \|\mathbf{E}\mathbf{x}\| \leq \|\mathbf{E}\| \|\mathbf{x}\|.$$

Hence $\|\mathbf{E}\| \geq 1$ and we have the following proposition.

PROPOSITION 4.7.1.   *If $\|\mathbf{E}\| < 1$, then $\mathbf{I} + \mathbf{E}$ is non-singular.*

Consider the solution $\mathbf{x}$ of the system of linear equations $\mathbf{A}\mathbf{x} = \mathbf{b}$. Let $\delta\mathbf{b}$ represent the vector of perturbations in $\mathbf{b}$ and let $\delta\mathbf{x}$ be the resulting vector of perturbations in the solution $\mathbf{x}$, so that:

$$\mathbf{A}(\mathbf{x} + \delta\mathbf{x}) = \mathbf{b} + \delta\mathbf{b}. \qquad (4.7.38)$$

Hence

$$\mathbf{A}\delta\mathbf{x} = \delta\mathbf{b}, \qquad \delta\mathbf{x} = \mathbf{A}^{-1}\delta\mathbf{b}, \qquad (4.7.39)$$

and

$$\|\delta\mathbf{x}\| = \|\mathbf{A}^{-1}\delta\mathbf{b}\| \leq \|\mathbf{A}^{-1}\| \|\delta\mathbf{b}\|. \qquad (4.7.40)$$

From $\mathbf{b} = \mathbf{Ax}$ we get

$$\|\mathbf{b}\| = \|\mathbf{Ax}\| \le \|\mathbf{A}\| \|\mathbf{x}\|, \tag{4.7.41}$$

whence

$$\frac{\|\delta\mathbf{x}\|}{\|\mathbf{x}\|} \le c(\mathbf{A}) \frac{\|\delta\mathbf{b}\|}{\|\mathbf{b}\|}. \tag{4.7.42}$$

The left-hand side of (4.7.42) represents the relative perturbation in $\mathbf{x}$, hence a relative perturbation in $\mathbf{b}$ is magnified at most $c(\mathbf{A})$ times.

Next we consider the perturbation $\delta\mathbf{x}$ in $\mathbf{x}$ due to a perturbation $\delta\mathbf{A}$ in $\mathbf{A}$, that is

$$(\mathbf{A} + \delta\mathbf{A})(\mathbf{x} + \delta\mathbf{x}) = \mathbf{b}. \tag{4.7.43}$$

As before, we assume that $\mathbf{A}$ is non-singular so that the inverse matrix $\mathbf{A}^{-1}$ exists [see I, Theorem 6.4.2]. The coefficients of $\delta\mathbf{A}$ must be sufficiently small so that $\mathbf{A} + \delta\mathbf{A}$ is also non-singular. Let

$$\mathbf{A} + \delta\mathbf{A} = \mathbf{A}(\mathbf{I} + \mathbf{A}^{-1}\delta\mathbf{A}) = \mathbf{A}(\mathbf{I} + \mathbf{E}),$$

where $\mathbf{E} = \mathbf{A}^{-1}\delta\mathbf{A}$. Hence, using Proposition 4.7.1, the condition for $\mathbf{A} + \delta\mathbf{A}$ to be non-singular is that

$$\|\mathbf{A}^{-1}\delta\mathbf{A}\| < 1. \tag{4.7.44}$$

From (4.7.43) we have

$$(\mathbf{A} + \delta\mathbf{A})\delta\mathbf{x} = -\delta\mathbf{Ax}. \tag{4.7.45}$$

Assuming that the condition (4.7.44) is satisfied, we take norms of (4.7.45) and use the basic norm properties, thus

$$\frac{\|\delta\mathbf{x}\|}{\|\mathbf{x}\|} \le \frac{c(\mathbf{A})}{1 - \|\mathbf{A}^{-1}\| \|\delta\mathbf{A}\|} \frac{\|\delta\mathbf{A}\|}{\|\mathbf{A}\|}. \tag{4.7.46}$$

The inequality (4.7.46) may be interpreted in the same way as (4.7.42); $c(\mathbf{A})$ expresses the greatest possible magnification of the relative perturbation in $\mathbf{A}$ or $\mathbf{b}$; it measures the sensitivity of the system of linear equations $\mathbf{Ax} = \mathbf{b}$. The system is said to be *ill-conditioned* if $c(\mathbf{A})$ is large, for example if $c(\mathbf{A}) > 100$.

EXAMPLE 4.7.7.   Let

$$\mathbf{A} = \begin{pmatrix} \frac{1}{2} & \frac{1}{3} & \frac{1}{4} & \frac{1}{5} \\ \frac{1}{3} & \frac{1}{4} & \frac{1}{5} & \frac{1}{6} \\ \frac{1}{4} & \frac{1}{5} & \frac{1}{6} & \frac{1}{7} \\ \frac{1}{5} & \frac{1}{6} & \frac{1}{7} & \frac{1}{8} \end{pmatrix}$$

$$\alpha_1(\mathbf{A}) = 0 \cdot 0000213$$

$$\alpha_4(\mathbf{A}) = 0 \cdot 977.$$

Since **A** is symmetric we have, using (4.7.37):

$$c(\mathbf{A}) = \alpha_4(\mathbf{A})/\alpha_1(\mathbf{A}) \doteq 40{,}000.$$

If

$$\mathbf{b} = (1, 1, 1, 1),$$

then

$$\mathbf{x} = (-20, 180, -420, 280)',$$

and if

$$\delta\mathbf{b} = (0\cdot0001, -0\cdot0001, -0\cdot0002, 0\cdot0001)',$$

then

$$\delta\mathbf{x} = (0\cdot4, -3\cdot1, 6\cdot0, -3\cdot3)'.$$

The relative perturbations in **b** and **x** are approximately equal to $0\cdot0002$ and to $0\cdot014$ respectively, giving a magnification $\doteq 70$. If

$$\mathbf{b} = (6\cdot95, 5\cdot02, 3\cdot96, 3\cdot28)'$$

then

$$\mathbf{x} = (8\cdot4, -1\cdot2, 12\cdot6, 0)'$$

and if

$$\delta\mathbf{b} = (0\cdot000001, -0\cdot00001, 0\cdot00002, -0\cdot00001)'$$

then

$$\delta\mathbf{x} = (0\cdot1, -0\cdot5, 0\cdot9, -0\cdot5).$$

In this case the relative perturbation in **b** and **x** are approximately equal to $0\cdot000003$ and to $0\cdot07$ respectively giving a magnification of the same order of magnitude as $c(\mathbf{A})$.

EXAMPLE 4.7.8.   We use the same **A**, the first **b** and the corresponding **x** as in Example 4.7.7. Let

$$\delta\mathbf{A} = \begin{pmatrix} \varepsilon & -\varepsilon & -\varepsilon & \varepsilon \\ -\varepsilon & -\varepsilon & \varepsilon & \varepsilon \\ -\varepsilon & \varepsilon & \varepsilon & -\varepsilon \\ \varepsilon & \varepsilon & -\varepsilon & \varepsilon \end{pmatrix}, \quad \text{where } \varepsilon = 0\cdot0000001.$$

Then

$$\delta\mathbf{x} = (0\cdot2, -1\cdot2, 2\cdot4, -1\cdot4)'.$$

The relative perturbation in **x** is $\doteq 0\cdot006$, the relative perturbation in **A** is $\doteq 0\cdot0000004$, giving a magnification of the same order of magnitude as $c(\mathbf{A})$.

4.8.   ITERATIVE METHODS FOR COMPUTING EIGENVALUES AND EIGENVECTORS

For a more detailed treatment and proofs of the results assumed in this section see Wilkinson (1965), Chapter 9. Let A be a real $n \times n$ matrix with eigenvalues

$$\alpha_1, \alpha_2, \ldots, \alpha_n \tag{4.8.1}$$

arranged in ascending order of magnitude, that is

$$|\alpha_1| \le |\alpha_2| \le \ldots \le |\alpha_n|, \tag{4.8.2}$$

and let

$$\mathbf{u}_1, \mathbf{u}_2, \ldots, \mathbf{u}_n \tag{4.8.3}$$

be the corresponding eigenvectors.

The *power method* starts with a vector $\mathbf{x}$ which can be arbitrary if no suitable approximation is available. Then the following computations are carried out repeatedly:

$$\mathbf{y} = \mathbf{A}\mathbf{x}, \tag{4.8.4}$$

$$\mathbf{x} = \frac{1}{k}\mathbf{y}. \tag{4.8.5}$$

The new value of $\mathbf{x}$ derived from (4.8.5) is used in (4.8.4) to compute a new value of $\mathbf{y}$ from (4.8.4), and so on. The scaling factor k in (4.8.5) is used to prevent the numbers involved in the computation from getting too large or too small. The most common methods for computing k are:

$$k = \pm\|\mathbf{y}\|_2 = \pm(\mathbf{y}'\mathbf{y})^{1/2} \tag{4.8.6}$$

or

$$k = \pm\|\mathbf{y}\|_\infty = \pm\max |y_i|. \tag{4.8.7}$$

This ensures that $\|\mathbf{x}\|_2 = 1$ or $\|\mathbf{x}\|_\infty = 1$. The sign is chosen to be the same as the sign of the component y_i of $\mathbf{y}$ which has the largest modulus. After repeated applications of (4.8.4) and (4.8.5), the vector $\mathbf{x}$ converges to $\mathbf{u}_n$ and k converges to α_n. The speed with which this convergence takes place after r iterations is the same as the speed with which

$$\left(\frac{\alpha_{n-1}}{\alpha_n}\right)^r \tag{4.8.8}$$

converges to zero.

EXAMPLE 4.8.1.   Let

$$\mathbf{A} = \begin{pmatrix} 6 & 3 & 1 \\ 3 & 2 & 1 \\ 1 & 1 & 1 \end{pmatrix}. \tag{4.8.9}$$

We start with $\mathbf{x} = (1, 1, 1)'$. To simplify the arithmetic we choose the scaling factor to be $k = y_3$ (the third component of $\mathbf{y}$); this ensures that x_3 (the third component of $\mathbf{x}$) remains 1. This differs from the formulae (4.8.6), (4.8.7) for k, but convergence still takes place. Table 4.8.1 gives the results of the first six

r	0	1	2	3	4	5	6
		10	22·68	33·58	34·751	34·907	34·926
$\mathbf{y}$		6	14·9	18·47	19·092	19·171	19·181
		3	6·3	7·61	7·840	7·869	7·873
	1	3·3	4·25	4·413	4·433	4·436	4·436
$\mathbf{x}$	1	2	2·36	2·427	2·435	2·436	2·436
	1	1	1	1	1	1	1

Table 4.8.1

iterations. This gives $\alpha_3 = 7·873$ and $\mathbf{u}_3 = (4·436, 2·436, 1)'$. The scaling factor is the third component of $\mathbf{y}$. The convergence is faster than usual because $\alpha_2 = 1$ whence $\alpha_2/\alpha_3 = 0·127$, and from (4.8.8) the approximation error tends to zero as $(0·127)^r$ tends to zero, giving nearly one extra digit of accuracy per iteration.

Formula (4.8.8) indicates the various snags which can occur. It may happen that $\alpha_{n-1} = \pm\alpha_n$ or $\alpha_{n-1} = \bar\alpha_n$ (complex conjugate); in which case there is no convergence. Often, when n is large, $|\alpha_{n-1}/\alpha_n|$ is very nearly 1, say 0·95 or 0·98; in which case convergence is very slow. Finally, the method only works for the dominant eigenvalue. See Wilkinson (1965), Chapter 9, for methods of overcoming these difficulties. On the other hand the method is simple; when $\mathbf{A}$ is very large or sparse it need not be stored explicitly, and $\mathbf{Ax}$ can be computed indirectly from the physics or geometry of the problem.

Next, we explain the *method of inverse iteration*. In principle this is the power method applied to $\mathbf{A}^{-1}$ instead of $\mathbf{A}$. The eigenvectors of $\mathbf{A}$ and $\mathbf{A}^{-1}$ are the same, but the eigenvalues of $\mathbf{A}^{-1}$ when arranged in ascending order of magnitude are [cf. (4.8.1)]:

$$1/\alpha_n, 1/\alpha_{n-1}, \ldots, 1/\alpha_2, 1/\alpha_1 \qquad (4.8.10)$$

[see I, Proposition 7.2.1(ii)]. Hence the inverse iteration method will yield $\alpha_1, \mathbf{u}_1$, instead of $\alpha_n, \mathbf{u}_n$ and the speed of convergence corresponding to (4.8.8) is

$$\left(\frac{1/\alpha_2}{1/\alpha_1}\right)^r = \left(\frac{\alpha_1}{\alpha_2}\right)^r. \qquad (4.8.11)$$

Multiplying by $\mathbf{A}^{-1}$ involves the solution of a system of n linear equations [see § 4.5], which is much more expensive than multiplying by $\mathbf{A}$; however this is more than compensated for by using a method for accelerating convergence which is not applicable to the power method. The eigenvalues of $\mathbf{A} - h\mathbf{I}$ are [see I, Proposition 7.2.1(i)]

$$\alpha_1 - h, \alpha_2 - h, \ldots, \alpha_n - h$$

where h is called *shift of origin*. The eigenvalues of $(\mathbf{A} - h\mathbf{I})^{-1}$ are

$$\frac{1}{\alpha_1 - h}, \frac{1}{\alpha_2 - h}, \ldots, \frac{1}{\alpha_n - h}.$$

If h is equal to an approximation to α_1 then these eigenvalues, when arranged in ascending order of magnitude become

$$\frac{1}{\alpha_n - h}, \ldots, \frac{1}{\alpha_2 - h}, \frac{1}{\alpha_1 - h} \tag{4.8.12}$$

where the last eigenvalue is very large, and formula (4.8.11) for the speed of convergence becomes

$$\left(\frac{\alpha_1 - h}{\alpha_2 - h}\right)^r. \tag{4.8.13}$$

This tends to zero very rapidly as h approaches α_1.

    We now describe more formally the method of inverse iteration. Starting with a given $\mathbf{x}$ and h (which is usually zero initially), compute $\mathbf{y}$ from the system of linear equations:

$$(\mathbf{A} - h\mathbf{I})\mathbf{y} = \mathbf{x}. \tag{4.8.14}$$

Determine a suitable scaling factor k such as (4.8.6) or (4.8.7) then compute

$$\mathbf{x} = \frac{1}{k}\mathbf{y}. \tag{4.8.15}$$

Usually k is an approximation to $1/(\alpha_1 - h)$, hence the best value for the next h, which should be an approximation to α_1, is given by the formula:

$$\text{new } h = \text{old } h + \frac{1}{k}. \tag{4.8.16}$$

Usually the computations (4.8.14), (4.8.15) and (4.8.16) have to be repeated only two or three times; as h approaches α_1, the system (4.8.14) becomes almost singular and k becomes very large [see the discussion preceding Example 4.6.4].

EXAMPLE 4.8.2.   We apply this method to the matrix of Example 4.8.1, but somewhat circumspectly. Thus we apply no shift in the first two iterations ($h = 0$). When we have a reasonable approximation $\mathbf{x}$ to $\mathbf{u}_1$ at the end of the second iteration we employ the Rayleigh–Ritz ratio (4.6.19) because the matrix is symmetric. This gives α_1 correct to 3 decimal places. Using this as a shift of origin we obtain the results

$$h = \alpha_1 = 0 \cdot 127017$$

$$\mathbf{x} = \mathbf{u}_1 = (-0 \cdot 39227, 1, 0 \cdot 69613)'$$

correct to the last decimal place, where h was computed by applying (4.6.19) again. Table 4.8.2 gives the calculation details.

r	0	1	2	3
y		1 -3 3	10 -26 19	-23,371 59,580 -41,481
k		1	-26	59,580
x	0 0 1	1 -3 3	-0·381 1·000 -0·730	-0·39227 1·00000 0·69613
h	0	0	0·127	0·127017

Table 4.8.2

To sum up, the convergence of the power method is usually slow, the shift of origin cannot be used to speed up convergence. The power method yields the largest eigenvalue and corresponding eigenvector whereas the method of inverse iteration yields the smallest eigenvalue and corresponding eigenvector. If the matrix is symmetric then the convergence of the method of inverse iteration can be speeded up still further by using the Rayleigh–Ritz ratio to compute the shift of origin instead of relying on (4.8.16).

4.9.   METHODS OF COMPUTING EIGENVALUES FOR TRIDIAGONAL MATRICES

The matrix $\mathbf{A}$ is said to be *tridiagonal* (or *codiagonal*) if it has zero elements except along the diagonal and the lower and upper codiagonals, that is elements adjacent to the diagonal, thus

$$\mathbf{A} = \begin{pmatrix} a_1 & c_1 & 0 & \cdot & 0 \\ b_1 & a_2 & c_2 & \cdot & 0 \\ \vdots & \vdots & \vdots & \vdots & c_{n-1} \\ 0 & 0 & \cdot & b_{n-1} & a_n \end{pmatrix} \tag{4.9.1}$$

where

$$\text{diag}(\mathbf{A}) = (a_1, a_2, \ldots, a_n),$$

$$\text{upper codiag}(\mathbf{A}) = (c_1, c_2, \ldots, c_{n-1}),$$

$$\text{lower codiag}(\mathbf{A}) = (b_1, b_2, \ldots, b_{n-1}).$$

Computing methods for tridiagonal matrices are frequently required either because they arise directly in applications or because general matrices are transformed to a tridiagonal form as a first step in matrix computations.

Let A_r denote the $r \times r$ leading submatrix of **A** corresponding to the first r rows and columns of **A** [see I (6.13.6)], and let

$$f_0(t) = 1, \qquad f_r(t) = \det(\mathbf{A}_r - t\mathbf{I}) \tag{4.9.2}$$

denote the characteristic polynomial of $\mathbf{A}_r$ [see I (7.3.1)], so that the eigenvalues of **A** are the roots of the equation $f_n(t) = 0$ [see I, Theorem 7.2.3]. To simplify the computations we can assume that **A** is *irreducible*, that is

$$b_i \neq 0, \qquad c_i \neq 0, \qquad (i = 1, 2, \ldots, n-1) \tag{4.9.3}$$

as otherwise **A** would be reducible and the problem of computing the eigenvalue of **A** can be reduced to computing those of smaller matrices [see I, Example 7.3.1].

Let $f_{r+1}(t) = \det(\mathbf{A}_{r+1} - t I)$. Because of the tridiagonal form, if we expand this determinant in terms of the last row or column [see I (6.11.3), (6.11.4)] we get [see I, § 14.12]

$$f_{r+1}(t) = (a_{r+1} - t)f_r(t) - b_r c_r f_{r-1}(t). \tag{4.9.4}$$

Since

$$f_0(t) = 1, f_1(t) = a_1 - t \tag{4.9.5}$$

we can apply (4.9.4) for $r = 1, 2, \ldots, n-1$.

We observe from (4.9.4) that the polynomials $f_r(t)$ are not altered if b_r, c_r are changed to b_r', c_r' provided that

$$b_r c_r = b_r' c_r'. \tag{4.9.6}$$

If

$$b_r c_r > 0 \qquad (r = 1, 2, \ldots, n-1) \tag{4.9.7}$$

the matrix **A** is said to be *quasisymmetric*, because we can take

$$b_r' = c_r' = (b_r c_r)^{1/2}. \tag{4.9.8}$$

Hence the eigenvalues of a quasisymmetric matrix, like those of a symmetric matrix, are real.

To simplify the computations we shall assume that

$$c_r = 1 \qquad (r = 1, 2, \ldots, n-1). \tag{4.9.9}$$

This is achieved by taking

$$b_r' = b_r c_r, c_r' = 1 \qquad (r = 1, 2, \ldots, n-1). \tag{4.9.10}$$

let

$$f_i = f_i(\alpha) \qquad (i = 0, 1, 2, \ldots, n); \tag{4.9.11}$$

then in this simplified form, (4.9.5) and (4.9.4) become:

$$f_0 = 1, f_1 = a_1 - \alpha$$
$$f_{r+1} = (a_{r+1} - \alpha)f_r - b_r f_{r-1} \qquad (r = 1, 2, \ldots, n-1). \qquad (4.9.12)$$

For quasisymmetric matrices, the polynomials

$$f_0(t), f_1(t), \ldots, f_n(t) \qquad (4.9.13)$$

form a *Sturm sequence*. For the properties of Sturm sequences [see I, Theorem 7.9.4 and Wilkinson (1965), p. 300]. In particular, we shall use the following property: Let $v(\alpha)$ the number of changes of signs in the sequence

$$f_0, f_1, f_2, \ldots, f_n \qquad (4.9.14)$$

defined by (4.9.11) and (4.9.12), and let

$$\alpha_1 < \alpha_2 \ldots < \alpha_n \qquad (4.9.15)$$

be the eigenvalues of **A**. Then $v(\alpha)$ is equal to the number of eigenvalues α_i to the left of (α); and if $a < b$, then $v(b) - v(a)$ gives the number of eigenvalues in the interval $[a, b]$. It also follows from the properties of Sturm sequences that the eigenvalues are distinct if **A** is irreducible.

This enables us to apply the bisection method [§ 5.3.1] to compute any eigenvalue α_i of a symmetric, or quasisymmetric, matrix **A**. We start with an interval $[a, b]$ which contains α_i; in the absence of any information we can take $a = -\|A\|_\infty$ and $b = +\|A\|_\infty$. Let $\alpha = \frac{1}{2}(a + b)$; use (4.9.12) to compute $v(\alpha)$; if $v(\alpha) < i$ then $\alpha < \alpha_i$ and we take

$$\text{new } a = \alpha,$$

$$\text{new } b = \text{old } b;$$

otherwise $\alpha \geq \alpha_i$ and we take

$$\text{new } a = \text{old } a,$$

$$\text{new } b = \alpha$$

and repeat. The computation stops when $b - a$ is less than the required accuracy.

Though simple, this algorithm is slow. We can substantially decrease the number of iterations by switching to Newton's method [see § 5.4.1] when the bisection method has isolated α_i, that is when $v(a) = i - 1$ and $v(b) = i$, so that $[a, b]$ contains only one eigenvalue α_i. If $\alpha = \frac{1}{2}(a + b)$ is an approximation to α_i then a much better approximation is given by

$$\alpha - f_n(\alpha)/f'_n(\alpha). \qquad (4.9.14)$$

To compute $f_n = f_n(\alpha)$ we use (4.9.12). To compute $f'_n = f'_n(\alpha)$ we use the following formula, derived by differentiating (4.9.12) with respect to α_i:

$$f'_{r+1} = (a_{r+1} - \alpha)f'_r - b_r f'_{r-1} - f_r \quad (r = 1, 2, \ldots, n-1) \left. \right\}$$

where

$$f'_0 = 0, \quad f'_1 = -1. \qquad (4.9.15)$$

This algorithm, called the *Sturm–Newton* algorithm, can be coded very efficiently by using the same loop to compute f_r and f'_r simultaneously for $r = 2, 3, \ldots, n$. Its convergence is very fast, but some caution is needed to check that the new α is still in $[a, b]$. This may not be the case if we accidentally hit a point where $f'_n(\alpha)$ is zero or nearly zero. The properties of Sturm sequences apply to the characteristic polynomials of the leading minors of any symmetric matrix but the recurrence relations (4.9.4) apply to tridiagonal matrices only.

We consider next a method, called the left–right, or *LR method*, for computing the eigenvalues of a general matrix which need not be tridiagonal nor symmetric, and the eigenvalues may not be distinct nor real. We construct a sequence of matrices

$$\mathbf{A}_1 = \mathbf{A}, \mathbf{A}_2, \mathbf{A}_3, \mathbf{A}_4, \ldots \qquad (4.9.16)$$

by the following series of transformations:

$$\mathbf{A}_{r+1} = Q(\mathbf{A}_r) = \mathbf{U}_r \mathbf{L}_r, \qquad (4.9.17)$$

where $\mathbf{L}_r, \mathbf{U}_r$ are computed by factorising

$$\mathbf{A}_r = \mathbf{L}_r \mathbf{U}_r, \qquad (4.9.18)$$

[see (4.2.10)]. We observe that Q is a similarity transformation which preserves the eigenvalues [see I, § 7.4] since

$$\mathbf{A}_{r+1} = \mathbf{U}_r \mathbf{L}_r = \mathbf{L}_r^{-1} \mathbf{L}_r \mathbf{U}_r \mathbf{L}_r \mathbf{L}_r^{-1} \mathbf{L}_r = \mathbf{L}_r^{-1}(\mathbf{L}_r \mathbf{U}_r) \mathbf{L}_r = \mathbf{L}_r^{-1} \mathbf{A}_r \mathbf{L}_r. \qquad (4.9.19)$$

We refer to Wilkinson (1965), pp. 487–492, for the properties of this method, of which the following is the most important:

Let the eigenvalues be arranged so that

$$|\alpha_1| \geq |\alpha_2| \geq \ldots \geq |\alpha_n| \qquad (4.9.20)$$

and let

$$\mathbf{L}_r = (l_{ij}^{(r)}), \qquad \mathbf{U}_r = (u_{ij}^{(r)}).$$

By definition $l_{ij}^{(r)} = 1$ when $i = j$ and 0 when $i < j$. For the lower diagonal elements it can be shown that under fairly general conditions [see § 2.3]

$$l_{ij}^{(r)} = O((\alpha_i/\alpha_j)^r). \qquad (4.9.21)$$

If the inequalities in (4.9.20) are strict, it follows that

$$\mathbf{L}_r \to \mathbf{I}. \tag{4.9.22}$$

The analysis needs slight modifications to deal with equal or complex conjugate eigenvalues; this will be dealt with later. It follows from (4.9.17) that

$$\mathbf{U}_r \to \mathbf{U}, \tag{4.9.23}$$

$$\mathbf{A}_r \to \mathbf{U}, \tag{4.9.24}$$

$$u_{ii}^{(r)} \to \alpha_i, \tag{4.9.25}$$

since the eigenvalues of a triangular matrix are equal to the diagonal element [see I, Example 7.3.1].

This method may appear to be impractical since the repeated factorization (4.9.18) is very expensive; but two modifications, namely shift of origin (similar to the one used in inverse iteration [see (4.8.13)]) and deflation make it very efficient. They are described below where we consider a special case of the *LR* method, called the *QD algorithm*, applicable to tridiagonal matrices. Let

$$\mathbf{A} = \begin{pmatrix} a_1 & 1 & 0 & \cdot & 0 \\ b_1 & a_2 & 1 & \cdot & 0 \\ 0 & \cdot & \cdot & \cdot & 1 \\ 0 & 0 & \cdot & b_{n-1} & a_n \end{pmatrix}. \tag{4.9.26}$$

We can show that if $\mathbf{A} = \mathbf{LU}$, then $\mathbf{L}$ and $\mathbf{U}$ have the form

$$\mathbf{L} = \begin{pmatrix} 1 & 0 & 0 & \cdot & \cdot \\ q_1 & 1 & 0 & \cdot & \cdot \\ 0 & q_2 & 1 & \cdot & \cdot \\ \cdot & \cdot & \cdot & \cdot & \cdot \\ 0 & 0 & \cdot & q_{n-1} & 1 \end{pmatrix}$$

$$\mathbf{U} = \begin{pmatrix} d_1 & 1 & 0 & \cdot & 0 \\ 0 & d_2 & 1 & 0 & \cdot \\ \cdot & \cdot & \cdot & \cdot & \cdot \\ \cdot & \cdot & 0 & d_{n-1} & 1 \\ 0 & \cdot & 0 & 0 & d_n \end{pmatrix} \tag{4.9.27}$$

and these shapes are preserved for all the matrices $\mathbf{A}_r$, $\mathbf{L}_r$, $\mathbf{U}_r$ involved in the iteration (4.9.17), (4.9.18).

The coefficients q_i and d_i are computed from the relation $\mathbf{A} = \mathbf{LU}$, which gives the algorithm

$$\left. \begin{array}{c} d_1 = a_1, \\ q_i = b_i/d_i, \quad d_{i+1} = a_{i+1} - q_i \qquad (i = 1, 2, \ldots, n-1). \end{array} \right\} \tag{4.9.28}$$

Let $A_1 = UL$, and let a_i^1, b_i^1 be the coefficients of A_1; they are computed from the relation $A_1 = UL$ by the algorithm:

$$a_i^1 = d_i + q_i, \qquad b_i^1 = d_{i+1}q_i \qquad (i = 1, 2, \ldots, n-1)\Big\}$$
$$a_n^1 = d_n. \tag{4.9.29}$$

We need computer memory space for a_i, b_i only, as d_i, q_i can overwrite a_i, b_i, and a_i^1, b_i^1 can overwrite d_i, q_i.

We now explain various refinements which make the QD algorithm very efficient. If follows from (4.9.21) and (4.9.25) that

$$a_n^{(r)} \to \alpha_n \tag{4.9.30}$$

and

$$b_{n-1}^{(r)} = O((\alpha_n/\alpha_{n-1})^r) \tag{4.9.31}$$

which tends to zero when $\alpha_{n-1} \neq \alpha_n$ because of the ordering (4.9.20). If h is an approximation to α_n, then we can apply a shift of origin h by replacing A by $A - hI$. This speeds up the convergence (4.9.31) and hence (4.9.30) considerably since, as we can verify,

$$\frac{|\alpha_n - h|}{|\alpha_{n-1} - h|} < \frac{|\alpha_n|}{|\alpha_{n-1}|}. \tag{4.9.32}$$

Hence

$$\left(\frac{|\alpha_n - h|}{|\alpha_{n-1} - h|}\right)^r \ll \left(\frac{|\alpha_n|}{|\alpha_{n-1}|}\right)^r. \tag{4.9.33}$$

The shift of origin involves a simple modification of the algorithm (4.9.28) and (4.9.29) which we indicate by the following replacements:

$$a_1 \to a_1 - h$$
$$b_i \to b_i/a_i, \qquad a_{i+1} \to a_{i+1} - h - b_i \qquad (i = 1, 2, \ldots, n-1) \tag{4.9.34}$$

and

$$a_i \to a_i + b_i + h, \qquad b_i \to a_{i+1}b_i \qquad (i = 1, 2, \ldots, n-1)$$
$$a_n \to a_n + h. \tag{4.9.35}$$

The transformation restores the shift of origin and ensures that all the matrices A_r remain similar to A.

Next we consider what happens when $|\alpha_n| = |\alpha_{n-1}|$. In this case b_{n-2} will tend to zero while, in general, b_{n-1} will not. A_r will tend to a partitioned matrix, that is, it becomes reducible [see I (7.3.14)], and the eigenvalues α_{n-1}, α_n are the roots of

$$\det \begin{pmatrix} a_{n-1} - t & 1 \\ b_{n-1} & a_n - t \end{pmatrix} = 0 \tag{4.9.36}$$

which is the quadratic equation

$$(a_n - t)(a_{n-1} - t) = b_{n-1}. \tag{4.9.37}$$

The eigenvalue α_n 'emerges', if b_{n-1} has converged to zero; when this happens we can drop the last row and column of A, before proceeding to compute α_{n-1}. Similarly, if $|\alpha_n| = |\alpha_{n-1}|$ then we can drop the last two rows and columns when b_{n-2} has converged to zero. This is called a *deflation* process. Generally, if b_r converges to zero before $b_{r+1}, \ldots, b_{n-1}$ (which may occur because the last $n - r + 1$ eigenvalues have equal or nearly equal modulus) then we can partition A_r into smaller matrices; this happens for Jordan blocks [see I (7.6.4)].

The combined refinements of shift of origin and deflation make the *QD* algorithm faster than the Sturm–Newton method, but the *QD* algorithm has to compute $\alpha_n, \alpha_{n-1}, \ldots, \alpha_{i+1}$ before it can compute α_i, whereas the Sturm–Newton method can compute α_i directly. Finally, the *QD* algorithm copes with general tridiagonal real matrices which may have complex eigenvalues, whereas the Sturm–Newton method applies to quasisymmetric matrices only.

Finally, the shift of origin and deflation refinements are applicable to the *LR* method also. This will be considered in section 4.11.

EXAMPLE 4.9.1.   We apply to *QD* algorithm with shift and deflation to the matrix

$$A = \begin{pmatrix} 4 & 1 & 0 & 0 \\ 0{\cdot}1 & 3 & 1 & 0 \\ 0 & 0{\cdot}1 & 2 & 1 \\ 0 & 0 & 0{\cdot}1 & 1 \end{pmatrix}. \tag{4.9.38}$$

The following Table 4.9.1 gives the details of the calculations. The required eigenvalues are 0·90449, 1·99566, 3·00434, 4·09951. The computation illustrates an important feature of this type of method, namely that the computation of α_4, which is the first eigenvalue to emerge, contributes to the computation of α_3, which in turn contributes to the computation of α_2 and so on.

r	0	1	2	3	4	5	6	7
a_1	4	4·02500	4·04880	4·06456	4·08059	4·08827	4·09548	4·09551
a_2	3	3·00861	3·01652	3·01714	3·01933	3·01157	3·00437	3·00434
a_3	2	2·01724	2·03000	2·01381	1·99558	1·99566		
a_4	1	0·94915	0·90468	0·90449				
b_1	0·1	0·07437	0·04962	0·03310	0·01605	0·00784	0·00003	
b_2	0·1	0·06610	0·03442	0·01824	0·00008			
b_3	0·1	0·04827	0·00021					
h	0	0·9	0·9	2	1·99	3	3	

Table 4.9.1

EXAMPLE 4.9.2. Having shown the algorithm at its best, we now show it at its worst by applying it to the same matrix reversed:

$$\mathbf{A} = \begin{pmatrix} 1 & 1 & 0 & 0 \\ 0{\cdot}1 & 2 & 1 & 0 \\ 0 & 0{\cdot}1 & 3 & 1 \\ 0 & 0 & 0{\cdot}1 & 4 \end{pmatrix}. \qquad (4.9.39)$$

Instead of 7 iterations, this time we need 21; the effect of the first 11 iterations is to 'drag' the smaller eigenvalues to the bottom right-hand side before convergence begins; during this preliminary 'dragging' stage the b_i coefficients become larger before they start converging to zero. A shift of origin during this preliminary stage would not help. We conclude from this example that the *QD* and *LR* type of algorithms can be made much more efficient if it is possible to rearrange the matrix so that the small diagonal elements are in the bottom right-hand corner. Table 4.9.2 gives some of the iterations.

r	0	1	5	10	11	12	13	14
a_1	1	1·10000	1·99601	2·78009	2·93557	3·15108	3·33591	3·45189
a_2	2	1·95263	2·04371	2·97695	3·04607	3·17820	3·29613	3·32899
a_3	3	2·98130	2·37176	3·21268	3·07076	2·76610	2·46346	2·31463
a_4	4	3·96607	3·58853	1·03029	0·94760	0·90462	0·90449	
b_1	0·1	0·1900	0·35211	0·43227	0·43869	0·41606	0·38691	0·36884
b_2	0·1	0·15513	1·01545	0·63372	0·67114	0·63379	0·47332	0·34450
b_3	0·1	0·13456	0·60067	0·24708	0·07836	0·00020		
h	0	0	0	0	0·9	0·9	0	0

r	15	16	17	18	19	20	21
a_1	3·55874	3·65549	3·84438	3·95811	3·99122	4·09505	4·09551
a_2	3·32905	3·30533	3·25606	3·14174	3·10862	3·00480	3·00434
a_3	2·20771	2·13469	1·99506	1·99566			
a_4							
b_1	0·34429	0·31271	0·21089	0·131048	0·10292	0·000498	
b_2	0·23604	0·15589	−0·00069				
b_3							
h	0	2	1·99	0	3	3	

Table 4.9.2

4.10. EIGENVALUES AND EIGENVECTORS OF SYMMETRIC MATRICES

The similarity transformation

$$Q(\mathbf{A}) = \mathbf{P}^{-1}\mathbf{A}\mathbf{P} \qquad (4.10.1)$$

is said to be orthogonal [see I, § 7.7] if

$$\mathbf{P}^{-1} = \mathbf{P}'. \tag{4.10.2}$$

Then, if $\mathbf{A}$ is symmetric, so is $Q(\mathbf{A})$. We seek a sequence of orthogonal transformations

$$\mathbf{A}_{r+1} = Q_r(\mathbf{A}_r) \qquad (r = 1, 2, \ldots, n-2) \tag{4.10.3}$$

where $\mathbf{A}_1 = \mathbf{A}$ and $\mathbf{A}_{n-1}$ is tridiagonal. We can then apply the methods of § 4.9 to compute the eigenvalues of $\mathbf{A}$. Since $\mathbf{A}_1$ is symmetric, so are the matrices $\mathbf{A}_2, \mathbf{A}_3, \ldots, \mathbf{A}_{n-1}$. The *Householder type* of orthogonal transformation is defined by

$$\mathbf{H} = \mathbf{I} - 2\mathbf{w}\mathbf{w}' \tag{4.10.4}$$

where

$$\mathbf{w} = (w_1, w_2, \ldots, w_n)', \tag{4.10.5}$$

and

$$\mathbf{w}'\mathbf{w} = 1, \tag{4.10.6}$$

that is, $\mathbf{w}$ is a unit vector. We shall now show that $\mathbf{H}$ is indeed an orthogonal matrix. For we deduce from (4.10.4) that

$$\mathbf{H}' = \mathbf{H}, \tag{4.10.7}$$

and using (4.10.6) we find that

$$\mathbf{H}^2 = \mathbf{H}'\mathbf{H} = \mathbf{H}\mathbf{H}' = \mathbf{I}. \tag{4.10.8}$$

Thus $\mathbf{H}$ is orthogonal, that is

$$\mathbf{H}^{-1} = \mathbf{H}'. \tag{4.10.9}$$

Let

$$\mathbf{x} = (x_1, x_2, \ldots, x_n)', \tag{4.10.10}$$

and

$$\mathbf{H}_r = \mathbf{I} - 2\mathbf{w}\mathbf{w}', \tag{4.10.11}$$

where

$$\mathbf{w} = (0, 0, \ldots, w_{r+1}, w_{r+2}, \ldots, w_n)', \tag{4.10.12}$$

that is, the first r elements of $\mathbf{w}$ are zero. It follows that the transformation

$$\mathbf{y} = \mathbf{H}_r\mathbf{x} \tag{4.10.13}$$

leaves unaltered the first r elements of $\mathbf{x}$. We wish to determine $\mathbf{H}_r$, that is $\mathbf{w}$, so that

$$\mathbf{y} = (x_1, x_2, \ldots, x_r, \rho, 0, 0, \ldots, 0)'. \tag{4.10.14}$$

The following formulae for the elements of $\mathbf{w}$ follow from (4.10.12) and (4.10.14). Let

$$\rho = \text{sign} (x_{r+1})(x_{r+1}^2 + x_{r+2}^2 + \ldots + x_n^2)^{1/2}, \tag{4.10.15}$$

and

$$s = (2\rho(\rho + x_{r+1}))^{1/2}. \tag{4.10.16}$$

Then

$$w_{r+1} = \frac{\rho + x_{r+1}}{s}, \tag{4.10.17}$$

$$w_i = \frac{x_i}{s} \quad (i = r+2, r+3, \ldots, n). \tag{4.10.18}$$

The transformations (4.10.3) will take the following form:

$$\mathbf{A}_{r+1} = \mathbf{H}_r \mathbf{A}_r \mathbf{H}_r \quad (r = 1, 2, \ldots, n-2) \tag{4.10.19}$$

where $\mathbf{A}_r$ has the form shown in Table 4.10.1: the submatrix corresponding to the first r rows and columns of $\mathbf{A}_r$ is symmetric tridiagonal, the submatrix corresponding to the last $n - r + 1$ rows and columns of $\mathbf{A}_r$ is symmetric full, all other elements of $\mathbf{A}_r$ are zero.

a_1	b_1	0	0	0	$\cdot$		$\cdot$		$\cdot$	$\cdot$	$\cdot$
b_1	a_2	b_2	0	0	$\cdot$		$\cdot$		$\cdot$	$\cdot$	$\cdot$
0	b_2	a_3	b_3	0	$\cdot$		$\cdot$		$\cdot$	$\cdot$	$\cdot$
0	0	b_3	a_4	b_4	$\cdot$		$\cdot$		$\cdot$	$\cdot$	$\cdot$
0	0	0	b_4	a_5	$\cdot$		$\cdot$		$\cdot$	$\cdot$	$\cdot$
$\cdot$	$\cdot$	$\cdot$	$\cdot$	$\cdot$	$\cdot$	b_{r-1}	0		$\cdot$		0
$\cdot$	$\cdot$	$\cdot$	$\cdot$	$\cdot$	b_{r-1}	a_r	$a_{r,r+1}^{(r)}$		$\cdot$		$a_{r,n}^{(r)}$
$\cdot$	$\cdot$	$\cdot$	$\cdot$	$\cdot$	0	$a_{r+1,r}^{(r)}$	$a_{r+2,r+1}^{(r)}$		$\cdot$		$a_{r+1,n}^{(r)}$
$\cdot$	$\cdot$	$\cdot$	$\cdot$	$\cdot$	0	$a_{n,r}^{(r)}$	$a_{n,r+1}^{(r)}$		$\cdot$		$a_{nn}^{(r)}$

Table 4.10.1

We seek to determine $\mathbf{H}_r$ in (4.10.18) so that $\mathbf{A}_{r+1}$ has a structure similar to that of $\mathbf{A}_r$ with $r+1$ instead of r. That is the submatrix corresponding to the first $r+1$ rows and columns of $\mathbf{A}_{r+1}$ is symmetric tridiagonal, the submatrix corresponding to the last $n - r$ rows and columns of $\mathbf{A}_{r+1}$ is full symmetric and all other elements of $\mathbf{A}_{r+1}$ are zero.

It follows from (4.10.10) to (4.10.14) that $\mathbf{H}_r$ is determined from (4.10.15) to (4.10.18) by taking

$$\mathbf{x} = (r+1)\text{-th column of } \mathbf{A}_r. \tag{4.10.20}$$

(The rth row and column of $\mathbf{A}_r$ are especially marked in Table 4.10.1.)

This method is used to determine $H_1, H_2, \ldots, H_{n-2}$ in turn by taking $r = 1, 2, \ldots, n-2$ in (4.10.15) to (4.10.19). The first and last H_r are slightly different but the general algorithm still applies.

We now describe various features of this method from the computer organization point of view. The sign of ρ in (4.10.15) is chosen so as to minimize the loss of accuracy when computing $\rho + x_{r+1}$ in (4.10.16) and (4.10.17). Since A_r remains symmetric, we store, compute and process only one half, e.g. the lower triangular half, of the matrix A; each A_r overwrites the previous one. The transformations H_r have to be stored for later use when computing eigenvectors (see (4.10.22) below); this is done most economically by storing the non-zero element of w_r mainly in the zero locations introduced by H_r in the rth column. This is because the zeros introduced by H_r are not altered by later transformations, a crucial point which does not often hold for other types of transformations. To compute HAH, where (H, A) are any of the (H_r, A_r) we proceed as follows: compute $p = Aw$, then $k = w'p$, $q = p - kw$; the required matrix is given by

$$HAH = A - 2wq' - 2qw',$$

that is

$$\text{new } a_{ij} = \text{old } a_{ij} - 2w_i q_j - 2w_j q_i. \tag{4.10.21}$$

Thus, the elements of H need not be computed or stored explicitly.

The matrix A_{n-1} is symmetric tridiagonal; its eigenvalues are the same as those of the original A and can be computed by any of the methods of section 4.9. To compute the eigenvectors of A_{n-1} we use the method of inverse iteration (4.8.14) to (4.8.15); because we can take h equal to an accurate eigenvalue, only two iterations are normally needed. If x is an eigenvector of A_{n-2}, then it follows from (4.10.19) that the corresponding eigenvector of A is

$$H_1 H_2 \ldots H_{n-2} x \tag{4.10.22}$$

that is, we apply the transformations H_r in reverse order; this is computed most conveniently by repeated application of the following formula:

$$Hx = (I - 2ww')x = x - 2(w'x)w. \tag{4.10.23}$$

4.11. EIGENVALUES AND EIGENVECTORS OF REAL UNSYMMETRIC MATRICES

We consider in more detail the transformation (4.10.9) which can be written in the form:

$$A_{n-1} = H_{n-2} \ldots H_2 H_1 A H_1 H_2 \ldots H_{n-2}; \tag{4.11.1}$$

consider the first step

$$B = H_1 A, \tag{4.11.2}$$

$$C = BH_1. \tag{4.11.3}$$

The left transformation $\mathbf{H}_1$ in (4.11.2) operates on the last $n-1$ rows and does not alter the first row; it has the effect of introducing zeros in the first column of $\mathbf{B}$ below position (2, 1), that is

$$b_{31} = b_{41} = \ldots = b_{n1} = 0. \tag{4.11.4}$$

The right transformation $\mathbf{H}_1$ in (4.11.3) operates on the last $n-1$ columns of $\mathbf{B}$ leaving the first column of $\mathbf{B}$, and in particular its zero elements, unaltered. If $\mathbf{A}$ is symmetric, the right transformation $\mathbf{H}_1$ in (4.11.3) will introduce zeros in the first row of $\mathbf{C}$ to the right of position (1, 2); but if $\mathbf{A}$ is unsymmetric, as we shall assume throughout this section, then, in general, all the elements of the first row will be non-zero. Generally the transformation will introduce zero elements below the lower codiagonal, that is, below position $(r+1, r)$ without affecting the previously introduced zeros. Hence $\mathbf{A}_{n-1}$ will have zero elements below the codiagonal. This is called the *Hessenberg* form. It is described by the *schematic* form:

$$\mathbf{A}_{n-1} = \begin{pmatrix} x & x & x & x & x \\ x & x & x & x & x \\ & x & x & x & x \\ & & x & x & x \\ & & & x & x \end{pmatrix}. \tag{4.11.5}$$

It is schematic in the sense that the letter x merely indicates a (possibly) non-zero position. We shall also take $n=6$ or 8 to illustrate the general case, thus we have taken $n=5$ in (4.11.5). These conventions will be used in the rest of this section to help describe the more involved algorithms.

Let us assume that $\mathbf{A}$ has already been reduced to the Hessenberg form, and let

$$\mathbf{A}_1 = \mathbf{A} = \mathbf{LU}, \tag{4.11.6}$$

$$\mathbf{A}_2 = \mathbf{UL}, \tag{4.11.7}$$

be the *LR* transformation (4.9.17), (4.9.18) applied to the Hessenberg matrix $\mathbf{A}$. Then it can be shown that $\mathbf{L}$ has zeros below the codiagonal and that $\mathbf{A}_2$ is also a Hesenberg matrix, thus *LR* transformations preserve the Hessenberg form, and the factorization (4.11.6) involves much less arithmetic for a Hessenberg matrix than for a full matrix. The shift of origin and deflation refinements described in section 4.8 and section 4.9 for the general *LR* method are applicable; suitable modifications described later [see (4.11.23)] can cope with complex or equal eigenvalues. Unfortunately the method cannot be safely applied to every matrix because it can involve very small or zero divisors, in fact it fails right at the start if a_{11} (the leading element of $\mathbf{A}$) is zero.

We now describe a more stable method, called the *QR method*. It is of the form

$$\mathbf{QA}_r = \mathbf{U}$$
$$\mathbf{A}_{r+1} = \mathbf{UQ} \tag{4.11.8}$$

where $\mathbf{A}_r$ and $\mathbf{A}_{r+1}$ have the Hessenberg form, $\mathbf{U}$ is upper triangular and $\mathbf{Q}$ involves $n-1$ steps $\mathbf{J}_1, \mathbf{J}_2, \ldots, \mathbf{J}_{n-1}$ eliminating in turn the $n-1$ elements $a_{21}, a_{32}, \ldots, a_{n,n-1}$ respectively. For example, when $n = 6$, $\mathbf{J}_2\mathbf{J}_1\mathbf{A}_r$ has the following schematic form:

$$\mathbf{J}_2\mathbf{J}_1\mathbf{A}_r = \begin{pmatrix} u & u & u & u & u & u \\ u & u & u & u & u \\ & q & y_1 & y_2 & y_3 \\ & p & x_1 & x_2 & x_3 \\ & & a & a & a \\ & & & a & a \end{pmatrix}, \tag{4.11.9}$$

where u indicates the elements of $\mathbf{U}$ computed so far, a indicates the elements of $\mathbf{A}_r$ so far unaltered, and p, q, x, y indicate the elements about to be processed by $\mathbf{J}_3$.

The transformations $\mathbf{J}_1, \mathbf{J}_2, \ldots, \mathbf{J}_{n-1}$ belong to an important type of orthogonal transformation, called *Jacobi transformation*, which represents a plane rotation. For example $\mathbf{J}_3$, described in (4.11.10) below, represents a rotation in the $(3, 4)$ plane and generally, $\mathbf{J}_i$ represents a rotation in the $(i, i+1)$ plane. Thus $\mathbf{J}_3$ has the following form:

$$\mathbf{J} = \begin{pmatrix} 1 \\ & 1 \\ & & c & s \\ & & -s & c \\ & & & & 1 \\ & & & & & 1 \end{pmatrix} \tag{4.11.10}$$

where

$$c^2 + s^2 = 1. \tag{4.11.11}$$

This condition leads to the relation $\mathbf{J}_3' = \mathbf{J}_3^{-1}$, that is, $\mathbf{J}_3$ is orthogonal [see I, § 7.7]. If we choose c, s as

$$c = q/r, \qquad s = p/r$$

where $(4.11.12)$

$$r = (p^2 + q^2)^{1/2}$$

then $\mathbf{J}_3\mathbf{J}_2\mathbf{J}_1\mathbf{A}_r$ takes the following schematic form:

$$\mathbf{J}_3\mathbf{J}_2\mathbf{J}_1\mathbf{A}_r = \begin{pmatrix} u & u & u & u & u & u \\ u & u & u & u & u \\ & r & y_1' & y_2' & y_3' \\ & & x_1' & x_2' & x_3' \\ & & a & a & a \\ & & & a & a \end{pmatrix} \tag{4.11.13}$$

where

$$x_i' = cx - sy_i,$$

and $(i = 1, 2, 3)$ (4.11.14)

$$y_i' = cy_i + sx_i.$$

Thus J_3 extends the upper triangular form by another row; it follows that

$$QA_r = J_5J_4J_3J_2J_1A_r = U \tag{4.11.15}$$

has the required triangular form.

We give next a typical stage of the backward transformation UQ. Let $UJ_1'J_2'$ have the schematic form

$$UJ_1'J_2' = \begin{pmatrix} a & a & x_1 & y_1 & u & u \\ a & a & x_2 & y_2 & u & u \\ & a & x_3 & y_3 & u & u \\ & & r & & u & u \\ & & & & u & u \\ & & & & & u \end{pmatrix}, \tag{4.11.16}$$

where a indicates the elements of the 'emerging' Hessenberg form, u the elements of the initial triangular form r, x, y the elements of the columns about to be processed by J_3'. Then $UJ_1'J_2'J_3'$ has the schematic form:

$$UJ_1'J_2'J_3' = \begin{pmatrix} a & a & x_1' & y_1' & u & u \\ a & a & x_2' & y_2' & u & u \\ & a & x_3' & y_3' & u & u \\ & & p & q' & u & u \\ & & & & u & u \\ & & & & & u \end{pmatrix}, \tag{4.11.17}$$

where

$$p = sr, \qquad q' = cr \tag{4.11.18}$$

(using the s, c of (4.11.12)),

$$\begin{aligned} x_i' &= cx_i + sy_i, \\ y_i' &= cy_i - sx_i, \end{aligned} \qquad (i = 1, 2, 3) \tag{4.11.19}$$

and the elements a, u are unaltered from (4.11.16). Comparing (4.11.17) with (4.11.16), we see that the Hessenberg form is extended by another column, and that

$$A_{r+1} = UQ = UJ_1'J_2'J_3'J_4'J_5'. \tag{4.11.20}$$

Because matrix multiplication is associative [see I (6.3.3)], the combined transformations (4.11.15) and (4.11.20) can be carried out in the following order:

$$A_{r+1} = J_5(\ldots (J_2(J_1(A_r)J_1')J_2')\ldots)J_5'. \tag{4.11.21}$$

The matrix $\mathbf{U}$ is not explicitly formed. But this order has a slight advantage for the computer organization: the coefficients c and s computed in (4.11.12) for the various transformations $\mathbf{J}_i$ have to be stored for later use in (4.11.20), whereas in (4.11.21) each pair of coefficients (c, s) can in turn overwrite the previous one.

After a number of repeated applications of (4.11.8), that is (4.11.21), the element a_{nn} of $\mathbf{A}_r$ will converge to α_n as the element $a_{n-1,n}$ tend to zero, where $\alpha_1, \alpha_2, \ldots, \alpha_n$ are the eigenvalues of $\mathbf{A} = \mathbf{A}_1$ arranged in descending order of magnitude as in (4.9.20). Convergence can be speeded up substantially by applying a shift of origin h [see (4.9.32), (4.9.33)] by using

$$\mathbf{Q}(\mathbf{A}_r - h\mathbf{I}) = \mathbf{U}$$
$$\mathbf{A}_{r+1} = \mathbf{U}\mathbf{Q} + h\mathbf{I} \tag{4.11.22}$$

instead of (4.11.8). This is usually done as soon as we have a rough estimate of α_n and we take this estimate to be the value of h. The deflation and partitioning methods, described for the QD algorithm after (4.9.37), are also applicable for the QR method.

If α_{n-1} and α_n are nearly equal, then $a_{n,n-1}$ will tend to zero very slowly and if $\alpha_{n-1} = \alpha_n$ or $\alpha_{n-1} = \bar{\alpha}_n$ then $a_{n,n-1}$ will not converge to zero at all. Usually in such cases the next lower codiagonal element $a_{n-1,n-2}$ will tend to zero before $a_{n,n-1}$. When this happens, that is when $a_{n-1,n-2}$ is sufficiently small we partition off the bottom right-hand 2×2 matrix given by (4.11.23) and process the two diagonal partitions separately, thus

$$\begin{pmatrix} a_{n-1,n-1} & a_{n-1,n} \\ a_{n,n-1} & a_{nn} \end{pmatrix}. \tag{4.11.23}$$

The eigenvalues of (4.11.23) are the roots of

$$t^2 - st + p = 0 \tag{4.11.24}$$

where

$$s = a_{n-1,n-1} + a_{nn},$$
$$p = a_{nn}a_{n-1,n-1} - a_{n,n-1}a_{n-1,n}. \tag{4.11.25}$$

Unfortunately the convergence of $a_{n-1,n-2}$ will be very slow and we cannot speed it up by a single shift of origin since $a_{n,n-1}$ does not converge to zero at all. To speed up the convergence we have to apply two shifts of origin h_1, h_2 taken to be the roots of (4.11.24). This double shift of origin can be applied as soon as $a_{n-1,n-2}$ starts to decrease in magnitude, which is much earlier than its ultimate convergence. Unfortunately h_1, h_2 will frequently be complex conjugate and the transformations (4.11.22) will involve complex arithmetic. However, it is possible to organise the double shift h_1, h_2 simultaneously so that no complex arithmetic is involved. The method for doing this is called the *double QR method*; we proceed to describe it in detail. Let

$$\mathbf{Q}_1(\mathbf{A} - h_1\mathbf{I}) = \mathbf{U}_1,$$
$$\mathbf{A}^{(1)} = \mathbf{U}_1\mathbf{Q}_1' + h_1\mathbf{I}, \tag{4.11.26}$$

and

$$\mathbf{Q}_2(\mathbf{A}^{(1)} - h_2\mathbf{I}) = \mathbf{U}_2,$$
$$\mathbf{A}^{(2)} = \mathbf{U}_2\mathbf{Q}_2' + h_2\mathbf{I}, \tag{4.11.27}$$

obtained by using first h_1 and then h_2 in (4.11.22). If h_1, h_2 are complex conjugate, $\mathbf{A}^{(2)}$ will be real even though the intermediate results in (4.11.26) and (4.11.27) are not. Let

$$\mathbf{W} = \mathbf{Q}_2\mathbf{Q}_1,$$
$$\mathbf{U} = \mathbf{U}_2\mathbf{U}_1, \tag{4.11.28}$$

then, by (4.11.26) and (4.11.27) we get that

$$\mathbf{W}(\mathbf{A} - h_1\mathbf{I})(\mathbf{A} - h_2\mathbf{I}) = \mathbf{U}$$
$$\mathbf{A}^{(2)} = \mathbf{W}\mathbf{A}\mathbf{W}'. \tag{4.11.29}$$

The matrix $\mathbf{A}$ is real and has one subdiagonal; the matrix

$$\mathbf{C} = (\mathbf{A} - h_1\mathbf{I})(\mathbf{A} - h_2\mathbf{I}) \tag{4.11.30}$$

is real and has 2 subdiagonals. Fortunately, we only need the first column c, of $\mathbf{C}$. Let

$$\mathbf{c} = (c_1, c_2, c_3, 0, 0, \dots)'. \tag{4.11.31}$$

Since h_1, h_2 are the roots of (4.11.24), we have that

$$h_1 + h_2 = s, \qquad h_1 h_2 = p \tag{4.11.32}$$

and the following formulae for c_1, c_2, c_3:

$$c_1 = a_{11}(a_{11} - s) + a_{12}a_{21} + p$$
$$c_2 = a_{21}(a_{11} + a_{22} - s) \tag{4.11.33}$$
$$c_3 = a_{21}a_{32}.$$

As in the case of the single QR method, $\mathbf{W}$ consists of $n-1$ stages $\mathbf{H}_1, \mathbf{H}_2, \dots, \mathbf{H}_{n-1}$ so that instead of (4.11.21) we have

$$\mathbf{A}_{r+1} = \mathbf{H}_{n-1}(\dots (\mathbf{H}_2(\mathbf{H}_1(\mathbf{A}_r)\mathbf{H}_1')\mathbf{H}_2') \dots)\mathbf{H}_{n-1}'. \tag{4.11.34}$$

Any orthogonal transformations which introduce the required zeros can be used. We shall use Householder transformations [see § 4.10], so that $\mathbf{H}'_i = \mathbf{H}_i$ (see (4.10.9)). Let $\mathbf{A}_r$ be 8×8 with the following schematic form:

$$\mathbf{A}_r = \mathbf{A}_r^{(1)} = \begin{pmatrix} a & a & a & a & a & a & a & a \\ a & a & a & a & a & a & a & a \\ & a & a & a & a & a & a & a \\ & & a & a & a & a & a & a \\ & & & a & a & a & a & a \\ & & & & a & a & a & a \\ & & & & & a & a & a \\ & & & & & & a & a \end{pmatrix}. \qquad (4.11.35)$$

The first transformation

$$\mathbf{A}_r^{(2)} = \mathbf{H}_1 \mathbf{A}_r^{(1)} \mathbf{H}_1 \qquad (4.11.36)$$

is slightly different from the rest: $\mathbf{H}_1$ is determined so that

$$\mathbf{H}_1(c_1, c_2, c_3, 0, 0, \ldots)' = (\rho, 0, 0, \ldots)'.$$

This may be achieved either by using the general method (4.10.15) to (4.10.18) or the simplified version (4.11.42) to (4.11.49) suitably modified. The result is described in the following schematic form, where the elements of the rows affected by the left transformation $\mathbf{H}_1$, that is row 1 to row 3, are underlined and the elements of the columns affected by the right transformation $\mathbf{H}_1$, that is column 1 to column 3, are primed. Note that non-zero elements are introduced in positions $(3, 1)$, $(4, 1)$ and $(4, 2)$, but these will be eliminated by $\mathbf{H}_2$ and $\mathbf{H}_3$.

$$\mathbf{A}_r^{(2)} = \begin{pmatrix} \underline{a}' & \underline{a}' & \underline{a}' & \underline{a} & \underline{a} & \underline{a} & \underline{a} & \underline{a} \\ \underline{a}' & \underline{a}' & \underline{a}' & \underline{a} & \underline{a} & \underline{a} & \underline{a} & \underline{a} \\ \underline{a}' & \underline{a}' & \underline{a}' & \underline{a} & \underline{a} & \underline{a} & \underline{a} & \underline{a} \\ a' & a' & a' & a & a & a & a & a \\ & & a & a & a & a & a & a \\ & & & a & a & a & a & a \\ & & & & a & a & a & a \\ & & & & & a & a & a \end{pmatrix}. \qquad (4.11.37)$$

The remaining transformations $\mathbf{H}_2, \mathbf{H}_3, \ldots, \mathbf{H}_7$ are very similar, and as an example we describe $\mathbf{H}_4$ in detail. We have

$$\mathbf{A}_r^{(5)} = \mathbf{H}_4 \mathbf{A}_r^{(4)} \mathbf{H}_4 \qquad (4.11.38)$$

where $\mathbf{A}_r^{(4)}$ has the following schematic form:

$$\mathbf{A}_r^{(4)} = \begin{pmatrix} a & a & a & a & a & a & a & a \\ a & a & a & a & a & a & a & a \\ & a & a & a & a & a & a & a \\ & & d_1 & a & a & a & a & a \\ & & d_2 & a & a & a & a & a \\ & & d_3 & a & a & a & a & a \\ & & & & & a & a & a \\ & & & & & & a & a \end{pmatrix} \qquad (4.11.39)$$

where $\mathbf{H}_4$ operates in the dimension (4 to 6) and is defined by

$$\mathbf{H}_4(a, a, a, d_1, d_2, d_3, 0, 0)' = (a, a, a, \rho, 0, 0, 0, 0)'. \qquad (4.11.40)$$

Let

$$\mathbf{H}_4 = \mathbf{I} - 2\mathbf{w}\mathbf{w}' \qquad (4.11.41)$$

[see (4.10.11)]. We could compute $\mathbf{w}$ by the general method (4.10.15) to (4.10.18). But we shall instead describe a more convenient formulation, which is a slight modification of (4.10.15) to (4.10.18). Let

$$\mathbf{w} = (0, 0, 0, s, sp_1, sp_2, 0, 0)'; \qquad (4.11.42)$$

then

$$\rho = \text{sign} (d_1)(d_1^2 + d_2^2 + d_3^2)^{1/2}, \qquad (4.11.43)$$

$$f = d_1/\rho, \qquad s = (1+f)^{1/2}, \qquad (4.11.44)$$

$$p_1 = (d_2/r + d_1), \qquad p_2 = (d_3/r + d_1). \qquad (4.11.45)$$

Let $\mathbf{r}_4, \mathbf{r}_5, \mathbf{r}_6$ be rows 4 to 6 of $\mathbf{A}_r$. Then the result of the left transformation $\mathbf{H}_4$ in (4.11.38) is as follows: let

$$\mathbf{y} = (1+f)(\mathbf{r}_4 + p_1\mathbf{r}_5 + p_2\mathbf{r}_6). \qquad (4.11.46)$$

Then

$$\text{new } \mathbf{r}_4 = \mathbf{r}_4 - \mathbf{y} \qquad (4.11.47)$$

$$\text{new } \mathbf{r}_5 = \mathbf{r}_5 - p_1\mathbf{y} \qquad (4.11.48)$$

$$\text{new } \mathbf{r}_6 = \mathbf{r}_6 - p_2\mathbf{y}. \qquad (4.11.49)$$

Note that s, defined in (4.11.44) and used in (4.11.42), does not occur in the actual computations (4.11.46) to (4.11.49) and need not be computed. Similar computations are carried out on columns 4 to 6 for the right transformation $\mathbf{H}_4$ in (4.11.38). The matrix $\mathbf{A}_r^{(5)}$ is the result of the complete transformation (4.11.38) and is given by the following schematic form, which uses the same convention as (4.11.37).

$$\mathbf{A}_r^{(5)} = \begin{pmatrix} a & a & a & a' & a' & a' & a & a \\ a & a & a & a' & a' & a' & a & a \\ & a & a & a' & a' & a' & a & a \\ & & \rho & \underline{a}' & \underline{a}' & \underline{a}' & \underline{a} & \underline{a} \\ & & \underline{d}_1' & \underline{a}' & \underline{a}' & \underline{a} & \underline{a} \\ & & \underline{d}_2' & \underline{a}' & \underline{a}' & \underline{a} & \underline{a} \\ & & \underline{d}_3' & a' & a' & a & a \\ & & & & & & a & a \end{pmatrix}. \qquad (4.11.50)$$

The transformation $\mathbf{H}_7$ is a special case in that it operates on two rows and columns, but is otherwise similar. After $\mathbf{H}_7$ the Hessenberg matrix $\mathbf{A}_{r+1}$ emerges. When $a_{n-1,n-2}$ becomes sufficiently small, we compute the eigenvalues α_{n-1}, α_n using (4.11.24), and deflate the $n \times n$ matrix to an $n-2 \times n-2$ matrix. If an earlier subdiagonal element converges to zero first then $\mathbf{A}$ is partitioned into two smaller Hessenberg matrices.

To compute the eigenvector corresponding to an eigenvalue α we use the method of inverse iteration (4.8.14) to (4.8.15) applied to the matrix in Hessenberg form. This involves much less arithmetic than the original full matrix. The remarks at the end of section 4.10 concerning the speed of convergence of the method of inverse iteration apply also in this case as does to formula (4.10.22) for finding the eigenvectors of the original full matrix.

The treatment of the double QR method in this section is based on Francis (1962).

4.12. MISCELLANEOUS METHODS

For simplicity we have so far considered real matrices and vectors only; however, most of the methods can be applied to complex matrices and vectors after only slight modification. Although computers can be programmed to cope with complex arithmetic very conveniently, it is best avoided when possible as it involves substantially more airthmetic. Complex arithmetic can be avoided if the matrix involved is real but this may require considerable ingenuity as the double QR method of the previous section shows. However, complex arithmetic is appropriate if the matrix involved is complex. To give an idea of the necessary modification we consider how to do this for the methods of section 4.10 concerned with symmetric matrices. Instead of symmetric matrices we deal with *Hermitian* matrices. $\mathbf{A}$ is said to be Hermitian if $\mathbf{A}^* = \mathbf{A}$, where $\mathbf{A}^*$ denotes the complex conjugate transpose of $\mathbf{A}$ [see I, § 6.7(viii)]. Instead of similarity transformations $\mathbf{P}$ satisfying the relation $\mathbf{P}^{-1} = \mathbf{P}'$, we use *unitary* transformations $\mathbf{U}$ which satisfy the relation $\mathbf{U}^{-1} = U^*$ [see I, § 6.7(x)]. The eigenvalues of a Hermitian matrix are real [see I, Theorem 7.8.1]. Householder transformations have the form $\mathbf{H} = \mathbf{I} - 2\mathbf{w}\mathbf{w}^*$. With the above modifications, all the methods of section 4.10 can be extended to Hermitian matrices.

The eigenvalue problem

$$\mathbf{A}\mathbf{x} = \alpha\mathbf{x} \qquad (4.12.1)$$

is a special case of the more general problem

$$\mathbf{A}\mathbf{x} = \alpha\,\mathbf{B}\mathbf{x} \qquad (4.12.2)$$

[see I, § 7.12]. That is, given $\mathbf{A}, \mathbf{B}$ find $\alpha, \mathbf{x}$ which satisfy (4.12.2). If $\mathbf{B}$ is non-singular [see I, Theorem 6.4.2], then this can be reduced to the form (4.12.1), because (4.12.2) is equivalent to

$$\mathbf{B}^{-1}\mathbf{A}\mathbf{x} = \alpha\mathbf{x}. \qquad (4.12.3)$$

Unfortunately (4.12.3) loses some of the useful properties which (4.12.2) may have. For example, in many applications $\mathbf{A}$ is symmetric and $\mathbf{B}$ is symmetric positive definite, but $\mathbf{B}^{-1}\mathbf{A}$ is not necessarily symmetric; in this case we use the following method: compute $\mathbf{U}$ using the Choleski method of section 4.2 so that

$$\mathbf{B} = \mathbf{U}'\mathbf{U}, \qquad (4.12.4)$$

where $\mathbf{U}$ is upper triangular. Then

$$\mathbf{V} = \mathbf{U}^{-1} \qquad (4.12.5)$$

is also upper triangular. Let

$$\mathbf{y} = \mathbf{U}\mathbf{x}, \qquad \mathbf{x} = \mathbf{V}\mathbf{y}. \qquad (4.12.6)$$

Then, using (4.12.4) to (4.12.6) we can write (4.12.2) in the form

$$\mathbf{V}'\mathbf{A}\mathbf{V}\mathbf{y} = \alpha\mathbf{y}. \qquad (4.12.7)$$

Hence

$$\mathbf{A}_1\mathbf{y} = \alpha\mathbf{y}, \qquad (4.12.8)$$

where

$$\mathbf{A}_1 = \mathbf{V}'\mathbf{A}\mathbf{V} \qquad (4.12.9)$$

is symmetric.

Thus we compute the eigenvalues and eigenvectors of $\mathbf{A}_1$ by the methods of section 4.10; the eigenvalues of $\mathbf{A}$ are the same as the eigenvalues of $\mathbf{A}_1$ and the eigenvectors of $\mathbf{A}$ are derived from the eigenvectors of $\mathbf{A}_1$ by means of (4.12.6).

The methods of direct inverse iterion of section 4.6 may also be used after a suitable modification. For example, the Rayleigh–Ritz method (4.6.19) to (4.6.22) is modified as follows: given an approximate eigenvector $\mathbf{x}$, compute

$$\alpha = \frac{\mathbf{x}'\mathbf{A}\mathbf{x}}{\mathbf{x}'\mathbf{B}\mathbf{x}}, \qquad (4.12.10)$$

solve for $\mathbf{y}$ the (nearly singular) system of linear equations:

$$(\mathbf{A} - \alpha\mathbf{I})\mathbf{y} = \mathbf{B}\mathbf{x}, \qquad (4.12.11)$$

compute [see I (7.12.5)]

$$\mathbf{x} = \mathbf{y}/(\mathbf{y}'\mathbf{B}\mathbf{y})^{1/2} \qquad (4.12.12)$$

and repeat from (4.12.10).

We have so far managed to avoid having to evaluate determinants; but as this is needed next, we indicate how it can be done most conveniently for a given matrix A. We compute L, U using (4.2.10) so that $A = LU$. Then [see I (6.10.4), (6.14.2)] we have that

$$\det A = \det LU = \det L \det U = u_{11}u_{22}\ldots u_{nn}. \qquad (4.12.13)$$

If A is symmetric positive definite, then we use the Choleski factorization of section 4.2, thus

$$A = U'U. \qquad (4.12.14)$$

It follows that

$$\det A = \det U' \det U = (u_{11}u_{22}\ldots u_{nn})^2. \qquad (4.12.15)$$

We consider next the problem of computing the eigenvalues and eigenvectors of very large matrices, for example when A is $n \times n$ where $n > 100$ or $n > 1000$. Often such matrices are sparse and have a particular structure of diagonal and codiagonal blocks which reflects the geometry of the physical problem, and which is used to store the elements of the matrix conveniently in the limited computer memory. Most of the similarity transformations used so far destroy the sparsity and special structure that make computer storage and manipulation possible. In such cases the methods of direct and inverse iterations of section 4.8 are most suitable. Since the original matrix A is not altered, it need not be stored explicitly as a matrix; all that is required is an algorithm which, given x, computes Ax, $A^{-1}x$ or $(A - hI)^{-1}x$.

Band matrices are a special form of sparse matrices which occur frequently in applications. The band is said to be of width $2k + 1$ if the elements of the matrix are zero except along the diagonal, k upper and k lower codiagonals, thus a tridiagonal matrix is a band matrix of width 3, but we are concerned with cases where for example $n = 1000$ and $k = 50$. If A_r is a symmetric positive definite band matrix of width $2k + 1$, then we can use the following combination of the Choleski and LR methods to compute the eigenvalues most conveniently. Let

$$A_r = U'U; \qquad (4.12.16)$$

then we deduce from the observation at the end of section 4.2 that U is an upper codiagonal matrix of width k. We observe that, as in the case of the LR method,

$$A_{r+1} = UU' \qquad (4.12.17)$$

is also symmetric positive definite, similar to A_1 and a band matrix of the same width $2k + 1$. As in the case of the LR method, it can be shown that

$$A_r \to \operatorname{diag}(\alpha_1, \alpha_2, \ldots, \alpha_n) \qquad (4.12.18)$$

where α_i are the real positive [see I, Theorem 7.9.1] eigenvalues of $A = A_1$ arranged in descending order. The order of convergence is like the order for the LR and QD methods of section 4.10, starting with α_n which is obtained as the limit of a_{nn}. The shift of origin and deflation refinements are also applicable.

We describe next a method for accelerating convergence further which is applicable only when $\mathbf{A}$ is symmetric positive definite. Let α be a root of the equation

$$f(x) = 0, \tag{4.12.19}$$

and let h_r, h_{r-1} be two reasonably good approximations to α. Then the secant or modified Newton method [see § 5.3.3] gives the following improved approximation to α:

$$h_{r+1} = \frac{h_{r-1}f_r - h_rf_{r-1}}{f_r - f_{r-1}}, \tag{4.12.20}$$

where $f_i = f(h_i)$. Let

$$f(h) = \det(\mathbf{A} - h\mathbf{I}) \tag{4.12.21}$$

and let

$$\mathbf{A} - h\mathbf{I} = \mathbf{U}'\mathbf{U} \tag{4.12.22}$$

be the Choleski factorization of $\mathbf{A} - h\mathbf{I}$, where

$$h < \alpha_n. \tag{4.12.23}$$

Then $\mathbf{A} - h\mathbf{I}$, like $\mathbf{A}$, is symmetric positive definite, and we can use (4.12.15) to compute the value of $f(h)$ in (4.12.21). We apply shifts of origin h_r to $\mathbf{A}$, so that (4.12.16) and (4.12.17) become

$$\mathbf{A}_r - h_r\mathbf{I} = \mathbf{U}'\mathbf{U} \tag{4.12.24}$$

$$\mathbf{A}_{r+1} = \mathbf{U}\mathbf{U}' + h_r\mathbf{I}. \tag{4.12.25}$$

For the next application of (4.12.24) and (4.12.25) we use a shift of origin h_{r+1} computed from (4.12.20) where f_r is computed by using (4.12.15).

The formula (4.12.20) is applicable to any reasonable function, but if, further, we assume that $f(x)$ is a polynomial and if α is the smallest root of the equation (4.12.19), then a graphical representation of the secant method will show that if

$$h_{r-1} < h_r < \alpha, \tag{4.12.26}$$

then

$$h_r < h_{r+1} < \alpha. \tag{4.12.27}$$

Hence h_{r+1} satisfies (4.12.23) and the Choleski factorization can be carried out in the next round of the application of (4.12.24) and (4.12.25), and we have the monotone increasing sequence

$$\ldots < h_{r-1} < h_r < h_{r+1} < \ldots < \alpha_n \tag{4.12.28}$$

with α_n as upper bound.

Since we could be dealing with very large band matrices we use this additional refinement to cut down the number of iterations. The method requires two reasonable estimates h_1, h_2 which satisfy the inequality

$$h_1 < h_2 < \alpha_n \qquad (4.12.29)$$

before we can start applying (4.12.20).

The combined effect of the various refinements will, in most cases, result in an average of two iterations per eigenvalue except possibly for α_n which is the first eigenvalue to be computed and which may need 3 or 4 iterations. The remaining eigenvalues $\alpha_{n-1}, \alpha_{n-2}, \ldots, \alpha_1$ computed in that order should require fewer iterations. In any case we need at least two iterations to apply (4.12.20). Deflation results in smaller and smaller matrices; when computing α_i we will be dealing with band matrices of dimension i.

I.M.K.

4.13. REFERENCES

Wilkinson (1965) is the authoritative work in this area, it lays particular importance on error analysis. For an introduction see Noble (1969) and Stewart (1973).

There are several well documented collections of algorithms maintained by computer manufacterers and large research laboratories. The documentation often includes a wealth of practical information. The following are also recommended: Wilkinson and Reinsch (1971); the collection of algorithms maintained by ACM, the American Association for Computing Machinery; and the collection of algorithms published by ICL and maintained by NAG Ltd. supported by the U.K. Computer Board for Universities and Research Councils.

BIBLIOGRAPHY

Bodewig, E. (1959). *Matrix Calculus*, North Holland Publishing Company, Amsterdam.

Francis, J. G. F. (1961, 1962). The QR Transformation, Parts I and II, *Computer J.* **4**, 265–271, 332–345.

Householder, A. (1957). A Survey of Some Closed Methods for Inverting Matrices, *J. Soc. Indust. Appl. Math.* **5**, 155–169.

Kron, G. (1963). *Diakoptics: the piecewise solution of large-scale systems*, MacDonald, London.

Noble, B. (1969). *Applied Linear Algebra*, Prentice Hall, Inc., Englewood Cliffs, N.J.

Stewart, G. W. (1973). *Introduction to Matrix Computation*, Academic Press, New York.

Wilkinson, J. H. (1963). *Rounding Errors in Algebraic Processes*, Notes on Applied Science No. 32, Her Majesty's Stationery Office, London.

Wilkinson, J. H. (1965). *The Algebraic Eigenvalue Problem*, Clarendon Press, Oxford.

Wilkinson, J. H. and Reinsch, C. (1971). *Handbook of Automatic Computation, Vol. II, Linear Algebra*, Springer-Verlag, Berlin.

PROGRAM

In the appendix there is a listing and example of a program (No. 5) to find the largest eigenvalue and associated eigenvector by the Power Method.

CHAPTER 5

Non-Linear Equations

5.1. INTRODUCTION

In Chapter 3 methods for the solution of systems of linear equations were described. The methods were of two kinds: direct and indirect; the former have the advantage that they will always find the solution (if one exists) but from a computational point of view the indirect methods are often more convenient, provided that convergence to the solution can be ensured.

For the equations to be considered in this chapter direct methods of solution will almost never be available. The simplest type of non-linear equation is a polynomial in a single variable [see I, § 14.1], viz.

$$a_n x^n + a_{n-1} x^{n-1} + \ldots + a_0 = 0 \quad (a_n \neq 0) \qquad (5.1.1)$$

(which is non-linear if $n \geq 2$) where the coefficients $a_n, a_{n-1}, \ldots, a_0$ are given. When $n \geq 5$ there is no formula for solving (5.1.1) and although formulae exist when $n = 3$ or $n = 4$ it is debatable whether or not they are worth using. When $n = 2$ the formula is, of course, well-known [see I (14.5.1)] and is

$$x = \frac{-a_1 \pm \sqrt{a_1^2 - 4a_0 a_2}}{2a_2}.$$

More generally a *non-linear equation* in a *single* variable is any equation of the type

$$f(x) = 0$$

where $f(x)$ is *not* expressible in the form

$$f(x) = ax + b$$

where a, b do not depend on x. The following are examples of non-linear equations:

(i) $x^2 + x - 1 = 0$
(ii) $x^3 - 8x - 10 = 0$
(iii) $2x^7 - 5x^4 + 1 = 0$
(iv) $3 \cos x + 5 \cos 3x - 7 \cos 5x = 0$

135

(v) $2x = e^{-x}$

(vi) $x \log x + 4x^2 = \sin x$.

For these equations direct methods of solution exist only for (i) and (ii); (iii) is a polynomial of degree >4 and (iv) can be transformed into a polynomial (of degree 5) by writing $t = \cos x$, since $\cos 3x$ and $\cos 5x$ can be expressed as polynomials in $\cos x$ [see V, § 1.2.3]. Equations (v) and (vi) are transcendental equations and cannot be transformed into polynomials. All these equations however can be solved by the methods to be described in this chapter. These methods are all *iterative* in nature and are of very wide applicability, but before we investigate them a brief look at iterative methods in general is desirable.

5.2. ITERATIVE METHODS

An iterative method for solving an equation involving one unknown, x, is one in which a sequence of approximations $\{x_n\}$ to the solution are computed using a formula of the type

$$x_n = F_n(x_0, x_1, \ldots, x_{n-1}). \tag{5.2.1}$$

The process is terminated when some criterion is satisfied, the most common criterion being

$$|x_n - x_{n-1}| < \varepsilon \tag{5.2.2}$$

for some pre-assigned number ε (such as $\frac{1}{2} \times 10^{-8}$, say).

In practice it is unlikely that the function $F_n(\cdot)$ on the right-hand side of (5.2.1) will involve more than 2 of the previous members of the sequence and these are likely to be x_{n-1} and x_{n-2} so that in effect the formulae used are likely to be of the form

$$x_n = F_n(x_{n-1}) \tag{5.2.3}$$

or

$$x_n = F_n(x_{n-1}, x_{n-2}). \tag{5.2.4}$$

If such a method is to be used there are 2 relevant questions: will the method converge to the solution of the equation [see I, § 1.2] and, if so, how rapidly?

Before we consider these questions in general let us look at an example.

EXAMPLE 5.2.1.   Use the iterative procedure

$$x_{n+1} = \frac{1}{2}\left(x_n + \frac{10}{x_n}\right)$$

starting with $x_0 = 1$ and stopping when $|x_n - x_{n-1}| < \frac{1}{2} \times 10^{-5}$.

Solution.   We carry 6 decimals at every stage so that the convergence criterion can be tested adequately.

Then we find that

$$x_0 = 1 \cdot 000000$$

$$x_1 = 5 \cdot 500000$$

$$x_2 = 3 \cdot 659091$$

$$x_3 = 3 \cdot 196005$$

$$x_4 = 3 \cdot 162456$$

$$x_5 = 3 \cdot 162278$$

$$x_6 = 3 \cdot 162278$$

and convergence to the required accuracy has occurred.

The value so found is $\sqrt{10}$ to 6 places of decimals, and it is a simple matter to verify that if the iterative procedure of the example converges, then it converges to $\sqrt{10}$. For if $x_{n+1} = x_n$ we have

$$x_n = \frac{1}{2}\left(x_n + \frac{10}{x_n}\right)$$

so that

$$\tfrac{1}{2}x_n = \frac{5}{x_n}.$$

That is, $x_n^2 = 10$ and so $x_n = \pm\sqrt{10}$, but since we started with $x_0 = +1$ all subsequent values will be positive and convergence will therefore be to $+\sqrt{10}$.

This iterative procedure is a special case of the *Newton–Raphson method*, which we shall discuss in section 5.4.1.

That not every iterative procedure converges is shown by the following example.

EXAMPLE 5.2.2.   Use the iterative procedure

$$x_{n+1} = x_n^2 - 1$$

(i) starting with $x_0 = 1$;
(ii) starting with $x_0 = 2$; in both cases continuing for 4 steps.

Solution

(i) $x_0 = 1$, $x_1 = 0$, $x_2 = -1$, $x_3 = 0$, $x_4 = -1$.
(ii) $x_0 = 2$, $x_1 = 3$, $x_2 = 8$, $x_3 = 63$, $x_4 = 3968$.

It is clear that we do not have convergence in either case although there is a solution to the equation $x = x^2 - 1$ which lies between 1 and 2 ($x \doteqdot 1 \cdot 61803$). Even if we begin with a starting value which is very close to the solution, such as $x_0 = 1 \cdot 618$, this procedure will not converge; the reason for this will become clear in section 5.5.

5.3. TWO-POINT ITERATIVE METHODS

Iterative methods in which each successive approximation is based upon the values of two earlier approximations are called *two-point methods*. Three such methods will be described; the first is:

5.3.1. The Method of Bisection

Suppose that we wish to solve

$$f(x) = 0 \qquad (5.3.1)$$

where $f(x)$ is some known function which is continuous and which is computable over some interval $[a, b]$. Suppose also that there are two numbers x_1, x_2 where $a \leq x_1 < x_2 \leq b$ such that $f(x_1)$ and $f(x_2)$ have opposite signs. It follows that there is at least one solution of (5.3.1) in the interval $[x_1, x_2]$. If we therefore regard x_1 and x_2 as two successive approximations to the solution how shall we find the next approximation, x_3? In the method of bisection we take

$$x_3 = \tfrac{1}{2}(x_1 + x_2). \qquad (5.3.2)$$

This will give us an approximation which is certain to be closer to the solution than at least one of x_1, x_2; we cannot be sure that it is closer than both, though it sometimes will be.

Having found the value of x_3 by using (5.3.2) we must now answer two questions:

(i) is x_3 sufficiently close to the solution of (5.3.1)?
if not,
(ii) how do we obtain our next approximation, x_4?

The answer to (i) will generally be found by applying some criterion; this might either be of the type (5.2.2) or of the type that we accept x_3 as the solution if

$$|f(x_3)| < \varepsilon \qquad (5.3.3)$$

where ε is, as before, some predetermined (small) number.

If the answer to (i) is 'Yes' then we accept x_3 as the solution; if it is 'No' we choose

$$x_4 = \tfrac{1}{2}(x_1 + x_3) \quad \text{if } f(x_3)f(x_1) < 0 \qquad (5.3.4)$$

$$x_4 = \tfrac{1}{2}(x_2 + x_3) \quad \text{if } f(x_3)f(x_2) < 0. \qquad (5.3.5)$$

Note that one, and only one, of (5.3.4) and (5.3.5) will be true, since $f(x_1)f(x_2) < 0$.

Having found x_4 the process is repeated to produce $x_5, x_6, \ldots$ until a value x_n, is found which is considered to be sufficiently close to the solution. The method is illustrated by the following example.

EXAMPLE 5.3.1.   Find the cube root of 2 to 4 d.p. by the method of bisection.

Solution. $2^{1/3}$ is the root of $x^3 - 2 = 0$ so we take $f(x) = x^3 - 2$. For 4 d.p. accuracy [see Definition 1.3.2] we set our criterion that we will stop when $|x_n - x_{n-1}| < \frac{1}{2} \times 10^{-4}$.
   ·Since $f(1) = -1$ and $f(2) = +6$ we take $x_1 = 1$, $x_2 = 2$ and bisection then gives $x_3 = 1 \cdot 5$. Since $f(x_3) = (1 \cdot 5)^3 - 2 = +1 \cdot 375$ we take

$$x_4 = \tfrac{1}{2}(x_3 + x_1) = 1 \cdot 25.$$

Now $f(x_4) = (1 \cdot 25)^3 - 2 = -0 \cdot 046875$ and so we take

$$x_5 = \tfrac{1}{2}(x_4 + x_3) = 1 \cdot 375$$

and so on. The table below shows the values obtained for the x_n to 5 d.p. from $n = 3$,

n	x_n	$f(x_n)$
3	1·5	+1·37500
4	1·25	−0·04688
5	1·375	+0·59961
6	1·3125	+0·26099
7	1·28125	+0·10330
8	1·26563	+0·02729
9	1·25782	−0·01001
10	1·26172	+0·00859
11	1·25977	−0·00071
12	1·26075	+0·00393
13	1·26026	+0·00160
14	1·26001	+0·00044
15	1·25989	−0·00014
16	1·25996	+0·00014
17	1·25992	−0·00000

Since $|x_{16} - x_{17}| \doteqdot 0 \cdot 00004 < \frac{1}{2} \times 10^{-4}$ the convergence criterion is satisfied. We take their mean value (1·25994) as the solution, or to 4 d.p., 1·2599. In this case x_{17}, as might be expected from the fact that $f(x_{17})$ is so small, is in fact correct to 5 d.p.
   In the example we saw that convergence to the solution occurs, but rather slowly, 15 iterations being required for convergence to 4 d.p. It is not difficult to see that slow convergence is a characteristic of this method and indeed we can even estimate how many iterations will be required to produce convergence to any given accuracy. For suppose that $|x_1 - x_2| = d$, then the solution lies in an interval of length d. After bisection we will have an interval of length $\frac{1}{2}d$ containing the solution, and after bisection again the interval will be of length $\frac{1}{4}d$. Continuing in this way we see that when we have reached the value x_n we

will have an interval of length

$$\frac{d}{2^{n-2}}$$

which contains the solution. If we want a solution which is accurate to k places of decimals it must lie in an interval of length at most $\frac{1}{2} \times 10^{-k}$. It therefore follows that this will occur when

$$\frac{d}{2^{n-2}} < \frac{1}{2} \times 10^{-k} \tag{5.3.6}$$

that is [using I (3.6.1)], when

$$n > 3 + \log_2 (d \times 10^k).$$

In the example above $d = 1$ and $k = 4$ so that convergence could be expected when

$$n > 3 + \log_2 (10{,}000) \doteqdot 16 \cdot 29$$

that is, $n \geq 17$, in agreement with our result.

5.3.2.   The Method of False Position

In Example 5.3.1 note the values of $|f(x_3)|$ and $|f(x_4)|$, the former is nearly 30 times greater than the latter. This indicates that the solution to the equation probably lies much closer to $x_4(1 \cdot 25)$ than to $x_3(1 \cdot 5)$ but the method of bisection makes no allowance for this. This partly explains the slow convergence of the method: it ignores vital information. Can we improve the rate of convergence by taking information of this kind into account?

Consider the case just mentioned. The ratio of $|f(x_3)|$ to $|f(x_4)|$ is 29·33 to 2 d.p., so instead of taking x_5 to have the value half way between x_3 and x_4 let us take it to be (say) 29/30ths of the way between x_3 and x_4, so that

$$x_5 = \frac{x_3 + 29x_4}{30} = \frac{1 \cdot 5 + 29 \times (1 \cdot 25)}{30} = 1 \cdot 2583 \text{ (to 4 d.p.).}$$

This is a much better estimate than the value for x_5 previously obtained, indeed it is better than x_6, x_7, x_8, x_9 and x_{10}. We therefore look a little more closely at this method.

The process which we have just carried out is a simple case of *inverse linear interpolation* which we have already discussed in Chapter 2. We may briefly re-state the method as follows: if a continuous function $f(x)$ takes values $f(x_1)$ and $f(x_2)$ at two points x_1, x_2 where $f(x_1)/f(x_2) < 0$ then $f(x) = 0$ at some point x_3 lying between x_1 and x_2 where

$$x_3 \doteqdot \frac{x_1 f(x_2) - x_2 f(x_1)}{f(x_2) - f(x_1)}. \tag{5.3.7}$$

In the example above $x_1 = 1 \cdot 5$, $x_2 = 1 \cdot 25$, $f(x_1) = 1 \cdot 375$ and $f(x_2) = -0 \cdot 04688$ so that, from (5.3.7)

$$x_3 = 1 \cdot 25824.$$

The method defined by (5.3.7) is known as *the Method of False Position*. It has been used for many centuries and is still sometimes known by its original Latin name of '*Regula Falsi*'.

Having found a new approximation, x_3, from x_1, x_2 by means of (5.37) we would then proceed to find a new approximation x_4 by using either

$$\text{(i)} \quad x_3 \text{ and } x_1 \text{ if } f(x_3) f(x_1) < 0 \qquad\qquad (5.3.8)$$

or

$$\text{(ii)} \quad x_3 \text{ and } x_2 \text{ if } f(x_3) f(x_2) < 0 \qquad\qquad (5.3.9)$$

and so on until a suitably accurate value x_n has been obtained.

We now rework Example 5.3.1 using the method of false position.

EXAMPLE 5.3.2.   Find the cube root of 2 to 4 d.p. by the method of false position.

Solution.   As in Example 5.3.1 we take $f(x) = x^3 - 2$, $x_1 = 1$, $x_2 = 2$ and will stop when $|x_n - x_{n-1}| < \frac{1}{2} \times 10^{-4}$. The results are shown below, to 5 d.p.

n	x_n	$f(x_n)$
1	1	$-1 \cdot 00000$
2	2	$+6 \cdot 00000$
3	$1 \cdot 14286$	$-0 \cdot 50729$
4	$1 \cdot 20968$	$-0 \cdot 22985$
5	$1 \cdot 23884$	$-0 \cdot 09873$
6	$1 \cdot 25116$	$-0 \cdot 04143$
7	$1 \cdot 25630$	$-0 \cdot 01721$
8	$1 \cdot 25843$	$-0 \cdot 00709$
9	$1 \cdot 25931$	$-0 \cdot 00291$
10	$1 \cdot 25967$	$-0 \cdot 00120$
11	$1 \cdot 25982$	$-0 \cdot 00048$
12	$1 \cdot 25988$	$-0 \cdot 00020$
13	$1 \cdot 25990$	$-0 \cdot 00010$

and since $|x_{13} - x_{12}| = 0 \cdot 00002 < \frac{1}{2} \times 10^{-4}$ we accept $1 \cdot 2599$ as the value of $2^{1/3}$ to 4 d.p.

In this case only 11 iterations were required to converge to 4 d.p. compared to 15 for the method of bisection. Had we started with $x_1 = 1 \cdot 5$, $x_2 = 1 \cdot 25$ the improvement over the method of bisection would have been even more striking for whereas that method as we saw in Example 5.3.1 required a further 13 iterations to converge the method of false position reaches the same accuracy in only 4 iterations as the table below shows

n	x_n	$f(x_n)$
1	1·5	1·37500
2	1·25	−0·04688
3	1·25824	−0·00799
4	1·25964	−0·00134
5	1·25987	−0·00024
6	1·25991	−0·00005

and since $|x_6 - x_5| < \frac{1}{2} \times 10^{-4}$ convergence has occurred.

These examples indicate that the method of false position is probably superior to the method of bisection in that it converges faster but there is no general rule which will tell us how much faster, for this depends in a fairly complex way upon the behaviour of the solution. In particular if $f(x)$ is convex between x_1 and x_2 [see IV, § 5.2.6] then all subsequent approximations to the solutions will be on one side of it. This happened in the example above, as can be seen from the fact that $f(x_i)$ is negative from a certain point onwards; under these circumstances convergence will generally be slow. The diagram below shows why this happens.

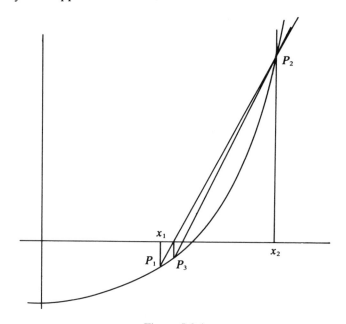

Figure 5.3.1

The shape of the graph $y = x^3 - 2$ between $x = x_1$ and $x = x_2$ is such that $x_3, x_4, \ldots$ all lie to the left of the solution and the point $P_2(x_2, f(x_2))$ is always one of the points used to compute each new approximation. It can be proved that this leads to slow convergence.

5.3.3. The Secant Method

We have just seen that the method of false position will converge to the solution from one side in certain (very common) cases and this leads to slow convergence. A variation on the method, known as the *Secant method* avoids this situation.

In the secant method we do not insist on using 2 points which lie on either side of the solution; in other words we are prepared to use inverse *extrapolation* or inverse *interpolation*, as required. The description of the secant method is therefore:

If x_{n-1}, x_n are 2 successive approximations to the solution of $f(x) = 0$ then choose

$$x_{n+1} = x_n - \frac{(x_n - x_{n-1})}{(f(x_n) - f(x_{n-1}))} f(x_n) \qquad (5.3.10)$$

as the next approximation. If $|x_{n+1} - x_n|$ is sufficiently small stop, and take x_{n+1} as the solution; otherwise repeat (5.3.10) using x_{n+1} in place of x_n and x_n in place of x_{n-1} to give a new approximation x_{n+2}.

EXAMPLE 5.3.3.   Repeat Example 5.3.2 starting with $x_1 = 1$, $x_2 = 2$ using the secant method.

Solution

n	x_n	$f(x_n)$
1	1	$-1 \cdot 00000$
2	2	$+6 \cdot 00000$
3	$1 \cdot 14286$	$-0 \cdot 50729$
4	$1 \cdot 20968$	$-0 \cdot 22985$
5	$1 \cdot 26504$	$+0 \cdot 02448$
6	$1 \cdot 25971$	$-0 \cdot 00100$
7	$1 \cdot 25992$	$-0 \cdot 00001$
8	$1 \cdot 25992$	$-0 \cdot 00001$

and the sequence has converged to 4 d.p.

As a further comparison of the two methods we use them to find that root of the quartic

$$x^4 + 3x^3 - 2x^2 - 12x - 8 = 0$$

which lies between 1 and 3. Starting with $x_1 = 1$ and $x_2 = 3$ the subsequent values are shown, in part, below

n	x_n (False Position)	x_n (Secant)
1	1·00000	1·00000
2	3·00000	3·00000
3	1·30508	1·30508
4	1·55749	1·55749
5	1·73821	2·69607
6	1·85310	1·78698
7	1·92028	1·90560
8	1·95758	2·02021
9	1·97767	1·99834
10	1·98831	1·99997
11	1·99390	2·00000
12	1·99682	2·00000

and the secant method has converged to the root, which is exactly 2, whereas the method of false position does not reach 5 figure accuracy until $n = 23$. Note that the estimates are the same up to $n = 4$ and that the secant method results are the inferior of the two sets until $n = 8$ after which it converges rapidly in contrast to the slow convergence of the other method.

In view of results such as these it might seem that the secant method should always be used in preference to the method of false position, but this is not so, for the former may fail to converge at all whereas the latter will always converge.

Computer programs for the two methods are almost identical; for the secant method the test to see if $f(x_n)f(x_{n-1}) < 0$ is suppressed. Consequently it is quite feasible to combine the two programs into one and arrange that the secant method is followed so long as the process seems to be converging but that if the process produces large changes in the sequence of approximate solutions then the program switches to using the method of false position.

Fortran programs for these two methods will be found in the Appendix.

There is one final point which should be mentioned. It is important to use the secant method in the form given by (5.3.10), viz.

$$x_{n+1} = x_n - \frac{(x_n - x_{n-1})f(x_n)}{(f(x_n) - f(x_{n-1}))}$$

rather than in the apparently simpler equivalent form

$$x_{n+1} = \frac{x_{n-i}f(x_n) - x_n f(x_{n-i})}{f(x_n) - f(x_{n-1})} \tag{5.3.11}$$

because as x_n approaches the solution, $f(x_n)$ and $f(x_{n-1})$ will both be close to zero and (5.3.10) will be numerically more unstable than (5.3.11).

5.4.   ONE-POINT METHODS

Iterative methods of the type

$$x_{n+1} = F(x_n) \tag{5.4.1}$$

are called *one-point methods*; they constitute a particularly important class and have been studied extensively. One-point methods are often to be preferred to two-point methods from a computational point of view provided that convergence can be ensured. The most celebrated one-point method, the Newton–Raphson, is a limiting form of the secant method so we shall discuss this first.

5.4.1.   The Newton–Raphson Method

In (5.3.10) let us put $x_{n-1} = x_n - h$, then $x_n - x_{n-1} = h$ and $f(x_{n-1}) = f(x_n - h) \doteqdot f(x_n) - hf'(x_n)$. If we now let $h \to 0$ we have, in the limit in place of (5.3.10)

$$x_{n+1} = x_n - \frac{f(x_n)}{f'(x_n)} \tag{5.4.1}$$

which is a one-point formula according to the definition above.

The formula defined by (5.4.1) is known as the *Newton–Raphson Formula*; it has a very simple graphical interpretation. In Figure 5.4.1 below the curve represents the function $y = f(x)$

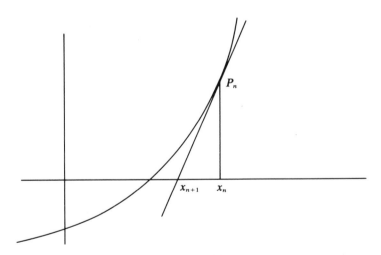

Figure 5.4.1

P_n is the point on the curve $(x_n, f(x_n))$. The tangent to the curve at P_n cuts the line $y = 0$ at x_{n+1}; then x_{n+1} is the new approximation to the solution given by the Newton–Raphson formula (5.4.1).

EXAMPLE 5.4.1.   Use the Newton–Raphson method to find the cube-root of 2 to 4 d.p. starting with $x_1 = 1$.

Solution.   As in Example 5.3.1 we note that $2^{1/3}$ satisfies $x^3 - 2 = 0$ so we take $f(x) = x^3 - 2$; then $f'(x) = 3x^2$ and 5.4.1 gives us the iterative formula

$$x_{n+1} = x_n - \frac{(x_n^3 - 2)}{3x_n^2}$$

or, more conveniently,

$$x_{n+1} = \frac{2}{3}\left(\frac{x_n^3 + 1}{x_n^2}\right).$$

Starting with $x_1 = 1$ the iteration produces

n	x_n
1	1·00000
2	1·33333
3	1·26389
4	1·25993
5	1·25992

and x_5 is correct to 4 d.p. (in fact it is correct to 5 d.p.). Comparison with the two-point methods shows that convergence is significantly faster than any of them.

So far we have only used the methods of this chapter to solve polynomials, but they are just as useful for solving other equations, as we now illustrate, first with a trigonometric equation (which could be transformed into a polynomial) and then with a transcendental equation.

EXAMPLE 5.4.2.   Use the Newton–Raphson method to solve $\sin 3x = \cos 2x$ starting with $x_1 = 0$ and working to 5 d.p.

Solution.   We take $f(x) = \sin 3x - \cos 2x$ so that $f'(x) = 3 \cos 3x + 2 \sin 2x$ and the Newton–Raphson formula is

$$x_{n+1} = x_n - \frac{(\sin 3x_n - \cos 2x_n)}{(3 \cos 3x_n + 2 \sin 2x_n)}.$$

The results are as follows

n	x_n
1	0·00000
2	0·33333
3	0·31388
4	0·31416
5	0·31416

The result is correct to 5 d.p. as we can verify easily since the solution of the equation is $x = \pi/10$ ($\doteq 0.314159$).

EXAMPLE 5.4.3.   Solve $x = e^{-x}$ to 5 d.p. starting with $x_1 = 0$ and using the Newton–Raphson method.

Solution.   We take $f(x) = x - e^{-x}$, so $f'(x) = 1 + e^{-x}$ and the iterative formula is

$$x_{n+1} = x_n - \left(\frac{x_n - e^{-x_n}}{1 + e^{-x_n}} \right)$$

which leads to

n	x_n
0	0·000000
1	0·500000
2	0·566311
3	0·567143
4	0·567143

and the result is already correct to 5 d.p.

Before turning to the general question of convergence of the Newton–Raphson method we show, by means of an example that there are cases for which it may fail to converge.

EXAMPLE 5.4.4.   Find the solution to 4 d.p. of

$$e^{-x^2} = \frac{x}{1 + x^2}$$

using the Newton–Raphson method

   (i) starting with $x_1 = -1$;
   (ii) starting with $x_1 = 0$.

Solution.   We take $f(x) = e^{-x^2} - x/(1 + x^2)$ so that

$$f'(x) = -2x\, e^{-x^2} - \frac{1}{(1 + x^2)} + \frac{2x^2}{(1 + x^2)^2}$$

and the calculations for case (i) then give

n	x_n	$f(x_n)$	$f'(x_n)$
1	−1·00000	+0·86788	+0·73576
2	−2·17957	+0·38767	+0·15111
3	−4·74505	+0·20178	+0·03891
4	−9·93086		

It is apparent that we do not have convergence in this case, or, to put it another way, the sequence $x_1, x_2, x_3, \ldots$ is decreasing rapidly to a limit of $-\infty$, which *is* a zero of $f(x)$, though not the one in which we are interested.

For case (ii) the results are

n	x_n	$f(x_n)$	$f'(x_n)$
1	0·00000	1·00000	−1·00000
2	1·00000	−0·13212	−0·73576
3	0·82043	+0·01976	−0·95382
4	0·84115	+0·00025	−0·92944
5	0·84142	0·00000	

and the process has converged, the solution being 0·8414 to 4 d.p.

The reason why the process fails to converge in case (i) but succeeds in case (ii) becomes clear if we look at the graph of $f(x)$. Evaluation of $f(x)$ and $f'(x)$ at $x = -2, -1, 0, +1$ and $+2$ gives the values (to 4 d.p.)

$x =$	−2	−1	0	1	2
$f(x) =$	0·4183	0·8679	1·0000	−0·1321	−0·3817
$f'(x) =$	0·1933	0·7358	−1·0000	−0·7358	+0·0467

Table 5.4.1

and these values show that $f(x)$ has a maximum at a point between −1 and 0 and a minimum at a point just below 2, in addition we can see that there is a zero between 0 and 1. Also $f(x) \to 0$ as $x \to +\infty$ and as $x \to -\infty$ and the graph of $y = f(x)$ is as shown in Figure 5.4.2.

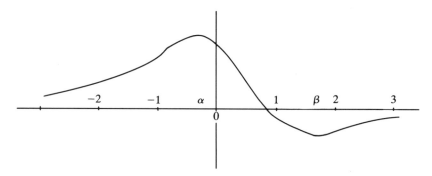

Figure 5.4.2

If α denotes the point between −1 and 0 at which $f(x)$ has its maximum value then convergence of the Newton–Raphson method will not occur if we choose as starting value a number $x_1 < \alpha$. For if $x_1 < \alpha$ and P_1 is the point $(x_1, f(x_1))$ on the curve $y = f(x)$ the tangent at P_1 meets the line $y = 0$ in a point $x_2 < x_1$ and the successive approximations $x_3, x_4, \ldots$ will then move further and further to the left.

A similar problem arises if we choose $x_1 > \beta$ where β is the value at which $f(x)$ has a minimum. In this case the sequence of approximations $x_2, x_3, \ldots$ would move further and further to the right approaching the zero of the function at $x = +\infty$. Thus if we take $x_1 = 2$, then our next approximation is

$$x_2 = x_1 - \frac{f(x_1)}{f'(x_1)}$$

$$= 2 + \frac{0 \cdot 3817}{0 \cdot 0467} \doteqdot 10 \cdot 173$$

from Table 5.4.1.

We can prove these assertions by noting that for large values of x, e^{-x^2} is sufficiently small to be ignored so that we have

$$f(x) \doteqdot \frac{-x}{1+x^2} \sim \frac{-1}{x} \quad \text{as } x \to \pm\infty$$

and

$$f'(x) \doteqdot \frac{x^2-1}{(x^2+1)^2} \sim \frac{1}{x^2} \quad \text{as } x \to \pm\infty$$

and the iterative formula

$$x_{n+1} = x_n - \frac{f(x_n)}{f'(x_n)}$$

becomes, approximately

$$x_{n+1} \doteqdot x_n + x_n = 2x_n.$$

We see therefore that in this case the Newton–Raphson method will certainly not converge unless we choose x to lie in the interval (α, β), but a glance at the graph will indicate that even if x_1 lies in this interval it is possible that x_2 will lie outside it, in which case convergence will still not occur. The condition $\alpha < x_1 < \beta$ is necessary but not sufficient for convergence, and in fact in order to ensure convergence x_1 must be chosen to lie in a certain sub-interval (γ, δ) of (α, β), that is we must have

$$\alpha < \gamma < x_1 < \delta < \beta.$$

The values of $\alpha, \beta, \gamma, \delta$ are approximately $-0 \cdot 38, -0 \cdot 13, 1 \cdot 47, 1 \cdot 84$ respectively as may be verified with a suitable computer program.

5.4.2. Convergence of the Newton–Raphson Method

The Newton–Raphson method is widely used for the solution of non-linear equations despite its two disadvantages, viz.

(i) it may not converge;
(ii) it is necessary to find $f'(x)$ and if $f(x)$ is a complicated function this may be a very tedious task.

Its advantage is, of course, that if convergence does occur it will be very rapid. The convergence of iterative methods such as (5.4.1) may be studied by making use of Taylor series [see IV, § 3.6]. If we let α be the exact solution of $f(x) = 0$ and suppose that

$$x_n = \alpha + \varepsilon_n, \qquad x_{n+1} = \alpha + \varepsilon_{n+1} \tag{5.4.2}$$

(so that ε_n, ε_{n+1} are the errors in our nth and $(n+1)$st approximations to the solution) then it is not difficult to prove that for the Newton–Raphson Method if ε_n is sufficiently small and $f'(\alpha) \neq 0$ then

$$\varepsilon_{n+1} \doteqdot \frac{\varepsilon_n^2 f''(\alpha)}{2f'(\alpha)}. \tag{5.4.3}$$

If $f'(\alpha) = 0$, $\varepsilon_{n+1} = A\varepsilon_n$ where $|A| < 1$, so convergence occurs, but slowly. In the example above $\alpha \doteqdot 0{\cdot}8414$, $f'(\alpha) \doteqdot -0{\cdot}9291$ and $f''(\alpha) \doteqdot 1{\cdot}5923$ so that once we are sufficiently close to the root the errors at consecutive stages are related approximately by

$$\varepsilon_{n+1} \doteqdot -0{\cdot}86\varepsilon_n^2. \tag{5.4.4}$$

If we look at x_2 and x_3 in case (ii) above we see that we have

$$x_2 = 1{\cdot}00000 = 0{\cdot}84142 + 0{\cdot}15858$$

$$x_3 = 0{\cdot}82043 = 0{\cdot}84142 - 0{\cdot}02099$$

so that $\varepsilon_2 = 0{\cdot}15858$ and $\varepsilon_3 = -0{\cdot}02099$ and therefore, in this case

$$\varepsilon_3 \doteqdot -0{\cdot}83\varepsilon_2^2$$

in good agreement with (5.4.4).

The relationship between ε_{n+1} and ε_n given by (5.4.3) is only valid when ε_n is 'sufficiently small'; whilst it is impossible to be more precise on the meaning of the phrase in general it may be interpreted as meaning that when x_n, x_{n+1} are sufficiently close to the solution for the process to be converging the relationship (5.4.3) is valid.

5.4.3. Order of Convergence of Iterative Methods

From (5.4.3) we see that the error at any stage of the Newton–Raphson method is proportional to the *square* of the error at the previous stage. For other methods the relationship between consecutive errors will not necessarily be of this form but if an iterative procedure has the property that the errors at the $(n + 1)$st and nth stages are related by an expression of the form

$$\varepsilon_{n+i} \doteqdot A\varepsilon_n^p \qquad (5.4.5)$$

then we say that the iterative procedure *converges with power p*.

The higher the value of p the faster the convergence. We have just seen that $p = 2$ for the Newton–Raphson method and we saw in (5.3.1) that $p = 1$ for the method of bisection. It can be proved that $p = 1$ for the method of false position and $p \doteqdot 1 \cdot 618$ for the secant method.

The fact that $p = 2$ for the Newton–Raphson method can be interpreted in a nice way, which is useful as a practical guide to how long convergence will take. For suppose ε_n is correct to k places of decimals then ε_{n+1} will be correct to about $2k$ places of decimals (the reason why we cannot be more precise is that we don't know the value of the constant A in (5.4.5) in general, but we hope that it is not too big). Thus, roughly speaking, the Newton–Raphson approximations double in accuracy at each stage.

5.4.4.   Halley's Method

It might be asked whether there exist methods which converge faster than the Newton–Raphson method, that is, methods for which $p > 2$. A number of such methods have been derived one of which, due to Edmund Halley, the astronomer, dates back to 1694; it is similar in principle to the Newton–Raphson but makes use of an extra term involving the second derivative [see IV, § 3.4] in the iterative formula which is

$$x_{n+1} = x_n - \frac{f_n}{f_n' - (f_n f_n'' / 2f_n')} \qquad (5.4.6)$$

where f_n, f_n', f_n'' respectively stand for $f(x_n), f'(x_n)$ and $f''(x_n)$.

It can be proved that (5.4.6) has *cubic* convergence (i.e. $p = 3$). More precisely: if α is the exact solution of $f(x) = 0$ and E_n, E_{n+1} are the errors at consecutive stages then

$$E_{n+1} \doteqdot \left[\frac{1}{4}\left(\frac{f''(\alpha)}{f'(\alpha)}\right)^2 - \frac{1}{6}\left(\frac{f'''(\alpha)}{f'(\alpha)}\right) \right](E_n)^3. \qquad (5.4.7)$$

The main disadvantage of (5.4.6) lies in the necessity of having to work out analytical expressions for $f'(x)$ and $f''(x)$. If $f(x)$ is a polynomial [cf. IV, Theorem 3.2.2] this is easily done, even symbolically, on a computer, but in other cases the difficulty in obtaining these expressions may outweigh the very rapid convergence.

Halley's formula simplifies considerably in the special case of the polynomial $f(x) = x^m - k$, when (5.4.6) reduces to

$$x_{n+1} = \left(\frac{(m-1)x_n^m + (m+1)k}{(m+1)x_n^m + (m-1)k}\right)x_n \qquad (5.4.8)$$

which is a very useful iterative formula for finding $k^{1/m}$.

EXAMPLE 5.4.5.   Using Halley's formula find $(59)^{1/7}$ to 9 d.p.

Solution.   In this case $m = 7$ and $k = 59$ so that (5.4.8) becomes

$$x_{n+1} = \left(\frac{6x_n^7 + 472}{8x_n^7 + 354}\right)x_n.$$

Starting from $x_0 = 1$ we obtain

$$x_1 = 1 \cdot 320441989$$
$$x_2 = 1 \cdot 655396497$$
$$x_3 = 1 \cdot 787160251$$
$$x_4 = 1 \cdot 790518643$$
$$x_5 = 1 \cdot 790518691, \quad \text{which is correct to 9 d.p.}$$

5.4.5.   Steffensen's Method

The simplicity of form and quadratic convergence of the Newton–Raphson method have made it very popular but, as remarked in section 5.4.2, the need to evaluate $f'(x)$ is sometimes a serious drawback. A method, due to Steffensen, which avoids the use of $f'(x)$ but retains quadratic convergence is therefore very useful in many cases.

Steffensen's method is a one-point method; the formula for iteration is

$$x_{n+1} = x_n - \frac{f^2(x_n)}{f(x_n + f(x_n)) - f(x_n)} \qquad (5.4.9)$$

and it can be proved that if convergence occurs it does so quadratically. From a computational point of view (5.4.9) is attractive for only two functional evaluations, $f(x_n)$ and $f(x_n + f(x_n))$ are required and (5.4.9) may be re-written in the form

$$y_n = f(x_n)$$

$$x_{n+1} = x_n - \frac{y_n^2}{f(x_n + y_n) - y_n} \qquad (5.4.10)$$

which is suitable either for a calculator or as the basis of an algorithm for a computer program.

As an illustration of Steffensen's method we re-work Example 5.4.3.

EXAMPLE 5.4.6. Solve $x = e^{-x}$ to 5 d.p. starting with $x_1 = 0$ and using Steffensen's method.

Solution. We put $f(x) = x - e^{-x}$ and (5.4.10) simplifies to

$$y_n = x_n - e^{-x_n}$$

$$x_{n+1} = x_n - \frac{y_n^2}{x_n - e^{-(x_n + y_n)}}$$

and leads to the table below (to 5 d.p.)

n	x_n	y_n	x_{n+1}
0	0	−1	0·36788
1	0·36788	−0·32432	0·54631
2	0·54631	−0·03277	0·56694
3	0·56694	−0·00032	0·56714

and the result is now correct to 5 d.p.

5.5. THE GENERAL ONE-POINT METHOD

The Newton–Raphson method is a particular case of the general one-point method which covers all iterative methods of the form

$$x_{n+1} = F(x_n) \tag{5.5.1}$$

where $F(x)$ is some function which depends, possibly in a complex way, upon the equation, $f(x) = 0$, that we wish to solve. Comparison of equations (5.4.1) and (5.5.1) shows that in the case of the Newton–Raphson method

$$F(x) = x - \frac{f(x)}{f'(x)}.$$

We begin our account of such methods by observing that for any given equation $f(x) = 0$ we can construct an infinite number of iterative methods of the type defined by (5.5.1). For since

$$f(x) = 0 \tag{5.5.2}$$

it follows that

$$x + f(x) = x$$

and, more generally, that for any constant $k(\neq 0)$

$$kx = kx + f(x). \tag{5.5.3}$$

If we now put (5.5.3) into the iterative form

$$kx_{n+1} = kx_n + f(x_n)$$

and divide by k we have

$$x_{n+1} = x_n + \frac{f(x_n)}{k} \qquad (5.5.4)$$

which defines an infinite class of iterative methods of the type defined by (5.5.1). If α is a solution to (5.5.2) then it is also a solution to (5.5.4) when we take $x_{n+1} = x_n$, but this does not necessarily mean that the method defined by (5.5.4) will converge to α. We illustrate this by an example

EXAMPLE 5.5.1.   Use the iterative formulae

(i) $x_{n+1} = 1 + 2x_n - x_n^2$;
(ii) $x_{n+1} = \frac{1}{2}(1 + 3x_n - x_n^2)$;

starting with $x_0 = 1$ and performing 4 iterations in each case.

Solution

(i) $x_0 = 1$, $x_1 = 2$, $x_3 = 1$, $x_4 = 2$,

(ii) $x_0 = 1$, $x_1 = 1 \cdot 5$, $x_2 = 1 \cdot 625$, $x_4 = 1 \cdot 6172$.

Clearly (i) cannot converge whereas (ii) looks as if it is converging quite quickly. If convergence is to occur then the solution must satisfy

$$x = 1 + 2x - x^3 \qquad \text{in case (i)}$$

and

$$x = \tfrac{1}{2}(1 + 3x - x^2) \qquad \text{in case (ii)}$$

and these are both equivalent to

$$x^2 - x - 1 = 0$$

which has the solutions $x = 1 \cdot 6180$ or $-0 \cdot 6180$ (to 4 d.p.). It is clear that (ii) is converging to the positive solution whilst (i) is oscillating about it indefinitely.

Thus we see that whilst (i) and (ii) are different iterative forms of the same equation (ii) converges whereas (i) does not. Even more strikingly, consider the iterative procedure

(iii) $x_{n+1} = 2 + 3x_n - 2x_n^2$

which is yet another iterative form of the same equation and gives rise to the values

$x_0 = 1$, $x_2 = 3$, $x_3 = -7$, $x_4 = -117$.

In this case the iterates are obviously diverging rapidly and the situation is not improved if we begin with an approximation which is quite close to the solution; thus, taking $x_0 = 1 \cdot 61803$ (which is correct to 5 d.p.) produces

$x_1 = 1 \cdot 61805$, $x_2 = 1 \cdot 61798$, $x_3 = 1 \cdot 61822$, $x_4 = 1 \cdot 61739$,

and we see that, far from converging, we are moving further away from the solution at each step.

It is clear from this Example that before using an iterative procedure such as (5.5.1) we must have some criterion for determining whether it will converge or not. Fortunately such a criterion exists as we shall see below, but first let us look at two diagrams associated with the procedures (ii) and (iii) above.

Both diagrams show the line $y = x$ and the curve $y = F(x)$ where $F(x) = \frac{1}{2}(1 + 3x - x^2)$ in Figure 5.5.1 and $F(x) = 2 + 3x - 2x^2$ in Figure 5.5.2.

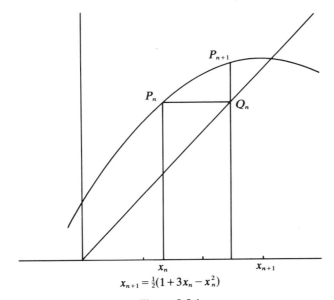

$$x_{n+1} = \tfrac{1}{2}(1 + 3x_n - x_n^2)$$

Figure 5.5.1

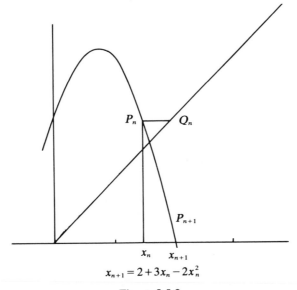

$$x_{n+1} = 2 + 3x_n - 2x_n^2$$

Figure 5.5.2

In both cases we start with an approximate solution x_n, the iterative formula then requires us to evaluate $F(x_n)$ so that we are led to the point P_n on the curve $y = F(x)$ where P_n is $(x_n, F(x_n))$. We now take $x_{n+1} = F(x_n)$, that is, we move from P_n parallel to the x-axis until we meet the line $y = x$ in a point Q_n which has co-ordinates (x_{n+1}, x_{n+1}). The process is now repeated, producing a new point P_{n+1} on the curve, and so on.

In Figure 5.5.1 the distance between consecutive points P_n, $P_{n+1}, \ldots$ steadily decreases and the points converge to the point of intersection of the line and the curve (i.e. to the solution); in Figure 5.5.2 the distance between consecutive points steadily increases and convergence does not occur.

It is not difficult to analyse the situation by elementary calculus, and this will lead us to the desired criterion.

Let the exact solution of (5.5.1) be $x = \alpha$, then

$$\alpha = F(\alpha).$$

Suppose that we have found an approximate solution x_n which differs from α by a small amount ε_n so that

$$x_n = \alpha + \varepsilon_n.$$

Then, from (5.5.1) we take x_{n+1} as our next approximation where

$$x_{n+1} = F(x_n) = F(\alpha + \varepsilon_n).$$

Now since ε_n is small

$$\frac{F(\alpha + \varepsilon_n) - F(\alpha)}{\varepsilon_n} \doteqdot F'(\alpha)$$

[see IV, Definition 3.1.1] and so

$$F(\alpha + \varepsilon_n) \doteqdot F(\alpha) + \varepsilon_n F'(\alpha)$$

or, since $F(\alpha) = \alpha$

$$x_{n+1} = F(\alpha + \varepsilon_n) \doteqdot \alpha + \varepsilon_n F'(\alpha)$$

and we see that if x_n is in error by an amount ε_n then x_{n+1} will be in error by an amount which is approximately $\varepsilon_n F'(\alpha)$. If therefore $|F'(\alpha)| < 1$ the new error will be smaller than the old one, that is x_{n+1} will be a better approximation than x_n to α. If however $|F'(\alpha)| > 1$, x_{n+1} will be a worse approximation than x_n. So we have proved:

CRITERION FOR CONVERGENCE. *If α is the exact solution of the equation $x = F(x)$ the iterative procedure $x_{n+1} = F(x_n)$ will converge to α if and only if x_0 is sufficiently close to α and $|F'(\alpha)| < 1$.*

Let us now test the three procedures in the Example above using this criterion, given that $\alpha = 1 \cdot 6180$.

(i) $F(x) = 1 + 2x - x^2, F'(x) = 2 - 2x$, so

$$F'(\alpha) \doteqdot -1 \cdot 2360;$$

(ii) $F(x) = \frac{1}{2}(1 + 3x - x^2), F'(x) = 1 \cdot 5 - x$, so

$$F'(\alpha) = -0 \cdot 1180;$$

(iii) $F(x) = 2 + 3x - 2x^2, F'(x) = 3 - 4x$, so

$$F'(\alpha) \doteqdot -3 \cdot 472,$$

and we conclude that (i) and (iii) will not converge whereas (ii) will.

Although this analysis is elementary it is quite rigorous, for if a procedure is to converge our approximations must eventually be arbitrarily close to the solution and the approximation to the value of $F'(\alpha)$ given by (5.5.1) becomes as accurate as we wish, and the criterion for convergence then applies.

In practice, of course, we do not know the value of α so that at first sight it seems that the criterion is not much use, but this is not so since $F(x)$ may have the property that $|F'(x)| < 1$ for a wide range of values (possibly infinite) of x, in which case convergence is assured provided only that α lies somewhere in this range, and this we can often detect easily enough.

5.6. SYSTEMS OF NON-LINEAR EQUATIONS

By a system of non-linear equations we mean a set of n equations in n unknowns

$$f_1(x_1, x_2, \ldots, x_n) = 0$$
$$f_2(x_1, x_2, \ldots, x_n) = 0$$
$$\vdots \qquad\qquad \vdots$$
$$f_n(x_1, x_2, \ldots, x_n) = 0$$

where not all of the functions f_i are linear in all of the variables x_j.

In general there is no analytical method for solving such a system and some iterative-type method must be employed. We illustrate one of the most commonly used methods by a simple example.

EXAMPLE 5.6.1. Solve $x^2 + y^2 = 5, x^2 - y^2 = 1$ using an iterative method starting at $(1, 1)$.

Solution. We suppose that we have an approximate solution (x_n, y_n) and that we wish to find a better one $(x_n + h_n, y_n + k_n)$ so that

$$(x_n + h_n)^2 + (y_n + k_n)^2 = 5$$
$$(x_n + h_n)^2 - (y_n + k_n)^2 = 1.$$

Expanding these equations and ignoring h_n^2 and k_n^2 (which we hope are small) we have

$$x_n^2 + 2h_n x_n + y_n^2 + 2k_n y_n \doteq 5$$

$$x_n^2 + 2h_n x_n - y_n^2 - 2k_n y_n \doteq 1$$

or:

$$2h_n x_n + 2k_n y_n = 5 - x_n^2 - y_n^2$$

$$2h_n x_n - 2k_n y_n = 1 - x_n^2 + y_n^2$$

which give:

$$h_n = \frac{3 - x_n^2}{2x_n}$$

$$k_n = \frac{2 - y_n^2}{2y_n}.$$

If we now start with $(x_0, y_0) = (1, 1)$ we find $h_0 = 1$, $k_0 = 0 \cdot 5$ so that $x_1 = 2$, $y_1 = 1 \cdot 5$ and convergence then continues as shown below;

n	x_n	y_n
0	1	1
1	2	1·5
2	1·75	1·417
3	1·7321	1·4142

The exact solution is easily obtained in this case, for addition of the 2 equations gives $2x^2 = 6$ or $x = \pm\sqrt{3} (= \pm 1 \cdot 7321)$, and subtraction gives $2y^2 = 4$ or $y = \pm\sqrt{2} (= \pm 1 \cdot 4142)$ and we see that we have converged to the solution lying in the first quadrant, to 4 d.p., in 3 steps.

The method used in this example is easily generalized. Consider the case of two non-linear equations in two unknowns

$$f_1(x, y) = 0$$

$$f_2(x, y) = 0$$

and suppose that we have an approximate solution (x_n, y_n). Let $(x_{n+1}, y_{n+1}) = (x_n + h_n, y_n + k_n)$ be the exact solution, then

$$f_1(x_{n+1}, y_{n+1}) = 0$$

$$f_2(x_{n+1}, y_{n+1}) = 0.$$

Now $f_1(x_{n+1}, y_{n+1}) = f_1(x_n + h_n, y_n + k_n)$ and by Taylor's theorem for two variables [see IV, Theorem 5.8.1]

$$f_1(x_n + h_n, y_n + k_n) \doteq f_1(x_n, y_n) + h_n \frac{\partial f_1}{\partial x}(x_n, y_n) + k_n \frac{\partial f_1}{\partial y}(x_n, y_n)$$

so that

$$h_n \frac{\partial f_1}{\partial x} + k_n \frac{\partial f_1}{\partial y} \doteq -f_1(x_n, y_n) \tag{5.6.1}$$

and similarly

$$h_n \frac{\partial f_2}{\partial x} + k_n \frac{\partial f_2}{\partial y} \doteq -f_2(x_n, y_n). \tag{5.6.2}$$

Equations (5.6.1) and (5.6.2), being linear in h_n and k_n, can be solved and give

$$h_n = \frac{f_2 \, \partial f_1/\partial y - f_1 \, \partial f_2/\partial y}{\partial f_1/\partial x \; \partial f_2/\partial y - \partial f_2/\partial x \; \partial f_1/\partial y} \tag{5.6.3}$$

and

$$k_n = \frac{f_2 \, \partial f_1/\partial x - f_1 \, \partial f_2/\partial x}{\partial f_1/\partial y \; \partial f_2/\partial x - \partial f_2/\partial y \; \partial f_1/\partial x} \tag{5.6.4}$$

where in (5.6.3) and (5.6.4) all functions and derivatives are evaluated at $x = x_n$, $y = y_n$. Thus in the example above

$$f_1(x, y) = 5 - x^2 - y^2$$
$$f_2(x, y) = 1 - x^2 + y^2$$

so that

$$\frac{\partial f_1}{\partial x} = -2x, \qquad \frac{\partial f_1}{\partial y} = -2y, \qquad \frac{\partial f_2}{\partial x} = -2x, \qquad \frac{\partial f_2}{\partial y} = 2y$$

and (5.6.3) gives

$$h_n = \frac{-2y_n(5 - x_n^2 - y_n^2) - 2y_n(1 - x_n^2 + y_n^2)}{-4x_n y_n - 4x_n y_n}$$

or

$$h_n = \frac{3 - x_n^2}{2x_n},$$

as we found before. The value for k_n is confirmed similarly.

The expressions given by (5.6.3) and (5.6.4) are rather involved and in practice it may be better to work from the linear equations (5.6.1) and (5.6.2), but we must remember of course that the numerical values of the coefficients change at each iteration. This approach is shown in the next example.

EXAMPLE 5.6.2.   Solve the equations

$$x^3 - y^2 + 1 = 0$$
$$x^2 - 2x + y^3 - 2 = 0$$

starting from $(x_0, y_0) = (1, 1)$.

Solution.   As before we suppose that we have an approximate solution (x_n, y_n) and try to obtain a better one $(x_{n+1}, y_{n+1}) = (x_n + h_n, y_n + k_n)$. Proceeding as before and ignoring h_n^2, k_n^2 etc. we obtain the linear equations

$$3x_n^2 h_n - 2y_n k_n = y_n^2 - x_n^3 - 1 \tag{5.6.5}$$

$$2(x_n - 1)h_n + 3y_n^2 k_n = 2 - y_n^3 - x_n^2 + 2x_n. \tag{5.6.6}$$

Substituting $x_0 = 1$, $y_0 = 1$ the equations become

$$3h_0 - 2k_0 = -1$$

$$3k_0 = 2$$

giving $k_0 = \frac{2}{3}$, $h_0 = \frac{1}{9}$ so that $(x_1, y_1) = (\frac{10}{9}, \frac{5}{3})$. With these values the equations become (to 4 d.p.)

$$3 \cdot 7037h_1 - 3 \cdot 3333k_1 = 0 \cdot 4060$$

$$0 \cdot 2222h_1 + 8 \cdot 3333k_1 = 1 \cdot 6420$$

which give $h_1 = -0 \cdot 0661$, $k_1 = -0 \cdot 1953$ so that $(x_2, y_2) = (1 \cdot 045, 1 \cdot 4714)$, and substitution of these values in (5.6.5) and (5.6.6) leads to the new equations

$$3 \cdot 2761h_2 - 2 \cdot 9428k_2 = 0 \cdot 0239$$

$$0 \cdot 0900h_2 + 6 \cdot 4951k_2 = -0 \cdot 1876.$$

We note that the absolute values of the numbers on the right hand side of the equations are decreasing rapidly which indicates that the iterations are converging and this is confirmed when we obtain the values of h_2 and k_2 which are: $h_2 = -0 \cdot 0184$, $k_2 = -0 \cdot 0286$ giving $(x_3, y_3) = (1 \cdot 0266, 1 \cdot 4428)$. The equations then become

$$3 \cdot 1617h_3 - 2 \cdot 8856k_3 = -0 \cdot 0003$$

$$0 \cdot 0532h_3 + 6 \cdot 2450k_3 = -0 \cdot 0041$$

which give $h_3 = -0 \cdot 0007$, $k_3 = -0 \cdot 0007$.

The approximate solution is now $(x_3, y_3) = (1 \cdot 0259, 1 \cdot 4421)$ and with these values we have

$$|y_3^2 - x_3^3 - 1| < 0 \cdot 0001$$

and

$$|2 - y_3^3 - x_3^2 + 2x_3| < 0 \cdot 0003.$$

A more accurate solution can only be obtained if we work to 5 d.p. at least. We therefore accept that the solution is approximately $(1 \cdot 0259, 1 \cdot 4421)$.

It is instructive to draw the two curves used in this example on graph paper and it will then be seen that the solution which we have found is one of three; the other solutions are close to $(-0 \cdot 466, 0 \cdot 948)$ and $(-0 \cdot 818, -0 \cdot 673)$ and the iterations would have converged to these solutions had we used appropriate starting values.

R.F.C.

REFERENCES

There have been many books and research papers devoted to numerical methods for solving non-linear equations of a *single* variable; two of the best texts are:

Householder, A. S. (1970). *The Numerical Treatment of a Single Nonlinear Equation*, McGraw-Hill, New York.

Traub, J. F. (1964). *Iterative Methods for the Solution of Equations*, Prentice-Hall, Englewood Cliffs, N.J.

The subject of *systems* of non-linear equations is less widely covered in books but

Ortega, J. and Rheinboldt, W. (1970). *Iterative Solution of Non-Linear equations in several variables*, Academic Press, New York.

can be recommended.

PROGRAMS

In the Appendix programs will be found for finding a root of an equation (1) by the Newton–Raphson Method, (2) by the Regula–Falsi or Secant Methods, (3) by the Method of Bisection.

CHAPTER 6

Curve Fitting and Approximation of Functions

6.1. INTRODUCTION

The problem of fitting a curve, or series of curves to data occurs frequently in all branches of the physical, biological and social sciences. The data may have been obtained experimentally or may be given by the values taken by a complicated function at specified points. If the data have been obtained experimentally they may be subject to experimental errors and this may have to be taken into account in our choice of curve; if the data have been obtained from the values of a known but complicated function our objective may be to find a simpler function of some particular type which fits the data to within some specified accuracy over a finite range.

No matter how the data have been obtained the problem of fitting a curve in two dimensions can be stated thus: given the values of a function at a set of points, S, in a finite one-dimensional interval $[a, b]$ to find a function of a single variable which, in some sense, provides a 'good' approximation to the given function at the points of S.

This statement of the problem raises several important questions:

(Q1) Is the set of points S continuous or discrete? If discrete does the set contain an infinite number of points, or only a finite number? Are the points equally spaced?

(Q2) Are there any restrictions placed on the form of the approximating function? In particular:
    (i) must it be continuous? [see IV, Definition 2.1.2];
    (ii) must it be differentiable up to some order ≥ 1? [see IV, Definition 2.10.1];
    (iii) must it be in the form of a polynomial and if so is there a limit to its degree? [see I, § 14.1];
    (iv) can a rational function be used? [see I, § 14.9];
    (v) can the interval $[a, b]$ be divided into sub-intervals $[a_i, a_{i+1}]$ with a different approximation function being used over each

163

sub-interval? If this is allowed must there be continuity and differentiability of these functions at the end-points of the sub-intervals (a_i)?

(Q3) What is meant by a 'good' approximation? In some way we must be able to decide whether one approximation is 'better' than another and this implies that we must have some measure of the closeness of fit of the approximating curve to the data. There is no unique measure appropriate to all circumstances, so we must decide in each case which measure to use.

Depending on the answers to these questions different approximation functions will be found.

We begin by considering the simplest case and we shall introduce concepts which will be required in the more complicated cases.

6.1.1. Straight Line Approximation at Discrete Points

We are given a (finite) set of points x_i $(i = 1, 2, \ldots, n)$—not necessarily equally spaced—and a set of values y_i $(i = 1, 2, \ldots, n)$ at these points. We wish to find a straight line $y = ax + b$ which passes 'as close as possible' (in some sense) to all the points $P_i(x_i, y_i)$.

Clearly if $n > 2$ we will not, in general, be able to find a line which passes *through* all the n points. When $x = x_i$ the line will obviously pass through the point $Q_i(x_i, ax_i + b)$ and so will 'miss' P_i by an amount $e_i = |y_i - (ax_i + b)|$. We must try to keep these numbers e_i fairly small since the e_i provide a measure of how well our line fits the points. There are however n numbers e_i so how do we decide that one line is better than another?

There are several possible approaches but among the more commonly used are:

Least Squares Approach: We choose that line that minimizes

$$\sum_{i=1}^{n} e_i^2$$

that is, that minimizes

$$S = \sum_{i=1}^{n} (y_i - ax_i - b)^2.$$

Perpendicular Least Squares Approach: In the least squares approach we are tacitly assuming that the values of the y_i are subject to error whereas the values of the x_i are exact. In some problems both x_i and y_i must be considered equally liable to error in which case the measure must take this into account. We shall examine one such measure in section 6.1.6.

Minimax Approach: We choose the line which has the smallest value of the largest absolute deviation $|e_i|$, that is, that minimizes $\max_{i=1,n} |e_i|$. That is, we find a, b so that we obtain

$$\min_{a,b} \ \max_{i=1,\dots,n} \ |y_i - ax_i - b|.$$

Initially we shall concentrate on the least squares method. The minimax approach will lead us to the use of Chebyshev polynomials and we will come to that in section 6.3.4.

6.1.2. Least Squares Fit of a Straight Line

Following the previous notation we wish to minimize

$$S = \sum_{i=1}^{n} (y_i - ax_i - b)^2$$

by appropriate choice of a, b. For a minimum we must have $\partial S/\partial a = \partial S/\partial b = 0$ [see IV, § 5.6] and so

$$\sum_{i=1}^{n} -2x_i(y_i - ax_i - b) = 0$$

$$\sum_{i=1}^{n} -2(y_i - ax_i - b) = 0$$

or:

$$\sum_{i=1}^{n} x_i(y_i - ax_i - b) = 0 \qquad (6.1.1)$$

and

$$\sum_{i=1}^{n} (y_i - ax_i - b) = 0. \qquad (6.1.2)$$

The second equation shows that the centre of gravity of the points P_i must lie on the line $y = ax + b$, for

$$(\textstyle\sum y_i) - a(\textstyle\sum x_i) - nb = 0$$

that is

$$\left(\frac{\sum y_i}{n}\right) = a\left(\frac{\sum x_i}{n}\right) + b$$

and

$$\left(\frac{\sum x_i}{n}, \frac{\sum y_i}{n}\right)$$

is the centre of gravity of the points $P_i(x_i, y_i)$.

Equations (6.1.1) and (6.1.2) can be solved for a, b, for

$$(\textstyle\sum x_i y_i) - a \textstyle\sum x_i^2 - b(\textstyle\sum x_i) = 0$$

and

$$\sum y_i - a \sum x_i - nb = 0$$

and so

$$n \sum x_i y_i - an \sum x_i^2 - nb \sum x_i = 0$$

and

$$(\sum x_i)(\sum y_i) - a(\sum x_i)^2 - nb \sum x_i = 0.$$

We now eliminate b and obtain

$$a\{n \sum x_i^2 - (\sum x_i)^2\} = n(\sum x_i y_i) - (\sum x_i)(\sum y_i)$$

so that

$$a = \frac{n(\sum x_i y_i) - (\sum x_i)(\sum y_i)}{n \sum x_i^2 - (\sum x_i)^2}. \tag{6.1.3}$$

Given this value of a the value of b can be found at once since the centre of gravity of the points lies on the line and so

$$b = \frac{1}{n}\{\sum y_i - a \sum x_i\}. \tag{6.1.4}$$

If we choose a, b to satisfy (6.1.3) and (6.1.4) we will have found an extremal value of S. This cannot be a maximum, for by choosing our line sufficiently far away from all the points P_i (and since there is a *finite* number of these points we can certainly move as far away from all of them as we wish) we can make S as large as we desire. Hence S can have no maximum. We have therefore found a minimum—as required.

EXAMPLE 6.1.1.   Find the least-squares line fit for the 3 points $(0, 1)$, $(1, 1)$, $(2, 2)$.

Solution.   We require $\sum x_i, \sum y_i, \sum x_i^2, \sum x_i y_i$ in the course of the calculation, we therefore form a table:

x_i	y_i	x_i^2	$x_i y_i$	
0	1	0	0	$n = 3$
1	1	1	1	
2	2	4	4	
Σ 3	4	5	5	

Hence, from (6.1.3)

$$a = \frac{3(5) - (3)(4)}{3(5) - (3)(3)} = \frac{3}{6} = \frac{1}{2}$$

and so, from (6.1.4)

$$b = \tfrac{1}{3}((4) - \tfrac{1}{2}(3)) = \tfrac{5}{6}.$$

Thus the line is

$$y = \tfrac{1}{2}x + \tfrac{5}{6}, \tag{6.1.5}$$

which passes through $(0, \tfrac{5}{6})$, $(1, \tfrac{4}{3})$, $(2, \tfrac{11}{6})$ with absolute errors of $\tfrac{1}{6}$, $\tfrac{1}{3}$ and $\tfrac{1}{6}$ in y_i respectively.

It is worth noting that the least squares line, (6.1.5), doesn't pass through *any* of the 3 points; Figure 6.1.1 shows its position; P_1, P_2 and P_3 are the points and C is their centre of gravity.

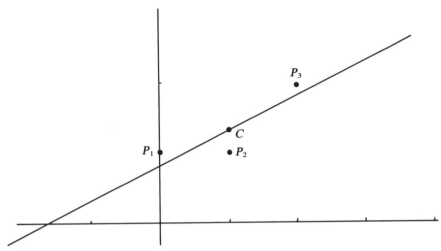

Figure 6.1.1

6.1.3.   Least Squares Approximation by a Polynomial

The least squares approximation by a straight line can be generalized in an obvious way to approximation by a polynomial. The method is the same in principle but the amount of work involved to find the polynomial is, of course, greater.

Suppose that we have n points $P_i(x_i, y_i)$ and we wish to find a polynomial

$$f(x) = a_0 + a_1 x + a_2 x^2 + \ldots + a_m x^m$$

which fits the points P_i as closely as possible in the least-squares sense. Of course if $n \leq m + 1$ we can find a perfectly fitting polynomial, for the Lagrange Interpolation polynomial will pass through all the n points [see § 2.3.4]. We therefore assume that $n \geq m + 2$. The least squares polynomial of degree m is then obtained by choosing $a_0, a_1, \ldots, a_m$ so that

$$S = \sum_{i=1}^{n} (f(x_i) - y_i)^2$$

is a minimum, that is, so that

$$S = \sum_{i=1}^{n} (a_0 + a_1 x_i + a_2 x_i^z + \ldots a_m x_i^m - y_i)^2$$

is a minimum. The $(m+1)$ necessary conditions are [see IV, § 5.6]

$$\frac{\partial S}{\partial a_k} = 2 \sum_{i=1}^{n} (a_0 + a_1 x_i + a_2 x_i^2 + \ldots + a_m x_i^m - y_i) x_i^k = 0 \qquad (k = 0, 1, \ldots, m).$$

We thus obtain a set of $(m+1)$ linear equations in the $(m+1)$ unknowns $a_0, a_1, a_2, \ldots, a_m$, viz.

$$a_0 \sum_{i=1}^{n} x_i^k + a_1 \sum_{i=1}^{n} x_i^{k+1} + a_2 \sum_{i=1}^{n} x_i^{k+2} + \ldots a_m \sum_{i=1}^{n} x_i^{k+m} = \sum_{i=1}^{n} x_i^k y_i$$

$$(k = 0, 1, 2, \ldots, m).$$

If we write $S_k = \sum_{i=1}^{n} x_i^k$, $V_k = \sum_{i=1}^{n} x_i^k y_i$ the equations can be written in the form:

$$\begin{aligned}
S_0 a_0 + S_1 a_1 + S_2 a_2 + \ldots + S_m a_m &= V_0 \\
S_1 a_0 + S_2 a_1 + S_3 a_2 + \ldots + S_{m+1} a_m &= V_1 \\
&\vdots \\
S_m a_0 + S_{m+1} a_1 + S_{m+2} a_2 + \ldots + S_{2m} a_m &= V_m
\end{aligned} \qquad (6.1.6)$$

These equations (6.1.6) are called the *normal equations*.

It can be proved that the normal equations in this case (i.e. polynomial approximation) have a unique solution which provides a least-squares minimum (see Atkinson p. 180). Thus, in theory, the solution of the equations (6.1.6) provides the desired least squares polynomial approximation of degree m at the n points $P_i(x_i, y_i)$. Unfortunately, as we shall see, unless m is rather small (say $m < 7$) least squares approximation by means of a polynomial is not likely to be very satisfactory unless a very large number of decimal places accuracy can be retained throughout the calculation.

A *least squares parabola* [see V, § 1.3.3] is frequently used to fit a 'smoothed' curve to 5 data points, as we illustrate in the next example.

EXAMPLE 6.1.2.   Fit a least squares parabola to the data given below; then compare the values of y obtained from the parabola with the original data

$$
\begin{array}{llllll}
x_i = 1 & 2 & 3 & 4 & 5 \\
y_i = 0.39 & 0.71 & 0.88 & 0.99 & 1.11.
\end{array}
$$

Solution.   Let the least-squares parabola be

$$y = a_0 + a_1 x + a_2 x^2 \qquad (6.1.7)$$

then, from equations (6.1.6) with $m = 2$,

$$S_0a_0 + S_1a_1 + S_2a_2 = V_0$$

$$S_1a_0 + S_2a_1 + S_3a_2 = V_1$$

$$S_2a_0 + S_3a_1 + S_4a_2 = V_2$$

where

$$S_n = \sum_{i=1}^{5} x_i^n \quad \text{and} \quad V_n = \sum_{i=1}^{5} x_i^n y_i.$$

Computation of these 8 values then gives us the equations

$$5a_0 + 15a_1 + 55a_2 = 4 \cdot 08$$

$$15a_0 + 55a_1 + 225a_2 = 13 \cdot 96$$

$$55a_0 + 225a_1 + 979a_2 = 54 \cdot 74$$

which may be solved by Gaussian elimination [see § 3.3.1] and the solutions (to 4 d.p.) are

$$a_0 = +0 \cdot 0701, \qquad a_1 = +0 \cdot 3691, \qquad a_2 = -0 \cdot 0329.$$

We now use these values in (6.1.7) and put $x = 1, 2, 3, 4, 5$ to produce the 'smoothed' values for y (to 2 d.p.), viz.

$$x_i = 1 \qquad 2 \qquad 3 \qquad 4 \qquad 5$$

$$y_i(\text{original}) = 0 \cdot 39 \quad 0 \cdot 71 \quad 0 \cdot 88 \quad 0 \cdot 99 \quad 1 \cdot 11$$

$$y_i(\text{smoothed}) = 0 \cdot 40 \quad 0 \cdot 68 \quad 0 \cdot 88 \quad 1 \cdot 02 \quad 1 \cdot 09.$$

We note that the smoothing parabola takes the same value (within the limits of the accuracy of the calculation) as the original data at $x = 3$; we shall see later that this might have been predicted and we shall also see that in cases where the data are given at equidistant points (as in Example 6.1.2) the calculation of the least squares parabola can be considerably simplified by a change of axes. We illustrate this by reworking Example 6.1.2 using the centre value $(x = 3)$ as origin, i.e. by writing $z = x - 3$ and computing y as a quadratic in z [see V, § 1.2.6].

EXAMPLE 6.1.3. Fit a least-squares parabola to the data given below

$$z_i = -2 \qquad -1 \qquad 0 \qquad 1 \qquad 2$$

$$y_i = \quad 0 \cdot 39 \qquad 0 \cdot 71 \quad 0 \cdot 88 \quad 0 \cdot 99 \quad 1 \cdot 11.$$

Solution. Let the parabola be

$$y = b_0 + b_1 z + b_2 z^2, \tag{6.1.8}$$

then the normal equations take the same form as before but with b_i replacing a_i.

The values of S_n and V_n are of course different since

$$S_n = \sum_{i=1}^{5} z_i^n \quad \text{and} \quad V_n = \sum_{i=1}^{5} z_i^n y_i.$$

We observe that $S_1 = S_3 = 0$ which considerably simplifies the equations which are

$$5b_0 \qquad\qquad\qquad + 10b_2 = 4\cdot08$$

$$10b_1 \qquad\qquad\qquad = 1\cdot72$$

$$10b_0 \qquad\qquad\qquad + 34b_2 = 7\cdot70$$

which lead to the solution

$$b_0 = 0\cdot8817, \qquad b_1 = 0\cdot1720, \qquad b_2 = -0\cdot0329.$$

The parabola is therefore

$$y = 0\cdot8817 + 0\cdot1720z - 0\cdot0329z^2 \tag{6.1.9}$$

or, in terms of the original variable x

$$y = 0\cdot8817 + 0\cdot1720(x-3) - 0\cdot0329(x-3)^2.$$

Putting $z = 0$ in (6.1.9) we obtain the value $0\cdot8817$ as our estimate corresponding to $x_i = 3$ and to 2 d.p. this agrees with the original data, as remarked above. We shall return to this in section 6.1.8.

6.1.4.   Least Squares Approximation to a Continuous Function

If we wish to find a least squares approximation to a continuous function our previous approach must be modified since the number of points (x_i, y_i) at which the approximation is to be measured is now infinite (and non-countable) [see I, § 1.6]. Therefore we cannot use a sum of the type $\sum (y_i - f(x_i))^2$ but must use a continuous measure, that is, an integral. If the interval of the approximation is $[a, b]$ so that $a \leq x_i \leq b$ for all points under consideration then we must minimize

$$\int_a^b (f(x) - y)^2 \, dx$$

[see IV, § 4.2] where $y = y(x)$ is our continuous function and $f(x)$ is our approximating function. Consider a simple example:

EXAMPLE 6.1.4.   Find the least squares straight line that provides the best fit to the curve $y = \sqrt{x}$ over the interval $0 \leq x \leq 1$.

Solution.   Let the line be $y = ax + b$. We must minimize

$$I(a, b) = \int_0^1 (\sqrt{x} - ax - b)^2 \, dx.$$

Expanding the integrand we obtain

$$I(a, b) = \int_0^1 (x - 2ax^{3/2} - 2bx^{1/2} + a^2x^2 + 2abx + b^2)\, dx$$

and integrating

$$I(a, b) = \left[\frac{1}{2}x^2 - \frac{4a}{5}x^{5/2} - \frac{4b}{3}x^{3/2} + \frac{a^2}{3}x^3 + abx^2 + b^2x \right]_0^1$$

so that finally

$$I(a, b) = \frac{1}{3}a^2 + b^2 + ab - \frac{4a}{5} - \frac{4b}{3} + \frac{1}{2}.$$

For a minimum we must have

$$\frac{\partial I(a, b)}{\partial a} = \frac{\partial I(a, b)}{\partial b} = 0$$

that is

$$\tfrac{2}{3}a + b = \tfrac{4}{5}$$

and

$$a + 2b = \tfrac{4}{3}$$

so that

$$\tfrac{1}{3}a = \tfrac{8}{5} - \tfrac{4}{3} = \tfrac{4}{15}$$

that is

$$a = \tfrac{4}{5}, \qquad b = \tfrac{4}{15}$$

and the least squares line is therefore $y = \tfrac{4}{15}(3x + 1)$. (A slightly simpler method for obtaining these equations is used in the more general case, below.)

It is easily seen that this line does produce a *minimum* value for $I(a, b)$ for consideration of the diagram shows that $I(a, b)$ cannot possibly have a maximum value, for we can move the line $y = ax + b$ arbitrarily far from the curve $y = \sqrt{x}$. Alternatively we can note that

$$I(a, b) = \frac{1}{3}a^2 + ab - \frac{4a}{5} + b^2 - \frac{4b}{3} + \frac{1}{2}$$

$$= \frac{1}{15}\left(a\sqrt{5} + \frac{3\sqrt{5}b}{2} - \frac{6}{\sqrt{5}} \right)^2 + \frac{1}{15}\left(\frac{b\sqrt{15}}{2} - \frac{2}{\sqrt{15}} \right)^2 - \frac{112}{225} + \frac{1}{2}.$$

Thus a minimum can only occur when both squared terms vanish, that is, when

$$\frac{b\sqrt{15}}{2} = \frac{2}{\sqrt{15}} \qquad \text{(giving } b = \tfrac{4}{15}\text{)}$$

and

$$a + \frac{3b}{2} = \frac{6}{5} \qquad \text{(giving } a = \tfrac{4}{5}\text{).}$$

The value of $I(a, b)$ at this minimum is then obviously

$$\tfrac{1}{2} - \tfrac{112}{225} = \tfrac{1}{450}.$$

The line $y = \tfrac{4}{15}(3x + 1)$ meets the curve $y = \sqrt{x}$ in two points $P(0 \cdot 1487,$
$0 \cdot 3856)$ and $Q(0 \cdot 7471, 0 \cdot 8643)$ as shown in Figure 6.1.2.

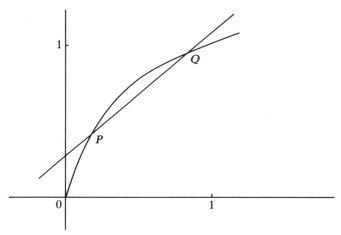

Figure 6.1.2

Returning now to the more general case, if we wish to find the least squares
approximation, using a polynomial of degree k to a continuous function y over
$[a, b]$ we must minimize

$$I(a_0, a_1, \ldots, a_k) = \int_a^b (y - a_0 - a_1 x - a_2 x^2 \ldots - a_k x^k)^2 \, dx.$$

Thus we require $\partial I / \partial a_i = 0$ $(i = 0, 1, \ldots, k)$. We derive the normal equations
in this case by differentiating under the integral sign, which is valid [see IV,
Theorem 4.7.3] and makes the algebra simpler;

$$\int_a^b -2x^r(y - a_0 - a_1 x - \ldots - a_k x^k) \, dx = 0 \qquad (r = 0, 1, \ldots, k).$$

The factor -2 can be ignored so that the normal equations can be written

$$\int_a^b x^r(y - a_0 - a_1 x - \ldots - a_k x^k) \, dx = 0 \qquad (r = 0, 1, 2, \ldots, k). \quad (6.1.10)$$

When the integrations have been performed we will have $(k + 1)$ linear
equations in the $(k + 1)$ unknowns $a_0, a_1, \ldots, a_k$ and these will (in general)
lead us to the required solution [see I, § 5.8].

EXAMPLE 6.1.5. Find the least squares quadratic, $ax^2 + bx + c$, which best fits the curve $y = \sqrt{x}$ over the interval $(0, 1)$. Compare the value of $I(a, b, c)$ with that obtained for $I(a, b)$ in Example 6.1.4.

Solution. The normal equations are

$$\int_0^1 (\sqrt{x} - ax^2 - bx - c)\, dx = 0 \qquad \text{i.e. } \tfrac{1}{3}a + \tfrac{1}{2}b + c = \tfrac{2}{3},$$

$$\int_0^1 x(\sqrt{x} - ax^2 - bx - c)\, dx = 0 \qquad \text{i.e. } \tfrac{1}{4}a + \tfrac{1}{3}b + \tfrac{1}{2}c = \tfrac{2}{5},$$

$$\int_2^1 x^2(\sqrt{x} - ax^2 - bx - c)\, dx = 0 \qquad \text{i.e. } \tfrac{1}{5}a + \tfrac{1}{4}b + \tfrac{1}{3}c = \tfrac{2}{7}.$$

The equations can be written

$$c + \tfrac{1}{2}b + \tfrac{1}{3}a = \tfrac{2}{3} \tag{i}$$

$$\tfrac{1}{2}c + \tfrac{1}{3}b + \tfrac{1}{4}a = \tfrac{2}{5} \tag{ii}$$

$$\tfrac{1}{3}c + \tfrac{1}{4}b + \tfrac{1}{5}a = \tfrac{2}{7} \tag{iii}$$

leading to

$$c + \tfrac{1}{2}b + \tfrac{1}{3}a = \tfrac{2}{3} \tag{i}$$

$$c + \tfrac{2}{3}b + \tfrac{1}{2}a = \tfrac{4}{5} \tag{iv}$$

$$c + \tfrac{3}{4}b + \tfrac{3}{5}a = \tfrac{6}{7} \tag{v}$$

(i) and (iv) give

$$\tfrac{1}{6}b + \tfrac{1}{6}a = \tfrac{2}{15} \tag{vi}$$

(iv) and (v) give

$$\tfrac{1}{12}b + \tfrac{1}{10}a = \tfrac{2}{35} \tag{vii}$$

or

$$\tfrac{1}{6}b + \tfrac{1}{5}a = \tfrac{4}{35} \tag{viii}$$

(vi) and (viii) give

$$\tfrac{1}{30}a = \tfrac{-2}{105} \quad \text{or} \quad a = -\tfrac{4}{7}$$

and so, from (vi)

$$b = \tfrac{4}{5} - a = \tfrac{48}{35}.$$

Finally, from (i)

$$c = \tfrac{2}{3} - \tfrac{1}{2}b - \tfrac{1}{3}a$$

$$= \frac{140 - 144 + 40}{210} = \frac{36}{210} = \frac{6}{35}.$$

The least squares quadratic is therefore

$$f(x) = \tfrac{1}{35}(-20x^2 + 48x + 6).$$

The value of $I(a, b, c)$ for these values of a, b, c can be found by integrating

$$\int_0^1 (\sqrt{x} - ax^2 - bx - c)^2 \, dx$$

which gives

$$\min I(a, b, c) = \tfrac{1}{2450}$$

which is less than one-fifth of the value of $\min I(a, b)$ found in Example 6.1.4.

6.1.5. Disadvantage of Least Squares Fitting with Polynomials of High Degree

When we used the least squares method to fit a polynomial of the second degree to $\sqrt{x}$ over 0, 1 we obtained a set of 3 linear equations in 3 unknowns which may be written in matrix form [see I, § 5.7] as

$$\begin{pmatrix} 1 & \tfrac{1}{2} & \tfrac{1}{3} \\ \tfrac{1}{2} & \tfrac{1}{3} & \tfrac{1}{4} \\ \tfrac{1}{3} & \tfrac{1}{4} & \tfrac{1}{5} \end{pmatrix} \begin{pmatrix} c \\ b \\ a \end{pmatrix} = \begin{pmatrix} \tfrac{2}{3} \\ \tfrac{2}{5} \\ \tfrac{2}{7} \end{pmatrix}.$$

When we fitted a straight line the equations were

$$\begin{pmatrix} 1 & \tfrac{1}{2} \\ \tfrac{1}{2} & \tfrac{1}{3} \end{pmatrix} \begin{pmatrix} b \\ a \end{pmatrix} = \begin{pmatrix} \tfrac{2}{3} \\ \tfrac{2}{5} \end{pmatrix}.$$

The form of these equations suggests that if we try to fit a polynomial of degree k to $\sqrt{x}$ over $[0, 1]$ we will arrive at a set of $(k+1)$ equations in $(k+1)$ unknowns $a_k, a_{k-1}, \ldots, a_1, a_0$ which will be representable in the form

$$\begin{bmatrix} 1 & \tfrac{1}{2} & \tfrac{1}{3} & \cdots & 1/(k+1) \\ \tfrac{1}{2} & \tfrac{1}{3} & & \cdots & 1/(k+2) \\ \vdots & & & & \vdots \\ 1/(k+1) & 1/(k+2) & & \cdots & 1/(2k+1) \end{bmatrix} \begin{bmatrix} a_0 \\ a_1 \\ \vdots \\ a_{k+1} \end{bmatrix} = \begin{bmatrix} 2/3 \\ 2/5 \\ \vdots \\ 2/(2k+3) \end{bmatrix}. \quad (6.1.11)$$

The matrix on the left-hand side of (6.1.11) is the principal minor $\mathbf{H}_{k+1}$ of order $(k+1)$ [see I, § 6.13] of the infinite *Hilbert Matrix*, one of the most notoriously *ill-conditioned* matrices in mathematics. (A matrix is said to be *ill-conditioned* if, when it has been normalised so that its largest element is 1, its inverse [see I, § 6.4] has very large elements.) The inverse matrices in the case of the un-normalised minors of order 2 and 3 are

$$\mathbf{H}_2^{-1} = \begin{pmatrix} 4 & -6 \\ -6 & 12 \end{pmatrix} \quad \text{and} \quad \mathbf{H}_3^{-1} = \begin{pmatrix} 9 & -36 & 30 \\ -36 & 192 & -180 \\ 30 & -180 & 180 \end{pmatrix}.$$

The size of the largest element in $\mathbf{H}_n^{-1}$ increases rapidly with n, for example in $\mathbf{H}_9^{-1}$ the largest element is about 3×10^{12}.

Now we observe that the matrix $\mathbf{H}_{k+1}$ is certain to be involved when we obtain the normal equations, using (6.1.10) for a polynomial approximation of degree k to *any* function $y = f(x)$, that is, it is not a special property of the function $y = \sqrt{x}$. For (6.1.10) may be written

$$\int_a^b x^r(a_0 + a_1 x + \ldots + a_k x^k)\, dx = \int_a^b x^r y\, dx \qquad (r = 0, 1, \ldots, k). \quad (6.1.12)$$

If we take the interval of integration as $(0, 1)$ for simplicity (which can always be done by a suitable change of variable [see IV, § 4.3]) (6.1.12) takes the form

$$\frac{a_0}{r+1} + \frac{a_1}{r+2} + \ldots + \frac{a_k}{r+k+1} = \int_0^1 x^r y\, dx \qquad (r = 0, 1, \ldots, k). \quad (6.1.13)$$

The integrals on the right are evaluated for the given function y and so produce $(k+1)$ numbers $b_0, b_1, \ldots, b_k$ and we then have $(k+1)$ linear equations in $(k+1)$ unknowns $a_0, a_1, \ldots, a_k$, viz.

$$a_0 + \frac{a_1}{2} + \frac{a_2}{3} + \ldots + \frac{a_k}{k+1} = b_0$$

$$\frac{a_0}{2} + \frac{a_1}{3} + \frac{a_2}{4} + \ldots + \frac{a_k}{k+2} = b_1$$

$$\vdots \qquad \qquad \vdots \qquad \vdots \qquad\qquad (6.1.14)$$

$$\frac{a_0}{k+1} + \frac{a_1}{k+2} + \ldots + \frac{a_k}{2k+1} = b_k$$

or $\mathbf{H}_{k+1}\mathbf{a}' = \mathbf{b}'$ where $\mathbf{a} = (a_0, a_1, \ldots, a_k)$, $\mathbf{b} = (b_0, b_1, \ldots, b_k)$ and $\mathbf{H}_{k+1}$ is the $(k+1)$st principal minor of the Hilbert Matrix.

Now $\mathbf{H}_n^{-1}$ contains very large elements even for moderately large values of n and so if we attempt to solve the linear equations (6.1.14) any round-off errors which occur during the calculation [see § 1.3] will produce greatly magnified errors in the solution. This can only be compensated by performing the calculations to an inconveniently (or impossibly) large number of decimal places of accuracy.

Thus in practice the least squares method of polynomial fitting using the method described above is unlikely to be satisfactory if n is even moderately large (say $n > 7$). It is however possible to approach the problem in a way that avoids the introduction of the Hilbert matrix but this method involves the use of orthonormal polynomials and so must be deferred until section 6.3.3.

Note that the left-hand side of the normal equations (6.1.13) are independent of y so that applications over $[0, 1]$ are easily carried out and we can simplify the work further by using inverse matrices $\mathbf{H}_2^{-1}$, $\mathbf{H}_3^{-1}$ etc. where relevant, viz.

EXAMPLE 6.1.6.   Find the best straight line fit in the least squares sense to e^x over the interval $[0, 1]$.

Solution.   Since $\int_0^1 e^x \, dx = e - 1$ and $\int_0^1 x e^x \, dx = 1$ the line is $y = ax + b$ where, using the inverse matrix $\mathbf{H}_2^{-1}$,

$$\binom{a}{b} = \begin{pmatrix} 4 & -6 \\ -6 & 12 \end{pmatrix} \binom{e-1}{1} \quad \text{i.e.} \quad \begin{array}{l} a = 4e - 10 \\ b = 18 - 6e \end{array}$$

that is, $y = 0.8731 + 1.6903x$. The value of the integral which measures the least square error $\int_0^1 (e^x - ax - b)^2 \, dx$ is then found to be 0.004 approximately.

6.1.6.   Least Squares Fit with Other Metrics

We return briefly to the consideration of the least squares fitting of a curve, $y = f(x)$ to a set of n discrete points $P_i(x_i, y_i)$. So far we have taken as our measure of the error

$$S = \sum_{i=1}^{n} f((x_i) - y_i)^2$$

and so we have tacitly assumed that the values x_i are not subject to error, but that the y_i are. This may be a reasonable assumption in some cases: for example we might measure some physical variable (temperature, position, velocity) at fixed intervals of time and so the times might be considered to be measured accurately but the variable subject to experimental error. In other cases, however, both the x_i and y_i might be subject to experimental error (for example, in astronomy, measuring the Right Ascension and Declination of a comet). In such cases a measure of the overall error ought to take into account experimental errors in both the x_i and y_i. We therefore require a different metric from the one above, that is, it is not satisfactory simply to measure the error at a point x_i, y_i from the curve $y = f(x)$ by

$$|f(x_i) - y_i|.$$

One possibility is to measure the distance of the point (x_i, y_i) from the *nearest* point of the curve $y = f(x)$. This is, unfortunately, a much more difficult problem to handle than the one used above. We illustrate the problem by considering the case where the approximating curve is just a straight line.

PROBLEM.   Given a set of n points $P_i(x_i, y_i)$ to find the straight line which minimises $\sum_{i=1}^{n} d_i^2$ where d_i is the perpendicular distance of P_i from the line.

Solution.   Let the line be $y = ax + b$. Then $d_i^2 = (y_i - ax_i - b)^2/(1 + a^2)$ [cf. V (1.2.29)]. We must therefore minimize

$$S_1 = \sum_{i=1}^{n} \frac{(y_i - ax_i - b)^2}{(1 + a^2)}. \tag{6.1.15}$$

The necessary conditions are, as usual,

$$\frac{\partial S_1}{\partial a} = \frac{\partial S_1}{\partial b} = 0.$$

The second condition is easily obtained and is

$$\frac{1}{(1+a^2)} \sum_{i=1}^{n} (y_i - ax_i - b) = 0. \tag{6.1.16}$$

The first condition is slightly more complex:

$$\sum_{i=1}^{n} \left\{ \frac{-2x_i(y_i - ax_i - b)}{1+a^2} - \frac{2a(y_i - ax_i - b)^2}{(1+a^2)^2} \right\} = 0$$

which reduces to

$$\frac{1}{(1+a^2)^2} \sum_{i=1}^{n} (y_i - ax_i - b)(ay_i + x_i - ab) = 0. \tag{6.1.17}$$

The factors $1/(1+a^2)$ and $1/(1+a^2)^2$ can be ignored provided we note that $a = \infty$ provides a possible minimum, although in practice this is most unlikely to arise since it corresponds to the degenerate solution $x = $ constant.

Equation (6.1.17) contains quadratic terms in a and b but since (6.1.16) is linear the problem of finding a and b is not so difficult as it might appear, as we now show.

Firstly, ignoring the factors $(1+a^2)^{-1}$ and $(1+a^2)^{-2}$, we note that (6.1.17) may be written

$$\sum_{i=1}^{n} (y_i - ax_i - b)(ay_i + x_i) = ab \sum_{i=1}^{n} (y_i - ax_i - b)$$

and, by (6.1.16), the right-hand side is zero so that (6.1.17) is equivalent to

$$\sum_{i=1}^{n} (y_i - ax_i - b)(ay_i + x_i) = 0. \tag{6.1.18}$$

Next let

$$X_1 = \sum_{i=1}^{n} x_i, \quad Y_1 = \sum_{i=1}^{n} y_i, \quad X_2 = \sum_{i=1}^{n} x_i^2, \quad Y_2 = \sum_{i=1}^{n} y_i^2 \quad \text{and} \quad P = \sum_{i=1}^{n} x_i y_i,$$

then (6.1.16) and (6.1.18) may be written

$$Y_1 - aX_1 - nb = 0 \tag{6.1.19}$$

and

$$aY_2 - aX_2 + P(1 - a^2) - b(aY_1 + X_1) = 0. \tag{6.1.20}$$

We now use (6.1.19) to substitute for b in (6.1.20) and so obtain a quadratic for a

$$a^2(X_1 Y_1 - nP) - a(Y_1^2 - X_1^2 - nY_2 + nX_2) - (X_1 Y_1 - nP). \tag{6.1.21}$$

Equation (6.1.21) will in general have two distinct real roots [see I, Proposition 14.5.4] and since the product of these roots is

$$-\frac{(X_1 Y_1 - nP)}{(X_1 Y_1 - nP)} = -1$$

the two lines corresponding to these roots are perpendicular [see V (1.2.2)]. One of the lines is the 'least perpendicular squares' line that we are seeking; it is not immediately obvious what the other line represents.

We also note that (6.1.19) may be written

$$\left(\frac{Y_1}{n}\right) = a\left(\frac{X_1}{n}\right) + b$$

so that, as in the case of the least-squares line [see § 6.1.2], the centre of gravity of the points lies on the line we are seeking and also on the line perpendicular to it. In view of this it will obviously simplify the calculations considerably if we move the (x, y) axes [see V, § 1.2.6] so that the centre of gravity of the points (x_i, y_i) becomes $(0, 0)$; this is achieved by writing

$$x_i' = x_i - \frac{X_1}{n} \tag{6.1.22}$$

$$y_i' = y_i - \frac{Y_1}{n}. \tag{6.1.23}$$

If we now put $X_1' = \sum_{i=1}^{n} x_i'$ etc. then $X_1' = Y_1' = 0$, equation (6.1.19) is satisfied with $b = 0$ (which we knew anyway since the line passes through the new origin) and (6.1.21) becomes

$$P'a^2 - (Y_2' - X_2')a - P' = 0 \tag{6.1.24}$$

where

$$P' = \sum_{i=1}^{n} x_i' y_i', \qquad X_2' = \sum_{i=1}^{n} (x_i')^2, \qquad Y_2' = \sum_{i=1}^{n} (y_i')^2.$$

When the values of a have been found from (6.1.24) it will be necessary to decide which of the two values provides the line we are seeking; if this is not obvious from the data the question can be decided by finding the values of S_1 as given by (6.1.15), using the two values of a and accepting that value which produces the smaller result.

EXAMPLE 6.1.7.   Find the least perpendicular squares line for the following data

$$x_i = 0\cdot1 \quad 0\cdot9 \quad 1\cdot6 \quad 3\cdot4$$

$$y_i = 1\cdot3 \quad 2\cdot4 \quad 3\cdot2 \quad 6\cdot3.$$

Solution.   We first find the centre of gravity of the points which is $(1\cdot5, 3\cdot3)$ so that referred to this point as origin the data becomes

$$x_i' = -1\cdot4 \quad -0\cdot6 \quad +0\cdot1 \quad +1\cdot9$$

$$y_i' = -2\cdot0 \quad -0\cdot9 \quad -0\cdot1 \quad +3\cdot0$$

and so

$$X_2' = \sum_{i=1}^{4} (x_i')^2 = 5\cdot94, \qquad Y_2' = \sum_{i=1}^{4} (y_i')^2 = 13\cdot82,$$

$$P' = \sum_{i=1}^{4} (x_i' y_i') = 9\cdot03.$$

Equation (6.1.24) then gives as the quadratic for a

$$a^2 - 0\cdot8726a - 1 = 0$$

the roots of which are $1\cdot5274$ and $-0\cdot6547$ (to 4 d.p.) [see I (14.5.1)]. Since the original data shows y_i increasing with x_i it is obvious that the positive value of a is the relevant one, we therefore take

$$y_i' = 1\cdot5274 x_i'$$

as our line based on the centre of gravity as origin. To convert to the original co-ordinate we put $y_i' = y_i - 3\cdot3$, $x_i' = x_i - 1\cdot5$ the line becomes

$$y_i - 3\cdot3 = 1\cdot5274(x_i - 1\cdot5)$$

or

$$L_1 : y = 1\cdot5274x + 1\cdot0089. \tag{6.1.25}$$

The line at right angles to (6.1.25) is

$$y_i' = -0\cdot6547 x_i$$

or, in original co-ordinates

$$L_2 : y = -0\cdot6547x + 4\cdot2805. \tag{6.1.26}$$

If we compute the two values of S_1 from (6.1.15) we find S_1 for $L_1 = 0\cdot0279$ and S_1 for $K_2 = 19\cdot7321$, so confirming our choice of L_1 as the required line.

Since L_2 arises from a solution of (6.1.16) and (6.1.17) it must have some significance and it is not difficult to prove that L_2 is the *greatest* perpendicular squares line which passes through the centre of gravity of the points and this will always be the case except when the quadratic (6.1.24) degenerates to an equation of the first degree, which happens if and only if $P' = 0$.

6.1.7.   Use of Weightings

It may happen that we are more confident of some of the values of y_i (in the classical least squares problem) than others. For example if several indepen-dent observers have recorded the position of a comet more or less simul-taneously the probable error in the co-ordinates is less than is the case for a single observer. In such cases we may wish to ensure that greater importance is attached to such observations than to others. This is achieved most simply by attaching a 'weighting factor' to each observation. Thus we minimize

$$S^1 = \sum_{i=1}^{n} w_i d_i^2$$

where the w_i are constants ('the weights') not all equal to 1. The problem is then solved exactly as before; the normal equations have coefficients which depend upon the w_i but they are no harder to solve.

6.1.8. Interpolation and Numerical Differentiation Using Least Squares

The method of least squares can be used to interpolate values in a table and also to estimate a numerical value for the derivative of a tabulated function.

Interpolation of a value for a smooth function is a relatively straightforward operation: we simply fit a polynomial (or other function, as desired) to a set of points surrounding the point of interpolation and then use this to estimate the value at the point (details will be found in § 2.3).

When the function is not smooth (due possibly to experimental errors) this approach is unlikely to lead to reliable results; in such cases the method of least squares is often used and one particular approach is to fit a least squares *parabola* to *five* consecutive points [see § 6.1.3]. We hope thereby to 'smooth out' experimental errors. The five points are chosen to lie on both sides of the point of interpolation. We have seen particular cases in Examples 6.2 and 6.3, and although the calculations in the general case are not very illuminating, if we take the case of 5 points equally spaced along the x-axis we can readily see what this method entails.

PROBLEM.   Fit a least squares parabola to the 5 points

$$P_{-2}(x_0-2h, y_{-2}), \quad P_{-1}(x_0-h, y_{-1}), \quad P_0(x_0, y_0),$$

$$P_1(x_0+h, y_1), \quad P_2(x_0+2h, y_2)$$

and then use it to *estimate* a value at P_0.

Solution.   Let $t = (x - x_0)/h$ then $t = -2, 1, 0, 1, 2$ at the 5 points.

Let the least squares parabola be $f(t) = a_0 + a_1 t + a_2 t^2$; the normal equations for determining the coefficients are

$$5a_0 \qquad\qquad + 10a_2 = \sum y_i$$
$$\qquad 10a_1 \qquad\qquad = \sum t_i y_i \qquad i = -2, -1, 0, 1, 2.$$
$$10a_0 \qquad\qquad + 34a_2 = \sum t_i^2 y_i$$

where $t_{-2} = -2, t_{-1} = -1, t_0 = 0, t_1 = 1, t_2 = 2$ which give:

$$35a_0 = 17 \sum y_i - 5 \sum t_i^2 y_i$$

that is

$$35a_0 = 17(y_{-2}+y_{-1}+y_0+y_1+y_2)-5(4y_{-2}+y_{-1}+y_1+4y_2)$$
$$= -3y_{-2}+12y_{-1}+17y_0+12y_1-3y_2$$
$$= 35y_0-3(y_{-2}-4y_{-1}+6y_0-4y_1+y_2)$$
$$= 35y_0-3\delta^4 y_0$$

that is

$$a_0 = y_0-\tfrac{3}{35}\delta^4(y_0) \tag{6.1.27}$$

where δ is the central difference operator defined by

$$\delta(z_n) = z_{n+\frac{1}{2}}-z_{n-\frac{1}{2}}$$

so that $\delta^2(z_n) = \delta(z_{n+\frac{1}{2}})-\delta(z_{n-\frac{1}{2}}) = z_{n+1}-2z_n+z_{n-1}$ etc. [see Definition 1.4.6].
Similarly we find

$$a_1 = \tfrac{1}{10}(-2y_{-2}-y_{-1}+y_1+2y_2) \tag{6.1.28}$$

and

$$a_2 = \tfrac{1}{14}(2y_{-2}-y_{-1}-2y_0-y_1+2y_2). \tag{6.1.29}$$

Now when we take $t=0$ we obtain $f(0)=a_0$ as our least-squares estimate at $x=x_0$, and so from (6.1.27) we see that our estimate differs from the value which we were given (y_0) by $\tfrac{3}{35}\delta^4 y_0$ and therefore if the fourth-order central differences are small the error at $x=x_0$ will also be small, and this explains our observation at the end of Example 6.1.2.

EXAMPLE 6.1.8.　The values of $\sqrt{n}$ for $n=10, 11, 12, 13, 14$ rounded to 2 places are shown. Use the method above to estimate a revised value for $\sqrt{12}$

$n = 10$	11	12	13	14
$\sqrt{n} \doteqdot$ 3·16	3·32	3·46	3·61	3·74.

Equation (6.1.27) gives

$$\sqrt{12} \doteqdot 3{\cdot}46 - \tfrac{3}{35}(3{\cdot}16 - 13{\cdot}28 + 20{\cdot}76 - 14{\cdot}44 + 3{\cdot}74)$$
$$= 3{\cdot}46 + \tfrac{0{\cdot}18}{35} \doteqdot 3{\cdot}4651.$$

The value of $\sqrt{12}$ to 4 d.p. is 3·4641 and so we see that in this case the smoothing effect of the least squares parabola has cut the error from 0·0041 to 0·0010, that is, in the value by a factor of 4.

The same method can be used for estimating the numerical value of a derivative at a point [see IV, Definition 2.9.1]. Numerical differentiation is always of doubtful validity and if the function is at all rough, estimates of the derivative obtained by differencing adjacent values are likely to be wildly inconsistent. We therefore use the 5-point parabola and estimate the derivative from that. We have already obtained the coefficients of this parabola above.

Since

$$f(t) = f\left(\frac{x - x_0}{h}\right) = a_0 + a_1 t + a_2 t^2$$

we have

$$\frac{df}{dx} = \frac{1}{h}\left(\frac{df}{dt}\right) = \frac{a_1 + 2a_2 t}{h} \qquad \text{[see IV (3.2.8)]}$$

and so our estimate of $y'(x_0)$ is

$$\frac{1}{h}\left(\frac{df}{dt}\right)_{t=0} = \frac{a_1}{h}$$

and so, using the value of a_1 from (6.1.28)

$$y'(x_0) \doteq \frac{1}{10h}(-2y_{-2} - y_{-1} + y_1 + 2y_2). \tag{6.1.30}$$

By way of a simple illustration we apply (6.1.30) to a case where we can check the accuracy of the result.

EXAMPLE 6.1.9.   Take $f(x) = \sqrt{x}$ and use the data of Example 6.1.8 and equafion (6.1.30) to estimate $f'(12)$.
   Equation (6.1.30) gives (since $h = 1$)

$$y'(12) \doteq \tfrac{1}{10}(-6 \cdot 32 - 3 \cdot 32 + 3 \cdot 61 + 7 \cdot 48) = 0 \cdot 145$$

And in fact $y'(12) = 1/(2\sqrt{12}) \doteq 0 \cdot 144$.

6.2.   VECTOR AND MATRIX NORMS

6.2.1.   Introduction

In approximation theory we continually find that we need to be able to assign some measure to how closely one function approximates to another. Thus in using the least squares method we used

$$S = \sum_{i=1}^{n} (f(x_i) - y_i)^2$$

as the measure of how closely the polynomial $f(x)$ approximated to the values y_i at the points x_i. Since we wished this measure to be as small as possible we chose $f(x)$ so that S was minimized. Similarly in the case of a continuous function we used

$$S = \int_a^b (f(x) - y)^2 \, dx$$

as the measure of the error.

We could of course have used other measures e.g. (in the continuous case)

$$S' = \int_a^b (f(x) - y)\, dx$$

or

$$S'' = \int_a^b |f(x) - y|\, dx.$$

It is clear that S' would not be a very good choice for we could have $S' = 0$ whilst $f(x)$ might be a very poor approximation to y. For example if $y = x$ over $[0, 1]$ and $f(x) = 1 - x$ over $[0, 1]$, $f(x)$ is a very bad approximation to y and yet $S' = 0$.

S'' is a more reasonable choice than S'. Analytically it is not so easy to handle as S since we need to split the integral into the sums and differences of several integrals over adjoining sub-intervals of $[a, b]$ [see IV, Theorem 4.1.2], the sign of the integrand being constant over each sub-interval, but it is sometimes used.

When we are approximating at a discrete set of points we have two vectors

$$(y_1, y_2, \ldots, y_n) \quad \text{and} \quad (f(x_1), f(x_2), \ldots, f(x_n)).$$

The difference between these two vectors gives the *error vector*

$$\mathbf{e} = (\varepsilon_1, \varepsilon_2, \ldots, \varepsilon_n) = (y_1 - f(x_1), y_2 - f(x_2), \ldots, y_n - f(x_n)).$$

How can we assign a measure to $\mathbf{e}$? So far we have used

$$\varepsilon_1^2 + \varepsilon_2^2 + \ldots + \varepsilon_n^2,$$

alternatives might be $\varepsilon_1 + \varepsilon_2 + \ldots + \varepsilon_n$ (which is not very good, for the reasons given above) and $|\varepsilon_1| + |\varepsilon_2| + \ldots + |\varepsilon_n|$ which is better, but difficult to use in practice.

If the measure of an error vector is to be any use it must possess some very simple properties. Any measure which possesses these properties we call a *norm*.

DEFINITION 6.2.1.   Let $\mathbf{x}, \mathbf{y}$ be vectors; then a measure $\|\mathbf{x}\|$ is said to be a *norm* if it satisfies:

   (i) $\|\mathbf{x}\| > 0$ unless $\mathbf{x} = \mathbf{0}$;
   (ii) $\|k\mathbf{x}\| = |k|\,\|\mathbf{x}\|$ for any *number k*;
   (iii) $\|\mathbf{x} + \mathbf{y}\| \leq \|\mathbf{x}\| + \|\mathbf{y}\|$.

(From (ii) it follows that $\|\mathbf{0}\| = 0$). Note that $\|\mathbf{e}\| = \varepsilon_1^2 + \varepsilon_2^2 + \ldots + \varepsilon_n^2$ is *not* a norm since (ii) is not satisfied, however $\|\mathbf{e}\|_2 = (\varepsilon_1^2 + \varepsilon_2^2 + \ldots + \varepsilon_n^2)^{1/2}$ *is* a norm.

6.2.2.   The Vector Norms $\|\mathbf{x}\|_p$

We define, for $p \geq 1$

$$\|\mathbf{x}\|_p = \left(\sum_{i=1}^n |x_i|^p \right)^{1/p}. \tag{6.2.1}$$

Note the restriction that $p \geq 1$; for $0 < p < 1$ the expression on the right-hand side of (6.2.1) is *not* a norm since condition (iii) is not satisfied (see Example 6.2.2).

For $p \geq 1$ it can be proved that $\|\mathbf{x}\|_p$ *is* a norm; conditions (i) and (ii) are obviously satisfied but to prove (iii) in the general case is non-trivial. We are, however, only interested in three values of p: $p = 1$, 2 and ∞, and in these cases (iii) is not difficult to prove.

The case $p = \infty$ requires some explanation; we assume $\mathbf{x} \neq \mathbf{0}$; let $|x_k| = \max_{i=1,\dots,n} |x_i|$ then, from (6.2.1)

$$\|\mathbf{x}\|_p = |x_k| \left(\sum_{i=1}^{n} \left| \frac{x_i}{x_k} \right|^p \right)^{1/p}. \tag{6.2.2}$$

All the terms except those equal to x_k (or $-x_k$) give rise to values inside the brackets on the right-hand side of (6.2.2) which are less than 1, the exceptional terms produce values equal to 1. As $p \to \infty$ all terms except the exceptional terms tend to zero; if there are m exceptional terms, then $m \geq 1$ and $m^{1/p} \to 1$ as $p \to \infty$ [see I, § 1.6] and so we have

$$\lim_{p \to \infty} \|\mathbf{x}\|_p = |x_k| = \max_{i=1,\dots,n} |x_i|.$$

Hence we define

$$\|\mathbf{x}\|_\infty = \max_{i=1,\dots,n} |x_i| \tag{6.2.4}$$

and the preceding argument provides the justification for this. If $\mathbf{x} = \mathbf{0}$, $\|\mathbf{x}\|_\infty = 0$ so that (6.2.4) is consistent with condition (ii).

The norm $\|\mathbf{x}\|_\infty$ is referred to under various names: *maximum norm, Chebyshev norm, infinity norm* or *uniform norm*.

The cases $p = 1$ and $p = 2$ require little comment:

$$\|\mathbf{x}\|_1 = \sum_{i=1}^{n} |x_i| \tag{6.2.5}$$

and

$$\|\mathbf{x}\|_2 = \left(\sum_{i=1}^{n} |x_i|^2 \right)^{1/2} \tag{6.2.6}$$

and we merely note that $\|\mathbf{x}\|_2$ is simply the *length* of the vector $\mathbf{x}$.

EXAMPLE 6.2.1.   If $\mathbf{x} = (1, 0, -2, 4, 3)$ what are the values of

$$\|\mathbf{x}\|_1, \|\mathbf{x}\|_2, \|\mathbf{x}\|_\infty?$$

Solution.

$$\|\mathbf{x}\|_1 = \sum_{i=1}^{5} |x_i| = 10$$

$$\|\mathbf{x}\|_2 = \left(\sum_{i=1}^{5} |x_i|^2 \right)^{1/2} = \sqrt{30}$$

$$\|\mathbf{x}\|_\infty = \max_{i=1,\ldots,5} |x_i| = 4.$$

EXAMPLE 6.2.2.   In 2 dimensions sketch the curves of unit norm $\|\mathbf{x}\|_p$ for $p = 1$, 2 and ∞, and also the curve $\|\mathbf{x}\|_{1/2} = 1$.

Solution.   Since we are dealing with 2 dimensions we have $\mathbf{x} = (x_1, x_2)$ and so the curves for $p = 1$, 2, and ∞ are

$$p = 1: \quad |x_1| + |x_2| = 1$$

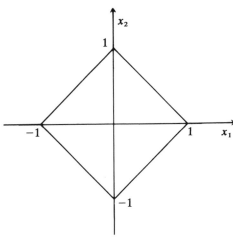

Figure 6.2.1

which is of the boundary of square of side $\sqrt{2}$ turned through $45°$;

$$p = 2: \quad x_1^2 + x_2^2 = 1$$

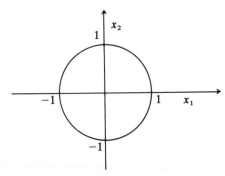

Figure 6.2.2

which is the boundary of the unit circle [see V (1.2.7)];

$$p = \infty: \quad \max\,(|x_1|, |x_2|) = 1$$

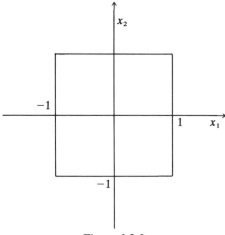

Figure 6.2.3

which is the boundary of the square of side 2.

For $p = \frac{1}{2}$, the curve is given by

$$(|x_1|^{1/2} + |x_2|^{1/2})^2 = 1$$

and so is the boundary of the astroidal region shown.

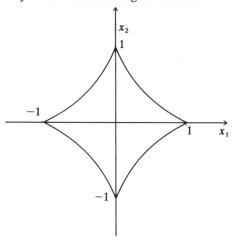

Figure 6.2.4

The astroidal region is non-convex, [see V, § 1.1.4], a property related to the fact that $\|x\|_p$ is not a norm for $0 < p < 1$. We can also verify directly that $\|x\|_{1/2}$ (say) is not a norm by taking $x = (1, 0)$, $y = (1, 1)$, then

$$\|x + y\|_{1/2} = (2^{1/2} + 1^{1/2})^2 = 3 + 2\sqrt{2} = 5 \cdot 828 \ldots$$

and

$$\|\mathbf{x}\|_{1/2} + \|\mathbf{y}\|_{1/2} = (1^{1/2} + 0^{1/2})^2 + (1^{1/2} + 1^{1/2})^2 = 5$$

and we see that condition (iii) for a norm is not satisfied.

Some simple properties of vector norms can be proved easily from the definitions; thus condition (iii) (usually known as *the triangle inequality*) gives us an *upper* bound for $\|\mathbf{x} + \mathbf{y}\|$; in the next example we prove a *lower* bound.

EXAMPLE 6.2.3

$$|\|\mathbf{a}\| - \|\mathbf{b}\|| \leq \|\mathbf{a} + \mathbf{b}\| \qquad\qquad (6.2.7)$$

and

$$|\|\mathbf{a}\| - \|\mathbf{b}\|| \leq \|\mathbf{a} - \mathbf{b}\|. \qquad\qquad (6.2.8)$$

Proof

$$\|\mathbf{a}\| = \|(\mathbf{a} + \mathbf{b}) + (-\mathbf{b})\|$$
$$\leq \|\mathbf{a} + \mathbf{b}\| + \|-\mathbf{b}\| \quad \text{by (iii)}$$

and

$$\|-\mathbf{b}\| = \|\mathbf{b}\| \quad \text{by (ii)}$$

hence

$$\|\mathbf{a}\| \leq \|\mathbf{a} + \mathbf{b}\| + \|\mathbf{b}\|$$

or

$$\|\mathbf{a}\| - \|\mathbf{b}\| \leq \|\mathbf{a} + \mathbf{b}\|. \qquad\qquad (1)$$

Interchanging **a** and **b** therefore:

$$\|\mathbf{b}\| - \|\mathbf{a}\| \leq \|\mathbf{a} + \mathbf{b}\| \qquad\qquad (2)$$

and so, combining (1) and (2)

$$|\|\mathbf{a}\| - \|\mathbf{b}\|| \leq \|\mathbf{a} + \mathbf{b}\|.$$

Replacing **b** by −**b** then gives

$$|\|\mathbf{a}\| - \|\mathbf{b}\|| \leq \|\mathbf{a} - \mathbf{b}\|$$

since $\|-\mathbf{b}\| = \|\mathbf{b}\|$.

This completes the proof.

(6.2.8) is often referred to as the *reverse triangle inequality*; one of its important uses is that it helps us to prove the following result:

THEOREM 6.2.1.   *If $\|\mathbf{x}\|$ is any vector norm on $\mathbf{x} = (x_1, \ldots, x_n)$, where n is finite, then $\|\mathbf{x}\|$ is a continuous function of each of the elements x_i[see IV, Definition 2.1.2].*

Proof.   Let $\mathbf{x}, \mathbf{y}$ be vectors in an n-dimensional vector space [see I, § 5.4]; let $\mathbf{e}_1, \mathbf{e}_2, \ldots, \mathbf{e}_n$ form a basis for the space, then

$$\mathbf{x} = \sum_{i=1}^{n} x_i \mathbf{e}_i, \qquad \mathbf{y} = \sum_{i=1}^{n} y_i \mathbf{e}_i,$$

the x_i, y_i being scalars, and so

$$(\mathbf{x} - \mathbf{y}) = \sum_{i=1}^{n} (x_i - y_i) \mathbf{e}_i.$$

Hence

$$\|\mathbf{x} - \mathbf{y}\| \le \sum_{i=1}^{n} |x_i - y_i| \|\mathbf{e}_i\| \le \|\mathbf{x} - \mathbf{y}\|_\infty \sum_{i=1}^{n} \|\mathbf{e}_i\|$$

but $\sum_{i=1}^{n} \|\mathbf{e}_i\| = C$ (a constant $< \infty$ since n is finite)

$$\|\mathbf{x} - \mathbf{y}\| \le C \|\mathbf{x} - \mathbf{y}\|_\infty.$$

Now from the reverse triangle inequality

$$|\|\mathbf{x}\| - \|\mathbf{y}\|| \le \|\mathbf{x} - \mathbf{y}\| \le C \|\mathbf{x} - \mathbf{y}\|_\infty$$

and it therefore follows that $\|\mathbf{x}\|$ is a continuous function of the elements of $\mathbf{x}$.   Q.E.D.

There is an important special case:

COROLLARY 6.2.2.   Put $\mathbf{y} = \mathbf{0}$ in Theorem 6.2.1, then

$$\|\mathbf{x}\| \le C \|\mathbf{x}\|_\infty \quad \text{for all } \mathbf{x},$$

that is, every norm is *bounded above* by a constant multiple of the ∞-norm.

It can also be proved that every norm is also *bounded below* by a constant multiple of the ∞-norm, so that for any norm $\|\mathbf{x}\|$ there exist constants C_1, C_2 such that

$$C_1 \|\mathbf{x}\|_\infty \le \|\mathbf{x}\| \le C_2 \|\mathbf{x}\|_\infty. \tag{6.2.9}$$

It then follows that if $\|\mathbf{x}\|_\alpha$ and $\|\mathbf{x}\|_\beta$ are *any* two norms defined on the same set of vectors there exist constants k_1 and k_2 such that, for all vectors $\mathbf{x}$

$$k_1 \|\mathbf{x}\|_\alpha \le \|\mathbf{x}\|_\beta \le k_2 \|\mathbf{x}\|_\alpha. \tag{6.2.10}$$

These results enable us to introduce the concept of convergence of a sequence of vectors $\{\mathbf{x}^{(1)}, \mathbf{x}^{(2)}, \ldots, \mathbf{x}^{(k)}, \ldots\}$ to a vector $\mathbf{x}$.

DEFINITION 6.2.2.   A sequence of vectors $\mathbf{x}^{(1)}, \mathbf{x}^{(2)}, \ldots, \mathbf{x}^{(k)} \ldots$ is said to *converge* to a vector $\mathbf{x}$ if and only if

$$\|\mathbf{x} - \mathbf{x}^{(k)}\| \to 0 \quad \text{as } k \to \infty.$$

In view of (6.2.10) convergence in one norm $\|\mathbf{x}\|_\alpha$ defined on an infinite number of elements occurs if and only if it also occurs in any other norm $\|\mathbf{x}\|_\beta$ defined on the same set of elements.

6.2.3.   Matrix Norms

It is also necessary for us to have some measure of the 'size' of a matrix and this leads us to the concept of *matrix norms*. These are based upon vector norms, viz.

DEFINITION 6.2.3.   With each vector norm $\|\mathbf{x}\|$ we associate a corresponding *matrix norm*

$$\|\mathbf{A}\| = \max \|\mathbf{A}\mathbf{x}\|/\|\mathbf{x}\| \qquad (\mathbf{x} \neq \mathbf{0}) \ldots . \tag{6.2.11}$$

Alternatively, since $\|k\mathbf{x}\| = |k|\,\|\mathbf{x}\|$, (6.2.11) can be written

$$\|\mathbf{A}\| = \max \|\mathbf{A}\mathbf{x}\| \qquad (\|\mathbf{x}\| = 1) \ldots . \tag{6.2.12}$$

It can be shown that with this definition the matrix norm has the following properties

  (i) $\|\mathbf{A}\| > 0$ unless $\mathbf{A} = \mathbf{0}$;
  (ii) $\|k\mathbf{A}\| = |k|\,\|\mathbf{A}\|$ for any number k;
  (iii) $\|\mathbf{A} + \mathbf{B}\| \leq \|\mathbf{A}\| + \|\mathbf{B}\|$;
  (iv) $\|\mathbf{A}\mathbf{B}\| \leq \|\mathbf{A}\|\,\|\mathbf{B}\|$;

in particular note that (iv) implies $\|\mathbf{A}^m\| \leq (\|\mathbf{A}\|)^m$ for positive integers m.

The matrix norm as defined by (6.2.11) or (6.2.12) clearly depends upon the particular vector norm $\|\mathbf{x}\|$ used in the definition. We therefore say that the matrix norm $\|\mathbf{A}\|$ is *subordinate* to the vector norm $\|\mathbf{x}\|$. Thus corresponding to the vector norms $\|\mathbf{x}\|_1$, $\|\mathbf{x}\|_2$ and $\|\mathbf{x}\|_\infty$ we have the three matrix norms $\|\mathbf{A}\|_1$, $\|\mathbf{A}\|_2$ and $\|\mathbf{A}\|_\infty$ and we can show that

$$\|\mathbf{A}\|_1 = \max_j \sum_i |a_{ij}| \tag{6.2.13}$$

$$\|\mathbf{A}\|_\infty = \max_i \sum_j |a_{ij}| \tag{6.2.14}$$

and

$$\|\mathbf{A}\|_2 = (\text{maximum eigenvalue of } \mathbf{A}^*\mathbf{A})^{1/2} \tag{6.2.15}$$

[see I, § 7.1] where $\mathbf{A}^*$ is the complex conjugate transpose of $\mathbf{A}$ [see I (6.7.11)]. $\|\mathbf{A}\|_2$ is sometimes called the *spectral norm*.

Proofs of these results will be found in Atkinson (section 7.3).

Another important matrix norm is the *Euclidean norm*

$$\|\mathbf{A}\|_E = \left(\sum_i \sum_j |a_{ij}|^2 \right)^{1/2}. \tag{6.2.16}$$

This norm is however not subordinate to any vector norm and this is a disadvantage. It is bounded above and below by multiples of the 2-norm $\|\mathbf{A}\|_2$, in fact

$$\|\mathbf{A}\|_2 \leq \|\mathbf{A}\|_E \leq n^{1/2}\|\mathbf{A}\|_2, \qquad (6.2.17)$$

and matrices exist such that both bounds are attained.

It is worth noting that (6.2.13) and (6.2.14) can be stated as

$\|\mathbf{A}\|_1 =$ greatest sum of the absolute values of the
elements in any *column* of $\mathbf{A}$;

$\|\mathbf{A}\|_\infty =$ greatest sum of the absolute values of the
elements in any *row* of $\mathbf{A}$.

EXAMPLE 6.2.4.   Which of the conditions (i) to (iv) is *not* satisfied by the proposed norm

$$\max_{i,j} |a_{ij}|?$$

Give a counter-example.

Solution.   Condition (iv) need not be satisfied for if

$$\|\mathbf{A}\| = \max_{i,j} |a_{ij}| \quad \text{and} \quad \|\mathbf{B}\| = \max_{i,j} |b_{ij}|$$

it may happen that *every* element of a certain row of $\mathbf{A}$ equals $\|\mathbf{A}\|$ and *every* element of a certain column of $\mathbf{B}$ equals $\|\mathbf{B}\|$ in which case we would have

$$\|\mathbf{AB}\| = n\|\mathbf{A}\|\|\mathbf{B}\|.$$

This case occurs if we take $\mathbf{A}$ to be the matrix with top row all ones, everywhere else zero, and $\mathbf{B} = \mathbf{A}'$ [see I, § 6.5].

EXAMPLE 6.2.5.   Let $\mathbf{X}_n = \mathbf{A}^{-1}(\mathbf{I} + \mathbf{E}_n)$ be an approximation to $\mathbf{A}^{-1}$ [see I, § 6.4] where $\|\mathbf{E}_n\|$ is small. Define

$$\mathbf{X}_{n+1} = \mathbf{X}_n(2\mathbf{I} - \mathbf{A}\mathbf{X}_n).$$

Show that $\mathbf{X}_n$ converges to $\mathbf{A}^{-1}$ if $\|\mathbf{E}_0\| < 1$.

Proof

$$\mathbf{X}_{n+1} = \mathbf{A}^{-1}(\mathbf{I} + \mathbf{E}_n)(\mathbf{I} - \mathbf{E}_n)$$
$$= \mathbf{A}^{-1}(\mathbf{I} - \mathbf{E}_n^2)$$

hence if we write

$$\mathbf{X}_{n+1} = \mathbf{A}^{-1}(\mathbf{I} + \mathbf{E}_{n+1})$$

then

$$\mathbf{E}_{n+1} = -\mathbf{E}_n^2$$

so that

$$\|\mathbf{E}_{n+1}\| = \|-\mathbf{E}_n^2\| = \|\mathbf{E}_n^2\| \leq (\|\mathbf{E}_n\|)^2.$$

If $\|\mathbf{E}_0\| < 1$ it follows that $\|\mathbf{E}_n\|$ decreases to zero as $n \to \infty$ so that $\mathbf{X}_n \to \mathbf{A}^{-1}$.

Q.E.D.

EXAMPLE 6.2.5.   A permutation matrix (a_{ij}) has the property that exactly one element in each row and exactly one element in each column equals 1, the rest being zero [see I, § 6.7(xi)].
    Prove that if $\mathbf{A}$ is a permutation matrix then

$$\|\mathbf{A}\|_1 = \|\mathbf{A}\|_\infty = \|\mathbf{A}\|_2.$$

What about $\|\mathbf{A}\|_E$?

Solution.   (i) Since each row and each column contain just one non-zero element and that element is equal to one it is clear that

$$\sum_{i=1}^n |a_{ij}| = 1 = \sum_{j=1}^n |a_{ij}|$$

and so

$$\|\mathbf{A}\|_1 = \|\mathbf{A}\|_\infty = 1.$$

(ii) Every permutation matrix $\mathbf{A}$ is orthogonal and so satisfies the equation $\mathbf{A}'\mathbf{A} = \mathbf{A}^*\mathbf{A} = \mathbf{I}$ (as $\mathbf{A}$ is real, $\mathbf{A}^* = \mathbf{A}'$) [see I, § 6.7(vii)]. It follows from (6.2.15) that

$$\|\mathbf{A}\|_2 = 1,$$

because all eigenvalues of $\mathbf{I}$ are equal to unity.
    Thus $\|\mathbf{A}\|_1 = \|\mathbf{A}\|_\infty = \|\mathbf{A}\|_2 = 1$.
    (iii) Evidently,

$$\|\mathbf{A}\|_E = \left(\sum_{j=1}^n \sum_{i=1}^n |a_{ij}|^2 \right)^{1/2} = n^{1/2},$$

and we see that the upper bound in the equality (6.2.17) is attained when $\mathbf{A}$ is a permutation matrix.

EXAMPLE 6.2.6.   Let

$$A = \begin{pmatrix} 1 & 0 & 0 \\ 1 & 1 & 1 \\ 0 & 0 & 0 \end{pmatrix}$$

then

$$\|A\|_1 = \text{greatest absolute column sum} = 2,$$

$$\|A\|_\infty = \text{greatest absolute row sum} = 3.$$

Next,

$$A'A = \begin{pmatrix} 2 & 1 & 1 \\ 1 & 1 & 1 \\ 1 & 1 & 1 \end{pmatrix}$$

whose characteristic polynomial is

$$t^3 - 4t^2 + 2t = t(t - 2 - \sqrt{2})(t - 2 + \sqrt{2})$$

so that

$$\|A\|_2 = 2 + \sqrt{2}.$$

6.2.4. Approximation to Continuous Functions Using the Norms L_1, L_2, L_∞

So far we have considered approximations to functions using least squares: this is equivalent to using the norm L_2:

$$\left(\int_0^1 (f(x) - p(x))^2 \, dx \right)^{1/2}.$$

Likewise we can consider approximations using the norm L_p:

$$\left(\int_0^1 |f(x) - p(x)|^p \, dx \right)^{1/p}$$

for values of p other than $p = 2$. The most important cases are $p = 1$ and $p = \infty$. The former is simply defined

$$L_1 : \quad \int_0^1 |f(x) - p(x)| \, dx$$

but is difficult to use.

L_∞ is the Chebyshev, or maximum, norm and is interpreted as

$$L_\infty : \quad \max_{0 \le x \le 1} |f(x) - p(x)|.$$

We shall deal with this later (6.3.4). However to illustrate the use of the L_1 norm:

EXAMPLE 6.2.7.   Find the best approximation by a constant, c, to $x^k (k > 0)$ over $[0, 1]$ such that $\int_0^1 |x^k - c| \, dx$ is minimized (i.e. in the L_1 sense).

Solution.   For $0 < x < c^{1/k}$, $|x^k - c| = c - x^k$. For $c^{1/k} < x < 1$, $|x^k - c| = x^k - c$. Hence we want to minimize

$$I(c) = \int_0^{c^{1/k}} (c - x^k) \, dx + \int_{c^{1/k}}^1 (x^k - c) \, dx$$

as a function of c. Now

$$I(c) = \left[cx - \frac{x^{k+1}}{k+1}\right]_0^{c^{1/k}} + \left[\frac{x^{k+1}}{k+1} - cx\right]_{c^{1/k}}^1$$

$$= 2c^{1+1/k}\left(\frac{k}{k+1}\right) - c + \frac{1}{k+1}$$

and so

$$\frac{\partial I}{\partial c} = 2\left(\frac{k+1}{k}\right)(c^{1/k})\frac{k}{k+1} - 1 = 2c^{1/k} - 1$$

$$= 0 \quad \text{if } c = (\tfrac{1}{2})^k$$

and

$$\frac{\partial^2 I}{\partial c^2} = \frac{2}{k}c^{(1/k)-1} > 0 \quad \text{since } k > 0.$$

Hence $\qquad\qquad c = (\tfrac{1}{2})^k$ gives a minimum.

(Note that the least squares solution is $c = 1/(k+1)$ which is the same as $(\tfrac{1}{2})^k$ if $k = 1$, but not otherwise.)

The result above is a special case of a more general result, viz.

EXAMPLE 6.2.8. If $f(x)$ is twice differentiable and monotonic [see IV, Definition 2.7.1] over $[a, b]$ the best constant approximation to $f(x)$ in the L_1 sense is $f((a+b)/2)$.

Proof. Suppose $y = C$ gives an approximation to $f(x)$ over $[a, b]$. Since $f(x)$ is monotone, $y = f(x)$ intersects $y = C$ at most once; for an L_1 minimum it must intersect at least once, hence $y = C$ intersects $y = f(x)$ just once in $[a, b]$, let it be at $x = x_1$; then $C = f(x_1)$. We may suppose $f(x)$ monotonic increasing without loss of generality.

We must minimise

$$I = \int_a^b |f(x) - C| \, dx.$$

Suppose

$$f(x) < C \quad \text{for } a \le x < x_1$$

then

$$f(x) > C \quad \text{for } x_1 < x \le b.$$

Hence $\qquad I = \int_a^{x_1} (C - f(x)) \, dx + \int_{x_1}^b (f(x) - C) \, dx$

$$= C(2x_1 - a - b) + F(b) + F(a) - 2F(x_1)$$

$$= f(x_1)(2x_1 - a - b) + F(b) + F(a) - 2F(x_1)$$

where

$$F(x) = \int^x f(t)\, dt.$$

Now

$$\frac{dI}{dx} = f'(x)(2x - a - b)$$

and for a minimum $dI/dx = 0$, but $f'(x_1) \neq 0$, hence $x_1 = (a + b)/2$. Now

$$\frac{d^2 I}{dx_1^2} = f''(x_1)(2x_1 - a - b) + 2f'(x_1)$$

and so at $x_1 = (a + b)/2$,

$$\frac{d^2 I}{dx_1^2} = 2f'(x_1) > 0$$

since $f'(x_1) > 0$.

Thus I has a minimum at $((a + b)/2)$.

6.3. APPROXIMATIONS USING ORTHOGONAL FUNCTIONS

6.3.1. Orthogonal Functions

When we considered the least squares fitting of polynomials to continuous functions over $[0, 1]$ in section 6.1.5 we saw that the calculations inevitably involved the inverse of a Hilbert matrix and that this makes the use of the method, as described in section 6.1.5, unsatisfactory when the degree of the polynomial is even moderately high. If the matrix of the normal equations had been well-conditioned the method would have been much more useful. A particularly satisfactory class of matrices are those having all off-diagonal elements equal to zero (i.e. the class of diagonal matrices [see I (6.7.3)]) and we shall see that it is possible to approach the least squares fitting of polynomials in such a way that instead of the Hilbert matrix for the coefficients we obtain a diagonal matrix. This is achieved by the use of *orthogonal polynomials* instead of simple power terms $(1, x, x^2, \ldots)$. Orthogonal polynomials are of fundamental importance in many branches of mathematics in addition to approximation theory and their applications are numerous but we shall be mainly concerned with two special cases, the polynomials of Legendre and of Chebyshev. More general applications are however easily worked out once the general principles have been understood.

We begin by defining orthogonal functions:

DEFINITION 6.3.1.   A system of real functions $\phi_0(x)$, $\phi_1(x)$, ... defined in an interval $[a, b]$ is said to be *orthogonal* in this interval if

$$\int_a^b \phi_m(x)\phi_n(x)\, dx = \begin{cases} 0 & (m \neq n) \\ \lambda_n & (m = n) \end{cases}.$$

If $\lambda_0 = \lambda_1 = \ldots = 1$ the system is said to be *normal*. An orthogonal system which is also normal is sometimes referred to as an *orthonormal system*.

Note that since $\phi_n(x)$ is real $\lambda_n \geq 0$ and we shall assume that each $\phi_n(x)$ is continuous and non-zero so that $\lambda_n > 0$.

The advantages offered by the use of orthogonal functions in approximation theory can now be made clear as follows. Suppose $\{\phi_n(x)\}$ is an orthogonal system and that $f(x)$ is any function and we wish to express $f(x)$ in the form

$$f(x) = C_0\phi_0(x) + C_1\phi_1(x) + \ldots + C_n\phi_n(x) + \ldots . \qquad (6.3.1)$$

Then

$$\int_a^b f(x)\phi_n(x)\, dx = C_n \int_a^b \phi_n^2(x)\, dx = C_n\lambda_n$$

since all the other terms on the right-hand side are zero and so

$$C_n = \frac{1}{\lambda_n} \int_a^b f(x)\phi_n(x)\, dx. \qquad (6.3.2)$$

Thus the coefficients C_n in (6.3.1) can be found. These coefficients C_n are called the *Fourier coefficients of* $f(x)$, with respect to the system $\{\phi_n(x)\}$.

EXAMPLE 6.3.1.   The best-known example of an orthogonal system is the trigonometrical system

$$1, \cos x, \sin x, \cos 2x, \sin 2x, \ldots$$

over the interval $[-\pi, \pi]$.

Since (see IV, example 4.3.12)

$$\int_{-\pi}^{\pi} \begin{Bmatrix} \cos nx \\ \sin nx \end{Bmatrix} \begin{Bmatrix} \cos mx \\ \sin mx \end{Bmatrix} dx = 0 \qquad (m \neq n)$$

and

$$\int_{-\pi}^{\pi} \cos nx \sin nx\, dx = 0$$

whereas

$$\int_{-\pi}^{\pi} \sin^2 nx\, dx = \int_{-\pi}^{\pi} \cos^2 nx\, dx = \pi$$

and, finally,

$$\int_{-\pi}^{\pi} 1^2\, dx = 2\pi$$

we have the values $\lambda_1 = 2\pi$, $\lambda_2 = \lambda_3 = \ldots = \pi$. It follows therefore that the system

$$\frac{1}{\sqrt{2\pi}}, \frac{\cos x}{\sqrt{\pi}}, \frac{\sin x}{\sqrt{\pi}}, \frac{\cos 2x}{\sqrt{\pi}}, \frac{\sin 2x}{\sqrt{\pi}}, \ldots$$

is orthogonal and normal.

DEFINITION 6.3.2.   Orthogonality with respect to a weight function. A series of functions $\{\phi_n(x)\}$ are said to be *orthogonal with respect to the weight function $w(x)$ over (a, b)* if

$$\int_a^b \phi_m(x)\phi_n(x)w(x)\,dx = \begin{cases} 0 & \text{if } m \neq n \\ \lambda_n & \text{if } m = n \end{cases}.$$

6.3.2.   The Legendre Polynomials

When we try to find good polynomial approximations to a given function $f(x)$ we are trying to represent $f(x)$ in the form

$$f(x) \doteqdot \sum_{k=0}^{n} C_k x^k \qquad (6.3.3)$$

which is of the form (6.3.1) with $\phi_k(x) = x^k$; unfortunately the set $1, x, x^2, \ldots$ is not orthogonal over any non-zero interval as may be seen at once since, for example

$$\int_a^b \phi_1(x)\phi_3(x)\,dx = \int_a^b x^4\,dx > 0$$

which contradicts the assertion that $\{x^k\}$ is orthogonal. It is however possible to construct a set of polynomials $P_0(x), P_1(x), P_2(x), \ldots, P_n(x), \ldots$ where $P_n(x)$ is of degree n which are orthogonal over the interval $[-1, +1]$, and from these a set of polynomials orthogonal over any given finite interval $[a, b]$ can be obtained.

The method for finding a set of polynomials which are orthogonal and normal over $[-1, +1]$ is a relatively simple one and we illustrate it by finding the first three such polynomials.

Let $q_0(x) = a_{00}$, $q_1(x) = a_{11}x + a_{10}q_0(x)$, $q_2(x) = a_{22}x^2 + a_{21}q_1(x) + a_{20}q_0(x)$ be three orthonormal polynomials of degrees 0, 1 and 2 over $[-1, +1]$. We proceed to evaluate the coefficients.

(1) Since $q_0(x)$ is orthonormal

$$\int_{-1}^{+1} q_0^2(x)\,dx = 1$$

hence $a_{00}^2 = \frac{1}{2}$.

There are therefore two possible solutions: $a_{00} = 1/\sqrt{2}$ and $a_{00} = -1/\sqrt{2}$.

We resolve the ambiguity (and all subsequent ambiguities) by choosing the sign that makes the leading coefficient of the polynomial positive.

Hence $a_{00} = 1/\sqrt{2}$, i.e.

$$q_0(x) = \frac{1}{\sqrt{2}}. \tag{6.3.4}$$

(2) $$\int_{-1}^{+1} q_1(x)q_0(x)\,dx = 0 \quad \text{and} \quad \int_{-1}^{+1} q_1^2(x)\,dx = 1$$

so that:

$$\frac{1}{\sqrt{2}} \int_{-1}^{+1} \left(a_{11}x + \frac{a_{10}}{\sqrt{2}}\right) dx = 0 \quad \text{which gives } a_{10} = 0$$

and

$$\int_{-1}^{+1} \left(a_{11}x + \frac{a_{10}}{\sqrt{2}}\right)^2 dx = 1 \quad \text{which gives } a_{11}^2 = \tfrac{3}{2};$$

we choose $a_{11} = \sqrt{\tfrac{3}{2}}$ so that

$$q_1(x) = x\sqrt{\tfrac{3}{2}}. \tag{6.3.5}$$

(3) $$\int_{-1}^{+1} q_2(x)q_0(x)\,dx = \int_{-1}^{+1} q_2(x)q_1(x)\,dx = 0$$

and

$$\int_{-1}^{+1} q_2^2(x)\,dx = 1.$$

The first condition tells us that $a_{22} = -3a_{20}$ and the second condition tells us that $a_{21} = 0$, and so the third condition reduces to

$$\int_{-1}^{+1} (a_{22}x^2 - \tfrac{1}{3}a_{22})^2\,dx = 1$$

which gives $a_{22} = \tfrac{3}{2}\sqrt{\tfrac{5}{2}}$ so that we have

$$q_2(x) = \sqrt{\frac{5}{2}}\left(\frac{3x^2 - 1}{2}\right). \tag{6.3.6}$$

This process may be continued and it will be found that

$$q_3(x) = \sqrt{\frac{7}{2}}\left(\frac{5x^3 - 3x}{2}\right) \tag{6.3.7}$$

$$q_4(x) = \frac{3}{\sqrt{2}}\left(\frac{35x^4 - 30x^2 + 3}{8}\right) \tag{6.3.8}$$

$$q_5(x) = \sqrt{\frac{11}{2}}\left(\frac{63x^5 - 70x^3 + 15x}{8}\right) \tag{6.3.9}$$

etc.

It will be observed that $q_n(x)$ apparently takes the form

$$q_n(x) = \sqrt{\frac{2n+1}{2}} P_n(x)$$

where $P_n(x)$ is a polynomial of degree n with rational coefficients [see I, § 14.1]. This is indeed the case and the polynomials $P_n(x)$ are known as the *Legendre Polynomials*. The Legendre polynomials may be defined in various ways and have many interesting properties [see IV, § 10.3], the more important of which are summarised in the following theorem, a proof of which may be found in Whittaker and Watson (Chapter XV).

THEOREM 6.3.1.   *If $P_n(x)$ is defined by Rodrigues' formula* [cf. IV, § 10.3.2]

$$P_n(x) = \frac{1}{2^n n!} \frac{d^n}{dx^n} (x^2 - 1)^n,$$

then $P_n(x)$ has the following properties

(i) $$\int_{-1}^{+1} P_n(x) P_m(x)\, dx = \begin{cases} 0 & \text{if } m \neq n \\ \dfrac{2}{2n+1} & \text{if } m = n; \end{cases}$$

(ii) *$P_n(x)$ satisfies the recurrence equation* [see I, § 14.13]

$$P_{n+1}(x) = \left(\frac{2n+1}{n+1}\right) x P_n(x) - \left(\frac{n}{n+1}\right) P_{n-1}(x);$$

(iii) *$P_n(x)$ satisfies the differential equation* [see IV, § 7.3]

$$(1 - x^2) P_n''(x) - 2x P_n'(x) + n(n+1) P_n(x) = 0.$$

It follows at once from (i) of the theorem that $\{P_n(x)\}$ forms an orthogonal, but not normal, set over $[-1, +1]$ with respect to the weight function $w(x) = 1$ and that

$$\{q_n(x)\} = \left\{ \sqrt{\frac{2n+1}{2}} P_n(x) \right\}$$

forms an orthonormal set, and so provides the set of polynomials we are seeking.

Property (ii) enables us to compute the polynomials quite easily. Thus since $P_0(x) = 1$ and $P_1(x) = x$

$$P_2(x) = (\tfrac{3}{2}) x(x) - \tfrac{1}{2}(1) = \tfrac{1}{2}(3x^2 - 1)$$

and

$$P_3(x) = \left(\frac{5}{3}\right) x \left(\frac{3x^2 - 1}{2}\right) - \left(\frac{2}{3}\right) x = \tfrac{1}{2}(5x^3 - 3x)$$

so that

$$q_2(x) = \sqrt{\frac{5}{2}}\left(\frac{3x^2-1}{2}\right), \qquad q_3(x) = \sqrt{\frac{7}{2}}\left(\frac{5x^3-3x}{2}\right)$$

in agreement with (6.3.6) and (6.3.7).

The more general problem of finding a set of polynomials $\{\phi_n(x)\}$ which are orthonormal with respect to a given weight function $w(x)$ over an interval $[a, b]$ is usually solved by a method known as the Gram–Schmidt orthogonalisation method. An explanation of this method is given in section 6.3.9.

6.3.3. Least Squares Approximation by Legendre Polynomials

Let $f(x)$ be any function defined over $[-1, +1]$ and let $L_n(x) = \sum_{k=0}^{n} a_k P_k(x)$ be a linear combination of Legendre polynomials; we shall now determine what values of the coefficients $\{a_k\}$ will make $L_n(x)$ the best approximation to $f(x)$ in the least squares sense over $[-1, +1]$. Our objective is to minimize

$$I(a_0, a_1, \ldots, a_n) = \int_{-1}^{+1} (f(x) - L_n(x))^2\, dx$$

and so, as in (6.1.4) we must have

$$\frac{\partial I}{\partial a_r} = 0 \qquad (r = 0, 1, \ldots, n)$$

which is equivalent to

$$\int_{-1}^{+1} P_r(x)(f(x) - \sum_{k=0}^{n} a_k P_k(x))\, dx = 0, \qquad (r = 0, 1, \ldots, n)$$

and by the orthogonality property of the Legendre polynomials this reduces to

$$a_r = \frac{2r+1}{2} \int_{-1}^{+1} f(x) P_r(x)\, dx. \tag{6.3.10}$$

When the coefficients $\{a_r\}$ have been found $L_n(x)$ can be re-arranged, if desired, as a polynomial in powers of x that is,

$$\sum_{k=0}^{n} a_k P_k(x) = \sum_{k=0}^{n} b_k x^k$$

which provides a solution to the least squares polynomial approximation problem of (6.1.4) and avoids the introduction of the Hilbert matrix (or a matrix of similar type). The evaluation of the integrals on the right-hand side of (6.3.10) may have to be done numerically but the same would also then be true in (6.1.4) since $P_k(x)$ is a polynomial.

EXAMPLE 6.3.2.   Find the fourth degree least squares polynomial to $|x|$ over $[-1, +1]$ by means of Legendre polynomials.

Solution.   Let the polynomial be

$$\sum_{k=0}^{4} a_k P_k(x);$$

then, from (6.3.10)

$$a_r = \frac{2r+1}{2} \int_{-1}^{+1} |x| P_r(x)\, dx \qquad (r = 0, 1, 2, 3, 4).$$

Hence

$$a_0 = \tfrac{1}{2} \int_{-1}^{+1} |x| \cdot 1\, dx = \tfrac{1}{2}$$

$$a_1 = \tfrac{3}{2} \int_{-1}^{+1} |x| \cdot x\, dx = 0$$

$$a_2 = \tfrac{5}{2} \int_{-1}^{+1} |x| \frac{(3x^2 - 1)}{2}\, dx = \tfrac{5}{2} \int_0^1 (3x^3 - x)\, dx = \tfrac{5}{8}$$

$$a_3 = \tfrac{7}{2} \int_{-1}^{+1} |x| \left(\frac{5x^3 - 3x}{2}\right) dx = 0$$

$$a_4 = \tfrac{9}{2} \int_{-1}^{+1} |x| \left(\frac{35x^4 - 30x^2 + 3}{8}\right) dx = \tfrac{9}{8} \int_0^1 (35x^5 - 30x^3 + 3x)\, dx = -\tfrac{3}{16}.$$

The required polynomial is therefore

$$\tfrac{1}{2} P_0(x) + \tfrac{5}{8} P_2(x) - \tfrac{3}{16} P_4(x). \qquad (6.3.11)$$

The expression (6.3.11) can be converted to normal polynomial form and will be found to be

$$\tfrac{1}{128}(15 + 210x^2 - 105x^4), \qquad (6.3.12)$$

which is therefore the least squares polynomial for $|x|$ over $[-1, +1]$.

The result may be verified directly by using (6.1.10); if the least squares polynomial is $\sum_{k=0}^{4} a_k x^k$ then the linear equations for the coefficients in this case are

$$a_0 \qquad + \frac{a_2}{3} \qquad + \frac{a_4}{5} \qquad = \tfrac{1}{2}$$

$$\frac{a_1}{3} \qquad + \frac{a_3}{5} \qquad = 0$$

$$\frac{a_0}{3} \qquad + \frac{a_2}{5} \qquad + \frac{a_4}{7} \qquad = \tfrac{1}{4}$$

$$\frac{a_1}{5} \qquad + \frac{a_3}{7} \qquad = 0$$

$$\frac{a_0}{5} \qquad + \frac{a_2}{7} \qquad + \frac{a_4}{9} \qquad = \tfrac{1}{6}.$$

We deduce at once that $a_1 = a_3 = 0$ and (rather more slowly) that

$$a_0 = \tfrac{15}{128}, \qquad a_2 = \tfrac{210}{128}, \qquad a_4 = -\tfrac{105}{128}$$

in agreement with (6.3.12).

6.3.4.   The Chebyshev Polynomials

We begin by considering the following problem. Suppose that $p_n(x) = x^n + a_{n-1}x^{n-1} + a_{n-2}x^{n-2} + \ldots + a_0$ is a polynomial with leading coefficient $= 1$. Let

$$\alpha_n = \sup_{-1 \le x \le 1} |p_n(x)|.$$

PROBLEM.   For what polynomial $p_n(x)$ is α_n as small as possible?

We solve the problem by introducing the Chebyshev polynomials and then show that these have the desired property.

DEFINITION 6.3.3.   The Chebyshev polynomial of degree n, $T_n(x)$, is defined by:

$$T_n(x) = \cos(n \arccos x).$$

In order to derive $T_n(x)$ we simply find the expressions for $\cos(ny)$ in terms of powers of $\cos(y)$ and then write $x = \cos(y)$. Thus:

$$T_0(x) = \cos(0y) = 1$$
$$T_1(x) = \cos(1y) = \cos y = x$$
$$T_2(x) = \cos(2y) = 2\cos^2 y - 1 = 2x^2 - 1$$
$$T_3(x) = \cos(3y) = 4\cos^3 y - 3\cos y = 4x^3 - 3x.$$

An alternative method for finding $T_n(x)$ will be proved below (Theorem 6.3.5).

LEMMA 6.3.2.   *In the interval $-1 \le x \le 1$ the Chebyshev polynomial $T_n(x)$ satisfies*

  (i) $-1 \le T_n(x) \le +1$;
  (ii) $|T_n(x)| = 1$ *at* $(n+1)$ *points* $x_0, x_1, \ldots, x_n$ *where* $x_\nu = \cos(\nu\pi/n)$, $\nu = 0, 1, 2, \ldots, n$ $(n \ge 1)$;
  (iii) $T_n(x_\nu) = (-1)^\nu$;
  (iv) *The leading term in* $T_n(x)$ *is* $2^{n-1}x^n$.

Proof.   (i) Since $T_n(x) = \cos(n \arccos x)$ and x is real and confined to the interval $-1 \le x \le 1$ it follows that $T_n(x) = \cos(z)$ where z is real.
  Hence $-1 \le T_n(x) \le 1$.

(ii) $|T_n(x)| = |\cos(z)| = 1$ implies that $z = k\pi$, hence $n \arccos(x) = k\pi$, $k = 0$, $\pm 1, \pm 2, \ldots$

Hence
$$x = \cos\left(\frac{k\pi}{n}\right),$$

and there are just $(n+1)$ distinct values of x in $[-1, 1]$ given by $x = 1$ and $x = \cos(\pm k\pi/n)$, $k = 1, 2, \ldots n$.

(iii) $T_n\left(\cos\dfrac{k\pi}{n}\right) = \cos\left(n\cos^{-1}\cos\left(\dfrac{k\pi}{n}\right)\right) = \cos(k\pi) = (-1)^k$.

(iv) $\cos(n\theta) = \mathrm{Re}(\cos\theta + i\sin\theta)^n$ [see I (2.7.29)]

$$= \cos^n\theta - \binom{n}{2}\cos^{n-2}\theta\sin^2\theta + \binom{n}{4}\cos^{n-4}\theta\sin^4\theta - \ldots$$

[see I (3.10.1)]

$$= \cos^n\theta - \binom{n}{2}\cos^{n-2}\theta(1-\cos^2\theta)$$

$$+ \binom{n}{4}\cos^{n-4}\theta(1-\cos^2\theta)^2 \ldots$$

Hence $\cos(n\theta) = A_n\cos^n\theta + \ldots,$

where

$$A_n = 1 + \binom{n}{2} + \binom{n}{4} + \ldots$$

$$= \tfrac{1}{2}\{(1+1)^n + (1-1)^n\} = \tfrac{1}{2}(2^n) = 2^{n-1}.$$ Q.E.D.

We can now prove the theorem which establishes the key role of the Chebyshev polynomials in min–max approximation.

THEOREM 6.3.3.   Of all polynomials of the nth degree with leading coefficient 1 the one with minimum maximum absolute value in the interval $[-1, 1]$ is $2^{1-n}T_n(x)$.

Proof.   (i) By virtue of (iv) in the lemma
$$2^{1-n}T_n(x) = x^n + \ldots$$

and so has leading coefficient 1. It is thus one of the class of polynomials under consideration.

(ii) By virtue of (i), (ii) and (iii) of the lemma
$$|2^{1-n}T_n(x)| \leq \tfrac{1}{2}^{n-1}   \text{ for } -1 \leq x \leq 1,$$

and equality holds at $(n+1)$ points $x_0, x_1, \ldots, x_n$.

(iii) Suppose that there exists a polynomial
$$p_n(x) = x^n + \ldots$$

such that

$$\sup_{-1 \le x \le 1} |p_n(x)| < \tfrac{1}{2}^{n-1}.$$

Consider $2^{1-n} T_n(x) - p_n(x)$—a polynomial of degree at most $(n-1)$, then

$$2^{1-n} T_n(x_0) - p_n(x_0) = \frac{1}{2^{n-1}} - p(x_0) > 0$$

$$2^{1-n} T_n(x_1) - p_n(x_1) = \frac{-1}{2^{n-1}} - p(x_1) < 0$$

$$\vdots$$

$$2^{1-n} T_n(x_n) - p_n(x_n) = \frac{(-1)^n}{2^{n-1}} - p(x_n) \begin{cases} >0 & \text{if } n \text{ is even} \\ <0 & \text{if } n \text{ is odd.} \end{cases}$$

Thus $2^{1-n} T_n(x) - p_n(x)$ changes sign at least n times in the interval $[-1, 1]$ and therefore has at least n roots, but this is impossible for a polynomial of degree $(n-1)$ unless it vanishes identically. Thus equality between $p_n(x)$ and $2^{1-n} T_n(x)$ must hold everywhere, that is, $2^{1-n} T_n(x)$ is the required polynomial.

Q.E.D.

The next important step is to establish that the Chebyshev polynomials are orthogonal with respect to the weight function $(1 - x^2)^{-1/2}$.

THEOREM 6.3.4

$$\int_{-1}^{+1} \frac{T_m(x) T_n(x)}{\sqrt{(1-x^2)}} \, dx = \begin{cases} 0 & m \ne n \\ \tfrac{1}{2}\pi & m = n \ne 0 \\ \pi & m = n = 0. \end{cases}$$

Proof. Let $x = \cos \theta$; then $dx = -\sin \theta \, d\theta$ and so

$$I_{m,n} = \int_{-1}^{+1} \frac{T_m(x) T_n(x)}{\sqrt{(1-x^2)}} \, dx = \int_0^\pi T_m(\cos \theta) T_n(\cos \theta) \, d\theta.$$

Now

$$T_m(\cos \theta) = \cos (m \cos^{-1} (\cos \theta)) = \cos (m\theta)$$

hence

$$I_{m,n} = \int_0^\pi \cos (m\theta) \cos (n\theta) \, d\theta$$

$$= \frac{1}{2} \int_0^\pi (\cos (m+n)\theta + \cos (m-n)\theta) \, d\theta$$

i.e.

$$I_{m,n} = \begin{cases} 0 & \text{if } m \ne n \\ \tfrac{1}{2}\pi & \text{if } m = n \ne 0 \\ \pi & \text{if } m = n = 0. \end{cases} \qquad \text{Q.E.D.}$$

The expression for $T_n(x)$ as a polynomial of degree n in x may be worked out directly from its definition but a much simpler method is available and is given by the following theorem.

THEOREM 6.3.5.   For $n \geq 1$

$$T_{n+1}(x) = 2xT_n(x) - T_{n-1}(x). \tag{6.3.13}$$

Proof.   In the identity [see V (1.2.20)]

$$\cos A + \cos B = 2 \cos \left(\frac{A-B}{2} \right) \cos \left(\frac{A+B}{2} \right)$$

put

$$A = (n+1) \arccos (x), \qquad B = (n-1) \arccos (x)$$

then

$$\cos ((n+1) \arccos (x)) = 2x \cos (n \arccos (x)) - \cos ((n-1) \arccos (x))$$

i.e.

$$T_{n+1}(x) = 2xT_n(x) - T_{n-1}(x). \qquad \text{Q.E.D.}$$

Since $T_0(x) = 1$, $T_1(x) = x$, (6.3.13) tells us that

$$T_2(x) = 2x(x) - 1 = 2x^2 - 1$$
$$T_3(x) = 2x(2x^2 - 1) - x = 4x^3 - 3x$$

and so on.

6.3.5.   Expansion as a Series of Chebyshev Polynomials

If we have a function $f(x)$ which we wish to approximate with a series of Chebyshev polynomials

$$f(x) \doteq \tfrac{1}{2}c_0 + c_1 T_1(x) + c_2 T_2(x) + \ldots + c_n T_n(x)$$

how can we find the coefficients c_j?

The theoretical method is to multiply $f(x)$ by $T_j(x)/\sqrt{(1-x^2)}$ and integrate over $[-1, 1]$, making use of the orthogonality property of $T_j(x)$. Thus:

$$\int_{-1}^{1} \frac{f(x)T_j(x)}{\sqrt{(1-x^2)}} \, dx \doteq \tfrac{1}{2}c_0 \int_{-1}^{1} \frac{T_j(x)}{\sqrt{1-x^2}} \, dx + \sum_{k=1}^{n} c_k \int_{-1}^{1} \frac{T_j(x)T_k(x)}{\sqrt{1-x^2}} \, dx.$$

The only term on the right which doesn't vanish is the one where $k = j$ and then we have

$$c_j = \frac{2}{\pi} \int_{-1}^{1} \frac{f(x)T_j(x)}{\sqrt{(1-x^2)}} \, dx. \tag{6.3.14}$$

The evaluation of the integral for c_j given by (6.3.14) will in general have to be done numerically and in such cases it is obviously important to ensure that the truncation error is sufficiently small or the accuracy available via the Chebyshev approximation to $f(x)$ will be reduced. In a few special cases the integral can be evaluated analytically and the problem of truncation error does not arise; the most important such case is when $f(x) = x^n$ ($n \geq 0$) and we shall deal with this case below; but first we look at an example where evaluation of (6.3.14) must be done numerically.

EXAMPLE 6.3.3. Find a cubic approximation to e^x by using Chebyshev polynomials.

Solution. Let the approximation be

$$e^x \doteq \tfrac{1}{2}c_0 T_0(x) + c_1 T_1(x) + c_2 T_2(x) + c_3 T_3(x)$$

then, from (6.3.14)

$$c_j = \frac{2}{\pi} \int_{-1}^{+1} \frac{e^x T_j(x)}{\sqrt{1-x^2}} \, dx \qquad (j = 0, 1, 2, 3).$$

The substitution $x = \cos \theta$ transforms the integral [see IV, § 4.3] to

$$c_j = \frac{2}{\pi} \int_0^\pi e^{\cos\theta} \cos(j\theta) \, d\theta \qquad (6.3.15)$$

which is better from a numerical point of view since the integrand no longer contains a singularity.

In evaluating integrals containing a periodic function [see IV, § 2.12] as a factor in the integrand it is usually best to make use of the simplest quadrature formulae, such as the midpoint rule or trapezium rule (see Chapter 7). By using either method the coefficients c_j can be evaluated for a series of decreasing step-sizes and the results compared, some confidence thus being established in the accuracy of the results. Thus using the midpoint rule with step-sizes $\pi/2^k$ ($k = 1, 2, 3, 4$) we obtain the following estimates for c_0

k	estimate	
1	2·521184	(6 d.p)
2	2·532131	(6 d.p)
3	2·53213176	(8 d.p)
4	2·53213176	(8 d.p)

and we conclude that $c_0 = 2 \cdot 53213176$ to 8 d.p.

The other coefficients are evaluated similarly and we find (to 8 d.p)

$$c_1 = 1 \cdot 13031821, \qquad c_2 = 0.27149534, \qquad c_3 = 0 \cdot 04433685$$

so that the required approximation is

$$e^x \doteqdot 1 \cdot 26606588 \, T_0(x) + 1 \cdot 13031821 \, T_1(x) + 0 \cdot 27149534 \, T_2(x)$$
$$+ 0 \cdot 04433685 \, T_3(x). \tag{6.3.16}$$

It is not necessary to re-order (6.3.16) in powers of x for this formula may be used directly for the computation of approximations to e^x by using (6.3.13); thus, taking $x = 0 \cdot 8$, for example: we have

$$T_0(0 \cdot 8) = 1, \qquad T_1(0 \cdot 8) = 0 \cdot 8$$

and so, from (6.3.13)

$$T_2(0 \cdot 8) = 2(0 \cdot 8)(0 \cdot 8) - 1 = 0 \cdot 28$$

and

$$T_3(0 \cdot 8) = 2(0 \cdot 8)(0 \cdot 28) - 0 \cdot 8 = -0 \cdot 352$$

and (6.3.16) then gives (rounded to 4 d.p)

$$e^{0 \cdot 8} \doteqdot 2 \cdot 2307.$$

The correct value to 4 d.p. is $2 \cdot 2255$.

By comparison the cubic approximation obtained by truncating the Taylor series for e^x [see IV (2.11.1)] after 4 terms gives $e^{0 \cdot 8} \doteqdot 2 \cdot 2053$ with an error almost 4 times as large. For small values of x however the Taylor series cubic will give better results e.g. at $x = 0 \cdot 2$:

$$(6.3.16) \text{ gives } e^{0 \cdot 2} \doteqdot 1 \cdot 2172 \quad (4 \text{ d.p.}).$$

The Taylor series cubic gives $e^{0 \cdot 2} \doteqdot 1 \cdot 2213$ and in fact $e^{0 \cdot 2} \doteqdot 1 \cdot 2214$ which illustrates the point that Chebyshev approximations do not necessarily produce the best approximations at any given point in the interval $[-1, +1]$ but they do guarantee to minimize the greatest error in the interval. In general it frequently happens that several approximation formulae are available and each will have its own advantages and disadvantages; in particular, different formulae may give the best results over different parts of the interval of approximation and it may require considerable analysis to decide which to use at any point.

We now return to the special case when $f(x) = x^n (n \geq 0)$ mentioned earlier. The importance of this case lies in its role in the method of economization which we shall discuss in section 6.3.6. The Chebyshev representations for x^n are very easily obtained by solving the Chebyshev polynomials successively, viz.

$$T_0(x) = 1, \quad \text{hence } 1 = T_0(x)$$

$$T_1(x) = x, \quad \text{hence } x = T_1(x)$$

$$T_2(x) = 2x^2 - 1 = 2x^2 - T_0(x), \quad \text{hence } x^2 = \tfrac{1}{2}(T_2(x) + T_0(x))$$

$$T_3(x) = 4x^3 - 3x = 4x^3 - 3T_1(x), \quad \text{hence } x^3 = \tfrac{1}{4}(T_3(x) + 3T_1(x))$$

and so on, the successive polynomials being obtained from the 3-term recurrence formula (6.3.13) proved earlier:

$$T_{n+1}(x) = 2xT_n(x) - T_{n-1}(x).$$

Thus

$$T_4(x) = 2xT_3(x) - T_2(x)$$
$$= 2x(4x^3 - 3x) - (2x^2 - 1)$$
$$= 8x^4 - 8x^2 + 1$$
$$= 8x^4 - 4(T_2(x) + T_0(x)) + T_0(x)$$

so that

$$x_4 = \tfrac{1}{8}(T_4(x) + 4T_2(x) + 3T_0(x)).$$

Since we can express x^k as a linear combination of $T_k(x), T_{k-1}(x), \ldots, T_0(x)$ we can convert any power series expansion of an arbitrary function $f(x)$ into an expansion in a series of Chebyshev polynomials.

EXAMPLE 6.3.4.   Convert the first 5 terms of the Taylor series expansion for e^x into Chebyshev polynomials.

Solution

$$e^x = 1 + x + \tfrac{1}{2}x^2 + \tfrac{1}{6}x^3 + \tfrac{1}{24}x^4 + \ldots$$
$$= T_0(x) + T_1(x) + \tfrac{1}{4}(T_2(x) + T_0(x)) + \tfrac{1}{24}(T_3(x) + 3T_1(x))$$
$$+ \tfrac{1}{192}(T_4(x) + 4T_2(x) + 3T_0(x))$$
$$= (1 + \tfrac{1}{4} + \tfrac{1}{64})T_0(x) + (1 + \tfrac{1}{8})T_1(x) + (\tfrac{1}{4} + \tfrac{1}{48})T_2(x)$$
$$+ \tfrac{1}{24}T_3(x) + \tfrac{1}{192}T_4(x)$$

i.e.

$$e^x \doteq \tfrac{81}{64}T_0(x) + \tfrac{9}{8}T_1(x) + \tfrac{13}{48}T_2(x) + \tfrac{1}{24}T_3(x) + \tfrac{1}{192}T_4(x).$$

Now we notice an interesting fact. If we truncate the Taylor series for e^x after the term $\tfrac{1}{6}x^3$ the error is $\tfrac{1}{24}x^4 + x^5/5! + \text{—} = \tfrac{1}{24}x^4(1 + (x/5) + \ldots)$ and in the range $[-1, +1]$ the maximum error will be at $x = 1$ and will exceed $\tfrac{1}{24}(\tfrac{6}{5}) \doteq 0{\cdot}05$.

If, on the other hand we truncate the Chebyshev series expansion above at the term $\tfrac{1}{24}T_3(x)$—which only involves terms up to and including x^3—the error will be $\tfrac{1}{192}T_4(x) + (x^5/5!) + \ldots$ and so, in the interval $-1 \leq x \leq 1$ we have

$$\left| e^x - \tfrac{81}{64}T_0(x) - \tfrac{9}{8}T_1(x) - \tfrac{13}{48}T_2(x) - \tfrac{1}{24}T_3(x) \right| = \left| \tfrac{1}{192}T_4(x) + \frac{x^5}{5!} + \ldots \right|.$$

Now $|T_4(x)| \leq 1$ and

$$\left| \frac{x^5}{5!} + \frac{x^6}{6!} + \ldots \right| < \frac{1}{5!}\left(1 + \frac{1}{6} + \frac{1}{6^2} + \ldots\right)$$

and so

$$\left|\tfrac{1}{192}T_4(x)+\frac{x^5}{5!}+\dots\right|<\tfrac{1}{192}+\tfrac{1}{120}(\tfrac{6}{5})\doteq 0\cdot 015.$$

Thus the maximum error on the interval has been reduced by a factor of more than 3 compared to that obtained by power series truncation. Thus we have found a *cubic*, namely,

$$\tfrac{81}{64}T_0(x)+\tfrac{9}{8}T_1(x)+\tfrac{13}{48}T_2(x)+\tfrac{1}{24}T_3(x) \tag{6.3.17}$$

which, on the interval $[-1, +1]$ gives a significantly better representation of e^x in the L_∞ sense, than the first 4 terms of its power series.

It is interesting to compare (6.3.17) and (6.3.16) since both are cubic approximations to e^x obtained by the use of Chebyshev polynomials. Converting the coefficients of (6.3.17) to decimal form we have the following comparison

	$T_0(x)$	$T_1(x)$	$T_2(x)$	$T_3(x)$
(6.3.16)	1·26606588	1·13031821	0·27149534	0·04433685
(6.3.17)	1·26562500	1·12500000	0·27083333	0·04166667

Since both cubics provide good approximations we would expect them to have similar coefficients but they are not identical because (6.3.16) is the approximation to e^x using the first 4 Chebyshev polynomials whereas (6.3.17) is based upon the Chebyshev equivalent of the first 5 terms of the Taylor Series for e^x 'economized' to a cubic.

The technique of 'economization' is a very useful one and can lead to significant improvements in the accuracy obtainable from a polynomial approximation to a power series. In the next section we present the technique in the general case and in passing see how (6.3.17) may be more easily obtained.

6.3.6. Economization of Power Series by Chebyshev Polynomials

The technique adopted in the example on e^x can be formalized as follows.

Suppose that we have a power series expansion of a function $f(x)$, about $x = 0$

$$f(x) = a_0 + a_1 x + a_2 x^2 + \dots + a_n x^n + R_{n+1}(x).$$

If we truncate at $a_n x^n$ we have

$$\text{Truncation error} = |f(x) - a_0 - a_1 x - \dots - a_n x^n| = |R_{n+1}(x)|.$$

Let the interval of approximation be $[-1, +1]$ and suppose that

$$\max_{-1\le x\le 1} |R_{n+1}(x)| = \varepsilon_1.$$

Suppose that we are prepared to accept a truncation error over $[-1, +1]$ of $(\varepsilon_1 + \varepsilon_2)$, can we then 'economize' on the degree of the approximating polynomial? Can we, for example, ignore the term $a_n x^n$ in the expansion above? If we do, the truncation error will be

$$|a_n x^n + R_{n+1}(x)| \le |a_n| + \varepsilon_1$$

and if $|a_n| < \varepsilon_2$ the economization has been satisfactorily achieved. If however $|a_n| > \varepsilon_2$ the direct truncation obtained by dropping $a_n x^n$ is not acceptable. We therefore seek an alternative form of the truncation using the representation of x^n by Chebyshev polynomials. Now

$$x^n = \frac{1}{2^{n-1}}(T_n(x) + S_{n-2}(x))$$

where $S_{n-2}(x)$ is a sum of Chebyshev polynomials of orders $(n-2), (n-4) \ldots$. Thus

$$a_n x^n = \frac{a_n}{2^{n-1}} T_n(x) + \frac{a_n}{2^{n-1}} S_{n-2}(x).$$

We now write

$$f(x) = a_0 + a_1 x + a_2 x^2 + \ldots + \frac{a_n}{2^{n-1}} S_{n-2}(x) + \frac{a_n}{2^{n-1}} T_n(x) + R_{n+1}(x)$$

and so

$$\left| f(x) - a_0 - a_1 x - a_2 x^2 \ldots - \frac{a_n}{2^{n-1}} S_{n-2}(x) \right| = \left| \frac{a}{2^{n-1}} T_n(x) + R_{n+1}(x) \right|$$

Hence, since $\max |T_n(x)| = 1$ on $-1 \le x \le 1$

$$\max_{-1 \le x \le +1} \left| f(x) - a_0 - a_1 x_2 - a_2 x^2 - \ldots - \frac{a_n}{2^{n-1}} S_{n-2}(x) \right| \le \frac{a_n}{2^{n-1}} + \varepsilon_1$$

and it is now only necessary that $a_n < 2^{n-1} \varepsilon_2$ for the truncation error to be satisfactory and our approximating polynomial has degree at most $(n-1)$.

If $a_n/2^{n-1}$ is much smaller than ε_2 the process can now be repeated to reduce the degree of the approximating polynomial still further.

We illustrate the method by:

EXAMPLE 6.3.5. Find an approximation to x^4 of degree 2 over $[-1, +1]$ using Chebyshev polynomials.

Solution. We express x^4 as a multiple of $T_4(x)$ and other Chebyshev polynomials.
    Now

$$x^4 = \tfrac{1}{8}(T_4 + 4T_2 + 3T_0);$$

whence

$$x^4 = \tfrac{1}{8}T_4(x) + \tfrac{1}{2}T_2(x) + \tfrac{3}{8}T_0(x)$$

$$= \tfrac{1}{8}T_4(x) + \tfrac{1}{2}(2x^2 - 1) + \tfrac{3}{8}(1)$$

i.e. $x^4 = \tfrac{1}{8}T_4(x) + x^2 - \tfrac{1}{8}.$

Hence the required polynomial is $x^2 - \frac{1}{8}$ and

$$\max_{-1 \le x \le 1} |x^4 - x^2 + \tfrac{1}{8}| = \max_{-1 \le x \le 1} \left| \frac{T_4(x)}{8} \right| = \tfrac{1}{8}.$$

The same idea of replacing x^4 by a linear combination of T_4, T_2, and T_0 and then ignoring T_4 itself leads us quickly to the 'economized' cubic approximation which we found in Example 6.3.4 above, for e^x.

For

$$\frac{x^4}{24} = \frac{1}{192} T_4(x) + \frac{x^2}{24} - \frac{1}{192},$$

as we just found. Hence

$$1 + x + \frac{x^2}{2} + \frac{x^3}{6} + \frac{x^4}{24} = (1 - \tfrac{1}{192}) + x + x^2(\tfrac{1}{2} + \tfrac{1}{24}) + \frac{x^3}{6} + \frac{T_4(x)}{192}.$$

Hence

$$e^x = \frac{191}{192} + x + \frac{13x^2}{24} + \frac{x^3}{6} + \frac{T_4(x)}{192} + \frac{x^5}{5!} + \cdots$$

so that, if $-1 \le x \le 1$,

$$\left| e^x - \frac{191}{192} - x - \frac{13x^2}{24} - \frac{x^3}{6} \right| < \frac{1}{192} + \frac{x^5}{5!} \left(1 + \frac{x}{6} + \cdots \right)$$

$$< \frac{1}{192} + \frac{x^5}{20(6 - x)} \qquad (-1 \le x \le 1)$$

[see IV, example 1.7.1. Since

$$\frac{191}{192} + x + \frac{13x^2}{24} + \frac{x^3}{6} = \frac{81}{64} T_0(x) + \frac{9}{8} T_1(x) + \frac{13}{48} T_2(x) + \frac{1}{24} T_3(x),$$

we see that we have obtained (6.3.17) by a rather easier method.

The coefficients of the economized polynomials found in this way may have less simple coefficients than the original truncated power series; but this is unimportant when the calculations are being done on a computer.

6.3.7. Use of Chebyshev Polynomials over Intervals Other than $[-1, 1]$

If we wish to find good L_∞-type approximations to a function $f(x)$ over an interval $[a, b]$ other than $[-1, 1]$ we can make a transformation of the variable

$$u = \frac{(a + b - 2x)}{(a - b)},$$

which maps the interval $a \le x \le b$ into the interval $-1 \le u \le 1$. The interval $[0, 1]$ is the most important special case. In this case the transformation is

$u = 2x - 1$. It is convenient to use a modified Chebyshev polynomial for approximations over $[0, 1]$, viz.

$$T_n^*(x) = T_n(u) = T_n(2x - 1).$$

The polynomials $T_n^*(x)$ can now easily be found:

$$T_0^*(x) = 1, \qquad T_1^*(x) = 2x - 1, \qquad T_2^*(x) = 8x^2 - 8x + 1,$$

$$T_3^*(x) = 32x^3 - 48x^2 + 18x - 1, \quad \text{etc.}$$

Note the relationship between T_n^* and T_{2n}

$$T_n^*(x) = T_{2n}(\sqrt{x}).$$

For approximations over $[0, 1]$ the polynomials $T_n^*(x)$ are used and clearly

$$\max_{0 \leq x \leq 1} |T_n^*(x)| = \max_{-1 \leq u \leq 1} |T_n(u)| = 1$$

so that the technique of economization and the analysis of the truncation error are carried out as before, but using $T_n^*(x)$ in place of $T_n(x)$.

EXAMPLE 6.3.6. Investigate the maximum truncation error over the interval $0 \leq x \leq 1$ of the cubic approximation for $\log(1 + x)$: $x - \frac{1}{2}x^2 + \frac{1}{3}x^3$. Use Chebyshev polynomials to economize to a linear function and investigate the maximum truncation error over $[0, 1]$ of this linear approximation.

Solution

$$\log(1 + x) \doteqdot x - \tfrac{1}{2}x^2 + \tfrac{1}{3}x^3.$$

The truncation error is

$$\tfrac{1}{4}x^4 - \tfrac{1}{5}x^5 + \tfrac{1}{6}x^6 \ldots < \frac{1}{4 \cdot 5} + \frac{1}{6 \cdot 7} + \frac{1}{8 \cdot 9} + \ldots$$

$$< \frac{1}{20} + \frac{1}{6^2} + \frac{1}{8^2} + \ldots$$

$$= \frac{1}{20} + \frac{1}{2^2}\left(\frac{\pi^2}{6} - 1 - \frac{1}{2^2}\right) \quad \text{[see IV, (20.5.21)]}$$

$$= \frac{1}{20} + \frac{4\pi^2 - 30}{96} \doteqdot \frac{1}{20} + \frac{1}{10} = 0 \cdot 15.$$

Turning now to the Chebyshev approximation, since the interval is $[0, 1]$ we use $T_n^*(x)$, and as we must replace $\frac{1}{3}x^3$ in the first instance we have

$$\tfrac{1}{3}x^3 = \tfrac{1}{3} \cdot \tfrac{1}{32}(T_3^*(x) + 48x^2 - 18x + 1)$$

$$= \tfrac{1}{96}T_3^*(x) + \tfrac{1}{2}x^2 - \tfrac{3}{16}x + \tfrac{1}{96}.$$

Hence

$$x - \tfrac{1}{2}x^2 + \tfrac{1}{3}x^3 = \tfrac{13}{16}x + \tfrac{1}{96} + \tfrac{1}{96}T_3^*(x)$$

the term in x^2 vanishing identically. We thus have an economized linear approximation to $\log(1+x)$ valid over $[0, 1]$, viz.

$$\log_e(1+x) \doteq \frac{1}{96} + \frac{13x}{16}.$$

The truncation error is $\tfrac{1}{96}T_3^*(x) + (x^4/4) - (x^5/5) + \ldots$ and so

$$|\text{Truncation error}| < \tfrac{1}{96} + 0.15 \doteq 0.16 \text{ to 2 d.p.}$$

6.3.8. Chebyshev Interpolation

If the values of a function $f(x)$ are known at a set of points $x_1 < x_2 < \ldots < x_n$ we can construct a polynomial of degree $(n-1)$ which takes the values $f(x_i)$ at x_i $(i = 1, \ldots, n)$. The polynomial is unique and can be found in various ways including the use of Newton's divided difference formula or Lagrange's method [see § 2.3.4]. Lagrange's formula is more cumbersome to use in practice but it has the advantage that we can write down the required polynomial explicitly, viz. [2.3.22]

$$p(x) = \sum_{j=1}^{n} f(x_j) \prod_{\substack{i=1 \\ i \neq j}}^{n} \frac{(x - x_i)}{(x_j - x_i)}. \qquad (6.3.18)$$

If $f(x)$ is not a polynomial of degree $\leq(n-1)$ the error when we use $p(x)$ for interpolation can be shown to be [see Atkinson, 1978 p. 110]

$$E(x) = \prod_{i=1}^{n} (x - x_i) \frac{f^{(n)}(\alpha)}{n!}$$

where α is some number between x_1 and x_n. If the values $x_1, x_2, \ldots, x_n$ have been fixed we can do nothing to minimize $E(x)$ but if we can choose *any* n points within a specified interval it may be worthwhile choosing them in a particular way as we now show.

Suppose, for simplicity, that we are interested in values of x lying in the interval $-1 \leq x \leq +1$ and that we are free to choose any n points $x_1, \ldots, x_n$ in this interval for use in the interpolation formula (6.3.18). Now

$$\prod_{i=1}^{n} (x - x_i) \qquad (6.3.19)$$

is a polynomial of degree n with leading coefficient 1 and we saw in Theorem 6.3.3 that of all such polynomials the one with the minimum maximum value is $2^{-(n-1)}T_n(x)$. It follows therefore that if we wish to minimize (6.3.18) we should choose the x_i so that

$$\prod_{i=1}^{n} (x - x_i) = \frac{T_n(x)}{2^{n-1}}$$

and this is equivalent to saying that we should choose $x_1, x_2, \ldots, x_n$ to be the n roots of $T_n(x)$, viz., take

$$x_m = \cos\left(\frac{2m-1}{2n}\pi\right) \qquad (m = 1, 2, \ldots, n). \qquad (6.3.20)$$

Having chosen the x_i in this way we then have the inequality that

$$\max_{-1 \le x \le 1} |E(x)| < \frac{|f^{(n)}(\alpha)|}{n!\,2^{n-1}}. \qquad (6.3.21)$$

This result may seem to be of limited usefulness since we have no idea of the value of α but this is a situation that occurs whichever interpolation method we use and the factor $n!\,2^{n-1}$ which appears in the denominator is larger than with any other method; had we used the first n terms of the Taylor series for $f(x)$ for example the corresponding error estimate would have been

$$\max_{-1 \le x \le 1} |E(x)| < \frac{|f^{(n)}(\beta)|}{n!} \qquad (6.3.22)$$

where β is some (unknown) number in $[-1, +1]$. Despite this apparent weakness bounds such as (6.3.21) and (6.3.22) may sometimes be useful in practice for we may be able to prove a worthwhile upper bound of the type

$$\max_{-1 \le x \le 1} |f^{(n)}(x)| < K$$

for some number K (possibly depending on n) and this will then enable us to place an upper bound on $|E(x)|$ which we can use. Thus if, say, $f(x) = \sin(x)$, $K = 1$ and if $f(x) = e^x$, $K = e$. In some cases however $f^{(n)}(x)$ is unbounded over $[-1, 1]$ even though $|f(x)|$ remains small and then no bound is available.

The main disadvantage of Chebyshev interpolation is the need to use the special values for x_i given by (6.3.20) rather than integral multiples of a step (such as $0 \cdot 1, 0 \cdot 2, \ldots$ etc.). The values (6.3.20) are however readily available in tables and books [see Abramovitz and Stegun] and although their use by hand is tedious they can be easily built into computer programs where their disadvantages are no longer of any consequence whilst their advantages remain.

If the interval of approximation is not $[-1, +1]$ but $[a, b]$ we must first make a change of variable as indicated in section 6.3.7.

EXAMPLE 6.3.7. Use Chebyshev interpolation to find a cubic polynomial approximation to $(1+x)^{1/2}$ over $[-1, 1]$.

Solution. The 4 Chebyshev interpolation points are $x_1 = \cos(\pi/8)$, $x_2 = \cos(3\pi/8)$, $x_3 = \cos(5\pi/8)$, $x_4 = \cos(7\pi/8)$. We note that $x_3 = -x_2$ and $x_4 = -x_1$. The cubic can therefore be simplified by combining terms involving (x_1 and x_4) and (x_2 and x_3), viz.

$$f(x) = (1+x_1)^{1/2}\frac{(x^2-x_2^2)(x+x_1)}{(x_1^2-x_2^2)(2x_1)} + (1-x_1)^{1/2}\frac{(x^2-x_2^2)(x-x_1)}{(x_1^2-x_2^2)(-2x_1)}$$

$$+ (1+x_2)^{1/2}\frac{(x^2-x_1^2)(x+x_2)}{(x_2^2-x_1^2)(2x_2)} + (1-x_2)^{1/2}\frac{(x^2-x_1^2)(x-x_2)}{(x_2^2-x_1^2)(-2x_2)}.$$

Computation of the coefficients then leads to the cubic (to 5 d.p.)

$$f(x) = 1\cdot01171 + 0\cdot49084x - 0\cdot21116x^2 + 0\cdot12947x^3.$$

Comparison of $f(x)$ with $(1+x)^{1/2}$ at $x = -\frac{1}{2}(\frac{1}{4})\frac{1}{2}$ gives

$x =$	$-0\cdot5$	$-0\cdot25$	0	$+0\cdot25$	$+0\cdot50$
$f(x) =$	$0\cdot69732$	$0\cdot87378$	$1\cdot01171$	$1\cdot12325$	$1\cdot22052$
$(1+x)^{1/2} =$	$0\cdot70711$	$0\cdot86603$	$1\cdot00000$	$1\cdot11803$	$1\cdot22475$
$\lvert error\rvert =$	$0\cdot00979$	$0\cdot00775$	$0\cdot01171$	$0\cdot00522$	$0\cdot00423$

Prediction of the maximum error in this case is pointless since $f^{iv}(x) = -\frac{15}{16}(1+x)^{-7/2}$ and so is unbounded at $x = -1$. Had we instead approximated $(1+x^2)^{1/2}$ over $[-1, 1]$ we would have obtained (to 5 d.p.)

$$(1+x^2)^{1/2} \doteqdot 1\cdot01051 + 0\cdot41115x^2$$

(the terms in x, x^3 vanish identically since we are dealing with an even function [see V, § 3.6.3]) and since $f^{iv}(x)$ in this case $= (12x - 3)/(1+x^2)^{7/2}$ we have $\max_{-1 \le x \le 1}\lvert f^{iv}(x)\rvert < 3$ so that from (6.3.21) the Chebyshev approximation has an error of at most

$$3/4!2^3 = \tfrac{1}{64} = 0\cdot015625$$

over $[-1, 1]$. In fact the maximum error in this case occurs at $x = 0$ and so is $0\cdot01051$.

6.3.9.   The Gram–Schmidt Orthogonalisation Method

In section 6.3.2 we obtained the first 3 Legendre polynomials directly by solving the problem of finding polynomials of degrees 0, 1 and 2 which are orthogonal over $[-1, +1]$ with respect to the weight function $w(x) = 1$. The more general problem of finding a set $\{\phi_n(x)\}$ of polynomials, where the degree of $\phi_n(x) = n$, which are orthonormal over an interval $[a, b]$ with respect to a weight function $w(x)$ is usually solved by a method known as the Gram–Schmidt method, which we now describe.

Let $(\phi_n, \phi_m) = \int_a^b w(x)\phi_n(x)\phi_m(x)\,dx$. Since $\{\phi_n(x)\}$ is to be orthonormal and $\phi_n(x)$ is to be a polynomial of degree n let $\phi_0(x) = a_{0,0}$ then we need

$$(\phi_0, \phi_0) = 1$$

that is

$$\int_a^b a_{0,0}^2\, w(x)\, dx = 1$$

so we take

$$\phi_0(x) = a_{0,0} = \left\{ \int_a^b w(x)\, dx \right\}^{-1/2}.$$

For $\phi_1(x)$ we first put

$$\psi_1(x) = x + a_{1,0}\phi_0(x)$$

and $\phi_1(x)$ will be a constant multiple of $\psi_1(x)$; now $(\phi_1, \phi_0) = 0$ and so $(\psi_1, \phi_0) = 0$, that is

$$(x, \phi_0) + a_{1,0}(\phi_0, \phi_0) = 0$$

and so

$$a_{1,0} = -(x, \phi_0) = -\int_a^b xw(x)\, dx \Big/ \left\{ \int_a^b w(x)\, dx \right\}^{1/2}.$$

If we now put

$$\phi_1(x) = \psi_1(x) / [(\psi_1, \psi_1)]^{1/2}$$

we have $(\phi_1, \phi_1) = 1$ and $(\phi_1, \phi_0) = 0$ so $\phi_1(x)$ is our required polynomial of degree one.

Suppose now that we have obtained a set of orthonormal polynomials $\phi_0(x), \phi_1(x), \ldots, \phi_{n-1}(x)$ and wish to find $\phi_n(x)$. We first put

$$\psi_n(x) = x^n + a_{n,n-1}\phi_{n-1}(x) + \ldots + a_{n,0}\phi_0(x)$$

and choose the coefficients so that $(\psi_n, \phi_k) = 0$ for $k = 0, 1, \ldots, n-1$, that is, we take

$$a_{n,k} = -(x^n, \phi_k).$$

If we now take

$$\phi_n(x) = \psi_n(x) / [(\psi_n, \psi_n)]^{1/2}$$

we see that $(\phi_n, \phi_n) = 1$ and $(\phi_n, \phi_k) = 0$ $(k = 0, 1, \ldots, n-1)$ so that the set $\{\phi_r\}$, $r = 0, 1, \ldots, n$ is orthonormal, as required.

EXAMPLE 6.3.8. Use the Gram–Schmidt process to find the first 3 polynomials which are orthonormal over $[0, \infty]$ with respect to e^{-x}.

Solution. We shall use the result that $\int_0^\infty x^n e^{-x}\, dx = n!$ [see IV, § 10.2]. In the notation used above

$$\phi_0(x) = a_{0,0} = \left\{ \int_0^\infty e^{-x}\, dx \right\}^{-1/2} = 1.$$

Putting $\psi_1(x) = x + a_{1,0}(1)$ we have, as shown above,

$$a_{1,0} = -\int_0^\infty x e^{-x} \, dx / 1 = -1$$

hence:

$$\dot{\psi}_1(x) = x - 1$$

and so

$$\phi_1(x) = (x - 1) \Big/ \left(\int_0^\infty (x - 1)^2 \, e^{-x} \, dx \right)^{1/2} = (x - 1).$$

Putting $\psi_2(x) = x^2 + a_{2,1}(x - 1) + a_{2,0}(1)$ we have

$$a_{2,1} = -(x^2, \phi_1) = -\int_0^\infty x^2(x - 1) \, e^{-x} \, dx = -4$$

$$a_{2,0} = -(x^2, \phi_0) = -\int_0^\infty x^2 e^{-x} \, dx = -2.$$

Hence

$$\psi_2(x) = x^2 - 4(x - 1) - 2 = x^2 - 4x + 2$$

and so:

$$\phi_2(x) = (x^2 - 4x + 2) \Big/ \left(\int_0^\infty (x^2 - 4x + 2)^2 \, e^{-x} \, dx \right)^{1/2}$$

$$= \tfrac{1}{2}(x^2 - 4x + 2).$$

Thus the first 3 polynomials are 1, $(x - 1)$ and $\tfrac{1}{2}(x^2 - 4x + 2)$. (These are in fact the first 3 *Laguerre polynomials* $L_0(x)$, $L_1(x)$, $L_2(x)$; it can be proved that

$$L_n(x) = \frac{1}{n! \, e^{-x}} \frac{d^n}{dx^n}(x^n e^{-x})$$

—the difference in sign, which gives $L_1(x) = (1 - x)$ according to this formula, being of no importance.)

6.4. NON-POLYNOMIAL APPROXIMATION

6.4.1. General Comments

In Chapter 2 we solved the problem of finding a polynomial of degree n which passes through $(n + 1)$ given points $P_1(x_1, y_1)$, $P_2(x_2, y_2), \ldots,$ $P_{n+1}(x_{n+1}, y_{n+1})$ and in section 6.1 we used the method of least squares to find a polynomial of degree n which best fits, in the least squares sense, either a set of more than $(n + 1)$ points or a continuous function over a finite interval. It sometimes happens however that we have reasons for believing that approximation by a polynomial is unlikely to be satisfactory; such circumstances arise for example if the function to be approximated grows too rapidly, or too slowly, as the variable increases or if the function or any of its derivatives becomes infinite or discontinuous at any point [see IV, § 2.3]. Examples of such functions are shown in the Figures below.

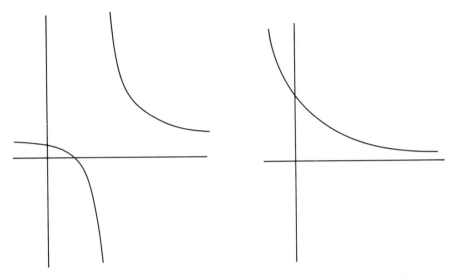

Figure 6.4.1: Function becomes infinite at $x = a$.

Figure 6.4.2: Function decreases asymptotically to zero.

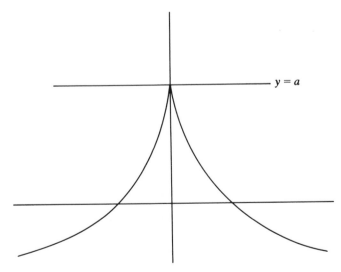

Figure 6.4.3: Function has a cusp at $x = 0$ [see V, § 3.7.3] (such a curve is given by $x^2 = (a - y)^3$).

In cases such as these approximation by functions other than polynomials might prove satisfactory. Such approximating functions include rational functions [see I, § 14.9], exponential or logarithmic functions [see IV, § 2.11], functions involving non-integral powers or combinations of all of these, or other functions having special properties (such as periodicity, when the trigonometric functions might be used [see IV, § 2.12]). If the analytic form of the function to be approximated is known suitable approximations might be found by analytic means, as we shall do for rational approximations in section 6.4.2 and as we did for Chebyshev approximations in section 6.3.4 but if the analytic form of the function is not known, that is if we only know its values at a set of discrete points, then there is no guaranteed route to a satisfactory approximation. In the case of discrete data, unless we have *a priori* reasons for expecting a particular form of solution, we can only examine the data, note any unusual properties and then try to construct a model to fit those properties. Some properties to look out for are listed below; in each case $F(x)$ is the unknown function and, except in 3, we shall assume that $F(x)$ contains only algebraic factors.

1. $F(x)$ *has a zero at a known point* $x = a$. Then $F(x) = (x-a)^s G(x)$ where $s > 0$ and $G(a) \neq 0$; note however that s is not necessarily an integer, but if $F(x)$ is single-valued s *is* an integer. $F(x)$ is said to have a zero of order s at $x = a$. If $F(x)$ is single-valued and

$$F(a) = F'(a) = \ldots = F^{(k)}(a) = 0, \qquad F^{(k+1)}(a) \neq 0$$

then $s = (k+1)$. If $F(x)$ is not single-valued and

$$F(a) = F'(a) = \ldots = F^{(k)}(a) = 0, \qquad F^{(k+1)}(a) = \infty$$

then s is non-integral and

$$k < s < (k+1).$$

2. $F(x)$ *is infinite at a known point* $x = b$. Unless $F(x)$ has an *essential singularity* (such as $e^{1/x}$ at $x = 0$) it may be written in the form

$$F(x) = \frac{G(x)}{(x-b)^s},$$

where $G(b)$ is neither zero nor infinite, for some number $s > 0$. It follows that the reciprocal function

$$\frac{1}{F(x)} = \frac{(x-b)^s}{G(x)}$$

has a zero of order s at $x = b$ and the remarks of the preceding section apply.

3. $F(x)$ *increases very slowly as* $x \to \infty$. By 'very slowly' we imply 'slower than any positive power of x' in which case it might be worthwhile comparing $F(x)$

with positive powers of log x for large values of x, i.e. study the ratio

$$\frac{F(x)}{(\log x)^s}$$

for various positive values of s as $x \to \infty$.

4. $F(x)$ *approaches a limit asymptotically as* $x \to \infty$. This case is frequently encountered both when $F(x)$ is given analytically (in a complicated form which we wish to approximate more simply) and when $F(x)$ is known only at discrete points. In either case the crucial question is 'How rapidly does $F(x)$ approach the limit?'. Let the limit be l then we are interested in

$$D(x) = l - F(x)$$

as $x \to \infty$. Since $D(x) \to 0$ it is worth investigating expressions such as $x^s D(x)$ $(s > 0)$ or $(\log x)^s D(x)$ $(s > 0$, if the convergence is very slow) or $e^{ax^s} D(x)$ $(a > 0, s > 0$, if the convergence is very fast). Alternatively it might be worth investigating the reciprocal of $D(x)$ as $x \to \infty$; this will become large and its rate of growth should tell us something about $F(x)$. This latter technique can be applied even in the case of discrete data but it is essential to realise that small changes in the values of $F(x)$ as it approaches its limit, l, produce increasingly larger changes in the reciprocal of $D(x)$ so that undue weight must not be given to values corresponding to larger values of x, where rounding errors in the values of $F(x)$ might be very misleading.

The simplest, and most commonly used, type of approximation apart from polynomial approximation is rational approximation. There are two distinct types of problem involving rational approximation:

(i) fitting a rational function to a given *function*;
(ii) fitting a rational function to given data.

In the next section we consider the first of these problems and in section 6.4.4 we consider the second.

6.4.2. Rational Approximations to Given Functions

Let $P_m(x)$ a polynomial of degree m and $Q_n(x)$ be a polynomial of degree n, and let

$$R_{m,n}(x) = \frac{P_m(x)}{Q_n(x)} ; \tag{6.4.1}$$

then we say that $R_{m,n}(x)$ is a *rational function of index* $(m + n)$.

We shall assume that $P_m(x)$ and $Q_n(x)$ have no common factor so that any zero of $P_m(x)$ is a zero of $R_{m,n}(x)$ and any zero of $Q_n(x)$ is a *pole* of $R_{m,n}(x)$ (i.e. causes $R_{m,n}(x)$ to be infinite) so that if we are interested in the behaviour of $R_{m,n}(x)$ over an interval $[a, b]$ on the real line and if $Q_n(x)$ has any zeros in $[a, b]$ then $R_{m,n}(x)$ becomes unbounded in the interval. If the interval $[a, b]$ is closed and finite and contains no zeros of $Q_n(x)$ then $R_{m,n}(x)$ remains bounded in the interval.

Rational approximations such as (6.4.1) are obviously more useful than polynomials if we wish to approximate to a function which becomes unbounded in a finite interval; but they are also useful for finding good approximations to functions which remain bounded for they are more general than polynomials and include them as a special case ($Q_n(x) \equiv 1$). We should however note that although a rational approximation may give a very good approximation to a bounded function within a specified finite interval it may also give a very bad approximation *outside* this interval, particularly as we approach a zero of $Q_n(x)$. We shall see examples of this; but first we must investigate how we can find rational approximations such as (6.4.1) to a given function $f(x)$. We observe at once however that even if m, n are specified in advance, representation of $f(x)$ in the form (6.4.1) is not unique; for the representation will depend upon the criterion adopted for deciding which is the 'best fit'—in the next section we shall use 'agreement with $f(x)$ and as many derivatives as possible' as the criterion but in section 6.3.4 the criterion was 'minimizing the maximum error at all points of the interval $[-1, +1]$'.

The number of coefficients available when we try to find a rational function of index $(m + n)$ is $(m + n + 2)$ but one of these is redundant since we can divide both numerator and denominator in (6.4.1) by any non-zero constant leaving $R_{m,n}(x)$ unchanged. Thus there are in fact only $(m + n + 1)$ coefficients to be evaluated and we may take the leading coefficient of $Q_n(x)$ to be 1 (say).

6.4.3. Padé Approximations

We suppose that we have a function $f(x)$ and that we wish to find a rational function $R_{m,n}(x)$ so that at $x = 0$ $R_{m,n}(x)$ and $f(x)$ and their derivatives are equal to as high an order as possible. If $n = 0$ (so that $Q_n(x) \equiv 1$) the solution is obvious, we take $P_m(x)$ to be equal to the first $(m + 1)$ terms of the Taylor series for $f(x)$ [see IV (2.10.17)]. If $n \neq 0$ the solution is not obvious and is found by solving a set of equations, as we now illustrate.

EXAMPLE 6.4.3.   Find a, b, c so that the rational function $R_{1,1}(x) = (a + bx)/(c + x)$ agrees with e^x and its derivatives at $x = 0$ to as high an order as possible.

Solution.   Consider

$$\frac{a + bx}{c + x} - e^x = \frac{(a + bx) - e^x(c + x)}{(c + x)}.$$

We try to make the numerator vanish at $x = 0$ to a high order. Now

$$(a + bx) - (c + x)(1 + x + \tfrac{1}{2}x^2 + \tfrac{1}{6}x^3 + \ldots)$$

$$= (a + bx) - (c + (c + 1)x + (\tfrac{1}{2}c + 1)x^2 + (\tfrac{1}{6}c + \tfrac{1}{2})x^3 + \ldots)$$

$$= (a - c) + (b - c - 1)x - (\tfrac{1}{2}c + 1)x^2 - (\tfrac{1}{6}c + \tfrac{1}{2})x^3 + \ldots.$$

We therefore take

$$a - c = 0$$

$$b - c - 1 = 0$$

$$\tfrac{1}{2}c + 1 = 0 \quad \text{i.e. } c = -2, b = -1, a = -2.$$

The coefficient of x^3 is $(\tfrac{1}{6}c + \tfrac{1}{2}) = +\tfrac{1}{6} \neq 0$. Thus

$$R_{1,1}(x) = \frac{-2-x}{-2+x} = \frac{2+x}{2-x} = \frac{1+\tfrac{1}{2}x}{1-\tfrac{1}{2}x} \tag{6.4.2}$$

and for small x we have

$$\frac{1+\tfrac{1}{2}x}{1-\tfrac{1}{2}x} - e^x \doteqdot \tfrac{1}{12}x^3$$

which is better than we obtain by truncating the Taylor series for e^x after the term in x^2, viz.

$$e^x \doteqdot 1 + x + \tfrac{1}{2}x^2 \tag{6.4.3}$$

with a truncation error of approximately $\tfrac{1}{6}x^3$ for small x. As x increases however the situation changes radically since $R_{1,1}(x)$ in this case has a pole at $x = 2$ whereas e^x does not. It is clear therefore that (6.3.2) cannot be expected to provide a good approximation to e^x except when x is small. The table below shows the errors when we use (6.4.2) and (6.4.3) for $x = 0 \cdot 2(0 \cdot 2)1 \cdot 0$

x	0·2	0·4	0·6	0·8	1·0
|Error (6.4.2)|	0·00082	0·00818	0·03502	0·10779	0·28172
|Error (6.4.3)|	0·00140	0·01182	0·04212	0·10554	0·21828

and we see that (6.4.3) is better than (6.4.2) from about $x = 0 \cdot 8$, though neither is worth using except for much smaller values of x. Better estimates can however be obtained from (6.4.2) by using the approximation in the form

$$e^x = (e^{x/n})^n \doteqdot \left(\frac{1 + (x/2n)}{1 - (x/2n)} \right)^n,$$

so that for example

$$e^{0 \cdot 8} = (e^{0 \cdot 2})^4 \doteqdot \left(\frac{1 \cdot 1}{0 \cdot 9} \right)^4 = 2 \cdot 23152,$$

a result which is in error by $0 \cdot 00598$, about 17 times less than the error when we used (6.4.2) directly with $x = 0 \cdot 8$.

Such approximations (but of higher index) are frequently used in computers and calculators for the computation of e^x.

A rational approximation $R_{m,n}(x)$ obtained in this way is called the (m, n)th *Padé approximation*. If we suppose that $f(x)$ has a Taylor expansion about $x = 0$ and that

$$f(x) = \sum_{n=0}^{\infty} c_n x^n$$

and we take

$$R_{m,n}(x) = \frac{P_m(x)}{Q_n(x)} = \frac{a + ax + \ldots + a_m x^m}{b_0 + b_1 x + \ldots + b_n x^n}$$

then we must choose the a_i and b_i so that the expression

$$(b_0 + b_1 x + \ldots + b_n x^n)(c_0 + c_1 x + c_2 x^2 + \ldots) - (a_0 + a_1 x + \ldots + a_n x^m) \quad (6.4.4)$$

has as high an order zero at $x = 0$ as possible. Thus we have a set of equations obtained by equating to zero the coefficients of (6.4.4). Since the coefficients c_i are known the equations are linear. Remembering that we can take $b_n = 1$ (or $b_0 = 1$, which is in some ways more convenient) we might hope to satisfy $(m + n + 1)$ equations in this way. Thus the first non-vanishing coefficient and hence the truncation error in (6.4.4) will in general be associated with x^{m+n+1} (e.g. $R_{1,1}(x)$ for e^x had error terms $O(x^3)$). Sometimes however the equations will have no solution and an approximation of the form $R_{m,n}(x)$ will not then exist. Another possibility is that the equations have a solution but $P_m(x)$ and $Q_n(x)$ happen to have a common factor so that $R_{m,n}(x)$ is of index lower than $(m + n)$. The first of these possibilities is illustrated by:

EXAMPLE 6.4.4.   Find a $(1, 1)$ Padé approximation to $\cos x$ near $x = 0$.

Solution.   We try

$$\frac{a + bx}{c + x} \doteq \cos x \doteq 1 - \frac{x^2}{2} + \frac{x^4}{24}$$

which leads to

$$(a + bx) - (c + x)(1 - \tfrac{1}{2}x^2 + \tfrac{1}{24}x^4 + \ldots) = O(x^3).$$

The equations are therefore:

$$a - c = 0$$

$$b - 1 = 0$$

$$-\tfrac{1}{2}c = 0$$

therefore $c = 0$, $a = 0$ and $b = 1$. Thus our $(1, 1)$ approximation turns out to be

$$\frac{0 + x}{0 + x} = \frac{1}{1} = 1$$

which is a $(0, 0)$ approximation! Hence no $(1, 1)$ Padé approximation exists in this case.

This phenomenon occurs because $\cos x$ is an *even* function of x [see V, § 3.6.3]. If instead we try $(a + bx^2)/(c + x^2)$ we have:

$$(a + bx^2) - (c + x^2)(1 - \tfrac{1}{2}x^2 + \tfrac{1}{24}x^4 \ldots) \doteq 0$$

and so:

$$a - c = 0$$

$$b - 1 + \tfrac{1}{2}c = 0$$

$$-\tfrac{1}{2} + \tfrac{1}{24}c = 0$$

therefore $c = 12$, $b = 5$, $a = 12$ giving

$$R_{2,2}(x) = \frac{12 - 5x^2}{12 + x^2} = \frac{1 - \tfrac{5}{12}x^2}{1 + \tfrac{1}{12}x^2}$$

with an error term $O(x^6)$.

6.4.4.   Rational Approximation to Given Values

If we are given the values of a function at a fixed set of points we can attempt to fit a rational function to them. In the special case where the rational function is chosen to be a polynomial the methods of Chapter 2 are suitable but if a non-polynomial is sought a different approach must be adopted.

We shall suppose that we have a finite set of (real or complex) distinct points $x_1, x_2, \ldots, x_e$ and a corresponding set of values $y_1, y_2, \ldots, y_e$. If any of the y_i are infinite we delete them from the list, together with the corresponding x_i but must remember to include a factor $(x - x_i)^{k_i}$, where k_i is a positive integer, in the denominator of the rational function for each such infinite y_i. The determination of k_i is not obvious; initially we take $k_i = 1$ and adjust later. Having said this we shall assume, for the present, that all y_i are finite.

We are seeking a rational function of the form (6.4.1), viz.

$$R_{m,n}(x) = \frac{P_m(x)}{Q_n(x)}$$

where $P_m(x)$ and $Q_n(x)$ are polynomials of degrees m, n respectively and such that

$$R_{m,n}(x_i) = y_i \qquad (i = 1, \ldots, e) \tag{6.4.5}$$

$R_{m,n}(x)$ contains $(m + n + 2)$ unknown coefficients but, as in (6.4.1), multiplication of both $P_m(x)$ and $Q_n(x)$ by an arbitrary (non-zero) constant leaves $R_{m,n}(x)$ unaltered so there are only $(m + n + 1)$ independent unknowns; the problem of finding these unknowns so that (6.4.5) is satisfied is called 'the Cauchy Interpolation Problem' since it was first considered by Cauchy in 1821.

At first sight the obvious way to solve (6.4.5) is write it in the form

$$R_{m,n}(x_i) = \frac{P_m(x_i)}{Q_m(x_i)} = y_i \qquad (i = 1, 2, \ldots, e)$$

so that

$$P_m(x_i) = y_i Q_n(x_i) \qquad (i = 1, 2, \ldots, e) \qquad (6.4.6)$$

and (6.4.6) is equivalent to a set of e linear equations in $(m+n+1)$ unknowns $(p_0, p_1, \ldots, p_m, q_1, q_2, \ldots, q_n)$ (note that we have taken $q_0 = 1$) which we will in general be able to solve if $e \le (m+n+1)$ [see I, § 5.10].

Unfortunately it is not necessarily true that (6.4.5) and (6.4.6) are equivalent for it may happen that for some particular x_i we have $Q_n(x_i) = 0$ in which case (6.4.6) is satisfied if $P_m(x_i) = 0$, but (6.4.5) is not. On the other hand any solution of (6.4.5) provides a solution of (6.4.6). We see therefore that (6.4.6) provides a problem which is not necessarily identical to the Cauchy inter-polation problem; it is therefore known as 'the Modified Cauchy Interpolation Problem'.

Providing we keep this reservation in mind (6.4.6) provides the simplest way of attempting to solve the Cauchy problem. It will rarely be the case that $Q(x_i) = 0$ so that we will usually find that the solution we obtain to (6.4.6) also satisfies (6.4.5). The example below illustrates the method which is, of course, similar to the method used in (6.4.2) to find the Padé approximation.

EXAMPLE 6.4.5.   Fit an $R_{2,1}(x)$ rational function to the following data

$$x = -1 \quad 0 \quad 1 \quad 3$$
$$y = \quad 2 \quad 1 \quad 2 \quad 6.$$

Solution.   Let $p_2(x) = p_0 + p_1 x + p_2 x^2$, $Q_1(x) = 1 + q_1 x$ then (6.4.6) gives 4 equations

$$
\begin{aligned}
p_0 - p_1 + p_2 &= 2 - 2q_1 \\
p_0 &= 1 \\
p_0 + p_1 + p_2 &= 2 + 2q_1 \\
p_0 + 3p_1 + 9p_2 &= 6 + 18q_1
\end{aligned}
$$

which lead to the solution $p_0 = 1$, $p_2 = 1$, $p_1 = \frac{2}{3}$, $q_1 = \frac{1}{3}$ giving

$$R_{2,1}(x) = \frac{1 + \frac{2}{3}x + x^2}{1 + \frac{1}{3}x} = \frac{3 + 2x + 3x^2}{3 + x}.$$

In this example $Q_1(x_i) \ne 0$ for any of the x_i in the set $(-1, 0, 1, 3)$. Had one of the x_i had the value -3 however the situation would have been different, and our solution would not necessarily have satisfied (6.4.5) as well as (6.4.6) this situation is illustrated in the following example.

EXAMPLE 6.4.6.   Fit an $R_{2,1}(x)$ approximation to the data

$$x = -2 \quad -1 \quad 0 \quad 2$$
$$y = \quad 0 \quad \quad 3 \quad 1 \quad 2.$$

Solution.   Let $p_2(x) = p_0 + p_1 x + p_2 x^2$, $Q_1(x) = 1 + q_1 x$ then the data leads to

$$p_0 - 2p_1 + 4p_2 = 0$$
$$p_0 - p_1 + p_2 \quad = 3 - 3q_1$$
$$p_0 \qquad\qquad = 1$$
$$p_0 + 2p_1 + 4p_2 = 2 + 4q_1.$$

From the first, third and fourth equations we deduce that $p_0 = 1$, $p_1 = \frac{1}{2} + q_1$, $p_2 = \frac{1}{2} q_1$ and so substituting in the second equation we find that we require

$$\tfrac{1}{2}(1 - q_1) = 3(1 - q_1)$$

which is satisfied if $q_1 = 1$. We therefore have

$$R_{2,1}(x) = \frac{1 + \frac{3}{2}x + \frac{1}{2}x}{1 + x} = \frac{(1 + x)(1 + \frac{1}{2}x)}{(1 + x)} = 1 + \tfrac{1}{2}x$$

so that $R_{2,1}(x)$ collapses to an $R_{1,0}(x)$ representation and although the function $(1 + \frac{1}{2}x)$ fits the data at $x = -2$, $x = 0$ and $x = 2$ it does not do so at $x = -1$ where it gives the value $y = \frac{1}{2}$. The value $y = 3$ at $x = 1$ is therefore unattainable by an $R_{2,1}(x)$ representation—indeed any value other than $y = \frac{1}{2}$ is unattainable. It follows that in this case the modified Cauchy problem is solvable and the Cauchy problem is not.

6.4.5. Continued Fractions

There is a method which has been used for centuries for obtaining good rational approximations to real numbers: the method of continued fractions. The method can also be used to obtain rational approximations to power series, as we shall see in section 6.4.7. In this section we introduce continued fractions of real numbers and in section 6.4.6 we summarise their main properties.

Let θ be any positive real number. We obtain a sequence of positive integers $a_0, a_1, a_2, \ldots$ from θ as follows: let $a_0 = [\theta]$, $\theta_0 = (\theta - a_0)^{-1}$ and for $n \geq 1$, $a_n = [\theta_{n-1}]$, $\theta_n = (\theta_{n-1} - a_n)^{-1}$ ($[x]$ denotes the integer part of x). Thus

$$\theta = a_0 + (\theta - a_0) = a_0 + \frac{1}{\theta_0}$$

$$\theta_0 = a_1 + (\theta_0 - a_1) = a_1 + \frac{1}{\theta_1}$$

$$\theta_1 = a_2 + (\theta_1 - a_2) = a_2 + \frac{1}{\theta_2}.$$

Hence $\theta = a_0 + \cfrac{1}{a_1 + \cfrac{1}{a_2 + \cfrac{1}{a_3 + \ldots}}}$

or, as it is written more conveniently:

$$\theta = a_0 + \cfrac{1}{a_1 +} \cfrac{1}{a_2 +} \cfrac{1}{a_3 +} \cdots \qquad (6.4.7)$$

We have therefore found the *unique* series of integers associated with θ when it is expressed in the form of a *continued fraction*—which is the name given to (6.4.7). The integers $a_0, a_1, a_2, \ldots$ are called the *partial quotients*.

EXAMPLE 6.4.7.   Find the continued fraction expansion of $\frac{17}{12}$.

Solution.

$$\tfrac{17}{12} = 1 + \tfrac{5}{12}, \; \tfrac{12}{5} = 2 + \tfrac{2}{5}$$

$$\tfrac{5}{2} = 2 + \tfrac{1}{2}, \; 2 = 2 + 0$$

$$\therefore \quad \tfrac{17}{12} = 1 + \cfrac{1}{2+} \cfrac{1}{2+} \cfrac{1}{2+}.$$

EXAMPLE 6.4.8.   Find the continued fraction expansion for $\sqrt{2}$.

Solution.   There are two ways of finding the continued fraction in this case. The first is to compute $\sqrt{2}$ to as many places of decimals as we think necessary and then find the expansion directly, viz. (to 4 d.p.)

$$\sqrt{2} = 1 \cdot 4142 = 1 + 0 \cdot 4142 = 1 + \cfrac{1}{2 \cdot 4142} \quad \text{etc.}$$

The second method, which can only be used in certain special cases (which include $\sqrt{2}$), is algebraic:

$$\sqrt{2} = 1 + (\sqrt{2} - 1) = 1 + \cfrac{1}{\sqrt{2} + 1} = 1 + \cfrac{1}{2 + (\sqrt{2} - 1)}.$$

Hence

$$\sqrt{2} = 1 + (\sqrt{2} - 1) = 1 + \cfrac{1}{2 + (\sqrt{2} - 1)}$$

$$= 1 + \cfrac{1}{2 + \cfrac{1}{2 + (\sqrt{2} - 1)}} \quad \text{etc.}$$

and so

$$\sqrt{2} = 1 + \cfrac{1}{2+} \cfrac{1}{2+} \cfrac{1}{2+} \cdots \qquad (6.4.8)$$

—a non-terminating continued fraction of cycle length 1.

It is easy to verify (6.4.8) directly. For let

$$y = \cfrac{1}{2+} \cfrac{1}{2+} \cfrac{1}{2+} \cdots .$$

Then $y = 1/(2+y)$ so that $y^2 + 2y - 1 = 0$ and so (since $y > 0$) $y = -1 + \sqrt{2}$. Hence

$$1 + \frac{1}{2+} \frac{1}{2+} \frac{1}{2+} = y + 1 = \sqrt{2}.$$

In general, given a real number in decimal form we will have to find its continued fraction by the direct method.

6.4.6. Approximation to Real Numbers by Continued Fractions

The importance of continued fractions lies in the fact that they provide (in a sense to be explained) the 'best' possible *rational* approximations to a real number θ.

Consider again the continued fraction for $\sqrt{2}$ given by (6.4.8). We take the successive approximations obtained when we truncate the expansion after the first, second, third terms, i.e. we have

$$1, 1 + \tfrac{1}{2}, 1 + \tfrac{1}{2+} \tfrac{1}{2}, \quad \text{etc.}$$

i.e. $1, \frac{3}{2}, \frac{7}{5}, \frac{17}{12}, \frac{41}{29}, \frac{99}{70}, \dots$ are the approximations. We convert each to decimal form and compare with the value for $\sqrt{2}$ which we take to be $1 \cdot 4142$.

p_n/q_n	Approximation	Error	$q_n^2 \times \lvert \text{Error} \rvert$
1/1	1·0000	−0·4142	0·4242
3/2	1·5000	+0·0858	0·3432
17/12	1·4167	+0·0025	0·3600
41/29	1·4138	+0·0004	0·3364
99/70	1·4142 (8)	−0·0000 (7)	0·3528

In the last column we form $q_n^2 \times \lvert \text{Error} \rvert$ where p_n/q_n is the nth approximation to $\sqrt{2}$. We observe that this column is reasonably constant. The rational approximations p_n/q_n obtained by truncating the continued fraction after n terms are called the *convergents* to θ.

The main properties of continued fractions are summarised in the following theorem.

THEOREM 6.4.1. *Let θ be a positive real number with continued fraction expansion*

$$= a_0 + \frac{1}{a_1+} \frac{1}{a_2+} \frac{1}{a_3+} \cdots$$

and successive convergents

$$\frac{p_0}{q_0}, \frac{p_1}{q_1}, \frac{p_2}{q_2}, \dots .$$

Then

(i) $a_n = 0$ *for some n if and only if θ is a rational number;*

(ii) *The a_i eventually becomes periodic (so that $a_{j+k} = a_j$ for some k and all $j \geq j_0$) if and only if θ is the root of a quadratic equation with integer coefficients [see I (14.5.1)].*

(iii) *Taking $p_0 = a_0$, $q_0 = 1$, $p_1 = a_0 a_1 + 1$, $q_1 = a_1$ the convergents are generated by*

$$p_n = a_n p_{n-1} + p_{n-2}$$

$$q_n = a_n q_{n-1} + q_{n-2}.$$

Furthermore p_n, q_n are relatively prime [see I, § 4.1.3] and

$$p_n q_{n-1} - q_n p_{n-1} = (-1)^{n-1}.$$

(iv) *The convergents p_n/q_n are the best possible rational approximations to θ in the sense that*

$$\left| \theta - \frac{p_n}{q_n} \right| < \left| \theta - \frac{p}{q} \right| \quad for\ all\ q < q_n.$$

(v) *If θ is irrational then an infinity of the successive convergents satisfy*

$$\left| \theta - \frac{p_n}{q_n} \right| < \frac{1}{q_n^2 \sqrt{5}}.$$

(The constant $\sqrt{5}$ cannot be replaced by any larger number because there are numbers, such as $\frac{1}{2}(\sqrt{5}-1)$ for which the statement above would then be false.)

For the proof of all these properties see books on the Theory of Numbers e.g. Hardy and Wright (Chapter 10).

Property (v) explains why we formed

$$q_n^2 \times |\text{Error}| = q_n^2 \left| \theta - \frac{p_n}{q_n} \right|,$$

for the theorem guarantees that

$$q_n^2 \left| \theta - \frac{p_n}{q_n} \right| < \frac{1}{\sqrt{5}}$$

infinitely often and it is also true (but extremely difficult to prove) that it is very unlikely that

$$q_n^2 \left| \theta - \frac{p_n}{q_n} \right|$$

becomes arbitrarily small.

EXERCISES.   (i) Take $e \doteq 2 \cdot 718281848$ and find the first 12 terms of the continued fraction for e. Can you see a pattern emerging? If so, what is it? [Solution 2; 1, 2, 1, 1, 4, 1, 1, 6, 1, 1, 8, 1, ①, . . . a change of last digit affects ①; there *is* a pattern; $e = 2$; 1, 2, 1, 1, $2n$ ($n = 2, 3, 4 \ldots$)].

(ii) Take $\pi \doteqdot 3 \cdot 141592654$ and find the first few terms of the continued fraction. Do you see a pattern? What are the first 4 convergents to π? How accurate is the fourth convergent? [Solution: there is no pattern; $\frac{3}{1}, \frac{22}{7}, \frac{333}{106}, \frac{355}{113}$; 6 d.p.]

6.4.7. Analytical Continued Fractions

Suppose we have a rational function of a single variable

$$R_{m,n}(x) = \frac{P_m(x)}{Q_n(x)}$$

where $P_m(x)$ and $Q_n(x)$ are polynomials of degrees m, n respectively with integer coefficients. By analogy with the continued fractions of rational numbers we can find an *analytical continued fraction* for $R_{m,n}(x)$. The method, although simple, is best illustrated by an example.

EXAMPLE 6.4.9. Express

$$\frac{x^4 - 5x^3 + 12x^2 - 11x + 2}{x^2 - 3x + 4}$$

as an analytic continued fraction

$$
\begin{array}{r}
x^2 - 2x + 2 \\
x^2 - 3x + 4\,\overline{\big)\ x^4 - 5x^3 + 12x^2 - 11x + 2} \\
\underline{x^4 - 3x^3 + 4x^2\ \ \ \ \ \ \ \ \ \ \ \ \ \ } \\
-2x^3 + 8x^2 - 11x + 2 \\
\underline{-2x^3 + 6x^2 - 8x\ \ \ \ \ \ \ \ } \\
2x^2 - 3x + 2 \\
\underline{2x^2 - 6x + 8} \\
3x - 6
\end{array}
$$

Hence

$$\frac{x^4 - 5x^3 + 12x^2 - 11x + 2}{x^2 - 3x + 4} = (x^2 - 2x + 2) + \frac{3x - 6}{x^2 - 3x + 4}$$

$$= (x^2 - 2x + 2) + \frac{3}{\dfrac{x^2 - 3x + 4}{x - 2}}$$

$$
\begin{array}{r}
x - 1 \\
x - 2\,\overline{\big)\ x^2 - 3x + 4} \\
\underline{x^2 - 2x\ \ \ \ \ \ } \\
-x + 4 \\
\underline{-x + 2} \\
2
\end{array}
\qquad \therefore \quad \frac{x^2 - 3x + 4}{x - 2} = (x - 1) + \frac{2}{(x - 2)}
$$

so that, finally,

$$\frac{x^4-5x^3+12x^2-11x+2}{x^2-3x+4}=(x^2-2x+2)+\frac{3}{(x-1)+}\ \frac{2}{(x-2)}. \qquad (6.4.9)$$

This continued fraction is of a more general kind than before in two respects, firstly it involves a variable (x), secondly it takes the form

$$a_0+\frac{b_0}{a_1+}\ \frac{b_1}{a_2+}\ \frac{b_2}{a_3+}\cdots,$$

where the b_i are not necessarily $=1$. The expansion is not, in fact unique. The b_i may be given arbitrary (real, non-zero) values, for example we could rewrite the expansion above as

$$(x^2-3x+2)+\frac{1}{\frac{1}{3}(x-1)+}\ \frac{\frac{2}{3}}{(x-2)}=(x-3x+2)+\frac{1}{\frac{1}{3}(x-1)+}\ \frac{1}{\frac{3}{2}(x-2)}.$$

Analytic continued fractions such as (6.4.9) may be useful for the repeated evaluation of a function. Not only will they frequently involve fewer multiplications and divisions than direct evaluation but, in the case of complicated (non-rational) functions truncation of the analytic continued fraction may provide good rational approximations which can then be used for approximate evaluation of the complicated function.

EXAMPLE 6.4.10. Given the analytic continued fraction:

$$e^x=1+\frac{x}{1-}\ \frac{x}{2+}\ \frac{x}{3-}\ \frac{x}{2+}\ \frac{x}{5-}\ \frac{x}{2+}\ \frac{x}{7-}$$

find the first four rational function approximations to e^x.

Note first that if

$$\theta=a_0+\frac{b_1}{a_1+}\ \frac{b_2}{a_2+}\ \frac{b_3}{a_3+}\cdots$$

then

$$p_0=a_0, \qquad q_0=1; \qquad p_1=a_0a_1+b_1, \qquad q_1=a_1$$

and that

$$p_n=a_np_{n-1}+b_np_{n-2}$$

$$q_n=a_nq_{n-1}+b_nq_{n-2}.$$

Hence since, for e^x: 1, $p_0=1$, $q_0=1$; $p_1=1+x$, $q_1=1$ we have:

$$p_2=2(1+x)-x(1)=x+2$$

$$q_2=2(1)-x(1)=2-x$$

and

$$p_3 = 3(x+2) + x(x+1) = x^2 + 4x + 6$$

$$q_3 = 3(2-x) + x(1) = 6 - 2x.$$

The first four rational functions are therefore

$$\frac{1}{1}, \frac{1+x}{1}, \frac{2+x}{2-x}, \frac{6+4x+x^2}{6-2x}.$$

Note that these are the Padé approximations $R_{0,0}$, $R_{1,0}$, $R_{1,1}$ and $R_{2,1}$ to e^x.

6.4.8. A Method of Deriving Continued Fractions from Series

There is a useful result which provides an identity between the terms of a finite series and a continued fraction. It is given by:

THEOREM 6.4.2

$$\sum_{k=1}^{n} \frac{1}{a_k} = \frac{1}{a_1-} \frac{a_1^2}{a_1+a_2-} \frac{a_2^2}{a_2+a_3-} \cdots \frac{a_{n-1}^2}{a_{n-1}+a_n}.$$

Proof. Taking $n=2$ we have the assertion that

$$\frac{1}{a_1} + \frac{1}{a_2} = \frac{1}{a_1-} \frac{a_1^2}{a_1+a_2} = \frac{a_1+a_2}{a_1 a_2} = \frac{1}{a_1} + \frac{1}{a_2}$$

so the theorem is true for $n=2$.

Suppose now that the theorem has been proved up to n. Then

$$\sum_{k=1}^{n+1} \frac{1}{a_k} = \sum_{k=1}^{n-1} \frac{1}{a_k} + \frac{1}{a_n} + \frac{1}{a_{n+1}} = \sum_{k=1}^{n-1} \frac{1}{a_k} + \frac{a_n + a_{n+1}}{a_n a_{n+1}}$$

$$= \sum_{k=1}^{n-1} \frac{1}{a_k} + \frac{1}{a_n a_{n+1}/(a_n + a_{n+1})}.$$

We now apply the theorem to the sum of the n terms $a_1^{-1}, a_2^{-1}, \ldots, a_{n-1}^{-1}$, $(a_n a_{n+1}/(a_n + a_{n+1}))^{-1}$ so that

$$\sum_{k=1}^{n+1} \frac{1}{a_k} = \frac{1}{a_1} \frac{a_1^2}{a_1+a_2-} \frac{a_2^2}{a_2+a_3-} \frac{a_{n-2}^2}{a_{n-2}+a_{n-1}} \frac{a_{n-1}^2}{-a_{n-1}+(a_n a_{n+1}/(a_n + a_{n+1}))}.$$

Now

$$\frac{a_{n-1}^2}{a_{n-1}+(a_n a_{n+1}/(a_n + a_{n+1}))} = \frac{a_{n-1}^2}{a_{n-1}+a_n - (a_n^2/(a_n + a_{n+1}))}.$$

and so

$$\sum_{k=1}^{n+1} \frac{1}{a_k} = \frac{1}{a_1-} \frac{a_1^2}{a_1+a_2-} \frac{a_2^2}{a_2+a_3-} \cdots \frac{a_n^2}{a_n + a_{n+1}}$$

and so the theorem is proved by induction.   Q.E.D.

EXAMPLE 6.4.11.　From the Taylor series for e^x

$$e^x = 1 + \frac{x}{1!} + \frac{x^2}{2!} + \frac{x^3}{3!} + \cdots$$

we can obtain an analytic continued fraction. For:

$$e^x = 1 + \frac{1}{x^{-1}} + \frac{1}{2x^{-2}} + \frac{1}{6x^{-3}} + \frac{1}{24x^{-4}} + \cdots$$

and so, from the theorem above

$$e^x = \frac{1}{1-} \frac{1}{1+x^{-1}-} \frac{x^{-2}}{x^{-1}+2x^{-2}-} \frac{4x^{-4}}{2x^{-2}+6x^{-3}-} \frac{36x^{-6}}{6x^{-3}+24x^{-4}-} \cdots$$

$$= \frac{1}{1-} \frac{x}{x+1-} \frac{x}{x+2-} \frac{4x}{2x+6-} \frac{36x}{6x+24-} \cdots$$

i.e.

$$e^x = \frac{1}{1-} \frac{x}{x+1-} \frac{x}{x+2-} \frac{2x}{x+3-} \frac{3x}{x+4-} \cdots \frac{kx}{x+k+1-}$$

$$\frac{1}{e^x} = 1 - \frac{x}{(x+1)-} \frac{x}{(x+2)-} \frac{2x}{(x+3)-} \frac{3x}{(x+4)-} \cdots$$

$$\frac{1-e^x}{e^x} = \frac{-x}{(x+1)-} \frac{x}{(x+2)-} \frac{2x}{(x+3)-} \frac{3x}{(x+4)-}$$

$$\frac{e^x-1}{e^x} = \frac{+x}{(x+1)-} \frac{x}{(x+2)-} \frac{2x}{(x+3)-}$$

$$\frac{e^x-1}{xe^x} = \frac{1}{(x+1)-} \frac{x}{(x+2)-} \frac{2x}{(x+3)-}$$

$$\frac{xe^x}{e^x-1} - (x+1) = \frac{-x}{(x+2)-} \frac{2x}{(x+3)-} \frac{3x}{(x+4)-}.$$

Put $x = -1$:

$$\frac{-1e^{-1}}{e^{-1}-1} = \frac{1}{1+} \frac{2}{2+} \frac{3}{3+} \cdots$$

or

$$\frac{1}{e-1} = \frac{1}{1+} \frac{2}{2+} \frac{3}{3+} \frac{4}{4+} \cdots.$$

6.4.9.　Fitting an Exponential Type Function to Given Values

Fitting functions of types other than polynomials or rational functions to given values may pose unpleasant problems in algebra; any special properties which the functions may possess should be exploited if the algebra is thereby simplified. One case which occurs occasionally where such simplification can be employed is that of fitting a negative exponential

$$f(x) = A e^{-\alpha x^\beta} \qquad (6.4.10)$$

where A, α, β are constants, to a set of points (x_i, y_i), $i = 1, \ldots, n$. The problem is considerably simplified if the points include one or more triads (x_i, x_j, x_k) say, such that $x_i x_k = x_j^2$ as the example below illustrates.

EXAMPLE 6.4.12.   The frequency of failure to observe a certain event during n trials has been predicted theoretically to be given by

$$f(n) = A n e^{-\alpha n^\beta}.$$

Estimate the values of A, α and β from the data

$n =$	10	20	30	40	50	60	70	80	90
$f(n) =$	476	446	403	363	328	298	272	249	229

where the values of n are given in thousands.

Solution.   If $f(n) \doteqdot A n e^{-\alpha n^\beta}$ then $n^{-1} f(n)$ is of type (6.4.10) and

$$\log_e \left(\frac{f(n)}{n} \right) \doteqdot \log A - \alpha n^\beta \qquad (6.4.11)$$

and so

$$\log_e \left(\frac{f(n)}{n} \right) - \log_e \left(\frac{f(m)}{m} \right) = \alpha (m^\beta - n^\beta) \qquad (6.4.12)$$

and

$$\log_e \left(\frac{f(m)}{m} \right) - \log_e \left(\frac{f(k)}{k} \right) = \alpha (k^\beta - m^\beta)$$

so that

$$\frac{\log_e \left(\frac{f(n)}{n} \right) - \log_e \left(\frac{f(m)}{m} \right)}{\log_e \left(\frac{f(m)}{m} \right) - \log_e \left(\frac{f(k)}{k} \right)} = \frac{m^\beta - n^\beta}{k^\beta - m^\beta}$$

or:

$$\frac{\log_e \left(\frac{mf(n)}{nf(m)} \right)}{\log_e \left(\frac{kf(m)}{mf(k)} \right)} = \frac{1 - \left(\frac{n}{m} \right)^\beta}{\left(\frac{k}{m} \right)^\beta - 1} \qquad (6.4.13)$$

The expression on the right-hand side of (6.4.13) takes a particularly simple form if

$$\frac{n}{m} = \frac{m}{k} (= t, \text{ say}),$$

that is if

$$nk = m^2; \tag{6.4.14}$$

for we can then find the value of β at once after which α can be found from (6.4.12) and finally A from (6.4.11).

In our data (6.4.14) is satisfied at $n = 10,000$, $m = 20,000$, $k = 40,000$, hence $t = \frac{1}{2}$ and (6.4.13) tells us that $\beta \doteqdot 0 \cdot 25$ and so, from (6.4.12) $\alpha \doteqdot 0 \cdot 4$, and then from (6.4.11), $A \doteqdot 2 \cdot 6$ and our conclusion based on the 3 points is that

$$f(n) \doteqdot 2 \cdot 6n \, e^{-0 \cdot 4n^{0 \cdot 25}}. \tag{6.4.15}$$

Since the solution has only used the data from 3 points it should now be checked against the remaining data points to see if the fit is satisfactory. In this particular example it is; if it were not, a different selection of 3 points should be chosen and new values of A, α and β found and checked. If values of A, α and β which give satisfactory results at all the points cannot be found the hypothesis (6.4.10) must be regarded as false (assuming the data are considered reliable).

If no three points satisfying (6.4.14) can be found the problem must be solved iteratively for (6.4.13) then takes the form

$$\frac{1 - t_1^{\beta}}{1 - t_2^{\beta}} = c$$

where t_1, t_2 and c are known and β is to be found; once β has been found, α and A are determined as before.

6.5. SPLINES AND HERMITE APPROXIMATIONS

6.5.1. Splines

One of the problems which frequently arises when we try to approximate to a function by means of a polynomial of high degree is that the polynomial turns out to have closely placed maxima and minima, thus giving it an undulatory (or 'wiggly') character. This is very undesirable if the polynomial is to be used for interpolation, and disastrous if it is to be used for numerical differentiation.

In 1945 I. J. Schoenberg* introduced the idea of approximating to functions by means of a series of polynomials over adjacent intervals with continuous derivatives at the end-points of the intervals. Such a set of polynomials he

---

* *Quarterly of Applied Mathematics*, **4** (1946), 45–99 and 112–141. The definition of a spline appears on p. 67.

called 'splines', pointing out that architects and designers had been using mechanical devices of this kind for years. In his paper Schoenberg explains: 'A spline is a simple mechanical device for drawing smooth curves. It is a slender flexible bar made of wood or some other elastic materials. The spline is placed on the sheet of graph paper and held in place at various points . . . '.

The mathematical equivalent of this 'flexible bar' is the *cubic spline* which has proved to be extremely useful for interpolation, numerical differentiation and integration and has been the subject of many research papers.

We begin by defining splines of degree k and then develop the special case of the cubic ($k = 3$).

DEFINITION 6.5.1.   A spline function $S(x)$ of degree k with n 'knots' (or 'nodes') $x_1 < x_2 < \ldots < x_n$ has the properties:

(i) $S(x)$ is given in the interval $[x_i, x_{i+1}]$ $i = 0, 1, \ldots, n$ (where $x_0 = -\infty$, $x_{n+1} = +\infty$) by a polynomial of degree at most k (in general, a different polynomial in each interval);

(ii) $S(x)$ and all its derivatives of orders $1, 2, \ldots, k-1$ are continuous on $(-\infty, +\infty)$.

In the case $k = 3$ the polynomials in each of the intervals are (at most) cubics and their first and second derivatives are continuous at the end points of the intervals. Such a set of polynomials form a *cubic spline*. We now see how such a set can be constructed.

Suppose that we are given a set of points $x_1 < x_2 < \ldots < x_n$, not necessarily equally spaced, and a set of values $f(x_1), f(x_2), \ldots, f(x_n)$ at these points. Take a particular interval $[x_i, x_{i+1}]$ and fit a cubic over the interval which satisfies the definition of a cubic spline. Since the cubic may differ from one interval to another let the cubic be

$$F_i(x) = a_0 + a_1 x + a_2 x^2 + a_3 x^3 \qquad (x_i \le x \le x_{i+1}). \qquad (6.5.1)$$

Equation (6.5.1) contains 4 unknowns. There are 2 obvious conditions: $F_i(x_i) = f(x_i)$, and $F_i(x_{i+1}) = f(x_{i+1})$. The remaining 2 conditions are obtained by choosing the coefficients so that the first and second derivatives of $F_i(x)$ at x_i are equal to the first and second derivatives of $F_{i-1}(x)$ at x_i, viz.

$$F_i'(x_i) = F_{i-1}'(x_i)$$

$$F_i''(x_i) = F_{i-1}''(x_i).$$

There remain special problems at x_1 and x_n, but we will deal with these later.

The conditions are now sufficient to determine the $(n-1)$ cubics which collectively constitute the cubic spline $S(x)$, viz.

$$S(x) = F_i(x) \quad \text{for } x_i \le x \le x_{i+1}.$$

How can we solve these equations? The simplest method is to note that $S''(x)$ is linear in x and is also continuous over the whole interval $[x_1, x_n]$; it is not, however differentiable. The graph of $S''(x)$ is therefore of the type:

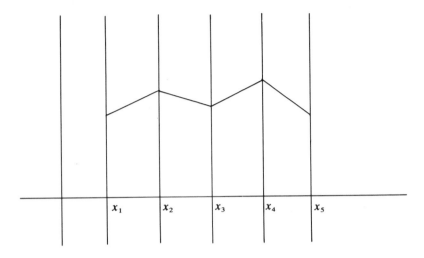

$S''(x)$ can therefore be represented in $[x_i, x_{i+1}]$ by a linear function which is easily seen to be

$$S''(x) = S''(x_i) + \frac{x - x_i}{x_{i+1} - x_i}(S''(x_{i+1})) - S''(x_i)). \qquad (6.5.2)$$

If we re-write (6.5.2) in the form

$$S''(x) = \frac{(x - x_i)}{(x_{i+1} - x_i)}S''(x_{i+1}) - \frac{(x - x_{i+1})}{(x_{i+1} - x_i)}S''(x_i)$$

and integrate twice then, putting $\Delta x_i = (x_{i+1} - x_i)$ as usual [see Definition 1.4.1] the result can be written

$$S(x) = \frac{(x - x_i)^3}{6\Delta x_i}S''(x_{i+1}) - \frac{(x - x_{i+1})^3}{6\Delta x_i}S''(x_i) + a(x - x_i) + b(x - x_{i+1}) \qquad (6.5.3)$$

since any expression of the form $Ax + B$ may be written as $a(x - x_i) + b(x - x_{i+1})$ for suitable choice of a, b provided $x_i \ne x_{i+1}$.

We now impose the conditions that $S(x_i) = f(x_i)$ and $S(x_{i+1}) = f(x_{i+1})$ so that, from (6.5.3)

$$f(x_i) = -\frac{(x_i - x_{i+1})^3}{6\Delta x_i}S''(x_i) + b(x_i - x_{i+1})$$

or

$$b = \frac{-f(x_i)}{\Delta x_i} + \frac{\Delta x_i}{6}S''(x_i) \qquad (6.5.4)$$

and

$$f(x_{i+1}) = \frac{(\Delta x_i)^3}{6\Delta x_i} S''(x_{i+1}) + a \, \Delta x_i$$

or

$$a = \frac{f(x_{i+1})}{\Delta x_i} - \frac{\Delta x_i}{6} S''(x_{i+1}). \tag{6.5.5}$$

Substituting from (6.5.4) and (6.5.5) in (6.5.3) gives, after slight re-arrangement of terms:

$$S(x) = \frac{S''(x_i)}{6} \left\{ \frac{(x_{i+1}-x)^3}{\Delta x_i} - \Delta x_i(x_{i+1}-x) \right\} + \frac{S''(x_{i+1})}{6} \left\{ \frac{(x-x_i)^3}{\Delta x_i} - \Delta x_i(x-x_i) \right\}$$

$$+ f(x_i)\left[\frac{x_{i+1}-x}{\Delta x_i}\right] + f(x_{i+1})\left[\frac{x-x_i}{\Delta x_i}\right] \tag{6.5.6}$$

this expression being valid for the interval $x_i < x < x_{i+1}$.

It is worth noting that if in (6.5.6) we ignore the two terms involving $S''(x_i)$ and $S''(x_{i+1})$ we obtain the approximation to $S(x)$

$$S(x) \doteqdot \frac{(x_{i+1}-x)f(x_i) + (x-x_i)f(x_{i+1})}{(x_{i+1}-x_i)} \; i$$

which is the result obtained from linear interpolation (Chapter 2). We see therefore that the terms involving $S''(x_i)$ and $S''(x_{i+1})$ can be regarded as correction terms obtained by using a cubic instead of a linear approximation. Before we can use (6.5.6) to determine $S(x)$ we must find the values $S''(x_i)$ and this we do by using the conditions that the 1st derivatives are continuous at the knots.

Differentiating (6.5.6) and putting $x = x_i$ we have

$$S'(x_i) = \frac{S''(x_i)}{6}\left\{ -\frac{3(x_{i+1}-x_i)^2}{\Delta x_i} + \Delta x_i \right\} - \frac{S''(x_{i+1})\,\Delta x_i}{6} - \frac{f(x_i)}{x_i} + \frac{f(x_{i+1})}{\Delta x_i}. \tag{6.5.7}$$

If we now replace i by $(i-1)$ in (6.5.6), differentiate and again put $x = x_i$ we obtain

$$S'(x_i) = \frac{S''(x_i)}{6}\left\{ \frac{3(x_i-x_{i-1})^2}{\Delta x_{i-1}} - \Delta x_{i-1} \right\} + S''(x_{i-1})\,\Delta x_{i-1} - \frac{f(x_{i-1})}{\Delta x_{i-1}} + \frac{f(x_i)}{\Delta x_{i-1}}. \tag{6.5.8}$$

The continuity of $S'(x)$ at x_i now requires that the expressions on the right of (6.5.7) and (6.5.8) are equal and this leads to the equation:

$$S''(x_{i-1})\,\Delta x_{i-1} + S''(x_i)\{2(x_{i+1}-x_{i-1})\} + S''(x_{i+1})\,\Delta x_i$$

$$= 6\left\{ \frac{f(x_{i+1})-f(x_i)}{\Delta x_i} - \frac{f(x_i)-f(x_{i-1})}{\Delta x_{i-1}} \right\}. \tag{6.5.9}$$

In the case where the x_i are evenly spaced (6.5.9) is simplified to

$$S''(x_{i-1}) + 4S''(x_i) + S''(x_{i+1}) = \frac{6(f(x_{i+1}) - 2f(x_i) + f(x_{i-1}))}{(\Delta(x_i))^2}. \quad (6.5.10)$$

The sets of $(n-1)$ equations, (6.5.9) and (6.5.10) contain $(n+1)$ unknowns $S''(x_j)$, $(j = 0, 1, \ldots, n)$ and in order to obtain a unique solution we must impose conditions on $S''(x_0)$ and $S''(x_n)$ and this is usually done by taking the spline in the intervals $[-\infty, x_0]$ and $[x_n, \infty]$ to be a straight line so that $S''(x_0) = S''(x_n) = 0$. This corresponds, in physical terms, to allowing the spline to assume its natural straight shape outside the intervals of approximation. The special $S(x)$ so determined is called the *natural cubic spline*.

Given these extra two conditions the equations (6.5.9) or (6.5.10) are now sufficient to determine the $S''(x_j)$ and so $S(x)$. The system of linear equations is of tri-diagonal form and such systems can be solved either by direct methods, such as Gaussian elimination or, if n is large, by indirect methods such as the Gauss–Seidel. Note that in the evenly-spaced case the coefficients of the equations, given by (6.5.10) are 1, 4, 1, so that the system is diagonally-dominant and the Gauss–Seidel method will converge. In the non-equally spaced case the same is still true for the coefficients are

$$(x_i - x_{i-1}), \ 2(x_{i+1} - x_{i-1}), \ (x_{i+1} - x_i)$$

and since

$$(x_{i+1} - x_{i-1}) = (x_{i+1} - x_i) + (x_i - x_{i-1})$$

the equations are again (easily) diagonally dominant (note that all expressions in brackets are >0).

EXAMPLE 6.5.1.   Fit a natural cubic spline to the data below and use it to estimate $f(55)$.

x	$f(x)$
25	5
36	6
49	7
64	8
81	9

We use (6.5.9) to form a set of linear equations for $S''(36)$, $S''(49)$, $S''(64)$ and we take $S''(25) = S''(81) = 0$. The equations are

$$(0) \cdot 11 + 2(24)S''(36) + 13S''(49) = 6\{\tfrac{1}{13} - \tfrac{1}{11}\}$$

$$13 \cdot S''(36) + 2(28)S''(49) + 15S''(64) = 6\{\tfrac{1}{15} - \tfrac{1}{13}\}$$

$$15 \cdot S''(49) + 2(32)S''(64) + 17(0) = 6\{\tfrac{1}{17} - \tfrac{1}{15}\}.$$

Rearranging these equations in a form suitable for Gauss–Seidel iteration they become (with coefficients to 6 d.p.):

$$S''(36) = -\tfrac{1}{48}[0 \cdot 083916 + 13S''(49)]$$

$$S''(49) = -\tfrac{1}{56}[0 \cdot 061538 + 13S''(36) + 15S''(64)]$$

$$S''(64) = -\tfrac{1}{64}[0 \cdot 047059 + 15S''(49)].$$

Starting at $S''(36) = S''(49) = S''(64) = 0$ and iterating produces the values

$$S''(36) = -0 \cdot 001594, \qquad S''(49) = -0 \cdot 000568, \qquad S''(64) = -0 \cdot 000602.$$

The point at which we wish to interpolate, $x = 55$, lies in the interval $[49, 64]$ so we must use the cubic appropriate to that interval, i.e. we use (6.5.6) with $x_{i+1} = 64$, $x_i = 49$, $x = 55$ and so obtain

$$S(55) = \frac{S''(49)}{6}\left[\frac{9^3}{15} - 15(9)\right] + \frac{S''(64)}{6}\left[\frac{6^3}{15} - 15(6)\right] + 7[\tfrac{9}{15}] + 8[\tfrac{6}{15}]$$

i.e.

$$S(55) = 0 \cdot 008179 + 0 \cdot 007585 + 7 \cdot 4 = 7 \cdot 415764.$$

So our estimate for $f(55)$ is $7 \cdot 415764$.

As remarked above the last two terms constitute the linear approximation, which therefore has the value

$$7(\tfrac{9}{15}) + 8(\tfrac{6}{15}) = 7 \cdot 4.$$

Since the function, $f(x)$, is in fact $\sqrt{x}$ we can check on the accuracy of the estimate for $\sqrt{55} = 7 \cdot 416198$ so our estimate using the cubic spline turns out to be correct to 3 d.p. The linear estimate is correct to only 1 d.p.

The result is satisfactory because we are working near the middle of a range of a smooth function with a small (absolute) value for its second derivative. Remember that we have taken $S''(25) = S''(81) = 0$. In fact $f''(x) = -\tfrac{1}{4}x^{-3/2}$ in this case so that $f''(25) = -0 \cdot 002$, $(81) \doteq -0 \cdot 00034$ and so our assumptions are not too far from the truth. Had we used the same method to interpolate say $\sqrt{5}$ using a table beginning at $x = 0$ the results would not have been so good since $f''(x)$ has a singularity at $x = 0$. The use of the natural cubic spline in cases where $f''(x)$ may be large at either end of the interval is not to be recommended.

A more typical example of the use of a spline is illustrated by

EXAMPLE 6.5.2.   Use a natural cubic spline based on the data below to interpolate values for $f(2 \cdot 7)$ and $f(3 \cdot 42)$.

$x =$	2·0	2·3	2·4	2·9	3·3	3·6	3·8
$f(x) =$	1·4094	2·0616	2·2619	3·1615	3·7845	4·2080	4·4728.

Solution.   There are 7 unknowns, $S''(2\cdot0)$, $S''(2\cdot3)$ etc. but we take $S''(2\cdot0) = S''(3\cdot8) = 0$ and so obtain 5 equations in 5 unknowns which we find from (6.5.9) to be, in matrix form.

$$\begin{pmatrix} 0\cdot8 & 0\cdot1 & 0 & 0 & 0 \\ 0\cdot1 & 1\cdot2 & 0\cdot5 & 0 & 0 \\ 0 & 0\cdot5 & 1\cdot8 & 0\cdot4 & 0 \\ 0 & 0 & 0\cdot4 & 1\cdot4 & 0\cdot3 \\ 0 & 0 & 0 & 0\cdot3 & 1\cdot0 \end{pmatrix} \begin{pmatrix} S''(2\cdot3) \\ S''(2\cdot4) \\ S''(2\cdot9) \\ S''(3\cdot3) \\ S''(3\cdot6) \end{pmatrix} = \begin{pmatrix} -1\cdot0260 \\ -1\cdot2228 \\ -1\cdot4502 \\ -0\cdot8750 \\ -0\cdot5260 \end{pmatrix}$$

These equations may be solved by the Gauss–Seidel method starting from $(0, 0, 0, 0, 0)$ and after 7 iterations the solutions converge (to 4 d.p.) to

$$S''(2.3) = -1\cdot1949, \qquad S''(2.4) = -0\cdot7007, \qquad S''(2\cdot9) = -0\cdot5250,$$

$$S''(3\cdot3) = -0\cdot3872, \qquad S''(3\cdot6) = -0\cdot4098.$$

For interpolation at $x = 2\cdot7$ we use equation (6.5.6) with $x_i = 2\cdot4$, $x_{i+1} = 2\cdot9$ which gives $S(2\cdot7) \doteqdot 2\cdot8199$ to 4 d.p. For interpolation at $x = 3\cdot42$ we use $x_i = 3\cdot3$, $x_{i+1} = 3\cdot6$ which gives $S(3\cdot42) \doteqdot 3\cdot9582$.

We can verify the accuracy of these results since

$$f(x) = 3 \log_e x + 1\cdot2x^{1/2} - 4\cdot1(1+x)^{-1/2}$$

so that $f(2\cdot7) = 2\cdot8201$ and $f(3\cdot42) = 3\cdot9579$. In each case therefore the interpolated values given by the cubic spline are correct to 3 d.p.

6.5.2.   Other End-Conditions for Splines

We saw earlier that the continuity conditions imposed upon the cubic spline left us with two values to be assigned, $S''(x_1)$ and $S''(x_n)$. The problem was resolved by taking both to be zero, thus giving the natural cubic spline. This apparently arbitrary choice may be justified in two ways

(i) on the 'flexible bar' interpretation the bar would take up a straight line position in $[-\infty, x_1]$ and in $[x_n, \infty]$, so that $S''(x) = 0$ corresponds to the physical reality;

(ii) by choosing $S''(x_0) = S''(x_n) = 0$ we minimize the integral

$$\int_{x_1}^{x_n} (S''(x))^2 \, dx$$

and so try to ensure that $S(x)$ has minimal oscillatory behaviour, which is desirable.

There are however other possibilities which are sometimes preferred when the underlying problem indicates that such is desirable. Two of the most common are:

(1) $S''(x_1), S''(x_n)$ *are specified*; if $S(x)$ is approximating to an analytic function [see IV, Definition 2.10.3] the true values of $S''(x_1), S''(x_n)$ may be available;

(2) $S(x)$ *is periodic*, so that $S'(x_1) = S'(x_n)$, $S(x_1) = S(x_n)$; this condition obviously only applies in appropriate circumstances; in general there is no justification for imposing such conditions.

6.5.3. Hermite Approximation

For approximation, or interpolation, of a function defined analytically a method due to Hermite is often useful. The method is superficially related to the spline method but in fact the two methods are very different because

(i) fitting of splines involves solving a system of linear equations to obtain numerical values for the second derivatives, $S''(x_i)$, whereas for Hermite interpolation the values of the first derivatives are given, and the second derivatives are not relevant;

(ii) splines are mainly used for fitting polynomials to *data*, Hermite polynomials are mainly used for approximating to *functions*.

The most commonly used Hermite approximation polynomial is the *cubic*, as in the case of splines and we shall discuss only this case in detail, the more general case can be analysed in a similar manner.

Suppose we have an analytic function $y = f(x)$ with values $f(x_i)$ and derivatives $f'(x_i)$ given at n points $x_i (i = 1, \ldots, n)$. Across each pair of adjacent points x_m, x_{m+1} we fit a cubic $p_m(x)$ such that

$$\left.\begin{matrix} p_m(x_m) = f(x_m), \; p'_m(x_m) = f'(x_m) \\ p_m(x_{m+1}) = f(x_{m+1}), \; p'_m(x_{m+1}) = f'(x_{m+1}) \end{matrix}\right\}. \qquad (6.5.11)$$

Since $p_m(x)$ contains 4 coefficients the 4 equations (6.5.11) determine it uniquely, and indeed the formula for $p_m(x)$ can be explicitly stated, viz.

$$p_m(x) = \left(1 + 2\frac{x - x_m}{x_{m+1} - x_m}\right)\left(\frac{x_{m+1} - x}{x_{m+1} - x_m}\right)^2 f(x_m)$$

$$+ \left(1 + 2\frac{x_{m+1} - x}{x_{m+1} - x_m}\right)\left(\frac{x - x_m}{x_{m+1} - x_m}\right)^2 f(x_{m+1})$$

$$+ \frac{(x - x_m)(x_{m+1} - x)^2}{(x_{m+1} - x_m)^2} f'(x_m) - \frac{(x - x_m)^2(x_{m+1} - x)}{(x_{m+1} - x_m)^2} f'(x_{m+1}). \qquad (6.5.12)$$

The error when we use (6.5.12) can be bounded by

$$\max_{x_m \leq x \leq x_{m+1}} |f(x) - p_m(x)| \leq \frac{(x_{m+1} - x_m)^4}{384} \max_{x_m \leq \alpha \leq x_{m+1}} |f^{(4)}(\alpha)|. \qquad (6.5.13)$$

A cubic Hermite approximation thus consists of a set of cubic polynomials, each defined over one interval, with continuity of the cubics and their first derivatives at all the nodes [cf. Definition 6.5.1].

EXAMPLE 6.5.3.   Use Hermite cubic interpolation to estimate the value of $\sqrt{55}$, taking $f(x) = \sqrt{x}$, $x_1 = 49$, $x_2 = 64$.

Solution. From (6.5.12) with $x_m = 49$, $x_{m+1} = 64$, $f(x_m) = 7$, $f'(x_m) = \frac{1}{14}$, $f(x_{m+1}) = 8$, $f'(x_{m+1}) = \frac{1}{16}$ we have the Hermite cubic approximation

$$f(x) \doteq \frac{(2x-83)(64-x)^2 \cdot 7}{15^3} + \frac{(143-2x)(x-49)^2 \cdot 8}{15^3} + \frac{(x-49)(64-x)^2}{15^2} \cdot \frac{1}{14}$$

$$- \frac{(x-49)^2(64-x)}{15^2} \cdot \frac{1}{16}.$$

Putting $x = 55$ yields the estimate

$$\sqrt{55} \doteq 7 \cdot 416286.$$

The correct value to 6 d.p., as mentioned in Example 6.5.1, is $7 \cdot 416198$, so the error is $0 \cdot 000088$ compared to an error of $0 \cdot 000434$ when we used the natural cubic spline in Example 6.5.1.

In general the errors when we use the Hermite cubic and the natural cubic spline on the same problem will not be very different for in both cases the error is proportional to h^4 (where h is the step between the two relevant nodes), although there is no estimate for the error in the spline quite so simple in form as (6.5.13).

We could have used (6.5.13) to obtain in advance an upper bound for the error in Example 6.5.3, viz.

$$\max_{49 \le x \le 64} |\sqrt{x} - H_2(x)| \le \frac{15^4}{384} \cdot \frac{15}{16} \cdot \frac{1}{7^7} \doteq 0 \cdot 00015$$

that is, about twice the error observed in this particular case, but small enough to have assured us that our result would be correct to (at least) 3 d.p. Had we wished to be certain of a result accurate to at least 5 d.p. (say) for $\sqrt{55}$ we would have had to ensure that the nodes were at most a distance h apart where

$$\frac{h^4}{384} \cdot \frac{15}{16} \cdot \frac{1}{7^7} < \frac{1}{2} \times 10^{-5}$$

which gives $h < 6 \cdot 4$ and this can be achieved by taking intermediate nodes at $53 \cdot 29$ $(= (7 \cdot 3)^2)$ and at $59 \cdot 29$ $(= (7 \cdot 7)^2)$. Repeating Example 6.5.3 with $x_m = 53 \cdot 29$ and $x_{m+1} = 59 \cdot 29$ we obtain the estimate $\sqrt{55} \doteq 7 \cdot 4162001$ (to 7 d.p.), an error of $0 \cdot 0000016$, so that our result *is* correct to 5 d.p. as predicted.

R.F.C.

REFERENCES

Abramovitz, M. and Stegun, I. (editors) (1964). *Handbook of Mathematical Functions*, National Bureau of Standards, U.S. Government Printing Office.

Atkinson, K. E. (1978). *An Introduction to Numerical Analysis*, Wiley, New York.

Hardy, G. H. and Wright, E. M. (1960). *An Introduction to the Theory of Numbers* (4th edition), Oxford University Press.

Whittaker, E. T. and Watson, G. N. (1946). *Modern Analysis* (4th edition), Cambridge University Press.

PROGRAMS

In the Appendix programs will be found which use the least square method (1) to fit a straight line, (2) to fit a parabola.

CHAPTER 7

Quadrature

7.1. INTRODUCTION

In this chapter we shall investigate techniques for the numerical evaluation of

$$I = \int_a^b f(x)\, dx$$

where a and b are constants and $f(x)$ is given either as a continuous differentiable function or discretely (i.e. the numerical value of $f(x)$ is given for specified values of x). It is always assumed that I exists and is finite [see IV, § 4.1].

7.1.1. The Riemann Sum

It is fairly natural to use the Riemann sum approach to integration to give an estimate for I. To recapitulate, we consider the interval $[a, b]$ to be subdivided at the points $x_0 = a, x_1, \ldots, x_n = b$. Then, if m_i is the least value of $f(x)$ in the interval (x_i, x_{i+1}) and M_i is the largest value, we have for the interval (x_i, x_{i+1})

$$b_i = m_i(x_{i+1} - x_i) \le \int_{x_i}^{x_{i+1}} f(x)\, dx \le M_i(x_{i+1} - x_i) = B_i. \qquad (7.1.1)$$

Summation over all the intervals gives

$$b = \sum_{i=0}^{n-1} m_i(x_{i+1} - x_i) \le \int_a^b f(x)\, dx \le \sum_{i=0}^{n-1} M_i(x_{i+1} - x_i) = B. \qquad (7.1.2)$$

Moreover, if we increase the number of divisions so that each $x_{i+1} - x_i \to 0$, the left-hand side and the right-hand side of (7.1.2) tend to the same limit. We can therefore take, for any subdivision, b or B as estimates for I. The method is easy to apply if the integrand $f(x)$ is known to be monotonic as in the following example [see IV § 2.7].

EXAMPLE 7.1.1. Evaluate $\int_1^2 dx'\, x$ by the Riemann sum method.

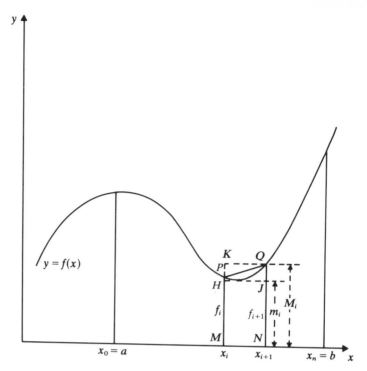

Figure 7.1.1

We divide the interval $[1, 2]$ into, say, ten equal intervals of width $0·1$. Since the function $1/x$ is a monotonic decreasing function of x we have, for the interval $[x_i, x_{i+1}]$, $m_i = 1/x_{i+1}$, $M_i = 1/x_i$. It follows from (7.1.2) that

$$b = \frac{1}{10}\left[\frac{1}{1·1}+\frac{1}{1·2}+\ldots+\frac{1}{2·0}\right] \leq \int_1^2 \frac{dx}{x} \leq \frac{1}{10}\left[\frac{1}{1·0}+\frac{1}{1·1}+\ldots+\frac{1}{1·9}\right] = B$$

i.e.

$$b = 0·66877 \leq \int_1^2 \frac{dx}{x} \leq 0·71877 = B.$$

We therefore have as approximations for $\int_1^2 dx/x$ either $0·66877$ or $0·71877$. These differ by $0·05$ so that there is a possibility of being in error by over 7% if either of these values is taken as an estimate for the integral. We note for future reference that the exact value of the integral is $\ln 2$ ($=0·693147$ to 6 decimal places) [cf. IV, Table 4.2.1].

7.1.2. Criteria for Good Quadrature Techniques

It is clear that the above technique is not a very good one from the practical point of view as we require good techniques to satisfy two important criteria.

1. The technique should give results of high accuracy.
2. The technique should require only a small number of function evaluations.

A further desirable feature of a technique is that it should give some measure of the error in the computational result although this may not always be possible.

The Riemann sum method just considered does not in most cases satisfy the criteria 1 and 2 but it does enable us to formulate other methods which can be improved upon to give these desirable properties. These basic methods will now be described.

7.1.3.   The Trapezium Rule

• If Figure 7.1.1 is examined then it is clear that (7.1.1) indicates that the area of the rectangle $HJNM$ ($= b_i$) is smaller than the area between the curve $y = f(x)$, the x-axis and the ordinates $x = x_i$ and $x = x_{i+1}$ ($= \int_{x_i}^{x_{i+1}} f(x)\, dx$) which in turn is smaller than the area of the rectangle $KQNM$ ($= B_i$). Thus, some area between b_i and B_i is likely to give a better approximation than either of these two quantities. Clearly, if one joins P to Q in Figure 7.1.1 then the area of the trapezium $PQNM$ lies between b_i and B_i. This area is

$$\tfrac{1}{2}(x_{i+1} - x_i)(f_i + f_{i+1}), \quad \text{where } f_i = f(x_i)$$

and we thus obtain the basic trapezium rule

$$\int_{x_i}^{x_{i+1}} f(x)\, dx \approx \tfrac{1}{2}(x_{i+1} - x_i)(f_i + f_{i+1}). \tag{7.1.3}$$

If, for each i, $x_{i+1} - x_i = h = (b - a)/n$ then h is called the step-length and summation over the whole range gives

$$\int_a^b f(x)\, dx \approx \tfrac{1}{2}h[f_0 + 2(f_1 + \ldots + f_{n-1}) + f_n] = I_T(n), \quad \text{say}. \tag{7.1.4}$$

This is known as the *composite trapezium rule*.

EXAMPLE 7.1.2.   Evaluate $\int_1^2 dx/x$ by the trapezium rule using

(a) 1 subdivision of the range,
(b) 2 subdivisions,
(c) 4 subdivisions,
(d) 10 subdivisions.

The answers we obtain are:

(a) $\displaystyle \int_1^2 \frac{dx}{x} \approx \frac{1}{2} \times 1 \left[\frac{1}{1} + \frac{1}{2} \right] = I_T(1) = 0 \cdot 75,$

(b) $\int_1^2 \dfrac{dx}{x} \approx \dfrac{1}{2} \times \dfrac{1}{2}\left[\dfrac{1}{1} + 2\times\dfrac{1}{1\cdot5} + \dfrac{1}{2}\right] = I_T(2)$

$\qquad = 0\cdot70833$ (to 5 decimal places),

(c) $\int_1^2 \dfrac{dx}{x} \approx \dfrac{1}{2} \times \dfrac{1}{4}\left[\dfrac{1}{1} + 2\left(\dfrac{1}{1\cdot25} + \dfrac{1}{1\cdot5} + \dfrac{1}{1\cdot75}\right) + \dfrac{1}{2}\right] = I_T(4)$

$\qquad = 0\cdot69702$ (to 5 decimal places),

(d) $\int_1^2 \dfrac{dx}{x} \approx \dfrac{1}{2} \times \dfrac{1}{10}\left[\dfrac{1}{1} + 2\left(\dfrac{1}{1\cdot1} + \dfrac{1}{1\cdot2} + \ldots + \dfrac{1}{1\cdot9}\right) + \dfrac{1}{2}\right]$

$\qquad = I_T(10) = 0\cdot69377$ (to 5 decimal places).

These numerical results appear to be tending to some limit in the region of $0\cdot69$ so that one would be fairly confident about quoting $I_T(10)$ as an approximation to the given integral. Since we can show that (see Hildebrand, 1974)

$$\int_{x_i}^{x_{i+1}} f(x)\,dx = \tfrac{1}{2}h[f_i + f_{i+1}] - \tfrac{1}{12}h^3 f''(\xi_i) \qquad (7.1.5)$$

where $h = x_{i+1} - x_i = (b-a)/n$ and $x_i < \xi_i < x_{i+1}$ (i.e. the approximation (7.1.3) is exact when $f(x)$ is a first order polynomial) it follows that

$$\int_a^b f(x)\,dx = I_T(n) - \tfrac{1}{12}h^3 \sum_{i=0}^{n-1} f''(\xi_i)$$

$$= I_T(n) - \tfrac{1}{12}h^2(b-a)f''(\xi), \qquad a < \xi < b. \qquad (7.1.6)$$

Thus

$$|\text{Error in trapezium rule}| < \tfrac{1}{12}h^2(b-a) \max_{a \le x \le b} |f''(x)|. \qquad (7.1.7)$$

In example 7.1.2(d) for instance, $h = 0\cdot1$, $b - a = 1$, $f(x) = 1/x$ giving $f''(x) = 2/x^3$ and $\max_{1 \le x \le 2} f''(x) = 2$, so that (7.1.7) informs us that the maximum truncation error in our computed result $I_T(10)$ is $0\cdot00167$, i.e. our result is only accurate to within 2 digits in the third decimal place. It follows that when stating the result at most 3 decimal places should be quoted and an indication should be given of the magnitude of the error. Thus we would write

$$\int_1^2 \dfrac{dx}{x} = 0\cdot693 \pm 0\cdot002.$$

It is worth noting that this is a substantially better result than that obtained by the method of Riemann sums although the same function evaluations were required.

In practice, as we might not be able to determine the derivatives of $f(x)$ easily, the following adaptive approach can be used to achieve a pre-determined accuracy. We select n and approximate $\int_{x_r}^{x_{r+1}} f(x)\,dx$ $(r = 0, \ldots, n-1)$ by the trapezium rule. For each r we then test if our result is sufficiently close to the trapezium rule estimate for

$$\int_{x_r}^{x_{r+\frac{1}{2}}} f(x)\,dx + \int_{x_{r+\frac{1}{2}}}^{x_{r+1}} f(x)\,dx \qquad (x_{r+\frac{1}{2}} = x_r + \tfrac{1}{2}h)$$

i.e. the difference between these two estimates is less in modulus than some prescribed constant ε. If, for some r, this criterion is not satisfied then the same test is applied to each of the subintervals $[x_r, x_{r+\frac{1}{2}}], [x_{r+\frac{1}{2}}, x_{r+1}]$, but now we insist that the difference between the two estimates should be less in modulus than $\frac{1}{2}\varepsilon$. This procedure is carried on until we achieve estimates for each interval $[x_r, x_{r+1}]$ which are in error by at most ε. One disadvantage of this approach is that a large number of function evaluations may be required and the Romberg technique of section 7.2 is generally preferable.

7.1.4.   The Mid-Point Rule

This method is similar to the trapezium rule in the sense that we are evaluating an area between b_i and B_i. In this case we assume the function values between x_i and x_{i+1} are constant and have value equal to the function value at $x_{i+\frac{1}{2}} = \frac{1}{2}(x_i + x_{i+1})$. Thus, we have the approximation (see Figure 7.1.2) which is known as the *basic mid-point rule*

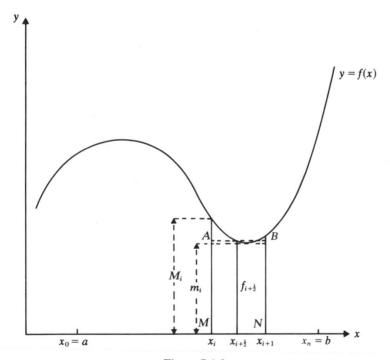

Figure 7.1.2

$$\int_{x_i}^{x_{i+1}} f(x)\, dx \approx \text{area } ABNM = (x_{i+1} - x_i) f_{i+\frac{1}{2}}, \qquad (7.1.8)$$

where

$$f_{i+\frac{1}{2}} = f(x_{i+\frac{1}{2}}) = f(x_i + \tfrac{1}{2}h); \qquad h = (x_{i+1} - x_i) = (b-a)/n.$$

It can be shown that (see Hildebrand, 1974) the approximation (7.1.8) is exact for functions $f(x)$ which are first degree polynomials or, more precisely, that

$$\int_{x_i}^{x_{i+1}} f(x)\, dx = h f_{i+\frac{1}{2}} + \tfrac{1}{24} h^3 f''(\xi_i) \qquad (7.1.9)$$

where

$$h = x_{i+1} - x_i = (b-a)/n \quad \text{and} \quad x_i < \xi_i < x_{i+1}.$$

Thus for the whole interval $[a, b]$

$$\int_a^b f(x)\, dx = h \sum_{i=0}^{n-1} f_{i+\frac{1}{2}} + \tfrac{1}{24} h^3 \sum_{i=0}^{n-1} f''(\xi_i),$$

$$= I_M(n) + \tfrac{1}{24} h^2 (b-a) f''(\xi), \qquad a < \xi < b \qquad (7.1.10)$$

where

$$I_M(n) = h \sum_{i=0}^{n-1} f_{i+\frac{1}{2}}.$$

This formula is known as the *composite mid-point rule*.

EXAMPLE 7.1.3.   Evaluate $\int_1^2 dx/x$ by the mid-point rule using

    (a) 1 subdivision of the range,
    (b) 2 subdivisions,
    (c) 4 subdivisions.

We obtain the following results:

(a) $\displaystyle \int_1^2 \frac{dx}{x} \approx 1 \left[\frac{1}{1\cdot5} \right] = I_M(1) = 0\cdot66667$ (to 5 decimal places),

(b) $\displaystyle \int_1^2 \frac{dx}{x} \approx \frac{1}{2}\left[\frac{1}{1\cdot25} + \frac{1}{1\cdot75} \right] = I_M(2) = 0\cdot68571$ (to 5 decimal places),

(c) $\displaystyle \int_1^2 \frac{dx}{x} \approx \frac{1}{4}\left[\frac{1}{1\cdot125} + \frac{1}{1\cdot375} + \frac{1}{1\cdot625} + \frac{1}{1\cdot875} \right] = I_M(4)$

$$= 0\cdot69122 \quad \text{(to 5 decimal places)}.$$

Note that the error in (c) is at most $\tfrac{1}{24}(\tfrac{1}{4})^2 \times 1 \times 2 = 0\cdot00521$ (in modulus) and thus

$$\int_1^2 \frac{dx}{x} = 0\cdot691 \pm 0\cdot005.$$

The adaptive approach used to get more precise results for the trapezium rule can also be applied in the case of the mid-point rule.

7.1.5. Simpson's Rule

In this method, too, we evaluate an area between b_i and B_i but, this time, we do not approximate the curve $y = f(x)$ by a straight line in the region $[x_i, x_{i+1}]$.

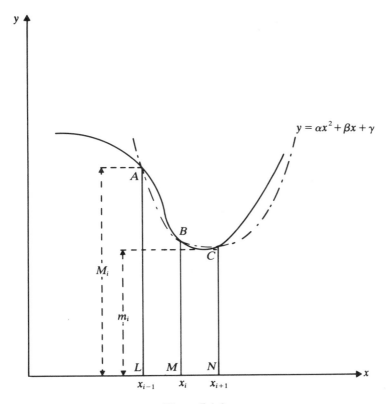

Figure 7.1.3

Instead, we consider the larger interval $[x_{i-1}, x_{i+1}]$ and we approximate $y = f(x)$ in this region by a parabola $y = \alpha x^2 + \beta x + \gamma$ which passes through the points $A(x_{i-1}, f_{i-1})$, $B(x_i, f_i)$, $C(x_{i+1}, f_{i+1})$ (Figure 7.1.3). We thus have the approximation

$$\int_{x_{i-1}}^{x_{i+1}} f(x)\, dx \approx \int_{x_{i-1}}^{x_{i+1}} (\alpha x^2 + \beta x + \gamma)\, dx$$

$$= 2h[\alpha(x_i^2 + \tfrac{1}{3}h^2) + \beta x_i + \gamma]$$

and, using the fact that A, B, C lie on the parabola, we obtain the *basic Simpson rule*

$$\int_{x_{i-1}}^{x_{i+1}} f(x)\,dx \approx \tfrac{1}{3}h[f_{i-1}+4f_i+f_{i+1}]. \qquad (7.1.11)$$

More precisely, it can be shown that

$$\int_{x_{i-1}}^{x_{i+1}} f(x)\,dx = \tfrac{1}{3}h[f_{i-1}+4f_i+f_{i+1}]-\tfrac{1}{90}h^5 f^{(\mathrm{iv})}(\xi_i) \qquad (7.1.12)$$

where

$$h = x_{i+1}-x_i = x_i-x_{i-1} = (b-a)/n \quad \text{and} \quad x_{i-1}<\xi_i<x_{i+1}$$

which indicates that the approximation (7.1.11) is exact for third order polynomials. To apply Simpson's rule to an interval $[a, b]$ we apply the basic rule (7.1.11) to the intervals (x_0, x_2), (x_2, x_4), ..., (x_{n-2}, x_n) where $n = 2m$, i.e. in this case we require $2m+1$ function evaluations. The formula for integration over the whole range is

$$\int_a^b f(x)\,dx = \tfrac{1}{3}h\!\left[f_0+2\sum_{i=0}^{m-1} f_{2i+1}+4\sum_{i=1}^{m-1} f_{2i}+f_{2m}\right]$$

$$-\tfrac{1}{90}h^5\sum_{i=0}^{m-1} f^{(\mathrm{iv})}(\xi), \qquad x_{2i}<\xi_i<x_{2i+2}$$

$$= I_S(m)-\tfrac{1}{180}h^4(b-a)f^{(\mathrm{iv})}(\xi), \qquad a<\xi<b \qquad (7.1.13)$$

where

$$I_S(m) = \tfrac{1}{3}h\!\left[f_0+2\sum_{i=0}^{m-1} f_{2i+1}+4\sum_{i=1}^{m-1} f_{2i}+f_{2m}\right].$$

This is known as the *composite Simpson rule*.

EXAMPLE 7.1.4.   Evaluate $\int_1^2 dx/x$ by means of Simpson's rule

  (a) with 3 function evaluations,
  (b) with 5 function evaluations.

The answers are:

(a) $$\int_1^2 \frac{dx}{x} \approx I_S(1) = \frac{1}{3}\times\frac{1}{2}\!\left[\frac{1}{1}+\frac{4}{1\cdot5}+\frac{1}{2}\right]$$

$$= 0\cdot69444 \quad \text{(to 5 decimal places)}$$

(b) $$\int_1^2 \frac{dx}{x} \approx I_S(2) = \frac{1}{3}\times\frac{1}{4}\!\left[\frac{1}{1}+\frac{4}{1\cdot25}+\frac{2}{1\cdot5}+\frac{4}{1\cdot75}+\frac{1}{2}\right]$$

$$= 0\cdot69325 \quad \text{(to 5 decimal places)}.$$

The error in these results can be estimated from formula (7.1.13).

The adaptive approach used earlier with the trapezium rule can be applied with Simpson's rule to give more accurate results.

7.1.6. Newton–Cotes Formulae

If the mid-point rule is recast in the form

$$\int_{x_i}^{x_{i+2}} f(x)\, dx \approx 2hf_{i+1}$$

then the trapezium, mid-point and Simpson rules belong to a class of formulae, called Newton–Cotes type formulae, which are of the form

$$\int_{x_0}^{x_n} f(x)\, dx \approx \sum_{i=0}^{n} \alpha_i f_i$$

where $x_i = x_0 + ih$, $f_i = f(x_i)$, $i = 0, \ldots, n$ and the α_i are constants. If α_0 and α_n are non-zero the formulae are termed closed formulae otherwise they are referred to as open formulae. The Newton–Cotes type formulae can also be used for quadrature but their main use has been in the integration of differential equations (see Chapter 8). A list of Newton–Cotes type formulae appears in Appendix 7.1. (Note that some authors, e.g. Davis and Rabinowitz, 1963, understand Newton–Cotes formulae to be those obtained by integrating the interpolating polynomial which passes through the points $(x_0, f_0), \ldots, (x_n, f_n)$.)

7.2. ROMBERG INTEGRATION

It will be seen from the examples in the preceding section that, as the number of function evaluations, n, increases, the computed values $I_T(n)$ (or $I_M(n)$ and $I_S(n)$) tend towards the correct value of $I = \int_1^2 dx/x$. This suggests that there is some relationship between $I - I_T(n)$ (or $I - I_M(n)$ and $I - I_S(n)$) and n which enables us to use Richardson's technique (2.4.1) to extrapolate and get an improved approximation to I. This is in fact the case and, in general, if $I = \int_0^1 f(x)\, dx$ where $f(x)$ and all its derivatives are finite, then it can be shown that

$$I = I_T(n) + \frac{A_T}{n^2} + \frac{B_T}{n^4} + \frac{C_T}{n^6} + \ldots \tag{7.2.1}$$

(or equivalently, $I = I_T(n) + A_T h^2 + B_T h^4 + C_T h^6 + \ldots$)

$$I = I_M(n) + \frac{A_M}{n^2} + \frac{B_M}{n^4} + \frac{C_M}{n^6} + \ldots \tag{7.2.2}$$

$$I = I_S(n) + \frac{B_S}{n^4} + \frac{C_S}{n^6} + \ldots \tag{7.2.3}$$

In the method of Romberg integration we systematically use the above formulae to eliminate in turn the coefficients of $1/n^2$, $1/n^4$, $1/n^6$, etc., by taking n to have the values 1, 2, 4, 8, ... We have, for example, using the trapezium rule

$$I = I_T(1) + A_T + B_T + C_T + \ldots \tag{7.2.4}$$

$$I = I_T(2) + \tfrac{1}{4}A_T + \tfrac{1}{16}B_T + \tfrac{1}{64}C_T + \ldots \tag{7.2.5}$$

$$I = I_T(4) + \tfrac{1}{16}A_T + \tfrac{1}{256}B_T + \tfrac{1}{4096}C_T + \ldots \tag{7.2.6}$$

$$I = I_T(8) + \tfrac{1}{64}A_T + \tfrac{1}{4096}B_T + \tfrac{1}{262144}C_T + \ldots \tag{7.2.7}$$

From (7.2.4) and (7.2.5) we get

$$I = I_T(1, 2) - \tfrac{1}{4}B_T - \tfrac{1}{16}(5C_T) - \ldots \tag{7.2.8}$$

and similarly

$$I = I_T(2, 4) - \tfrac{1}{64}B_T - \tfrac{1}{1024}(5C_T) - \ldots \tag{7.2.9}$$

$$I = I_T(4, 8) - \tfrac{1}{1024}B_T - \tfrac{1}{65536}(5C_T) - \ldots \tag{7.2.10}$$

where

$$I_T(r, 2r) = \tfrac{1}{3}[4I_T(2r) - I_T(r)], \qquad r = 1, 2, 3.$$

Next, we eliminate B_T between successive equations and we get

$$I = I_T(1, 2, 4) + \tfrac{1}{64}C_T + \ldots \tag{7.2.11}$$

and

$$I = I_T(2, 4, 8) + \tfrac{1}{4096}C_T + \ldots \tag{7.2.12}$$

where

$$I_T(r, 2r, 4r) = \tfrac{1}{15}[16I_T(2r, 4r) - I_T(r, 2r)]$$

and, proceeding further,

$$I_T(r, 2r, \ldots, 2^k r) = \frac{1}{2^{2k}-1}[2^{2k}I_T(2r, \ldots, 2^k r) - I_T(r, \ldots, 2^{k-1}r)]. \tag{7.2.13}$$

Further elimination can be carried out if required. We then set up the tableau

$$
\begin{array}{llll}
I_T(1) & & & \\
I_T(2) & I_T(1, 2) & & \\
I_T(4) & I_T(2, 4) & I_T(1, 2, 4) & \\
I_T(8) & I_T(4, 8) & I_T(2, 4, 8) & I_T(1, 2, 4, 8)
\end{array}
$$

If the entries above any upward sloping line tend to a limit then that limit is the value of $I = \int_0^1 f(x)\, dx$.

EXAMPLE 7.2.1.   Evaluate $I = \int_1^2 dx/x$ by Romberg integration.

From the values of $I_T(1)$, $I_T(2)$, $I_T(4)$ obtained in Example 7.1.2 we can form the tableau

$$
\begin{array}{ccc}
0\cdot75000 & & \\
& 0\cdot69444 & \\
0\cdot70833 & & 0\cdot69317 \\
& 0\cdot69325 & \\
0\cdot69702 & &
\end{array}
$$

Since the two adjacent values $0\cdot69325$ and $0\cdot69317$ are very close together this gives us some confidence in assuming that $0\cdot69317$ is a good approximation to I. Better results can, of course, be obtained if we evaluate $I_T(8)$ and the corresponding tableau entries $I_T(4, 8)$, $I_T(2, 4, 8)$, $I_T(1, 2, 4, 8)$ but this requires us to evaluate the integrand a further four times.

Identical results hold for the mid-point rule with

$$I_M(r, 2r) = \tfrac{1}{3}[4I_M(2r) - I_M(r)], \qquad r = 1, 2, 4, \ldots$$

$$I_M(r, 2r, 4r) = \tfrac{1}{15}[16I_M(2r, 4r) - I_M(r, 2r)], \qquad r = 1, 2, 4, \ldots$$

and, in general,

$$I_M(r, 2r, \ldots, 2^k r) = \frac{1}{2^{2k} - 1}[2^{2k} I_M(2r, \ldots, 2^k r) - I_M(r, \ldots, 2^{k-1} r)]. \quad (7.2.14)$$

However, there is an advantage in using the trapezium rule as fewer function evaluations will be required at subsequent stages of the computation since, with $h = 2^{-n}$, $2^n + 1$ function evaluations are required for the trapezium rule as compared with $2^{n+1} - 1$ function evaluations for the mid-point rule.

For Simpson's rule we have the slight variation

$$I_S(r, 2r) = \tfrac{1}{15}[16I_S(2r) - I_S(r)], \qquad r = 1, 2, 4, \ldots$$

$$I_S(r, 2r, 4r) = \tfrac{1}{63}[64I_S(2r, 4r) - I_S(r, 2r)], \qquad r = 1, 2, 4, \ldots$$

and so on. It is worth remarking that the quantity $I_T(r, 2r)$ is identical with $I_S(r)$ and thus Simpson's rule can be considered as the first stage of the Romberg extrapolation of the trapezium rule. In some applications, e.g. if the interval of integration is greater than 1, it may be advisable to start the tableau at $I(2)$ or $I(4)$. For the procedure to be adopted if the integrand $f(x)$ is singular at the end of the range (i.e. $f(x) \to \infty$ as $x \to 0$ or $f^{(n)}(x) \to \infty$ as $x \to 0$, where $f^{(n)}(x)$ denotes some derivative of $f(x)$) see section 7.6.

7.2.1. The Variation of Lyness and McHugh

A number of variations of the extrapolation procedure of the previous section exist. We mention briefly that due to Lyness and McHugh, 1963, as we shall require this later when we consider the evaluation of multiple integrals.

Lyness and McHugh substitute the values $n = 1, 2, 3, 4, \ldots$ in, say, (7.2.1) and proceed to eliminate A_T, B_T, C_T, $\ldots$ obtaining the formulae given in

Appendix 7.2(a). For instance, we can show that

$$I = I_T(1, 2, 3) + \tfrac{1}{36}C_T + \ldots$$

where

$$I_T(1, 2, 3) = \tfrac{1}{24}I_T(1) - \tfrac{16}{15}I_T(2) + \tfrac{81}{40}I_T(3). \qquad (7.2.15)$$

EXAMPLE 7.2.2.   Evaluate $I = \int_1^2 dx/x$ by the Lyness and McHugh extrapolation technique.

We have previously evaluated

$$I_T(1) = 0 \cdot 75, \qquad I_T(2) = 0 \cdot 70833,$$

and

$$I_T(3) = \tfrac{1}{2} \times \tfrac{1}{3}[1 + 2 \times \tfrac{3}{4} + 2 \times \tfrac{3}{5} + \tfrac{1}{2}] = 0 \cdot 7.$$

Thus

$$I \approx I_T(1, 2, 3) = 2 \cdot 025(0 \cdot 7) - 1 \cdot 06667(0 \cdot 70833) + 0 \cdot 04167(0 \cdot 75)$$
$$= 0 \cdot 69319.$$

7.3. GAUSSIAN QUADRATURE

The methods of the previous sections required the evaluation of the integrand at equal intervals. However, it is immaterial to an electronic computer whether a function is evaluated at $x = 1$ or $x = 0 \cdot 5771$ and it is therefore reasonable to seek approximations for $\int_a^b f(x)\, dx$ of the form

$$\int_a^b f(x)\, dx \approx \sum_{i=1}^n \alpha_i f(x_i)$$

where the α_i are constants and the abscissae x_i are values of x between a and b which are not equally spaced.

A whole class of such approximations exist of the form (see Cohen, 1973)

$$\int_a^b w(x)f(x)\, dx = \sum_{i=1}^n \alpha_i f(x_i) + R_n \qquad (7.3.1)$$

where $w(x)$ is a positive weight function in $[a, b]$ i.e. $w(x) \geq 0$ for all $x \in [a, b]$, x_i are the roots of the polynomial equation $p_n(x) = 0$ where $p_n(x)$ is a member of the sequence of polynomials $\{p_m(x)\}$ which are orthogonal over $[a, b]$ with respect to the weight function $w(x)$ i.e. $\int_a^b w(x)p_m(x)p_n(x)\, dx = 0$, $m \neq n$ and R_n is the error in the approximation which is proportional to $f^{(2n)}(\xi)$, $a < \xi < b$.

Since the $(2n)$th derivative of a polynomial of degree $2n-1$ or less is zero, it follows that the approximation in (7.3.1) is exact for such polynomials. It should be noted that once we have a table of the coefficients $\alpha_1, \ldots, \alpha_n$ and the abscissae $x_1, \ldots, x_n$ corresponding to a particular weight function $w(x)$ over an interval $[a, b]$ then only n function evaluations of a given polynomial $p(x)$, of degree $2n-1$ or less, are required to evaluate

$$\int_a^b w(x)p(x)\, dx$$

exactly. Had we used a Newton–Cotes formula we would, in general, have required about $2n$ function evaluations to produce an exact result.

7.3.1. Gauss–Legendre Quadrature

In this application of Gaussian quadrature we are interested in evaluating

$$\int_{-1}^1 f(x)\, dx$$

and we can show that

$$\int_{-1}^1 f(x)\, dx = \sum_{k=1}^n \alpha_k f(x_k) + R_n \tag{7.3.2}$$

where $x_1, \ldots, x_n$ are the roots of the Legendre polynomial equation $P_n(x) = 0$ [see IV, § 10.3.2],

$$\alpha_k = 2/\{[P'_n(x_k)]^2(1-x_k^2)\}, \qquad k = 1, 2, \ldots, n$$

$$R_n = 2^{2n+1}(n!)^4 f^{(2n)}(\xi)/(2n+1)[(2n)!]^3, \qquad -1 < \xi < 1.$$

The values of x_k and α_k, correct to 10 decimal places are given in Appendix 7.3(a) for $n = 1(1)6$ (n from 1 to 6 in steps of 1). More extensive tables can be found in Abramowitz and Stegun, 1965.

EXAMPLE 7.3.1. Evaluate $I = \int_{-1}^1 (x^4 + x^2 + 1)\, dx$ by Gauss–Legendre quadrature.

Employing the 3-point formula in Appendix 7.3(a) we get

$$I \approx 0{\cdot}555556(0{\cdot}774597^4 + 0{\cdot}774597^2 + 1) + 0{\cdot}888889(1{\cdot}0)$$
$$+ 0{\cdot}555556([-0{\cdot}774597]^4 + [-0{\cdot}774597]^2 + 1)$$
$$= 3{\cdot}066670.$$

The exact value of the integral is $3\frac{1}{15} = 3 \cdot 066667$ (to six decimal places) and the agreement between the computed value and the exact value is expected in view of the fact that the third-order formula is exact for polynomials of degree ≤ 5. The error can be accounted for by the rounding up of the coefficients α_k and rounding during the calculation.

EXAMPLE 7.3.2.   Evaluate $I = \int_1^2 dx/x$ by Gauss–Legendre quadrature.

Before we can apply the Gauss–Legendre quadrature formula it is necessary to transform the interval of integration to $[-1, 1]$. The transformation $x = \frac{1}{2}[(b - a)u + (b + a)]$ produces the result

$$\int_a^b f(x)\, dx = \frac{1}{2}(b - a) \int_{-1}^1 f[\tfrac{1}{2}(b - a)u + \tfrac{1}{2}(b + a)]\, du. \qquad (7.3.3)$$

Application of result (7.3.3) with $b = 2$, $a = 1$ gives

$$I = \int_1^2 \frac{dx}{x} = \int_{-1}^1 \frac{du}{3 + u}.$$

With the Gauss–Legendre 3-point formula we have the approximation

$$I \approx \frac{0 \cdot 555556}{2 \cdot 225403} + \frac{0 \cdot 888889}{3 \cdot 0} + \frac{0 \cdot 555556}{3 \cdot 774597} = 0 \cdot 693122.$$

Since

$$|f^{(6)}(\xi)| = \left| \frac{6!}{(3 + \xi)^7} \right| < \frac{6!}{2^7}, \qquad -1 < \xi < 1$$

it follows that

$$|R_3| < \frac{2^7 (3!)^4}{7(6!)^3} \times \frac{6!}{2^7} = \frac{1}{2800}$$

and the maximum truncation error incurred in using the 3-point formula is at most 4 units in the fourth decimal place.

In a large number of examples it may not be easy to differentiate the function $f(x)$ to obtain an estimate for the truncation error. In fact, $f(x)$ may not even be differentiable [see IV, § 3.1]. In this event we have two courses open to us to check the validity of the computed result.

   Strategy 1:   To use a higher order Gauss–Legendre quadrature formula to check results

and

   Strategy 2:   To apply the same Gauss–Legendre quadrature formula to sub-intervals of the range.

EXAMPLE 7.3.3.   Evaluate $\int_1^2 dx/x$ by Strategy 1.

If we denote the result of using an n-point formula by I_n, then

$I_3 = 0\cdot693122$ (see Example 7.3.2)

$$I_4 = 0\cdot347855\left[\frac{1}{2\cdot138864} + \frac{1}{3\cdot861136}\right] + 0\cdot652145\left[\frac{1}{2\cdot660019} + \frac{1}{3\cdot339981}\right]$$

$$= 0\cdot693146$$

$$I_5 = 0\cdot236927\left[\frac{1}{2\cdot093820} + \frac{1}{3\cdot906180}\right]$$

$$+ 0\cdot478629\left[\frac{1}{2\cdot461531} + \frac{1}{3\cdot538469}\right] + 0\cdot568889\left[\tfrac{1}{3}\right]$$

$$= 0\cdot693147.$$

The close agreement between I_4 and I_5 suggests that the value for I_5 is a good approximation to I. However, even with I_4 and I_5 consistent it is still worthwhile to use a higher order formula to check, as it has been pointed out by Clenshaw and Curtis (1960) that the 3- and 4-point formulae, when applied to evaluate the integral

$$J = \int_{-1}^{1} \frac{dx}{x^4 + x^2 + 0\cdot9}$$

give the consistent results $1\cdot585026$ and $1\cdot585060$, from which we might have drawn the conclusion that $1\cdot585$ is the correct value of J to four significant figures. That this is not the case can be seen by using the 5-point Gauss–Legendre formula. In fact, the exact value of J to 6 decimal places is $1\cdot582233$.

The advantage of adopting the approach of Strategy 1 is the certain knowledge (Cheney, 1966) that, with exact arithmetic, higher order approximations to

$$I = \int_a^b w(x)f(x)\,dx \qquad (a, b \text{ finite})$$

converge to I.

Strategy 2 can be applied in a number of ways. The first, and perhaps most obvious, consists of dividing the interval into two equal parts and applying the same order formula to each part. This is illustrated in Example 7.3.4.

EXAMPLE 7.3.4.   Evaluate $I = \int_1^2 dx/x$ by subdivision of the interval.

We have

$$I = \int_1^{1 \cdot 5} \frac{dx}{x} + \int_{1 \cdot 5}^2 \frac{dx}{x}$$

$$= \int_{-1}^1 \frac{du}{5 + u} + \int_{-1}^1 \frac{du}{u + 7}.$$

Employing the 3-point formula to evaluate each of these integrals we find

$$I \approx 0 \cdot 405464 + 0 \cdot 287682 = 0 \cdot 693146.$$

The process of subdivision is repeated as often as is necessary until the approximate integral over the sub-interval (a_k, b_k) agrees to specified accuracy, with the sum of the approximate integrals for $(a_k, \frac{1}{2}[a_k + b_k])$ and $(\frac{1}{2}[a_k + b_k], b_k)$. One disadvantage is that previously calculated function values are not utilised. Robinson (1971) has overcome this by using a 3-point formula initially and arranging for a subdivision into three parts in each of which the original abscissa x_k $(k = 1, 2, 3)$ is the mid-point. Thus the interval $(-1, 1)$ is divided up into the three intervals $(-1, -\alpha)$ $(-\alpha, \alpha)$, $(\alpha, 1)$ where $\alpha = 2\sqrt{(3/5)} - 1$. The 3-point formula is then applied to each of these intervals.

EXAMPLE 7.3.5.* Evaluate $I = \int_1^2 dx/x$ by Robinson's technique.

$$I = \int_{-1}^1 \frac{du}{3 + u}$$

$$= \int_{-1}^{-\alpha} \frac{du}{3 + u} + \int_{-\alpha}^{\alpha} \frac{du}{3 + u} + \int_{\alpha}^1 \frac{du}{3 + u}$$

$$= \beta \int_{-1}^1 \frac{dt}{3 - \gamma + \beta t} + \alpha \int_{-1}^1 \frac{dt}{3 + \alpha t} + \beta \int_{-1}^1 \frac{dt}{3 + \gamma + \beta t}$$

where

$$\alpha = 2\sqrt{(3/5)} - 1, \qquad \gamma = \sqrt{(3/5)}, \qquad \beta = 1 - \gamma.$$

Applying the 3-point formula to each of these integrals we find

$$I \approx 0 \cdot 203270 + 0 \cdot 370303 + 0 \cdot 119574 = 0 \cdot 693147.$$

The process can be carried further in an obvious manner, the calculations for each subinterval proceeding until the approximate value obtained by the direct use of the 3-point rule is equal, to specified accuracy, to the value obtained using the Robinson subdivision for that interval.

7.3.2.   The Correction Terms for Gauss–Legendre Quadrature and the Application to Romberg Integration

In the previous section it was pointed out that the error, R_n, in the n-point Gauss–Legendre formula was proportional to the $(2n)$th derivative of the integrand evaluated at some point in $(-1, 1)$. Alternatively, it can be shown that the error term in the n-point formula is of the form

$$R_n = c_n[f^{(2n-1)}(1) - f^{(2n-1)}(-1)] + c_{n+1}[f^{(2n+1)}(1) - f^{(2n+1)}(-1)]$$

$$+ \ldots + c_r[f^{(2r-1)}(1) - f^{(2r-1)}(-1)] + \ldots \tag{7.3.4}$$

where $f^{(k)}(\pm 1)$ denotes the kth derivative evaluated at ± 1 and the c_k are constants. Appendix 7.3(b) contains a table of the c_k for given n. It can be seen from this Appendix that we have, for example,

$$\int_{-1}^{1} f(x)\, dx = f\left(-\frac{1}{\sqrt{3}}\right) + f\left(\frac{1}{\sqrt{3}}\right) + \tfrac{1}{270}[f'''(x)]_{-1}^{1}$$

$$- \tfrac{4}{8505}[f^{(v)}(x)]_{-1}^{1} + \tfrac{17}{340200}[f^{(vii)}(x)]_{-1}^{1}. \tag{7.3.5}$$

If the derivative of $f(x)$ is easily obtainable then an improved approximation for the integral can be obtained by computing successive correction terms as in the following example.

EXAMPLE 7.3.6.   Evaluate

$$I = \int_{-1}^{1} \frac{du}{3+u}\left(= \int_{1}^{2} \frac{dx}{x}\right)$$

using formula (7.3.5).

For the given integral

$$f(u) = (3+u)^{-1}, \qquad f^{(r)}(u) = (-1)^r r!(3+u)^{-r-1}.$$

The Gauss–Legendre 2-point formula gives the approximation

$$f\left(-\frac{1}{\sqrt{3}}\right) + f\left(\frac{1}{\sqrt{3}}\right) = 0.69230_8$$

$$\text{first correction term} = 0.00130_2$$

$$\text{second correction term} = -0.00086_8$$

$$\text{third correction term} = 0.00098_0.$$

We would like each correction term to be much smaller than its predecessor but it will be observed that the third correction term is in fact larger than its predecessor. The formula (7.3.5) is an example of an asymptotic expansion [see § 7.7] and it is typical of such expansions for terms to decrease and then increase again. The procedure usually adopted is to terminate the expansion at the last decreasing term which is taken with a weight factor $\tfrac{1}{2}$. Applying this to the present data

$$I = 0.69230_8 + 0.00130_2 - \tfrac{1}{2}(0.00086_8)$$

$$= 0.69317_6.$$

If we had used a different example, such as the evaluation of $\int_{-1}^{1} e^x \, dx$, we would have found that the first 3 correction terms decreased and we would not have encountered the above difficulties.

In many instances it will be impossible to determine the derivatives and hence the correction terms. Nevertheless, (7.3.4) can still be of use to determine an improved approximation to I, since it forms the basis by which Gauss–Legendre formulae can be adapted for Romberg integration. If we divide the range of integration into k equal parts and use the Gauss–Legendre n-point formula in each part it can be shown that

$$I = I(k) + \frac{A}{k^{2n}} + \frac{B}{k^{2n+2}} + \frac{C}{k^{2n+4}} + \dots \qquad (7.3.6)$$

where $I(k)$ denotes the approximation using k intervals and $A, B, C, \dots$ are constants. The Romberg tableau can be set up in the manner that was demonstrated in section 7.2, eliminating $A, B, C, \dots$ in turn.

EXAMPLE 7.3.7.   Evaluate

$$I = \int_{1}^{2} \frac{dx}{x} \left(= \int_{-1}^{1} \frac{du}{3+u} \right)$$

by Romberg integration using a Gauss–Legendre 2-point formula.

We have

$$I(1) = 0 \cdot 69230_8 \qquad \text{(see Example 7.3.6)}$$

$$I(2) = \int_{1}^{1 \cdot 5} \frac{dx}{x} + \int_{1 \cdot 5}^{2} \frac{dx}{x} = \int_{-1}^{1} \frac{du}{5+u} + \int_{-1}^{1} \frac{du}{7+u}$$

$$= 0 \cdot 69307_7.$$

Similarly,

$$I(4) = \int_{1}^{1 \cdot 25} \frac{dx}{x} + \int_{1 \cdot 25}^{1 \cdot 5} \frac{dx}{x} + \int_{1 \cdot 5}^{1 \cdot 75} \frac{dx}{x} + \int_{1 \cdot 75}^{2} \frac{dx}{x}$$

$$= \int_{-1}^{1} \frac{du}{9+u} + \int_{-1}^{1} \frac{du}{11+u} + \int_{-1}^{1} \frac{du}{13+u} + \int_{-1}^{1} \frac{du}{15+u}$$

$$= 0 \cdot 69314_2.$$

The Romberg tableau is

0·692308		
0·693077	0·693128	
0·693142	0·693146₃	0·693146₆

From the tableau we find $I \approx 0 \cdot 693147$.

7.3.3. Gauss–Laguerre Quadrature

Here we are concerned with the evaluation of integrals of the form

$$\int_0^\infty e^{-x} f(x)\, dx$$

where $f(x)$ is a function which possesses a bounded $(2n)$th derivative. It can be shown that

$$\int_0^\infty e^{-x} f(x)\, dx = \sum_{k=1}^n \alpha_k f(x_k) + R_n \qquad (7.3.7)$$

where $x_1, x_2, \ldots, x_n$ are the roots of the Laguerre polynomial equation $L_n(x) = 0$, [see IV, § 10.5.1] and

$$\alpha_k = (n!)^2 / [x_k \{L_n'(x_k)\}^2], \qquad k = 1, \ldots, n;$$

$$R_n = (n!)^2 f^{(2n)}(\xi)/(2n)!, \qquad 0 < \xi < \infty$$

The values of x_k and α_k are given for $n = 1(1)6$ in Appendix 7.3(c).

EXAMPLE 7.3.8. Evaluate $I = \int_3^\infty e^{-x} x^{1/2}\, dx$.

We first have to transform the range into $(0, \infty)$ by the substitution $x = u + 3$. Then

$$I = e^{-3} \int_0^\infty e^{-u} (u + 3)^{1/2}\, du = e^{-3} I_1, \quad \text{say.}$$

If the 3-point Gauss–Laguerre formula is used to evaluate I_1 we get

$$I_1 \approx 0{\cdot}711093 \,(3{\cdot}415775)^{1/2} + 0{\cdot}278518 \,(5{\cdot}294280)^{1/2}$$

$$+ 0{\cdot}010389 \,(6{\cdot}289945)^{1/2}$$

$$= 1{\cdot}981133.$$

Hence $I \approx 0{\cdot}098635$ and the error in I is $e^{-3} R_3$ where

$$e^{-3} |R_3| < \frac{7e^{-3}}{2304\sqrt{3}} = 0{\cdot}000088.$$

We also require some strategy to check the validity of computed results as we often cannot determine the truncation error. Two courses present themselves in this case.

Strategy 1. The use of higher order Gauss–Laguerre formulae to check results.

Strategy 2: The integral is recalculated by using a Gauss–Legendre formula to evaluate $\int_0^a e^{-x} f(x)\, dx$ $(a = 2$, say) and the Gauss–Laguerre formula is used (in the manner of Example 7.3.8) to estimate $\int_a^\infty e^{-x} f(x)\, dx$.

7.3.4. Gauss–Hermite Quadrature

In Appendix 7.3(d) abscissae and weights are given for the evaluation of $\int_{-\infty}^{\infty} \exp(-x^2) f(x)\, dx$ in the form

$$\int_{-\infty}^{\infty} e^{-x^2} f(x)\, dx = \sum_{1}^{n} \alpha_k f(x_k) + R_n. \tag{7.3.8}$$

Here, x_k $(k = 1, \ldots, n)$ are the roots of the Hermite polynomial equation $H_n(x) = 0$, [see IV, § 10.5.2]

$$\alpha_k = 2^{n+1} n! \sqrt{\pi} / [H'_n(x_k)]^2, \qquad k = 1, \ldots, n$$

and

$$R_n = n! \sqrt{\pi}\, f^{(2n)}(\xi) / 2^n (2n)!, \qquad -\infty < \xi < \infty.$$

Strategies, similar to those given in section 7.3.3, can be devised to check the validity of the computed results.

7.3.5. Pseudo-Gaussian Quadrature

In a number of applications it is desirable to construct formulae with the property that the integrand $f(x)$ is evaluated at some fixed points, say the ends of the range, and at intermediate points which are chosen to minimise the number of function evaluations required. Such formulae may be termed *pseudo-Gaussian*, and one of the most important formulae of this type is that due to Lobatto, viz.

$$\int_{-1}^{1} f(x)\, dx = \alpha f(-1) + \alpha f(1) + \sum_{k=1}^{n-2} \alpha_k f(x_k) + R_n, \qquad n \geq 3 \tag{7.3.9}$$

where the x_k are the roots of the polynomial equation $P'_{n-1}(x) = 0$ ($P_n(x)$ is the Legendre polynomial of degree n [see IV, § 10.3.2])

$$\alpha = 2/n(n-1); \qquad \alpha_k = \alpha / [P_{n-1}(x_k)]^2$$

and

$$R_n = -\frac{2^{2n-1} n(n-1)^3 [(n-2)!]^4}{(2n-1)[(2n-2)!]^3} f^{(2n-2)}(\xi), \qquad -1 < \xi < 1.$$

The abscissae and weights for Lobatto quadrature are tabulated for $n = 3(1)6$ in Appendix 7.3(e).

By insisting that the integrand is evaluated at ± 1 we lose two degrees of freedom and a Lobatto n-point formula is only exact for polynomials of degree $2n - 3$, as compared with a Gauss–Legendre n-point formula which is exact for polynomials of degree $2n - 1$. However, some benefit does accrue when it is known that the function $f(x)$ is zero at the ends of the range. In this case a $(n+2)$-point Lobatto formula requires n function evaluations and is exact for polynomials of degree $2n + 1$, whereas the Gauss–Legendre n-point formula, which also requires n function evaluations, is only exact for polynomials of degree $2n - 1$.

EXAMPLE 7.3.9. Evaluate $I = \int_{-1}^{1} \cos \frac{1}{2}\pi x \, dx$ using the 5-point Lobatto rule (the exact value of I is $4/\pi = 1 \cdot 27324$ to 5 decimal places).

Since the integrand vanishes at the end points, the 5-point formula requires only 3 function evaluations and we find

$$I \approx \tfrac{49}{90} \cos \tfrac{1}{2}\pi(-0 \cdot 654654) + \tfrac{32}{45} \cos 0 + \tfrac{49}{90} \cos \tfrac{1}{2}\pi(0 \cdot 654654)$$

$$= 1 \cdot 27325 \quad \text{(to 5 decimal places)}.$$

The Gauss–Legendre 3-point formula, with the same number of function evaluations, produces

$$I \approx 1 \cdot 27412.$$

Finally, we note that strategies, similar to those employed for Gauss–Legendre quadrature can be used to assess the numerical accuracy of the results.

7.4. CHEBYSHEV METHODS

In section 6.3 it is demonstrated that a truncated Chebyshev series is a powerful way of approximating many functions in $(-1, 1)$ in that fewer terms of the truncated Chebyshev series are required for a good approximation than for the corresponding truncated Taylor series. Since a Chebyshev series is easily integrated, the main requirement of the method of this section is the establishment of a Chebyshev approximation to a function $f(x)$.

There are two methods for obtaining a Chebyshev approximation of degree n to $f(x)$. In the first it is assumed that the approximation and $f(x)$ assume the same values at the points

$$x_k = \cos\left(\frac{2k+1}{n+1} \frac{\pi}{2}\right), \qquad k = 0, 1, \ldots, n.$$

These points are called *collocation points*. If we denote the approximation by $A_1(x)$, we find

$$A_1(x) = \sum_{r=0}^{n}{}' b_r T_r(x) = \tfrac{1}{2} b_0 T_0(x) + b_1 T_1(x) + \ldots + b_n T_n(x) \qquad (7.4.1)$$

where the prime denotes that the first term has to be multiplied by the factor $\frac{1}{2}$ and each b_r is defined by

$$b_r = \frac{2}{n+1} \sum_{k=0}^{n} f(x_k) T_r(x_k). \qquad (7.4.2)$$

Provided that the coefficients in the infinite Chebyshev series for $f(x)$ are negligible after the term a_{n+3} it can be shown that

$$\left| \int_{-1}^{1} (f(x) - A_1(x))\, dx \right| \sim \begin{cases} \dfrac{2}{n^2} |a_{n+2}|, & \text{if } n \text{ even} \\[2mm] \dfrac{1}{n^2}\{|a_{n+1}| - 2|a_{n+3}|\}, & \text{if } n \text{ odd} \end{cases}$$

so that approximately

$$\int_{-1}^{1} f(x)\, dx \sim \int_{-1}^{1} A_1(x)\, dx = \sum_{r=0}^{n'} \int_{-1}^{1} b_r T_r(x)\, dx.$$

Since [see IV, § 10.5.3]

$$\int T_r(x)\, dx = \begin{cases} T_1(x), & \text{if } r = 0 \\[1mm] \frac{1}{4} T_2(x), & \text{if } r = 1 \\[1mm] \dfrac{1}{2}\left\{ \dfrac{T_{r+1}(x)}{r+1} - \dfrac{T_{r-1}(x)}{r-1} \right\}, & \text{if } r > 1, \end{cases}$$

it follows that

$$\int_{-1}^{1} T_r(x) = \begin{cases} 0, & \text{if } r \text{ odd} \\[1mm] \left(\dfrac{1}{r+1} - \dfrac{1}{r-1} \right), & \text{if } r \text{ even.} \end{cases}$$

Consequently, after a little manipulation, we obtain

$$\int_{-1}^{1} f(x)\, dx = \sum_{r=0}^{[\frac{1}{2}n]} B_{2r+1} = B_1 + B_3 + B_5 + \ldots \tag{7.4.3}$$

where $[\frac{1}{2}n]$ denotes the greatest integer $\leq \frac{1}{2}n$,

$$B_{2s+1} = b_{2s}/(2s+1), \qquad s = [\tfrac{1}{2}n]$$

and

$$B_{2r+1} = (b_{2r} - b_{2r+2})/(2r+1), \qquad r = 0, 1, \ldots, s-1.$$

The second method is very similar to the first, the main difference being that the collocation points are now taken to be $x_k = \cos(k\pi/n)$. The approximation to $f(x)$ is $A_2(x)$, where

$$A_2(x) = \sum_{r=0}^{n''} c_r T_r(x)$$

$$= \tfrac{1}{2} c_0 T_0(x) + c_1 T_1(x) + \ldots + c_{n-1} T_{n-1}(x) + \tfrac{1}{2} c_n T_n(x), \tag{7.4.4}$$

the double prime indicating that the first and last terms in the summation have to be multiplied by the factor $\frac{1}{2}$. Each c_r is defined by

$$c_r = \frac{2}{n} \sum_{k=0}^{n} {}'' f(x_k) T_r(x_k), \tag{7.4.5}$$

where the double prime has the same meaning as above. For this approximation

$$\left| \int_{-1}^{1} (f(x) - A_2(x))\, dx \right| \sim \begin{cases} \dfrac{8}{n^3}|a_{n+2}|, & \text{if } n \text{ is even} \\[2mm] \dfrac{4}{n^3}\{|a_{n+1}| + 3|a_{n+3}|\}, & \text{if } n \text{ is odd} \end{cases}$$

so that approximately

$$\int_{-1}^{1} f(x)\, dx \sim \int_{-1}^{1} A_2(x)\, dx = \sum_{r=0}^{[\frac{1}{2}n]} C_{2r+1} \tag{7.4.6}$$

where

$$C_{2s+1} = \alpha c_{2s}/(2s+1), \qquad C_{2s-1} = (c_{2s-2} - \alpha c_{2s})/(2s-1)$$

with

$$s = [\tfrac{1}{2}n], \qquad \alpha = \begin{cases} \tfrac{1}{2}, & \text{if } 2s = n,\, n \geq 2 \\ 1, & \text{if } 2s \neq n \end{cases}$$

and

$$C_{2r+1} = (c_{2r} - c_{2r+2})/(2r+1), \qquad r = 0, 1, \ldots, s-2.$$

There is some slight advantage in using the second method and we shall now demonstrate its use.

EXAMPLE 7.4.1.   Evaluate $I = \int_1^2 dx/x$.

We first have to transform the range of integration to $(-1, 1)$ and by application of (7.3.3) we have

$$\int_1^2 \frac{dx}{x} = \int_{-1}^1 \frac{du}{u+3}.$$

Taking $n = 2$, we find

$$c_0 = \tfrac{1}{2}f(x_0)T_0(x_0) + f(x_1)T_0(x_1) + \tfrac{1}{2}f(x_2)T_0(x_2); \qquad x_0 = -x_2 = 1, \qquad x_1 = 0,$$
$$= \tfrac{1}{2}\cdot\tfrac{1}{4} + \tfrac{1}{3} + \tfrac{1}{2}\cdot\tfrac{1}{2} = \tfrac{17}{24}.$$

Similarly,

$$c_2 = \tfrac{1}{2}\cdot\tfrac{1}{4} - \tfrac{1}{3} + \tfrac{1}{2}\cdot\tfrac{1}{2} = \tfrac{1}{24}.$$

Thus

$$C_3 = \tfrac{1}{2}\cdot\tfrac{1}{24}/3 = \tfrac{1}{144}$$
$$C_1 = (\tfrac{17}{24} - \tfrac{1}{2}\cdot\tfrac{1}{24}) = \tfrac{11}{16}$$

and

$$I \sim (\tfrac{11}{16} + \tfrac{1}{144}) = \tfrac{25}{36} = 0{\cdot}69444.$$

When $n = 4$,

$$c_0 = \tfrac{1}{2}[\tfrac{1}{2}.\tfrac{1}{4} + 1/(3 + 1/\sqrt{2}) + \tfrac{1}{3} + 1/(3 - 1/\sqrt{2}) + \tfrac{1}{2}.\tfrac{1}{2}] = \tfrac{577}{816} = 0.70710$$

$$c_2 = \tfrac{1}{2}[\tfrac{1}{2}.\tfrac{1}{4} - \tfrac{1}{3} + \tfrac{1}{2}.\tfrac{1}{2}] = \tfrac{1}{48} = 0.02083$$

$$c_4 = \tfrac{1}{2}[\tfrac{1}{2}.\tfrac{1}{4} - 1/(3 + 1/\sqrt{2}) + \tfrac{1}{3} - 1/(3 - 1/\sqrt{2}) + \tfrac{1}{2}.\tfrac{1}{2}] = \tfrac{1}{816} = 0.00122.$$

Accordingly,

$$C_5 = \tfrac{1}{2} \times 0.00122/5 = 0.00012$$

$$C_3 = (0.02083 - \tfrac{1}{2} \times 0.00122)/3 = 0.00674$$

$$C_1 = (0.70710 - 0.02083)/1 = 0.68627.$$

It follows that

$$I \sim C_1 + C_3 + C_5 = 0.69313,$$

which is already a good approximation for I.

7.5. SPECIAL INTEGRALS

In previous sections we have presented techniques for the evaluation of $\int_a^b f(x)\, dx$, a, b finite, which work satisfactorily provided that we take a suitable number of function evaluations and the integral is fairly well-behaved (in the sense that the integrand is never infinite at any point of the range and is fairly smooth, without too many maxima and minima [see IV, § 2.10]).

In this section we shall consider how to evaluate infinite integrals of the form $\int_0^\infty g(x)\, dx$ [see IV, § 4.6]. If $g(x) = e^{-x} f(x)$ then the Gauss–Laguerre formulae of section 7.3.3 should be employed. We shall also be concerned with the computation of oscillatory integrals of the form $\int_a^b f(x) \cos mx\, dx$ and $\int_a^b f(x) \sin mx\, dx$.

7.5.1. Salzer's Extrapolation Method

Let $I(na) = \int_0^{na} g(x)\, dx$, $n = 1, 2, \ldots, k$; $I(0) = 0$. We evaluate $I(na)$ from the relation

$$I(na) = I([n-1]a) + \int_{(n-1)a}^{na} g(x)\, dx, \qquad n = 1, 2, \ldots, k$$

where the integral $\int_{(n-1)a}^{na} g(x)\, dx$ is evaluated by some suitable quadrature formula, say a Gauss 5-point rule. We thus have a sequence of approximations $f(1), f(\tfrac{1}{2}), \ldots, f(1/n)$ where $f(1/n) = I(na)$. Finally, the Lagrange interpolation polynomial [see § 2.3.4], $p(x)$ is constructed which passes through the points $(1, f(1)), (\tfrac{1}{2}, f(\tfrac{1}{2})), \ldots, (1/k, f(1/k))$. Assuming that the interpolation polynomial $p(x)$ is fairly representative of the function $f(x)$ throughout the interval $(0, 1/k)$ we can estimate $I = \int_0^\infty g(x)\, dx = f(0)$ by computing $p(0)$.* The method is illustrated in Example 7.5.1.

---

* Alternatively, Levin's u-transformation can be applied. See Appendix 7.5.

EXAMPLE 7.5.1. Evaluate $I = \int_0^\infty dx/(1+x^2)$ (exact answer $= \frac{1}{2}\pi \approx$ 1·570796 [see IV, Table 4.2.1]).

We have the following results (correct to 6 decimal places)

$$f(1) = I(1) = \int_0^1 \frac{dx}{1+x^2} \approx 0.785398$$

$$f(\tfrac{1}{2}) = I(2) = I(1) + \int_1^2 \frac{dx}{1+x^2} = 1.107149$$

$$f(\tfrac{1}{3}) = I(3) = I(2) + \int_2^3 \frac{dx}{1+x^2} = 1.249046.$$

The Lagrange interpolation polynomial which passes through the points $(1, f(1))$, $(\tfrac{1}{2}, f(\tfrac{1}{2}))$, $(\tfrac{1}{3}, f(\tfrac{1}{3}))$ is

$$p(x) = \frac{(x - \tfrac{1}{2})(x - \tfrac{1}{3})}{(1 - \tfrac{1}{2})(1 - \tfrac{1}{3})} \times 0.785398 + \frac{(x - 1)(x - \tfrac{1}{3})}{(\tfrac{1}{2} - 1)(\tfrac{1}{2} - \tfrac{1}{3})} \times 1.107149$$

$$+ \frac{(x - 1)(x - \tfrac{1}{2})}{(\tfrac{1}{3} - 1)(\tfrac{1}{3} - \tfrac{1}{2})} \times 1.249046$$

and when $x = 0$ we find

$$p(0) = 0.5 \times 0.785398 - 4 \times 1.107148 + 4.5 \times 1.249045$$

$$= 1.584810.$$

Thus,

$$I \approx 1.584810.$$

We haven't got a particularly good result at this stage but greater accuracy can be obtained by computing $I(4)$ and $I(5)$ and determining the 4- and 5-point interpolation polynomials. If the extrapolation process is valid then there should be some consistency between the extrapolated results obtained from the 3-point, 4-point and 5-point interpolation polynomials. [See also section 7.7.2].

7.5.2. Oscillatory Integrals

Over a finite range $[a, b]$ a result of Filon (1928) enables us to evaluate $\int_a^b f(x) \sin mx \, dx$ and $\int_a^b f(x) \cos mx \, dx$. Filon's formula is

$$\int_a^b f(x) \sin (mx + \varepsilon) \, dx = h(\alpha_1 S_1 + \alpha_2 S_2 + \alpha_3 S_3) + R_n \qquad (7.5.1)$$

where

$$h = (b-a)/2n; \qquad \theta = mh,$$

$$\alpha_1 = (\theta^2 + \tfrac{1}{2}\theta \sin 2\theta - 2 \sin^2 \theta)/\theta^3,$$

$$\alpha_2 = 2[\theta(1+\cos^2 \theta) - \sin 2\theta]/\theta^3,$$

$$\alpha_3 = 4[\sin \theta - \theta \cos \theta]/\theta^3,$$

$$S_1 = f_0 \cos (ma + \varepsilon) - f_{2n} \cos (mb + \varepsilon),$$

$$S_2 = \sum_{r=1}^{n-1} f_{2r} \sin (mx_{2r} + \varepsilon) + \tfrac{1}{2}f_0 \sin (mx_0 + \varepsilon) \qquad (7.5.2)$$

$$+ \tfrac{1}{2}f_{2n} \sin (mx_{2n} + \varepsilon)$$

$$S_3 = \sum_{r=1}^{n} f_{2r-1} \sin (mx_{2r-1} + \varepsilon)$$

and, for the case $\varepsilon = 0$,

$$R_n = \tfrac{1}{12}h^3(b-a)\left(1 - \frac{1}{16 \cos \tfrac{1}{4}mh}\right) \sin \tfrac{1}{2}mh \ f^{(\mathrm{iv})}(\xi), \qquad a < \xi < b.$$

To avoid severe cancellation it is best to use the following formulae to evaluate $\alpha_1, \alpha_2, \alpha_3$ for small θ.

$$\alpha_1 = \frac{2\theta^3}{45} - \frac{2\theta^5}{315} + \frac{2\theta^7}{4725} - \cdots$$

$$\alpha_2 = \frac{2}{3} + \frac{2\theta^2}{15} - \frac{4\theta^4}{105} + \frac{2\theta^6}{567} - \cdots$$

$$\alpha_3 = \frac{4}{3} - \frac{2\theta^2}{15} + \frac{\theta^4}{210} - \frac{\theta^6}{11340} + \cdots.$$

EXAMPLE 7.5.2.   Evaluate $\int_0^1 x^5 \sin 2x$ using Filon's formula.

If we take $n = 5$, $\varepsilon = 0$, then the quantities in formula (7.5.1) are $h = 0\cdot1$ ($\theta = 0\cdot2$)

$$\alpha_1 = 0\cdot000354$$

$$\alpha_2 = 0\cdot671939$$

$$\alpha_3 = 1\cdot328008$$

$$S_1 = 0\cdot416147$$

$$S_2 = 0\cdot862135$$

$$S_3 = 0\cdot768342.$$

It follows that

$$\int_0^1 x^5 \sin 2x \, dx \approx 0.15998.$$

The remainder R_n is at most 1.2×10^{-3} in modulus so that the quoted result is in error by at most 2 units in the third decimal place. The exact value of the integral is $-\frac{7}{4} \cos 2 - \frac{5}{8} \sin 2 = 0.15995$ to 5 decimal places.

Note that to evaluate $\int_a^b f(x) \cos ms \, dx$ we have to put $\varepsilon = \frac{1}{2}\pi$ in (7.5.1).

Some new methods have recently been developed to integrate oscillatory functions and the interested reader should consult Littlewood and Zakian (1976).

Consider, next, the evaluation of integrals of the form

$$I = \int_0^\infty f(x) \sin mx \, dx,$$

where $f(x)$ is a monotonic decreasing function of x [see IV § 2.7 and Longman, 1956]. Then I can be expressed in the form

$$I = \int_0^{\pi/m} f(x) \sin mx \, dx + \int_{\pi/m}^{2\pi/m} f(x) \sin mx \, dx + \dots$$

$$+ \int_{n\pi/m}^{(n+1)\pi/m} f(x) \sin mx \, dx + \dots$$

$$= I_0 - I_1 + I_2 - I_3 + \dots \tag{7.5.3}$$

where

$$I_n = \int_0^{\pi/m} f\left(x + \frac{n\pi}{m}\right) \sin mx \, dx.$$

As $f(x)$ is a monotonic decreasing function of x the terms I_n are decreasing in magnitude and the evaluation of I thus consists of determining the sum of an alternating series*. This can be achieved by *Euler's method* which is as follows. We compute $I_0, I_1, \dots, I_k$, for some suitable k, say $k = 12$, by using one of the quadrature techniques of previous sections. The differences ΔI_0, $\Delta^2 I_0$, etc., are then computed [see § 1.4] and, finally, we form the sum

$$\tfrac{1}{2}I_0 - \tfrac{1}{4}\Delta I_0 + \tfrac{1}{8}\Delta^2 I_0 - \dots + \tfrac{1}{2}(-\tfrac{1}{2}\Delta)^r I_0 + \dots. \tag{7.5.4}$$

The sum (7.5.4) converges to I much faster than the partial sums in (7.5.3). The method is illustrated in Example 7.5.3.

EXAMPLE 7.5.3. Evaluate

$$I(t) = \int_0^\infty \frac{x^{1/2} \sin tx}{1+x} \, dx \quad \text{for } t = 1.$$

* We can also use Levin's t-transformation. See Appendix 7.5.

This integral has a singularity at $x = 0$ due to the presence of the term $x^{1/2}$ and, therefore,

$$I_0 = \int_0^{\pi} \frac{x^{1/2} \sin tx}{1+x}\, dx$$

requires a special technique to evaluate it. I_0 has been evaluated as a sum of

$$\int_0^1 \frac{x^{1/2} \sin tx}{1+x}\, dx + \int_1^{\pi} \frac{x^{1/2} \sin tx}{1+x}\, dx = 0 \cdot 217611 + 0 \cdot 733332$$

$$= 0 \cdot 950943,$$

where the first integral has been evaluated by Fox's method (§ 7.6.1), and the second integral and also $I_1, \ldots, I_{12}$ have been computed using a Gauss–Legendre 5-point rule. We find

$$I_0 - I_1 + I_2 - I_3 + I_4 - I_5 = 0 \cdot 950943 - 0 \cdot 762325 + 0 \cdot 634101$$

$$- 0 \cdot 553415 + 0 \cdot 497109 - 0 \cdot 455031$$

$$= 0 \cdot 311382.$$

The sum of the remaining terms in the alternating series is found by constructing a difference table from the calculated values of I_n ($n > 5$).

n	I_{n+5}		Δ^2		Δ^4		Δ^6	$\frac{1}{2}(-\frac{1}{2}\Delta)^n$
1	0·422070							0·211035
		−26714	4497					
2	0·395356	−22217		−1132				0·006679
			3365		360			
3	0·373139	−18852	2593	−772	228	−132	52	0·000562
4	0·354287	−16259		−544	148	−80		0·000071
			2049					
5	0·338028	−14210	1653	−396				0·000011
6	0·323818	−12557						0·000002
7	0·311261							0·0000004
							Sum =	0·218360

The terms in the difference table which contribute to the sum (7.5.4) are underlined, and the individual terms in the sum are given at the side of the difference table. The final result is

$$I(1) \approx 0 \cdot 311382 + 0 \cdot 218360 = 0 \cdot 529742.$$

This answer is in error by 1 unit in the sixth decimal place. The function $I(t)$ can be expressed in terms of Fresnel integrals, as has been shown in Cohen (1969) [see § 7.7.4].

It should be noted that oscillatory integrals such as $\int_0^{\infty} J_0(x)\, dx$, where $J_0(x)$ is the Bessel function of order zero, can be evaluated by the above technique. The integral is now subdivided as

$$\int_0^{\lambda_1} J_0(x)\,dx + \int_{\lambda_1}^{\lambda_2} J_0(x)\,dx + \int_{\lambda_2}^{\lambda_3} J_0(x)\,dx + \dots$$

where $\lambda_1, \lambda_2, \dots$ are the consecutive zeros of $J_0(x)$ [see IV, § 10.4].

7.6. SINGULAR INTEGRALS

In previous sections it was assumed that the integrand $f(x)$ was finite and differentiable throughout the range of integration $[a, b]$. If $f(x)$ is not finite, or one of its derivatives is infinite, at some point in $[a, b]$ then the methods of previous sections will yield unsatisfactory results. Thus, for example, the trapezium rule with $h = \frac{1}{10}$ gives

$$\int_0^1 x^{1/2}\,dx = \tfrac{1}{10}[\tfrac{1}{2}\sqrt{0} + \sqrt{0\cdot1} + \dots + \sqrt{0\cdot9} + \tfrac{1}{2}\sqrt{1}]$$

$$= 0\cdot6605 \quad \text{(to 4 decimal places)}.$$

This is in error by 1% but, if we did not know that the exact result was $\frac{2}{3}$, the error estimate in (7.1.6) indicates the possibility of a large error. Further, it appears that the trapezium rule cannot be employed to approximate

$$\int_0^1 x^{-1/2}\,dx \quad \text{(exact answer} = 2)$$

since the integrand is infinite at $x = 0$. Other quadrature formulae fare equally badly. For example, the Gauss–Legendre 3-point formula yields

$$\int_0^1 x^{1/2}\,dx = \tfrac{1}{2}\int_{-1}^1 [\tfrac{1}{2}(u+1)]^{1/2}\,du$$

$$\approx \frac{1}{2\sqrt{2}}\{0\cdot555556\,(0\cdot474767 + 1\cdot332140) + 0\cdot888889 \times 1\}$$

$$= 0\cdot6692 \quad \text{(to 4 decimal places)}$$

while

$$\int_0^1 x^{-1/2}\,dx = \tfrac{1}{2}\int_{-1}^1 [\tfrac{1}{2}(u+1)]^{-1/2}\,du$$

$$\approx \frac{1}{\sqrt{2}}\{0\cdot555556\,(2\cdot106298 + 0\cdot750672) + 0\cdot888889 \times 1\}$$

$$= 1\cdot7509.$$

Clearly, on the assumption that $\int_a^b f(x)\,dx$ is always a convergent integral, we have to adopt some new approach to evaluate singular integrals accurately and, in this section, we describe some of the techniques used.

7.6.1. Fox's Extension of the Method of Romberg Integration

Fox (1967) has shown that for a function $f(x)$ with a weak singularity at $x = 0$, i.e. one for which $f(0) = 0$ but for which a derivative of $f(x)$ is infinite at $x = 0$,

$$I = \int_0^1 f(x)\, dx$$

$$= I_T(n) + Ah^2 + Bh^4 + Ch^6 + \dots$$

$$+ \tfrac{1}{12}h^2 f'(h) - \tfrac{1}{12}h^3 f''(h) + \tfrac{29}{720}h^4 f'''(h) - \dots \qquad (7.6.1)$$

where $A, B, C, \dots$ are constants and the notation is the same as section 7.1.3. If $f(x) = x^{1/2}$, for example, we find

$$h^2 f'(h) = \tfrac{1}{2}h^{3/2}$$

$$h^3 f'(h) = -\tfrac{1}{4}h^{3/2}$$

$$h^4 f'''(h) = \tfrac{3}{8}h^{3/2}, \quad \text{etc.};$$

so that the correct formula for the application of the Romberg process when the integrand is $x^{1/2}$ is

$$I = I_T(n) + Kh^{3/2} + Ah^2 + Bh^4 + Ch^6 + \dots. \qquad (7.6.2)$$

As in section 7.2 we evaluate $I_T(2), I_T(4), I_T(8)$, etc. and, from these computed values, we eliminate $K, A, B, \dots$ the coefficients of the lowest powers of h being eliminated first. Thus, for example, we can form $I_T(2^r, 2^{r+1})$ such that

$$I_T(2^r, 2^{r+1}) = [2\sqrt{2}I_T(2^{r+1}) - I_T(2^r)]/(2\sqrt{2} - 1)$$

and

$$I = I_T(2^r, 2^{r+1}) + A'h^2 + B'h^4 + \dots \qquad (h = 1/2^r).$$

Similarly we can form

$$I_T(2^r, 2^{r+1}, 2^{r+2}) = \tfrac{1}{3}\{4I_T(2^{r+1}, 2^{r+2}) - I_T(2^r, 2^{r+1})\}$$

which satisfies

$$I = I_T(2^r, 2^{r+1}, 2^{r+2}) + B''h^4 + \dots,$$

and so on. The Romberg tableau for the evaluation of $\int_0^1 x^{1/2}\, dx$ is

0·603554	0·665012		
0·643283	0·666250	0·666663	
0·658130	0·666562	0·666667	0·666667
0·663581			

Because the elements on the sloping line are fairly consistent we are justified in thinking that 0·666667 is a good answer for $\int_0^1 x^{1/2}\, dx$.

In the error expansion (7.6.2) the term $Kh^{3/2}$ was introduced in addition to the normal terms occurring in the Romberg expansion (7.2.1). It follows from

(7.6.1) that if we are integrating

$$\int_0^1 x^{r+1/2} \, dx, \qquad r = 0, 1, 2, \ldots$$

then the extra term introduced into the Romberg expansion is of the form $K'h^{r+3/2}$. This result enables us to integrate functions of the form

$$\int_0^1 x^{1/2} f(x) \, dx$$

where $f(x)$ is of the form $a_0 + a_1 x + a_2 x^2 + \ldots$. The appropriate Romberg expansion is

$$I = I_T(n) + Ah^{3/2} + Bh^2 + Ch^{5/2} + Dh^{7/2} + Eh^4 + \ldots \qquad (7.6.3)$$

the term in $h^{r+3/2}$ only occurring if the coefficient of $x^{r+1/2}$ is not zero.

EXAMPLE 7.6.1. Evaluate

$$J = \int_0^1 \frac{x^{1/2} \sin x}{1+x} \, dx.$$

This integral was required in Example 7.5.3. The relevant expansion in this case is

$$I = I_T(n) + Bh^2 + Ch^{5/2} + Dh^{7/2} + Eh^4 + \ldots$$

and we can form the Romberg tableau

2	0·218186	0·218157		
4	0·218164	0·217701	0·217603	
8	0·217817	0·217626	0·217610	0·217611
16	0·217674			

From the tableau we deduce $J \approx 0·217611$.

Similar results to (7.6.1) hold for the mid-point and Simpson rules and also for Gauss–Legendre formulae. An important point to note is that the mid-point rule and the Gauss–Legendre formulae can be used to evaluate integrals such as

$$\int_0^1 x^{-1/2} f(x) \, dx, \qquad \int_0^1 \ln x f(x) \, dx$$

for functions $f(x)$ which are differentiable, although these integrals cannot be evaluated directly by the trapezium rule* or Simpson's rule. A fuller list of Romberg formulae is given in Appendix 7.2(b).

---

* The integrals can however be evaluated if we use the modified form of the trapezium rule $h(\tfrac{3}{2}f_1 + f_2 + \ldots + f_{N-1} + \tfrac{1}{2}f_N)$ in conjunction with the appropriate error expansion in powers of h (Appendix 7.2(b)).

EXAMPLE 7.6.2.   Evaluate $I = \int_0^1 x^{-1/2}\,dx$ using

(a) the mid-point rule
(b) the Gauss–Legendre 2-point formula.

For (a) the appropriate Romberg error formula is

$$I = I_M(n) + Ah^{1/2} + Bh^2 + Ch^4 + Dh^6 + \ldots .$$

The Romberg tableau is

1·414214			
1·577350	1·971197		
1·698844	1·992156	1·999142	
1·786461	1·997987	1·999930$_6$	1·999983

From the tableau we deduce that $I \approx 1\cdot99998$. For (b) the appropriate error formula is

$$I = I_G + Ah^{1/2} + Bh^4 + Ch^6 + \ldots$$

where I_G denotes the sum of the applications of the Gauss 2-point rule to each of the n subdivisions. The Romberg tableau is

1·650680		
1·752800	1·999337$_5$	
1·825187	1·999945$_7$	1·999986

From the tableau we find $I \approx 1\cdot99999$.

7.6.2.   Gaussian Quadrature Formulae

The theory of Gaussian quadrature formulae was developed for any positive weight function. Thus formulae of the type given in section 7.3 can be constructed for singular integrals such as

$$\int_0^1 x^{-1/2} f(x)\,dx, \qquad \int_0^1 -\ln x\; f(x)\,dx, \qquad \int_0^\infty x^{1/2} e^{-x} f(x)\,dx, \quad \text{etc.}$$

Thus, for example,

$$\int_0^1 x^{-1/2} f(x)\,dx = \sum_{k=1}^n \alpha_k f(x_k) + R_n \qquad (7.6.4)$$

where x_k is a root of the polynomial equation $P_{2n}(\sqrt{x}) = 0$, α_k is twice the weight coefficient in the $(2n)$-point Gauss–Legendre formula and

$$R_n = \frac{2^{4n+1}[(2n)!]^3}{(4n+1)[(4n)!]^2} f^{(2n)}(\xi), \qquad 0 < \xi < 1.$$

This is a n-point Gaussian quadrature formula which is exact for polynomials of degree $2n - 1$ or less. The abscissae and weights are tabulated for $n = 1(1)6$ in Appendix 7.3(f).

EXAMPLE 7.6.3.   Evaluate

$$J = \int_0^1 \frac{x^{1/2} \sin x}{1+x}\, dx.$$

This integral has been evaluated by other means in Example 7.6.1. For the application of the rule in (7.6.4) we have $f(x) = x \sin x/(1+x)$. The 4-point formula given in Appendix 7.3(f) yields

$$J \approx 0\cdot725368 \times \frac{0\cdot033648 \sin 0\cdot033648}{1+0\cdot033648} + 0\cdot627413 \times \frac{0\cdot276184 \sin 0\cdot276184}{1+0\cdot276184}$$

$$+ 0\cdot444762 \times \frac{0\cdot634677 \sin 0\cdot634677}{1+0\cdot634677} + 0\cdot202457$$

$$\times \frac{0\cdot922157 \sin 0\cdot922157}{1+0\cdot922157}$$

$$= 0\cdot21761 \quad \text{(to 5 decimal places)}.$$

7.6.3.   Other Techniques for Singular Integrals

The earlier methods for singular integrals were specifically designed for dealing with the singularity. An alternative approach is to remove the singularity. The simplest approach is to make some substitution which transforms the integrand into a regular function. For instance, the substitution $x = u^2$ transforms $\int_0^1 x^{-1/2} f(x)\, dx$ into $\int_0^1 2f(u^2)\, du$ while the substitution $x = \sin \theta$ transforms

$$\int_{-1}^1 \frac{f(x)\, dx}{\sqrt{(1-x^2)}}$$

into $\int_{-\frac{1}{2}\pi}^{\frac{1}{2}\pi} f(\sin \theta)\, d\theta$ [see IV, Theorem 4.3.2]. The transformed integrals can be evaluated by means of the Gauss–Legendre quadrature formulae or some other basic quadrature rules.

Another technique is to subtract out the singularity using known results. For instance [see IV, (2.11.7)],

$$I = \int_0^{\frac{1}{2}\pi} \log \sin x\, dx = \int_0^{\frac{1}{2}\pi} \log (\sin x/x)\, dx + \int_0^{\frac{1}{2}\pi} \log x\, dx$$

$$= I_1 + I_2, \quad \text{say}.$$

The function $\sin x/x$ is regular in the interval $[0, \frac{1}{2}\pi]$ and does not vanish there so that I_1 can be evaluated by the methods of previous sections. The exact value of I_2 is known to be $\frac{1}{2}\pi[\ln(\frac{1}{2}\pi) - 1]$ and thus I can be determined by adding the approximation for I_1 to I_2 [see IV, (4.3.2)].

Yet another device is to remove the singularity to infinity. For instance, Takahasi and Mori, 1973 have shown that integrals of the form

$$I = \int_{-1}^{1} (1-x)^{\alpha}(1+x)^{\beta}f(x)\, dx, \qquad \alpha, \beta > -1$$

can be evaluated by first making the transformation

$$x = \operatorname{erf} u = \frac{2}{\sqrt{\pi}} \int_{0}^{u} \exp(-t^2)\, dt$$

so that

$$I = \frac{2}{\sqrt{\pi}} \int_{-\infty}^{\infty} f(\operatorname{erf} u)\, e^{-u^2}(1 - \operatorname{erf} u)^{\alpha}(1 + \operatorname{erf} u)^{\beta}\, du.$$

The trapezium rule is then employed to evaluate I. Thus,

$$I \approx I_T = \frac{2}{\sqrt{\pi}} h \sum_{n=-\infty}^{\infty} f(\operatorname{erf} nh) \exp(-n^2 h^2)(1 - \operatorname{erf} nh)^{\alpha}(1 + \operatorname{erf} nh)^{\beta}.$$

Because of the factor $\exp(-n^2 h^2)$ the terms in the summation become negligible for moderate n (depending on the choice of h). The quantity $\operatorname{erf} nh$ can be computed by means of an expansion, see Abramowitz and Stegun (1965). For good results it is advisable to determine the value of $\operatorname{erf} nh$ using double length arithmetic as there is severe cancellation in working out $(1 - \operatorname{erf} nh)$, $n \gg 0$ and $(1 + \operatorname{erf} nh)$, $n \ll 0$.

7.6.4. Cauchy Principal Values

In some problems, in aeronautics for example, it is sometimes required to find the Cauchy principal value [see IV, Def. 4.6.3] of integrals of the form

$$\int_{-1}^{1} \frac{f(x)}{x}\, dx = \lim_{\eta \to 0} \left\{ \int_{-1}^{-\eta} \frac{f(x)}{x}\, dx + \int_{\eta}^{1} \frac{f(x)}{x}\, dx \right\}.$$

Piessens (1970) has shown that even order Gauss–Legendre quadrature formulae work well on such examples and

$$\mathcal{P} \int_{-1}^{1} \frac{f(x)}{x}\, dx \approx \sum_{i=1}^{2m} \alpha_i f(x_i)/x_i$$

where x_i and α_i are the abscissae and weights in the Gauss–Legendre formula of order $2m$.

EXAMPLE 7.6.4.   Estimate

$$I = \mathcal{P} \int_{-1}^{1} \frac{e^x}{x}\, dx.$$

The Gauss–Legendre 4-point formula gives

$$I \approx 0{\cdot}652145\left[\frac{e^{0{\cdot}339981}}{0{\cdot}339981}-\frac{e^{-0{\cdot}339981}}{0{\cdot}339981}\right]$$

$$+0{\cdot}347855\left[\frac{e^{0{\cdot}861136}}{0{\cdot}861136}-\frac{e^{-0{\cdot}861136}}{0{\cdot}861136}\right]$$

$$= 2{\cdot}11450.$$

This answer is correct to 5 decimal places.

7.6.5.   Integration of Discontinuous Functions

In some practical situations we are required to integrate discontinuous functions. There is no difficulty in determining numerical approximations as the theory of Gaussian integration shows that high order Gaussian formulae will produce good results. It is usually much better, however, to sub-divide the range of integration so that the integrand is continuous in each subdivision, and to apply a low order formula to evaluate the integral over each subinterval. We illustrate the idea by means of a simple example.

EXAMPLE 7.6.5.   Evaluate $I = \int_{-1}^{1} f(x)\, dx$ where

$$f(x)=\begin{cases}0, & -1\le x<0\\ e^{x}, & 0<x\le 1\end{cases}$$

(a) by means of a Gauss–Legendre 6-point formula
(b) by dividing the range of integration and using a Gauss–Legendre 2-point formula for each subinterval.

(a) The Gauss–Legendre 6-point formula produces the approximation

$$I \approx 0{\cdot}467914(0+e^{0{\cdot}238619})+0{\cdot}360762(0+e^{0{\cdot}661209})$$

$$+0{\cdot}171324(0+e^{0{\cdot}932470})$$

$$\approx 1{\cdot}728154.$$

(b) Consider the interval $(-1, 1)$ subdivided into the two intervals $(-1, 0)$ and $(0, 1)$. As $f(x)$ is a constant in $(-1, 0)$ the Gauss–Legendre quadrature formula is exact over this interval and $\int_{-1}^{0} f(x)\, dx = 0$. In the interval $(0, 1)$,

$$\int_{0}^{1} f(x)\, dx = \tfrac{1}{2}\int_{-1}^{1} f(\tfrac{1}{2}u + \tfrac{1}{2})\, du$$

$$\approx \frac{1}{2}\left[\exp\left(\frac{1}{2}-\frac{1}{2\sqrt{3}}\right)+\exp\left(\frac{1}{2}+\frac{1}{2\sqrt{3}}\right)\right]$$

$$\approx 1{\cdot}717896.$$

Thus

$$\int_{-1}^{1} f(x)\, dx = \int_{-1}^{0} f(x)\, dx + \int_{0}^{1} f(x)\, dx$$

$$\approx 1{\cdot}717896.$$

The exact value of this integral is known to be $e - 1$ ($= 1{\cdot}718282$ to 6 decimal places) and clearly the strategy (b) which only involved 4 function evaluations is more effective than strategy (a) which required 6 function evaluations.

It is worth noting that if a point of discontinuity coincides with an abscissa of a Gauss–Legendre formula (and this would be the case if we had used a 5-point formula in (a)) then the value of the function at the abscissa is taken to be the average of the values of $f(x)$ on either side of the discontinuity. Thus, for the integral in Example 7.6.5 we have, using the Gauss–Legendre 5-point formula

$$I \approx 0{\cdot}568889 f(0) + 0{\cdot}478629(0 + e^{0{\cdot}538469}) + 0{\cdot}236927(0 + e^{0{\cdot}906180})$$

where

$$f(0) = \lim_{\eta \to 0} \tfrac{1}{2}[f(0 + \eta) + f(0 - \eta)]$$

$$= \tfrac{1}{2}(0 + e^{0}) = 0{\cdot}5.$$

Thus $I \approx 1{\cdot}690878$.

A further example of the superiority of strategy (b) is given in Example 10.2.9.

7.7.   MISCELLANEOUS TECHNIQUES

The methods presented in previous sections were well-suited to automatic computation on an electronic digital computer. In some practical situations an answer is required quickly and there is no computer available. Alternatively, the function $f(x)$ might be presented in a table with no indication as to the form of the function. Some other case might arise which will make it desirable to employ a different technique to those discussed previously. In this section we will present some techniques which have been profitably used in the past, in the circumstances we have just outlined, and are still worthy of attention.

7.7.1.   Finite Difference Methods

These methods can, in general, be derived from the integration of interpolation formulae. To give an example (see Cohen, 1973), consider Bessel's interpolation formula [see § 2.3.5] in the form

$$f_p = f(x_0 + ph)$$

$$= f_0 + p\delta_{1/2} + \frac{1}{2!} p(p-1)\mu\delta^2_{1/2} + \frac{1}{3!} p(p-1)(p-2)\delta^3_{1/2}$$

$$+ \frac{1}{4!}(p+1)p(p-1)(p-2)\mu\delta^4_{1/2} + \frac{1}{5!}(p+1)p(p-1)(p-2)(p-\tfrac{1}{2})\delta^5_{1/2} + R$$

$$(7.7.1)$$

where

$$R = \frac{1}{6!}(p+2)(p+1)p(p-1)(p-2)(p-3)h^6 f^{(vi)}(\xi), \qquad x_{-2} < \xi < x_3.$$

Since

$$\int_{x_0}^{x_1} f(x)\, dx = h \int_0^1 f_p\, dp$$

it follows from (7.7.1) that

$$\int_{x_0}^{x_1} f(x)\, dx = h(f_0 + \tfrac{1}{2}\delta_{1/2} - \tfrac{1}{12}\mu\delta^2_{1/2} + \tfrac{11}{720}\mu\delta^4_{1/2}) + R_1^*$$

$$= h[\tfrac{1}{2}(f_0 + f_1) - (\tfrac{1}{12}\mu\delta - \tfrac{11}{720}\mu\delta^3)(f_1 - f_0)] + R_1^* \qquad (7.7.2)$$

where

$$R_1^* = \frac{-191}{60480} h^7 f^{vi}(\eta), \qquad x_{-2} < \eta < x_3.$$

Now, suppose we want to evaluate $\int_{x_0}^{x_N} f(x)\, dx$. Repeated application of (7.7.2) gives

$$\int_{x_0}^{x_N} f(x)\, dx = \int_{x_0}^{x_1} f(x)\, dx + \int_{x_1}^{x_2} f(x)\, dx + \ldots + \int_{x_{N-1}}^{x_N} f(x)\, dx$$

$$= h[\tfrac{1}{2}(f_0 + f_1) + \tfrac{1}{2}(f_1 + f_2) + \ldots + \tfrac{1}{2}(f_{N-1} + f_N)$$

$$- (\tfrac{1}{12}\mu\delta - \tfrac{11}{720}\mu\delta^3)\{(f_1 - f_0) + (f_2 - f_1) + \ldots + (f_N - f_{N-1})\}] + \sum_1^N R_i^*$$

which simplifies to (using the notation of (7.1.4)

$$\int_{x_0}^{x_N} f(x)\, dx = I_T(N) - (\tfrac{1}{12}\mu\delta - \tfrac{11}{720}\mu\delta^3)(f_N - f_0) + R^* \qquad (7.7.3)$$

where $R^* = \sum_1^N R_i^*$. Formula (7.7.3) may be considered as the trapezium rule with difference correction terms. Its application is demonstrated by the following example.

EXAMPLE 7.7.1. Evaluate $I = \int_1^2 dx/x$.

We form the difference table below

i	x_i	f_i		δ^2		δ^4		δ^6
-3	0·7	1·428571						
			-178571					
-2	0·8	1·250000		39682				
			-138889		-11904			
-1	0·9	1·111111		27778		4328		
			-111111		-7576		-1803	
0	1·0	1·000000		20202		2525		834
			-90909		-5051		-969	
1	1·1	0·909091		15151		1556		409
			-75758		-3495		-560	
2	1·2	0·833333		11656		996		232
			-64102		-2499		-328	
3	1·3	0·769231		9157		668		117
			-54945		-1831		-211	
4	1·4	0·714286		7326		457		78
			-47619		-1374		-133	
5	1·5	0·666667		5952		324		43
			-41667		-1050		-90	
6	1·6	0·625000		4902		234		25
			-36765		-816		-65	
7	1·7	0·588235		4086		169		28
			-32679		-647		-37	
8	1·8	0·555556		3439		132		2
			-29240		-515		-35	
9	1·9	0·526316		2924		97		15
			-26316		-418		-20	
10	2·0	0·500000		2506		77		2
			-23810		-341		-18	
11	2·1	0·476190		2165		-282		
			-21645					
12	2·2	0·454545		1883				
			-19762					
13	2·3	0·434783						

Table 7.7.1: Differences of the function $1/x$.

From the table we deduce that

$$I_T(N) = 0·693771_5; \qquad \tfrac{1}{2}\mu\delta f_N = -0·002088_6; \qquad \tfrac{1}{12}\mu\delta f_0 = -0·008417_5$$
$$\tfrac{11}{720}\mu\delta^3 f_N = -0·000005_8; \qquad \tfrac{11}{720}\mu\delta^3 f_0 = -0·000096_5.$$

It follows that

$$I \approx 0·693771_5 + 0·000208_9 - 0·000841_8 - 0·000000_6 + 0·000009_7$$
$$= 0·693147_7.$$

Apart from the rounding error incurred in the calculation, which is small because of the factor $h = 0·1$, account has not been taken of the error term R^*. The derivative in the error term R_1 satisfies approximately

$$h^6 f^{(vi)}(\eta) \approx \delta_0^6$$

and thus

$$R_1^* \approx -\frac{191}{60480} h\delta_0^6.$$

Treating the other error terms similarly we deduce

$$R^* \approx \sum_0^N \frac{-191}{60480} h\delta_i^6$$

$$\approx 0·6 \quad \text{(in the sixth decimal place).}$$

The net effect is that the error in the above quoted result is very small and is only of the order of 1 unit in the sixth decimal place.

One disadvantage of formula (7.7.3) is that tabular values outside the interval (x_0, x_N) are required to calculate the central differences involved. If only discrete data is available it is quite likely that the function values at x_{-3}, x_{-2}, x_{-1}, x_{11}, x_{12}, x_{13} are not known. Formula (7.7.3) cannot then be used. However, $\int_{x_0}^{x_N} f(x)\,dx$ can still be estimated by means of *Gregory's formula* which requires the knowledge of forward and backward differences instead of central differences. Gregory's formula when curtailed at fifth differences is

$$\int_{x_0}^{x_N} f(x)\,dx = I_T(N) - h[\tfrac{1}{2}(\nabla f_N - \Delta f_0) + \tfrac{1}{24}(\nabla^2 f_N + \Delta^2 f_0)$$

$$+ \tfrac{19}{720}(\nabla^3 f_N - \Delta^3 f_0) + \tfrac{3}{160}(\nabla^4 f_N + \Delta^4 f_0)$$

$$+ \tfrac{863}{60480}(\nabla^5 f_N - \Delta^5 f_0)] + R. \tag{7.7.4}$$

EXAMPLE 7.7.2. Evaluate $I = \int_1^2 dx/x$ using Gregory's formula.

If we take $h = 0\cdot1$ then the differences which are required in Gregory's formula are the underlined entries in Table 7.7.1. Thus

$$I \approx 0\cdot693771_5 - 0\cdot000538_3 - 0\cdot000075_3 - 0\cdot000007_9$$

$$- 0\cdot000002_1 - 0\cdot000000_4$$

$$= 0\cdot693147_5.$$

A large selection of finite difference formulae for the evaluation of $\int_{x_0}^{x_N} f(x)\,dx$ are given in Interpolation and Allied Tables (1956) but the two presented in this section are probably the most useful in practice for numerical quadrature. In § 2.3 other formulae are listed which facilitate the solution of ordinary differential equations.

7.7.2. Series Expansions and Transformations

Some integrals are conveniently computed by expanding the integrand in series form and integrating term by term. Consider, for instance, the complete elliptic integral [see IV, § 10.1.4]

$$I = \int_0^{\frac{1}{2}\pi} \frac{dx}{\sqrt{(1 - \tfrac{1}{4}\sin^2 x)}}.$$

Using the binomial expansion we can express the integrand in the form [see IV, § 1.10]

$$(1 - \tfrac{1}{4}\sin^2 x)^{-1/2} = 1 + \tfrac{1}{8}\sin^2 x + \tfrac{3}{128}\sin^4 x + \tfrac{5}{1024}\sin^6 x + \dots$$

so that

$$I = \int_0^{\frac{1}{2}\pi} (1 + \tfrac{1}{8}\sin^2 x + \tfrac{3}{128}\sin^4 x + \tfrac{5}{1024}\sin^6 x + \dots)\,dx$$

and, by application of Wallis' formula [IV, Example 4.4.1]

$$\int_0^{\frac{1}{2}\pi} \sin^{2n} x \, dx = \frac{1.3.5\ldots(n-1)}{2.4.6\ldots(2n)} \cdot \frac{\pi}{2},$$

we find

$$I = \frac{\pi}{2}(1 + \tfrac{1}{16} + \tfrac{9}{1024} + \tfrac{25}{16384} + \ldots)$$

$$= 1\cdot68575 \quad \text{(to 5 decimal places)}.$$

An alternative procedure is to note that the complete elliptic integral is of the form

$$I = \int_0^{\frac{1}{2}\pi} \frac{dx}{\sqrt{[a_n^2 \cos^2 x + b_n^2 \sin^2 x]}}.$$

If we make a *Landen transformation*, that is, we let

$$\sin t = \frac{a_{n+1} \sin 2x}{\sqrt{[a_n^2 \cos^2 x + b_n^2 \sin^2 x]}}$$

$$\cos t = \frac{c_{n+1} + a_{n+1} \cos 2x}{\sqrt{[a_n^2 \cos^2 x + b_n^2 \sin^2 x]}}$$

$$(7.7.5)$$

where $a_{n+1} = \tfrac{1}{2}(a_n + b_n)$, $b_{n+1} = \sqrt{(a_n b_n)}$, $c_{n+1} = \tfrac{1}{2}(a_n - b_n)$, then it can be shown that

$$I = \int_0^{\frac{1}{2}\pi} \frac{dt}{\sqrt{[a_{n+1}^2 \cos^2 t + b_{n+1}^2 \sin^2 t]}}$$

If this transformation is applied repeatedly the sequences $\{a_n\}, \{b_n\}$ tend rapidly to the same limit l [see IV, § 1.2] so that, in the limit,

$$I = \int_0^{\frac{1}{2}\pi} \frac{dt}{\sqrt{[l^2 \cos^2 t + l^2 \sin^2 t]}} = \int_0^{\frac{1}{2}\pi} \frac{dt}{l} = \frac{\pi}{2l}.$$

For the given example we have the sequence of values

$$a_0 = 1, \qquad\qquad b_0 = \sqrt{(1 - \tfrac{1}{4})} = 0\cdot8660254$$

$$a_1 = 0\cdot9330127, \qquad b_1 = 0\cdot9306048$$

$$a_2 = 0\cdot9318087, \qquad b_2 = 0\cdot9318080$$

$$a_3 = 0\cdot9318084, \qquad b_3 = 0\cdot9318084.$$

Thus

$$\int_0^{\frac{1}{2}\pi} \frac{dx}{\sqrt{(1 - \tfrac{1}{4}\sin^2 x)}} = \frac{\pi}{2 \times 0\cdot9318084} = 1\cdot685750 \quad \text{(to 6 decimal places)}.$$

This transformation method is particularly well-suited to evaluating complete elliptic integrals of the form

$$\int_0^{\frac{1}{2}\pi} \frac{dx}{\sqrt{(1-\sin^2\alpha \sin^2 x)}}$$

where $\sin^2\alpha$ is very close to 1. Expansion in series is not efficient for this problem as a large number of terms of the series have to be evaluated. When $\sin^2\alpha = 0.9999$, i.e. we are evaluating

$$\int_0^{\frac{1}{2}\pi} \frac{dx}{\sqrt{(1-0.9999\sin^2 x)}}$$

we obtain the sequence of values

$$a_0 = 1, \qquad\qquad b_0 = \sqrt{(1-0.9999)} = 0.01$$

$$a_1 = 0.505, \qquad\quad b_1 = 0.1$$

$$a_2 = 0.3025, \qquad\quad b_2 = 0.224722$$

$$a_3 = 0.263611, \qquad b_3 = 0.260727$$

$$a_4 = 0.262169, \qquad b_4 = 0.262165$$

$$a_5 = 0.262167, \qquad b_5 = 0.262167.$$

The value of the integral thus equals

$$\pi/2a_5 = 5.99159 \quad \text{(to 5 decimal places)}.$$

In section 7.6.3 transformations were applied to remove the singularity in singular integrands. It is worth mentioning one further transformation, $u = 1/x$, as this enables one to relate $\int_1^\infty f(x)\,dx$ with $\int_0^1 g(u)\,du$, which is sometimes more manageable from a computational point of view. Thus, for example,

$$\int_1^\infty \frac{dx}{1+x^2} = \int_1^0 \frac{-du/u^2}{1+(1/u)^2} = \int_0^1 \frac{du}{1+u^2}$$

[cf. § 7.5.1].
    We end this section with another example on series expansion.

EXAMPLE 7.7.3.   Evaluate $I = \int_0^1 x^x\,dx$.

The integral can be expressed in the form [see IV, § 2.11]

$$x^x = e^{x\ln x}$$

$$= 1 + x\ln x + \frac{(x\ln x)^2}{2!} + \ldots + \frac{(x\ln x)^n}{n!} + \ldots.$$

Since $\int_0^1 (x \ln x)^n dx$ is known, by integration by parts, to have the value $(-1)^n n!/(n+1)^{n+1}$ [see IV, Ex. 4.3.3] it follows that

$$\int_0^1 x^x dx = 1 - \frac{1}{2^2} + \frac{1}{3^3} - \frac{1}{4^4} + \ldots$$

$$= 0.783431 \quad \text{(to 6 decimal places).}$$

7.7.3. Asymptotic Expansions

In the previous section we were required to expand the integrand for small values of the variable x. There are many occasions when an expansion for large x is desirable. This expansion, to be of any use, will be in powers of $1/x$. Suppose we are given a function $f(x)$ and we can determine $S_n(x)$,

$$S_n(x) = A_0 + \frac{A_1}{x} + \frac{A_2}{x^2} + \ldots + \frac{A_n}{x^n}, \tag{7.7.6}$$

such that

$$R_n(x) = f(x) - S_n(x) \tag{7.7.7}$$

and, for fixed n,

$$\lim_{1x1 \to \infty} x^n R_n(x) = 0. \tag{7.7.8}$$

Then $S(x)$,

$$S(x) = A_0 + \frac{A_1}{x} + \frac{A_2}{x} + \ldots + \frac{A_n}{x^n} + \ldots$$

is referred to as the *asymptotic expansion* of $f(x)$ [see also IV, § 2.15] and we write

$$f(x) \sim S(x).$$

We illustrate the procedure for developing an asymptotic expansion in the following example.

EXAMPLE 7.7.4.   Estimate $I = \int_4^\infty e^{-\frac{1}{2}x^2} dx$.

Consider the more general problem of determining

$$f(t) = \int_t^\infty e^{-\frac{1}{2}x^2} dx.$$

We can write the integral in the form

$$\int_t^\infty x^{-1}(xe^{-\frac{1}{2}x^2}) dx = \frac{1}{t} e^{-\frac{1}{2}t^2} - \int_t^\infty x^{-2} e^{-\frac{1}{2}x^2} dx$$

by integration by parts [see IV, § 4.3]. This latter integral is equivalent to

$$\int_t^\infty x^{-3}(x e^{-\frac{1}{2}x^2}) dx = \frac{1}{t^3} e^{-\frac{1}{2}t^2} - \int_t^\infty \frac{3}{x^4} e^{-\frac{1}{2}x^2} dx.$$

Proceeding in the same fashion we obtain

$$\int_t^\infty e^{-\frac{1}{2}x^2}\,dx = e^{-\frac{1}{2}t^2}\left(\frac{1}{t}-\frac{1}{t^3}+\frac{3}{t^5}-\ldots+\frac{(-1)^n(2n)!}{2^n n!\,t^{2n+1}}\right)+R_n(t),$$

where

$$R_n(t)=\frac{(-1)^{n+1}(2n+2)!}{2^{n+1}(n+1)!}\int_t^\infty \frac{e^{-\frac{1}{2}x^2}}{x^{2n+2}}\,dx$$

$R_n(t)$ satisfies the condition

$$\lim_{|t|\to\infty} t^{2n+1}R_n(t)=0,$$

as the dominant term in $R_n(t)$ behaves like $\exp\left(-\tfrac{1}{2}t^2\right)$. We therefore assert that

$$f(t)=\int_t^\infty e^{-\frac{1}{2}x^2}\,dx \sim e^{-\frac{1}{2}t^2}\left(\frac{1}{t}-\frac{1}{t^3}+\frac{3}{t^5}-\ldots+\frac{(-1)^n(2n)!}{2^n n!\,t^{2n+1}}+\ldots\right).$$

The problem now, for any given value of t (in our case $t=4$), is to determine how many terms are needed to obtain a good approximation for $f(t)$. It will be noted that, for fixed t, the terms

$$(-1)^n(2n)!/2^n n!\,t^{2n+1}$$

decrease in magnitude and then increase. Thus in the present case we have

$$f(4)\sim e^{-8}(0\cdot 25-0\cdot 015625+0\cdot 002930-0\cdot 000916$$

$$+0\cdot 000401-0\cdot 000226+0\cdot 000156-0\cdot 000127$$

$$+0\cdot 000120-0\cdot 000127+\ldots).$$

Let

$$T_7=0\cdot 25-\ldots+0\cdot 000156-0\cdot 000127=0\cdot 236593$$

and

$$T_8=T_7+0\cdot 000120=0\cdot 236713.$$

Since

$$f(4)=e^{-8}T_7+R_7=S_7+R_7,\quad \text{say}$$

and

$$f(4)=e^{-8}T_8+R_8=S_8+R_8,$$

and R_7 and R_8 have opposite signs, it is clear that the true value of $f(4)$ lies between S_7 and S_8. A reasonable approximation to the true value of $f(4)$ is thus $\frac{1}{2}(S_7+S_8)=0\cdot 236653e^{-8}$. Since this differs by at most $0\cdot 00006e^{-8}$ from either S_7 or S_8 we can assert that

$$I=\int_4^\infty e^{-\frac{1}{2}x^2}\,dx \approx (0\cdot 236653\pm 0\cdot 00006)e^{-8}.$$

It should be pointed out that, in most cases where asymptotic expansions are required, we would normally be dealing with values of t which are much greater than 4.

7.7.4.   Use of Transforms

To evaluate certain integrals a parameter can be introduced which enables us, by means of Laplace or Fourier transforms, to express the given integral in a form which is very suitable for computation. Consider, for example,

$$I = \int_0^\infty \frac{x^{1/2} \sin x}{1+x} \, dx.$$

We introduce the parameter t and consider the evaluation of the integral

$$I(t) = \int_0^\infty \frac{x^{1/2} \sin tx}{1+x}, \qquad t > 0. \tag{7.7.9}$$

If we can evaluate $I(t)$ then I follows by putting $t = 1$. Now if $\bar{I}(s)$ denotes the Laplace transform of $I(t)$ [see IV, § 13.4.1] we have

$$\bar{I}(s) = \int_0^\infty e^{-st} I(t) \, dt,$$

$$= \int_0^\infty e^{-st} \left(\int_0^\infty \frac{x^{1/2} \sin tx}{1+x} \, dx \right) dt,$$

$$= \int_0^\infty \frac{x^{1/2}}{1+x} \left(\int_0^\infty e^{-st} \sin tx \, dt \right) dx,$$

the inversion of the order of integration being permissible if $\mathrm{Re}(s) > 0$. Thus

$$\bar{I}(s) = \int_0^\infty \frac{x^{3/2}}{(1+x)(s^2+x^2)} \, dx$$

and by a well known result in the theory of contour integration (Whittaker and Watson, 1962, p. 118) we get

$$\bar{I}(s) = \frac{\pi}{s^2+1} + \frac{\pi}{\sqrt{2}} \frac{s^{1/2}(s-1)}{s^2+1}. \tag{7.7.10}$$

It now remains to determine the function whose Laplace transform is given by (7.7.10). Since the Laplace transform of the function $\pi \sin t$ is $\pi/(s^2+1)$ we need only concern ourselves with the function

$$\bar{f}(s) = \frac{s^{1/2}(s-1)}{s^2+1} = s^{1/2} - s^{-1/2} \left\{ \frac{s}{s^2+1} + \frac{1}{s^2+1} \right\}.$$

Now $\mathscr{L}\{1/\sqrt{(\pi t)}\} = s^{1/2}$, so [see IV, § 13.4.4] by the convolution theorem for transforms

$$f(t) = (\pi t)^{-1/2} - \int_0^t [\{\cos(t-u) + \sin(t-u)\}/\sqrt{(\pi u)}]\, du$$

$$= (\pi t)^{-1/2} - 2\pi^{-1/2}[(\cos t + \sin t)C(\sqrt{t}) + (\sin t - \cos t)S(\sqrt{t})]$$

where $C(t)$, $S(t)$ are the Fresnel integrals

$$C(t) = \int_0^t \cos(u^2)\, du, \qquad S(t) = \int_0^t \sin(u^2)\, du.$$

Thus we have, for all $t > 0$,

$$I(t) = \pi \sin t + \sqrt{(\pi/2t)} - \sqrt{(2\pi)}(\cos t + \sin t)C(\sqrt{t}) + (\sin t - \cos t)S(\sqrt{t}).$$

$$(7.7.11)$$

The functions $C(\sqrt{t})$ and $S(\sqrt{t})$ can be evaluated for moderate t from the series expansions [see IV, (2.12.1), (2.12.2)]

$$C(\sqrt{t}) = t^{1/2}\left(1 - \frac{t^2}{5(2!)} + \frac{t^4}{9(4!)} - \ldots + (-1)^n \frac{t^{2n}}{(4n+1)(2n)!} + \ldots\right)$$

$$S(\sqrt{t}) = t^{1/2}\left(\tfrac{1}{3}t - \frac{t^3}{7(3!)} + \frac{t^5}{11(5!)} - \ldots + (-1)^n \frac{t^{2n+1}}{(4n+3)(2n+1)!} + \ldots\right).$$

For large t we have

$$C(\sqrt{t}) = \tfrac{1}{2}\sqrt{(\tfrac{1}{2}\pi)} - \int_{\sqrt{t}}^{\infty} \cos(u^2)\, du,$$

with a similar result for $S(\sqrt{t})$. The latter integral can be evaluated by an asymptotic expansion. In fact, we find

$$I(t) \sim (\pi/2t)^{1/2}\left(\frac{1}{2t} + \frac{3}{4t^2} - \frac{15}{8t^3} - \ldots\right).$$

The reader should check that for $t = 1$ (for instance) this approach produces a result close to the value obtained previously in Example 7.5.3. $I(t)$ can also be determined from (7.7.10) by a method due to Talbot, (1979).

7.8. THE EVALUATION OF MULTIPLE INTEGRALS

7.8.1. Integration over Hypercubes

Some of the rules developed for the integration of functions of one variable can be extended to enable us to integrate functions of several variables. For instance, consider the numerical integration of

$$I' = \int_a^b \int_c^d g(x', y')\, dx'\, dy' \qquad (a, b, c, d \text{ finite}).$$

Since we can make the transformation $x = (x'-a)/(b-a)$, $y = (y'-c)/(d-c)$; [see IV, § 6.2] there is no loss of generality if we consider

$$I = \int_0^1 \int_0^1 f(x, y) \, dx \, dy,$$

where the region of integration is the unit square $0 \leq x, \ y \leq 1$. The simplest mid-point approximation to I is

$$I_1 = f(\tfrac{1}{2}, \tfrac{1}{2}).$$

Alternatively, we can subdivide the square $0ABC$ into the squares S_1, S_2, S_3, S_4 (Figure 7.8.1) and obtain

$$I = \iint_{S_1} f(x, y) \, dx \, dy + \iint_{S_2} f(x, y) \, dx \, dy + \iint_{S_3} f(x, y) \, dx \, dy + \iint_{S_4} f(x, y) \, dx \, dy$$

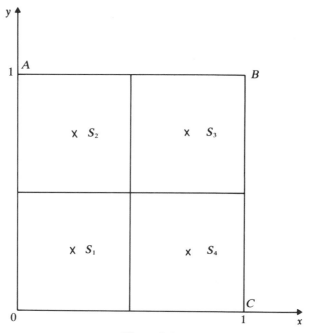

Figure 7.8.1

so that

$$I \approx I_2 = \tfrac{1}{4}[f(\tfrac{1}{4}, \tfrac{1}{4}) + f(\tfrac{1}{4}, \tfrac{3}{4}) + f(\tfrac{3}{4}, \tfrac{1}{4}) + f(\tfrac{3}{4}, \tfrac{3}{4})].$$

Further subdivisions will produce approximations to I which, in general, are more accurate. Thus if the unit square is subdivided into squares of side $h(=1/k)$ the kth order mid-point rule approximation to I, denoted by I_k, is given by

$$I_k = h^2 \sum_{s=0}^{k-1} \sum_{r=0}^{k-1} f(rh + \tfrac{1}{2}h, sh + \tfrac{1}{2}h). \qquad (7.8.1)$$

For integration over a m-dimensional hypercube, that is estimating

$$I = \int_0^1 \int_0^1 \cdots \int_0^1 f(x_1, \ldots, x_m)\, dx, \ldots dx_m$$

the corresponding result to (7.8.1) is

$$I \approx I_k = h^m \sum_{r_1=0}^{k-1} \cdots \sum_{r_m=0}^{k-1} f(r_1 h + \tfrac{1}{2}h, \ldots, r_m h + \tfrac{1}{2}h). \qquad (7.8.2)$$

Similar results to (7.8.1) and (7.8.2) can be obtained from the one dimensional trapezium and Simpson rules.

We can also extend Gaussian quadrature formulae to deal with the integration of functions of several variables. For instance, given the third order (fifth degree) formula for 1 variable

$$\int_{-1}^1 g(x)\, dx \approx \tfrac{5}{9}g(-\sqrt{\tfrac{3}{5}}) + \tfrac{8}{9}g(0) + \tfrac{5}{9}(\sqrt{\tfrac{3}{5}}),$$

then

$$\int_{-1}^1 \int_{-1}^1 f(x, y)\, dx\, dy \approx \tfrac{25}{81}f(-\sqrt{\tfrac{3}{5}}, -\sqrt{\tfrac{3}{5}}) + \tfrac{40}{81}f(-\sqrt{\tfrac{3}{5}}, 0) + \tfrac{25}{81}f(-\sqrt{\tfrac{3}{5}}, \sqrt{\tfrac{3}{5}})$$

$$+ \tfrac{40}{81}f(0, -\sqrt{\tfrac{3}{5}}) + \tfrac{64}{81}f(0, 0) + \tfrac{40}{81}f(0, \sqrt{\tfrac{3}{5}})$$

$$+ \tfrac{25}{81}f(\sqrt{\tfrac{3}{5}}, -\sqrt{\tfrac{3}{5}}) + \tfrac{40}{81}f(\sqrt{\tfrac{3}{5}}, 0) + \tfrac{25}{81}f(\sqrt{\tfrac{3}{5}}, \sqrt{\tfrac{3}{5}}) \qquad (7.8.3)$$

is a third order (fifth degree) formula for two variables. More generally, for the kth order formula in one dimension

$$\int_{-1}^1 g(x)\, dx \approx \sum_{i=1}^k \alpha_i g(\xi_i)$$

the corresponding formula in m dimensions is

$$\int_{-1}^1 \cdots \int_{-1}^1 f(x_1, \ldots, x_m) \approx \sum_{i_1-1}^k \cdots \sum_{i_m=1}^k \alpha_{i_1} \ldots \alpha_{i_m} f(\xi_{i_1}, \ldots, \xi_{i_m}) \qquad (7.8.4)$$

With all the formulae cited above, which are all termed *product* formulae, a formula requiring k function evaluations for the integration of a function of one variable requires k^m function evaluations for the integration of a function of several variables. Thus the extended third order Gaussian formula requires 59049 function evaluations to estimate the integral of a function of 10 variables.

EXAMPLE 7.8.1.   Evaluate $\int_0^1 \int_0^1 e^{-xy}\, dx\, dy$ by the following methods

(a) the mid-point rule with 9 function evaluations
(b) the Gauss formula (7.8.3)

(The exact answer to 5 decimal places is 0·79660.)

(a) The mid-point rule gives (after collection of like terms)

$$\int_0^1 \int_0^1 e^{-xy}\, dx\, dy \approx I_3$$

where

$$I_3 = \tfrac{1}{9}[e^{-1/36} + 2\,e^{-1/12} + 2\,e^{-5/36} + e^{-1/4} + 2\,e^{-5/12} + e^{-25/36}]$$

$$= 0·79444_2.$$

(b) Before we can apply the Gauss quadrature formula we have to transform the range of integration for each variable to $(-1, 1)$ [see IV, § 6.2]. The approximation we use is thus

$$\int_0^1 \int_0^1 f(x, y)\, dx\, dy \approx \tfrac{1}{324}[25f(\alpha, \alpha) + 40f(\alpha, \tfrac{1}{2}) + 25f(\alpha, \beta)$$

$$+ 40f(\tfrac{1}{2}, \alpha) + 64f(\tfrac{1}{2}, \tfrac{1}{2}) + 40f(\tfrac{1}{2}, \beta)$$

$$+ 25f(\beta, \alpha) + 40f(\beta, \tfrac{1}{2}) + 25f(\beta, \beta)]$$

where $\alpha = \tfrac{1}{2}(1 - \sqrt{(\tfrac{3}{5})})$, $\beta = \tfrac{1}{2}(1 + \sqrt{(\tfrac{3}{5})})$, so that

$$\int_0^1 \int_0^1 e^{-xy}\, dx\, dy \approx \tfrac{1}{324}[25\,e^{-\alpha^2} + 80e^{-\frac{1}{2}\alpha} + 50\,e^{-\alpha\beta}$$

$$+ 64\,e^{-\frac{1}{4}} + 80\,e^{-\frac{1}{2}\beta} + 25\,e^{-\beta^2}]$$

$$= 0·79660.$$

The results indicate the superiority of the Gaussian approximation over the mid-point rule, which was to be expected in view of the corresponding 1 variable results. However, if the integrand is differentiable, we can use the Romberg technique to improve the answers obtained from simple mid-point estimates. For instance, in the above example,

$$I_1 = e^{-1/4} = 0·778801$$

$$I_2 = \tfrac{1}{4}(e^{-1/16} + 2\,e^{-3/16} + e^{-9/16}) = 0·791814$$

$$I(1, 2) = \tfrac{1}{3}(4I_2 - I_1) = 0·796152$$

and using the Lyness and McHugh variation [see § 7.2.1] we get

$$I(1, 2, 3) = 2·025I_3 - 1·066667I_2 + 0·041667I_1$$

$$= 0·79659.$$

At this stage it is worth remarking that the natural extension of Romberg integration is not a very practical technique for the numerical evaluation of multiple integrals. For, given a m-dimensional integral

$$I = \int_0^1 \cdots \int_0^1 f(x_1, \ldots, x_m) \, dx_1 \ldots dx_m$$

the application of the Romberg technique to approximations of order 1, 2, 4 and 8 requires approximately

$$1^m + 2^m + 4^m + 8^m$$

function evaluations. For $m = 5$ this already totals 33825. The Lyness and McHugh variation, using the approximations of orders 1, 2, 3 and 4 requires a total of

$$1^m + 2^m + 3^m + 4^m$$

function evaluations which only totals 1300 when $m = 5$. The Lyness and McHugh variation therefore has the advantage of economy of function evaluations since both the above methods eliminate terms in h^2, h^4, and h^6 in the expansion of $I - I_k$. (Note, however, that the error in the Romberg results will, in general, be smaller than the Lyness and McHugh results.)

The extrapolation formulae of Romberg or Lyness and McHugh type are examples of *non-product formulae* and they are called this because they cannot be derived by direct extension from 1-dimensional formulae. Non-product Gaussian formulae also exist and an example is provided by *Radon's fifth degree formula* which only requires seven function evaluations to approximate

$$\int_{-1}^1 \int_{-1}^1 f(x, y) \, dx \, dy$$

as compared with the 9 required by the product formula encountered previously. Radon's formula is

$$\int_{-1}^1 \int_{-1}^1 f(x, y) \, dx \, dy \approx \tfrac{5}{9}[f(r, s) + f(r, -s) + f(-r, s) + f(-r, -s)]$$

$$+ \tfrac{20}{63}[f(0, t) + f(0, -t)] + \tfrac{8}{7}f(0, 0) \qquad (7.8.5)$$

where $r = \sqrt{\tfrac{3}{5}}$, $s = \sqrt{\tfrac{1}{3}}$, $t = \sqrt{\tfrac{14}{15}}$.

Unfortunately, non-product formulae are not, in general, easy to construct. A considerable amount of research is currently in progress to try and establish such formulae for 2 and higher dimensions, particularly formulae requiring a smaller number of function evaluations than the corresponding product formula.

7.8.2. Integration over General Regions

When one moves from one dimension to higher dimensions we are often faced with the problem of integrating over some region which need not necessarily be a hypercube. For instance, we may wish to evaluate

$$J = \iiint\limits_{S_3} \exp(x_1 x_2 x_3)\, dx_1\, dx_2\, dx_3$$

where S_3 denotes the unit sphere in three dimensions

$$S_3: \quad x_1^2 + x_2^2 + x_3^2 \leq 1.$$

A very crude way to evaluate J is to write

$$J = \int_{-1}^{1} \int_{-1}^{1} \int_{-1}^{1} f(x_1, x_2, x_3)\, dx_1\, dx_2\, dx_3$$

where

$$f(x_1, x_2, x_3) = \begin{cases} \exp(x_1 x_2 x_3), & (x_1, x_2, x_3) \in S_3 \\ 0, & (x_1, x_2, x_3) \notin S_3 \end{cases}$$

We are now in a position to estimate J using, say, a third degree Gaussian product formula. We have

$$J \approx f(r, r, r) + f(r, r, -r) + f(r, -r, r) + f(r, -r, -r)$$
$$+ f(-r, r, r) + f(-r, r, -r) + f(-r, -r, r) + f(-r, -r, -r),$$

where $r = 1/\sqrt{3}$. Thus $J \approx 8\cdot14861$. As no point outside the sphere S_3 has been used in the computation the above result effectively gives the value of the integral of $\exp(x_1 x_2 x_3)$ over the hypercube $C_3: -1 \leq x_1, x_2, x_3 \leq 1$. Thus a more realistic result would be obtained by multiplying the stated answer by the ratio of the volume of S_3 to the volume of C_3, i.e. by $\frac{4}{3}\pi/8$. If we do this we obtain the modified answer

$$J \approx (1\cdot01833)\tfrac{4}{3}\pi$$

which compares favourably with the answers obtained later (Example 7.8.2) using two tailor-made formulae.

It is best however to derive product formulae directly for each given region, where this is possible. For instance, for the sphere S_3 we can make the transformation [see IV, § 6.4]

$$x_1 = r \cos \theta_2 \cos \theta_1$$

$$x_2 = r \cos \theta_2 \sin \theta_1$$

$$x_3 = r \sin \theta_2$$

so that we have the following equivalent result

$$\iiint_{S_3} f(x_1, x_2, x_3)\, dx_1\, dx_2\, dx_3 = \int_{-1}^{1} \int_{-\pi/2}^{\pi/2} \int_{-\pi/2}^{\pi/2} f(r, \theta_1, \theta_2)|r|^2 \cos\theta_2\, dr\, d\theta_1\, d\theta_2$$

$$(7.8.6)$$

and the right-hand side is expressible as an integral over a standard hypercube by the substitutions $\theta_1 = \frac{1}{2}\pi\phi_1$, $\theta_2 = \frac{1}{2}\pi\phi_2$. Thus

$$J = \iiint_{S_3} \exp(x_1 x_2 x_3)\, dx_1\, dx_2\, dx_3$$

$$= \tfrac{1}{4}\pi^2 \int_{-1}^{1} \int_{-1}^{1} \int_{-1}^{1} r^2 \exp(r^3 \cos^2 \tfrac{1}{2}\pi\phi_2 \sin \tfrac{1}{2}\pi\phi_2 \cos \tfrac{1}{2}\pi\phi_1 \sin \tfrac{1}{2}\pi\phi_1)$$

$$. \cos \tfrac{1}{2}\pi\phi_2\, dr\, d\phi_1\, d\phi_2.$$

This can be approximated in the usual way by inserting the appropriate abscissae and weights. Yet another approach which could be adopted is to consider the term $|r|^2$ in (7.8.6) as the weight function for integration with respect to r, 1 as the weight function for integration with respect to θ_1, and $\cos\theta_2$ as the weight function for integration with respect to θ_2. This leads to another approximate formula.

In Appendix 7.4 we give a number of formulae for integration over hypercubes and spheres. A more extensive list of formulae can be obtained by consulting Stroud (1971) who also quotes results for other n-dimensional regions. Stroud's book also provides a large amount of theoretical detail about the construction of formulae and error estimates.

To conclude this section we apply the two relevant formulae in Appendix 7.4 to evaluate J.

EXAMPLE 7.8.2. Evaluation of $J = \iiint_{S_3} \exp(x_1 x_2 x_3)\, dx_1\, dx_2\, dx_3$ by means of formulae E, F (Appendix 7.4).

From formula E we have

$$J \approx \tfrac{1}{6} \cdot \tfrac{4}{3}\pi[e^0 + e^0 + e^0 + e^0 + e^0 + e^0].$$

$$= \tfrac{4}{3}\pi.$$

Formula F yields

$$J \approx \tfrac{1}{8} \cdot \tfrac{4}{3}\pi[4 \exp(1/5\sqrt{5}) + 4 \exp(-1/5\sqrt{5})],$$

$$= \tfrac{4}{3}\pi \cosh(1/5\sqrt{5})$$

$$= 1\cdot004003(\tfrac{4}{3}\pi).$$

7.9. SUMMARY

There are many integrals which can be evaluated in closed form, such as

$$\int_0^t \frac{dx}{x^4+1} = \tfrac{1}{8}\sqrt{2} \ln\left(\frac{t^2+\sqrt{2t+1}}{t^2-\sqrt{2t+1}}\right) + \tfrac{1}{4}\sqrt{2} \tan^{-1}\left(\frac{\sqrt{2t}}{1-t^2}\right).$$

However, the degree of difficulty in evaluating the right-hand side for given t is equal to, if not greater than, the degree of difficulty in estimating the integral by a numerical method. Other integrals, for example,

$$\int_0^1 e^{-x^2}\, dx$$

cannot be evaluated in closed form. Again, one might be confronted with integrating a function which is only known from experimental data. These examples indicate the need for using numerical methods to evaluate definite integrals. But which numerical method should we use? No clear cut answer can be given to this question. Some analysis is first required to decide the nature of the integrand. If we are satisfied that the integrand is a well-behaved analytic function, and the range of integration is finite, then elementary integration formulae such as the trapezium rule or Gauss–Legendre quadrature formula are probably adequate, especially if an adaptive approach (§ 7.1.3, § 7.3.1) or a Romberg type approach (§ 7.2, § 7.3.2) is used, although in some instances the Chebyshev technique (§ 7.4) is superior.

Quite often the integrand is not well-behaved in the sense that the integrand, or one of its derivatives, is infinite at an end point of the range. In such a case a special analysis will probably be required to obtain a good numerical answer (§ 7.6). Again, a special approach will be required when the integrand is oscillatory (§ 7.5.2), or when the range of integration is infinite (§§ 7.3.3, 7.3.4, 7.5.1). All these points have been dealt with to some extent in the text but the reader requiring to know more about numerical integration is advised to consult some of the classic texts in the field e.g. Davis and Rabinowitz (1963) and Krylov (1962). The reader may also find it beneficial to consult Stroud and Secrest (1966) who give extensive tables of various Gaussian quadrature formulae.

All the above remarks refer to the numerical evaluation of integrals of a single variable. The numerical evaluation of multiple integrals is a much harder problem and we have only been able to sketch some simple techniques. A comprehensive account of Gaussian methods, for a variety of regions, is given in Stroud (1971). A probabilistic technique, known as the Monte Carlo method, has proved competitive for certain types of nuclear reactor problems and further information about this method can be found in Hammersley and Handscomb (1964). Expansion of the integrand is another approach and this technique has been successful for integrals such as

$$\int_0^\pi \int_0^\pi \int_0^\pi \frac{1}{1 - \cos x \cos y \cos z}\, dx\, dy\, dz$$

(Byrnes *et al.* 1969).

In conclusion it will be seen that the choice of formula is largely a matter of experience and while some methods, such as Gaussian methods, are very robust one can find many integrals (of one variable or several variables) for which other techniques work better, and these should never be overlooked.

A.M.C.

REFERENCES

Abramowitz, M. and Stegun, I. A. (1964). *Handbook of Mathematical Functions*, Dover Publications Inc.

Byrnes, J. S., Podgor, S. and Zachary, W. W. (1969). On the Evaluation of Certain Lattice Green's Functions, *Proc. Camb. Phil. Soc.*, **66**, 377–380.

Cheney, E. W. (1966). *Introduction to Approximation Theory*, McGraw-Hill Book Co.

Clenshaw, C. W. and Curtis, A. R. (1960). A Method for Numerical Integration on an Automatic Computer, *Num. Math.*, **2**, 197–205.

Cohen, A. M. (ed.) (1973). *Numerical Analysis*, McGraw-Hill Book Co. (UK) Ltd.

Cohen, A. M. (1969). *Computational Physics Conference*, H.M.S.O. Vol. 2, Paper 36.

Davis, P. J. and Rabinowitz, P. (1963). *Numerical Integration*, Blaisdell Publishing Co.

Filon, L. N. G. (1928). On a Quadrature Formula for Trigonometric Integrals, *Proc. Roy. Soc. Edinburgh*, **49**, 38–47.

Fox, L. (1967). Romberg Integration for a class of Singular Integrals, *Computer J.*, **10**, 87–93.

Hammersley, J. M. and Handscomb, D. C. (1964). *Monte Carlo Methods*, Methuen.

Hildebrand, F. B. (1974). *Introduction to Numerical Analysis* (2nd edition), McGraw-Hill Book Co., 93, 94.

Interpolation and Allied Tables, H.M.S.O. (1956).

Krylov, V. I. (1962). *Approximate Calculation of Integrals*, The Macmillan Co. (translation by A. H. Stroud).

Levin, D. (1973). Development of Non-Linear Transformations for Improving Convergence of Sequences, *Intern. J. Computer Math.*, **3**, 371–388.

Littlewood, R. K. and Zakian, V. (1976). Numerical Evaluation of Fourier Integrals, *J. Inst. Maths. Applics.*, **18**, 331–339.

Longman, I. M. (1956). Note on a Method for Computing Infinite Integrals of Oscillatory Functions, *Proc. Camb. Phil. Soc.*, **52**, 764–768.

Lyness, J. N. and McHugh, B. J. J. (1963). Integration over Multi-Dimensional Hypercubes I. A Progressive Procedure, *Computer J.*, **6**, 264–270.

Piessens, R. (1970). Numerical Evaluation of Cauchy Principal Values of Integrals, *BIT*, **10**, 476–480.

Robinson, I. G. (1971). Adaptive Gaussian Integration, *Aust. Comput. J.*, **3**, 126–129.

Stroud, A. H. (1971). *Approximate Calculation of Multiple Integrals*, Prentice-Hall, Inc.

Stroud, A. H. and Secrest, D. (1966). *Gaussian Quadrature Formulas*, Prentice-Hall Inc.

Takahasi, H. and Mori, M. (1973). Quadrature Formulas obtained by Variable Transformation, *Numer. Math.*, **21**, 206–219.

Talbot, A. (1979). The Accurate Numerical Inversion of Laplace Transforms, *J. Inst. Maths. Applics.* **23**, 97–120.

Whittaker, E. T. and Watson, G. N. (1962). *A Course of Modern Analysis* (4th edition), C.U.P. 118.

APPENDIX 7.1

A Selection of Newton–Cotes Type Formulae

Formula	Correction	Type
$\displaystyle\int_{x_0}^{x_1} f(x)\,dx = hf_0$	$\frac{1}{2}h^2 f^{(i)}(\xi_1)$	P
$\displaystyle\int_{x_0}^{x_1} f(x)\,dx = \frac{1}{2}h(f_0+f_1)$	$-\frac{1}{12}h^3 f^{(ii)}(\xi_2)$	C
$\displaystyle\int_{x_0}^{x_2} f(x)\,dx = 2hf_1$	$\frac{1}{3}h^3 f^{(ii)}(\xi_3)$	P
$\displaystyle\int_{x_0}^{x_2} f(x)\,dx = \frac{1}{3}h(f_0+4f_1+f_2)$	$-\frac{1}{90}h^5 f^{(iv)}(\xi_4)$	C
$\displaystyle\int_{x_0}^{x_3} f(x)\,dx = \frac{3}{4}h(f_0+3f_2)$	$\frac{3}{8}h^4 f^{(iii)}(\xi_5)$	P
$\displaystyle\int_{x_0}^{x_3} f(x)\,dx = \frac{3}{8}h(f_0+3f_1+3f_2+f_3)$	$-\frac{3}{80}h^5 f^{(iv)}(\xi_6)$	C
$\displaystyle\int_{x_0}^{x_4} f(x)\,dx = \frac{4}{3}h(2f_1-f_2+2f_3)$	$\frac{14}{45}h^5 f^{(iv)}(\xi_7)$	P
$\displaystyle\int_{x_0}^{x_4} f(x)\,dx = \frac{1}{3}h(f_0+4f_1+2f_2+4f_3+f_4)$	$-\frac{1}{45}h^5 f^{(iv)}(\xi_8)$	C

P and C indicate that the formulae can be used respectively as predictor and corrector formulae for the solution of ordinary differential equations.

APPENDIX 7.2(a)

Coefficients w_{kr} in the Lyness and McHugh Extrapolation Formulae

r \ k	1	2	3	4	5
1	1·0000000	−0·3333333	0·0416667	−0·0027778	0·0011157
2		1·3333333	−1·0666667	0·3555555	−0·0677249
3			2·0250000	−2·6035714	1·4645089
4				3·2507937	−5·7791887
5					5·3822889

$$I_T(1, 2, \ldots, k) = \sum_{r=1}^{k} w_{kr} I_T(r)$$

or

$$I_M(1, 2, \ldots, k) = \sum_{r=1}^{k} w_{kr} I_M(r).$$

APPENDIX 7.2(b)

Error terms present in the evaluation of the integrals $I_1, \ldots, I_4$.

$$I_1 = \int_0^1 f(x)\, dx \qquad\qquad I_2 = \int_0^1 x^{1/2} f(x)\, dx$$

$$I_3 = \int_0^1 x^{-1/2} f(x)\, dx \qquad\qquad I_4 = \int_0^1 \ln x f(x)\, dx$$

$I_1 - I_T(n)$	h^2	h^4	h^6	$\ldots$			
$I_1 - I_M(n)$	h^2	h^4	h^6	$\ldots$			
$I_1 - I_S(n)$	h^4	h^6	h^8	$\ldots$			
$I_1 - I_{Gk}(n)$	h^{2k}	h^{2k+2}	h^{2k+4}	$\ldots$			
$I_2 - I_T(n)$	$h^{3/2}$	h^2	$h^{5/2}$	$h^{7/2}$	h^4	$\ldots$	
$I_2 - I_M(n)$	$h^{3/2}$	h^2	$h^{5/2}$	$h^{7/2}$	h^4	$\ldots$	
$I_2 - I_{G2}(n)$	$h^{3/2}$	$h^{5/2}$	$h^{7/2}$	h^4	$h^{9/2}$	$\ldots$	
$I_3 - I_M(n)$	$h^{1/2}$	$h^{3/2}$	h^2	$h^{5/2}$	$h^{7/2}$	h^4	$\ldots$
$I_3 - I_{G2}(n)$	$h^{1/2}$	$h^{3/2}$	$h^{5/2}$	$h^{7/2}$	h^4	$h^{9/2}$	$\ldots$
$I_4 - I_M(n)$	h	h^2	$h^2 \ln h$	h^3	h^4	$h^4 \ln h$	$\ldots$
$I_4 - I_{G2}(n)$	h	h^2	h^3	h^4	$h^4 \ln h$	$\ldots$	

($f(x)$ denotes an analytic function such that $f^{(n)}(0) \neq 0$, all n, and I_{Gk} denotes the Gauss–Legendre k-point rule.)

APPENDIX 7.3(a)

Table of Abscissae and Weight Coefficients for Gauss–Legendre Quadrature

n	x_k	α_k
1	0·00000 00000	2·00000 00000
2	±0·57735 02692	1·00000 00000
3	±0·77459 66692	0·55555 55556
	0·00000 00000	0·88888 88889
4	±0·86113 63116	0·34785 48451
	±0·33998 10436	0·65214 51549
5	±0·90617 98459	0·23692 68851
	±0·53846 93101	0·47862 86705
	0·00000 00000	0·56888 88889
6	±0·93246 95142	0·17132 44924
	±0·66120 93865	0·36076 15730
	±0·23861 91861	0·46791 39346

APPENDIX 7.3(b)

Coefficients in the Error Expansion for the n-point Gauss–Legendre Formula

n \ k	1	2	3	4
1	$\frac{1}{6}$	$\frac{-7}{360}$	$\frac{31}{15120}$	$\frac{-127}{2419200}$
2	—	$\frac{1}{270}$	$\frac{-4}{8505}$	$\frac{17}{340200}$
3	—	—	$\frac{1}{31500}$	$\frac{-7}{1620000}$

APPENDIX 7.3(c)

Table of Abscissae and Weight Coefficients for Gauss–Laguerre Quadrature

n	x_k	α_k
1	1·00000 00000	1·00000 00000
2	0·58578 64376	0·85355 33906
	3·41421 35624	0·14644 66094
3	0·41577 45568	0·71109 30099
	2·29428 03603	0·27851 77336
	6·28994 50829	0·01038 92565
4	0·32254 76896	0·60315 41043
	1·74576 11012	0·35741 86924
	4·53662 02969	0·03888 79085
	9·39507 09123	0·00053 92947
5	0·26356 03197	0·52175 56106
	1·41340 30591	0·39866 68111
	3·59642 57710	0·07594 24497
	7·08581 00059	0·00361 17587
	12·64080 08443	0·00002 33700
6	0·22284 66042	0·45896 46739
	1·18893 21017	0·41700 08308
	2·99273 63261	0·11337 33821
	5·77514 35691	0·01039 91975
	9·83746 74184	0·00026 10172
	15·98287 39806	0·00000 08985

APPENDIX 7.3(d)

Table of Abscissae and Weight Coefficients for Gauss–Hermite Quadrature

n	x_k	α_k
1	0·00000 00000	1·77245 38509
2	±0·70710 67812	0·88622 69255
3	±1·22474 48714	0·29540 89752
	0·00000 00000	1·18163 59006
4	±1·65068 01239	0·08131 28354
	±0·52464 76233	0·80491 40900
5	±2·02018 28705	0·01995 32421
	±0·95857 24646	0·39361 93231
	0·00000 00000	0·94530 87205
6	±2·35060 49737	0·00453 00099
	±1·33584 90740	0·15706 73203
	±0·43607 74119	0·72462 95952

APPENDIX 7.3(e)

Table of Abscissae and Weight Coefficients for Lobatto Quadrature

n	x_k	α_k
3	±1·00000 00000	0·33333 33333
	0·00000 00000	1·33333 33333
4	±1·00000 00000	0·16666 66667
	±0·44721 35955	0·83333 33333
5	±1·00000 00000	0·10000 00000
	±0·65465 36707	0·54444 44444
	0·00000 00000	0·71111 11111
6	±1·00000 00000	0·06666 66667
	±0·76505 53239	0·37847 49563
	±0·28523 15165	0·55485 83770

APPENDIX 7.3(f)

Table of Abscissae and Weight Coefficients for the Evaluation of $\int_0^1 x^{-1/2} f(x)\, dx$

n	x_k	α_k
1	0·33333 33333	2·00000 00000
2	0·11558 71100	1·30429 03097
	0·74155 57472	0·69570 96903
3	0·05693 91160	0·93582 78691
	0·43719 78528	0·72152 31461
	0·86949 93949	0·34264 89848
4	0·03364 82681	0·72536 75668
	0·27618 43139	0·62741 32918
	0·63467 74762	0·44476 20689
	0·92215 66085	0·20245 70726
5	0·02216 35688	0·59104 84494
	0·18783 15677	0·53853 34386
	0·46159 73615	0·43817 27250
	0·74833 46284	0·29890 26983
	0·94849 39263	0·13334 26886
6	0·01568 34066	0·49829 40916
	0·13530 00117	0·46698 50731
	0·34494 23794	0·40633 48534
	0·59275 01277	0·32015 66571
	0·81742 80133	0·21387 86520
	0·96346 12787	0·09435 06728

APPENDIX 7.4

Formulae for the evaluation of $I = \int \ldots \int f(x_1, \ldots, x_n) \, dx_1 \ldots dx_n$ **over the cube** $C_n: -1 \leq x_1, \ldots, x_n \leq 1$ **and the sphere** $S_n: x_1^2 + \ldots + x_n^2 \leq 1$

Formula	Description	Region	Degree	Number of points
A	Product trapezoidal rule	C_n	1	2^n
B	Product Gauss rule	C_n	3	2^n
C	Product Gauss rule	C_n	5	3^n
D	Radon's Fifth Degree Formula	C_2	5	7
E	Hammer and Stroud formula	S_n	3	$2n$
F	Third Degree formula	S_n	3	2^n

$$A: \quad I \approx \sum f(r_1, \ldots, r_n), \qquad r_i = \pm 1, i = 1, \ldots, n.$$

$$B: \quad I \approx \sum f(r_1, \ldots, r_n), \qquad r_i = \pm 1/\sqrt{3}, i = 1, \ldots, n.$$

$$C: \quad I \approx \sum A_{i_1} A_{i_2} \ldots A_{i_n} f(r_{i_1}, r_{i_2}, \ldots, r_{i_n})$$

where each subscript i_k denotes 1, 2 or 3 $(k = 1, \ldots, n)$ and $A_1 = \frac{5}{9}$, $A_2 = \frac{8}{9}$, $A_3 = \frac{5}{9}$; $r_1 = -\sqrt{\frac{3}{5}}$, $r_2 = 0$, $r_3 = \sqrt{\frac{3}{5}}$

$$D: \quad I \approx \frac{8}{7} f(0, 0) + \frac{20}{63} f(0, \pm\sqrt{\tfrac{14}{15}}) + \frac{5}{9} f(\pm\sqrt{\tfrac{3}{5}}, \pm\sqrt{\tfrac{1}{3}})$$

$$E: \quad I \approx (V/2n)[f(r, 0, \ldots, 0) + f(0, r, 0, \ldots 0) + \ldots + f(0, \ldots, 0, r)]$$

$$V = 2\pi^{n/2}/[n\Gamma(\tfrac{1}{2}n)], \qquad r = \pm\sqrt{[n/(n+2)]} \quad [\text{see IV, § 10.2}]$$

$$F: \quad I \approx 2^{-n} V \sum f(r_1, \ldots, r_n),$$

$$V = 2\pi^{n/2}/[n\Gamma(\tfrac{1}{2}n)], \qquad r_i = \pm\sqrt{[1/(n+2)]}, \qquad i = 1, 2, \ldots, n$$

APPENDIX 7.5

The Levin *t*- and *u*-transformations.

Let S_n be the *n*th partial sum of the convergent alternating series

$$u_1 - u_2 + u_3 - \ldots + (-1)^{n-1} u_n + \ldots, \quad u_n > 0 (S_0 = 0)$$

then we define the t_{2n} transformation to be

$$t_{2n}(S) = \frac{\sum\limits_{j=0}^{2n} (-1)^j \binom{2n}{j} \left(\dfrac{j+1}{2n+1}\right)^{2n-1} \dfrac{S_{j+1}}{\Delta S_j}}{\sum\limits_{j=0}^{2n} (-1)^j \binom{2n}{j} \left(\dfrac{j+1}{2n+1}\right)^{2n-1} \dfrac{1}{\Delta S_j}}, \qquad \Delta S_j = S_{j+1} - S_j.$$

As *n* increases $t_{2n}(S) \to$ the sum of the alternating series. Thus from only the first 5 terms in the alternating series in Example 7.5.3 we obtain

$$t_4(S) = 0 \cdot 529753.$$

If S_n denotes the *n*th partial sum of the convergent series of positive terms

$$u_1 + u_2 + u_3 + \ldots + u_n + \ldots. \qquad (S_0 = 0)$$

then we define the u_{2n+1} transformation to be

$$\frac{\sum\limits_{j=0}^{n+1} (-1)^j \binom{n+1}{j} \left(\dfrac{j+1}{2n+2}\right)^{2n-1} \dfrac{S_{j+1}}{\Delta S_j}}{\sum\limits_{j=0}^{n+1} (-1)^j j \binom{n+1}{j} \left(\dfrac{j+1}{2n+2}\right)^{2n-1} \dfrac{1}{\Delta S_j}} \qquad \Delta S_j = S_{j+1} - S_j$$

As *n* increases $u_{2n+1}(S) \to$ the sum of the series. Thus with the values obtained in Example 7.5.1 and the additional term $S_4 = I(4) = 1 \cdot 325818$ we obtain

$$u_3(S) = 1 \cdot 570368.$$

See Levin (1973) for further details.

CHAPTER 8

Ordinary Differential Equations

8.1. INTRODUCTION: EXISTENCE OF ANALYTIC SOLUTIONS

Differential Equations first arose as a consequence of the invention of the differential calculus by Newton and Leibnitz [see IV, Chapter 3]; up to the year 1800 considerable progress was made developing methods for finding analytical solutions to certain classes of ordinary differential equations. Since that date, however, few fundamentally new results have been obtained in this area [see IV, Chapter 7]. With the general availability of powerful computing facilities, much of the current research in this field is devoted to the construction of reliable methods for the determination of numerical solutions. Differential equations arise wherever the differential calculus is applied to describe a model, and they commonly occur in problems in engineering, physics, chemistry, economics and biology.

For example, Newton's law of cooling states that the rate of loss of heat from a liquid is proportional to the excess temperature of the liquid over that of its surroundings. If T_s denotes the temperature of the surroundings, and T is the temperature of the liquid at time t, and k is a constant of proportionality, then Newton's law can be stated in the form

$$\frac{dT}{dt} = -k(T - T_s).$$

The mathematical model is completed by a statement about the starting conditions of the experiment, which typically takes the form $T = T_0$ (some initial temperature) at time $t = 0$.

Another example of a differential equation arising in practice is that describing the behaviour of an electrical circuit in which a generator is producing a voltage $E_0 \sin \omega t$ and is connected in series through a condenser of capacitance C, a wire of resistance R and a coil of induction L. The charge q on the condenser will vary with time, and its value at any time t is given by the differential equation

$$L\frac{d^2q}{dt^2} + R\frac{dq}{dt} + \frac{q}{C} = E_0 \sin \omega t,$$

307

where, as in the above example, some information about the initial conditions is needed to complete the model.

Chemical reactions provide yet another instance of the use of differential equations and a particular example is the bimolecular reaction, where substances A and B form molecules of type C. The concentration x of C at a time t is given by the differential equation

$$\frac{dx}{dt} = k(a-x)(b-x)$$

where k is a constant and a and b are the original concentrations of A and B respectively.

All these 3 examples of differential equations can be solved in analytic form, that is, the solution can be expressed as a mathematical formula. For example, the concentration x of substance C is

$$x = \frac{a(1-e^{(a-b)kt})}{(1-(a/b)e^{(a-b)kt})}$$

from which it is possible, given the values of a, b and k, to calculate the value of x at any time t. However, in many other problems such simple solutions do not exist, and accordingly numerical methods are used.

8.1.1. First Order Equations: Initial-Value Problem

The equation

$$\frac{dy}{dx} = f(x, y) \qquad\qquad (8.1.1)$$

is called an ordinary differential equation of the first order [see IV, §§ 7.2 and 7.10]. Provided that the function $f(x, y)$ satisfies certain appropriate conditions (these are given below, see (8.1.3), and are almost always satisfied in physically realistic models), the solution of equation (8.1.1) is of the form $y = g(x)$ and can be represented as a family of curves in the (x, y) plane, as shown in Figure 8.1.1.

In general there is one and only one solution curve through each point in the (x, y)-plane, and a unique solution to equation (8.1.1) is specified by an initial condition of the form

$$y(x_0) = y_0. \qquad\qquad (8.1.2)$$

This effectively selects which solution curve is being calculated, and an accurate, stable numerical method would yield this solution, starting at the point (x_0, y_0) and tracing out the section PQ indicated by the heavier line. The solution then is uniquely determined by the differential equation (8.1.1) and the initial condition (8.1.2) and indeed these two equations constitute the *initial value problem* for a first-order differential equation.

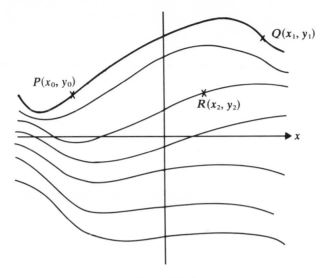

Figure 8.1.1

Figure 8.1.1 also clearly indicates that it is not always possible to require a solution of (8.1.1) to pass through 2 (or more) arbitrary points; for example the points P and R lie on different solution curves and no solution curve exists which passes through both these points. In cases where $f(x, y)$ becomes indeterminate at particular points (for example $f(x, y) = y/x$ at the origin) these results are not valid, and it is necessary to make a careful study of the equation in the immediate vicinity of the point. Because such cases arise rarely in problems encountered in practice their analysis will not be made here, but the interested reader is referred to Chapter 7 of Volume IV or to Stoker (1950) for a full discussion.

EXAMPLE 8.1.1.   The differential equation $dy/dx = -y$ has the general solution $y = A e^{-x}$, where A is an arbitrary constant [see IV, § 7.2]. The family of solution curves are shown in Figure 8.1.2 for a variety of values of A.

The solution to the initial-value problem

$$\frac{dy}{dx} = -y, \qquad y(0) = 1$$

is the curve MN, having the equation $y = e^{-x}$ which corresponds to the case $A = 1$ in the general solution given above.

The existence of a solution to the initial-value problem is governed by the following Theorem.

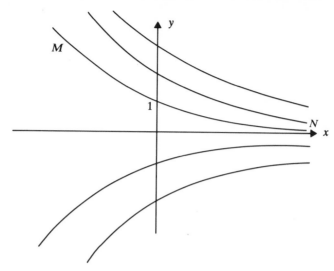

Figure 8.1.2: Solutions of the differential equation $dy/dx = -y$.

THEOREM 8.1.1.   *Let the function $f(x, y)$ be continuous for $a \leq x \leq b$ [see IV, § 5.1] and let there exist a constant L such that*

$$|f(x, y) - f(x, z)| \leq L|y - z| \qquad (8.1.3)$$

for all $x \in [a, b]$ and all real y and z. Then the initial-value problem

$$\frac{dy}{dx} = f(x, y), \qquad y(x_0) = y_0$$

where $x_0 \in [a, b]$, possesses a unique solution $y = y(x)$ for $x \in [a, b]$.

The relation (8.1.3) is called a *Lipschitz condition* on the function $f(x, y)$. A proof of this theorem will not be given here but the reader is referred to Phillips and Taylor (1973). The importance of this result is that it establishes the existence of solutions to a wide class of first-order initial-value problems, and not just to the relatively few which possess explicit, analytic solutions. As an example, let us apply this to the simple differential equation given above. Here

$$f(x, y) = -y$$

and hence

$$|f(x, y) - f(x, z)| = |y - z| \quad \text{for all } x$$

so that the inequality (8.1.3) is satisfied for any $L \geq 1$.

8.1.2.   Second Order Equations: Initial-Value and Boundary-Value Problems

The equation

$$\frac{d^2y}{dx^2} = f\left(x, y, \frac{dy}{dx}\right) \tag{8.1.4}$$

where f is any function, is an ordinary differential equation of the second order. Thus

$$\frac{d^2y}{dx^2} = x + y + \frac{dy}{dx}$$

and

$$\frac{d^2y}{dx^2} = xy^2 + y\left(\frac{dy}{dx}\right)^2 - y^3$$

are both examples of second-order differential equations.

To specify a unique solution to a second-order differential equation, two conditions are required. Suppose these are of the form

$$y(x_0) = y_0, \qquad \left.\frac{dy}{dx}\right|_{x_0} = y_0' \tag{8.1.5}$$

where y_0 and y_0' are given constants, and $dy/dx|_{x_0}$ denotes the value of dy/dx when $x = x_0$.

Then in general for physically realistic systems these conditions together with equation (8.1.4), specify a unique solution to the differential equation, again referred to as the *initial-value problem*. For example the second-order differential equation

$$\frac{d^2y}{dx^2} = x + y$$

has the solution $y = A e^x + B e^{-x} - x$, where A and B are arbitrary constants [see IV, Ex. 7.4.3]. If it is required that the solution satisfies the initial conditions

$$y(0) = 0 \quad \text{and} \quad \left.\frac{dy}{dx}\right|_0 = 1$$

then it follows that

$$A + B = 0 \quad \text{and} \quad A - B - 1 = 1.$$

These two equations have the solution $A = 1$, $B = -1$ so that the unique solution satisfying the initial conditions is

$$y = e^x - e^{-x} - x.$$

However, for second (and higher)-order systems other conditions than those specified in (8.1.5) are possible and these give rise to further classes of problems. For example the conditions

$$y(x_0) = y_0, \qquad y(x_1) = y_1 \tag{8.1.6}$$

(where y_0 and y_1 are given constants) together with the differential equations (8.1.4) again in general specify a unique solution, and this is referred to as the *boundary-value problem* for the second-order system. Thus supposing we require the solution of the differential equation

$$\frac{d^2 y}{dx^2} = x + y$$

which satisfies the conditions

$$y(0) = 0 \quad \text{and} \quad y(1) = 1,$$

the general solution is, as above,

$$y = A \, e^x + B \, e^{-x} - x.$$

If we now apply the boundary conditions we obtain the following equations which must be satisfied by A and B

$$A + B = 0$$

$$Ae + \frac{B}{e} - 1 = 1.$$

These have the solution

$$A = -B = \frac{2e}{e^2 - 1}$$

so that the solution of this boundary-value problem is

$$y = \frac{2e}{e^2 - 1} (e^x - e^{-x}) - x.$$

Other conditions, which also define a boundary-value problem, are

$$A \frac{dy}{dx}\bigg|_{x_0} + By(x_0) = \alpha_0$$

$$C \frac{dy}{dx}\bigg|_{x_1} + Dy(x_1) = \alpha_1$$

although these are less commonly encountered in practice than either the conditions (8.1.5) or (8.1.6).

Lastly the *eigenvalue problem* should be mentioned at this point; as a simple illustration let us consider the differential equation

$$\frac{d^2 y}{dx^2} + k^2 y = 0 \tag{8.1.7}$$

with boundary conditions $y(0) = 0$, $y(1) = 0$.
The solution of this differential equation is

$$y = A \sin kx + B \cos kx \tag{8.1.8}$$

where, again, A and B are arbitrary constants [see IV, Ex. 7.4.1].
The two boundary conditions give

$$B = 0 \quad \text{and} \quad A \sin k + B \cos k = 0.$$

The second condition is accordingly

$$A \sin k = 0$$

which leads to two possibilities, either $A = 0$ or $\sin k = 0$.
If $A = 0$, then the solution is identically zero, but if $\sin k = 0$ then $k = s\pi$, where $s = \pm 1, \pm 2, \pm 3, \ldots$. Provided that k has one of these values the solution to the problem is of the form

$$y = A \sin kx$$

where the value of A is still arbitrary. These values of k are *eigenvalues* and the solution is termed the *eigensolution*. A fuller explanation of these topics is given in IV, § 7.8.

In practice, if k is very close to one of these eigenvalues, the numerical solution to the differential equation (8.1.7) can become severely distorted due to the presence of the unwanted eigensolution. Thus a knowledge of the mathematical properties of the equation is important whenever a numerical solution is contemplated.

8.1.3. Higher Order Equations

The equation

$$\frac{d^n y}{dx^n} = f\left(x, y, \frac{dy}{dx}, \frac{d^2 y}{dx^2}, \ldots, \frac{d^{n-1} y}{dx^{n-1}}\right) \tag{8.1.9}$$

is an ordinary differential equation of the nth order since it involves differential coefficients up to, but not higher than the nth order. A commonly occurring fourth-order ordinary differential equation arises in elasticity theory. If the loading on a beam is a given function $w(x)$, where x is the distance measured along the beam, and its flexural rigidity is EI, then the displacement, y, of the beam at any point is given by the differential equation

$$\frac{d^2}{dx^2}\left(EI \frac{d^2 y}{dx^2}\right) = w$$

or, writing this in the form of equation (8.1.9),

$$\frac{d^4 y}{dx^4} = \frac{w}{EI} - \frac{2}{EI} \frac{d(EI)}{dx} \frac{d^3 y}{dx^3} - \frac{1}{EI} \frac{d^2(EI)}{dx^2} \frac{d^2 y}{dx^2}.$$

The equation requires four conditions to determine a unique solution, while then nth order equation (8.1.9) requires in general n conditions to specify a unique solution. If all the conditions apply at $x = x_0$, then this defines an initial-value problem; when conditions at more than one value of x are involved, a boundary-value problem is specified. As in the second-order case, eigensolutions can arise, and this emphasizes the need to understand the mathematical nature of the problem before embarking upon extensive numerical working.

Although most of the differential equations considered in this chapter are first-order equations, this is not a restriction on the equations which can be solved, since it is always possible to express an nth order equation as a set of n first-order equations [see IV, § 7.9.1]. If, in the example given above, we write

$$y = y_1, \qquad \frac{dy_1}{dx} = y_2, \qquad \frac{dy_2}{dx} = y_3, \qquad \frac{dy_3}{dx} = y_4$$

then

$$\frac{d^2 y}{dx^2} = \frac{d}{dx}\left(\frac{dy}{dx}\right) = \frac{d}{dx}\left(\frac{dy_1}{dx}\right) = \frac{dy_2}{dx} = y_3$$

and similarly

$$\frac{d^3 y}{dx^3} = y_4.$$

The fourth-order equation can accordingly be written in the form

$$\frac{dy_4}{dx} = \frac{w}{EI} - \frac{2}{EI} \frac{d(EI)}{dx} y_4 - \frac{1}{EI} \frac{d^2(EI)}{dx^2} y_3$$

In general, the nth-order equation can be written as a system of n first-order simultaneous equations

$$\frac{dy_1}{dx} = y_2$$

$$\frac{dy_2}{dx} = y_3$$

$$\vdots \qquad\qquad (8.1.10)$$

$$\frac{dy_{n-1}}{dx} = y_n$$

$$\frac{dy_n}{dx} = f(x, y_1, y_2, \ldots, y_n)$$

where y_1 has been written for y. The methods described can easily be generalized to cater for such a system, and explicit formulations of the nth-order problem will be given where appropriate.

8.1.4. Linear Equations

A linear differential equation of order n has the general form

$$f_0(x)\frac{d^n y}{dx^n}+f_1(x)\frac{d^{n-1}y}{dx^{n-1}}+\ldots+f_{n-1}(x)\frac{dy}{dx}+f_n(x)y = F(x). \quad (8.1.11)$$

The general solution of this equation has the form

$$y = Y(x)+ \sum_{i=1}^{n} \alpha_i y_i(x) \quad (8.1.12)$$

where $Y(x)$ is a particular solution of (8.1.11), α_i $(1 \le i \le n)$ are arbitrary constants to be determined by applying the initial and/or boundary conditions and $y_i(x)$ are independent solutions to the equation

$$f_0(x)\frac{d^n y}{dx^n}+f_1(x)\frac{d^{n-1}}{dx^{n-1}}y\ldots+f_{n-1}(x)\frac{dy}{dx}+f_n(x)y = 0. \quad (8.1.13)$$

[see IV, § 7.3].

Consider, for example, the linear second-order differential equation

$$\frac{d^2 y}{dx^2}+6\frac{dy}{dx}+8y = 35\, e^{3x}. \quad (8.1.14)$$

A particular solution to this equation is $y = e^{3x}$, as can be verified by direct substitution. If we now consider the equation

$$\frac{d^2 y}{dx^2}+6\frac{dy}{dx}+8y = 0 \quad (8.1.15)$$

then two independent solutions to this are

$$y = e^{-2x} \quad \text{and} \quad y = e^{-4x}.$$

The general solution of (8.1.14) is

$$y = e^{3x}+\alpha_1 e^{-2x}+\alpha_2 e^{-4x}$$

where α_1 and α_2 are arbitrary constants, corresponding to the solution to the nth-order equation given by (8.1.12). In fact, if $Y(x)$ is a particular solution to equation (8.1.14), and $Y_1(x)$ is any solution to (8.1.15), then

$$y = Y(x)+\alpha Y_1(x)$$

where α is an arbitrary constant, is also a solution of (8.1.14) and it is sometimes useful to exploit this fact in the numerical solution to a linear problem. However, even with linear problems, analytic methods are seldom adequate in themselves to provide the scientist or engineer with the solution to a particular problem, while in the case of non-linear systems numerical methods in general offer the only way of providing a solution. Consider for example the 3-body problem which is concerned with the motion of three particles under the action of Newton's law of gravitation. If the masses of these particles are m_i and their positions at time t are $\{x_i(t), y_i(t)\}$ $(i = 1, 2, 3)$ then the equations governing their motion in a plane can be written as

$$\frac{d^2 x_1}{dt^2} = \frac{(x_2 - x_1)m_2 G}{r_{12}^3} + \frac{(x_3 - x_1)m_3 G}{r_{13}^3}$$

$$\frac{d^2 y_1}{dt^2} = \frac{(y_2 - y_1)m_2 G}{r_{12}^3} + \frac{(y_3 - y_1)m_3 G}{r_{13}^3}$$

together with two similar pairs of equations for particles 2 and 3, where G is the gravitational constant and $r_{ij} = \{(x_i - x_j)^2 + (y_i - y_j)^2\}^{1/2}$ is the distance between the particles i and j at time t.

Various methods for obtaining numerical solutions to these equations have been successfully devised, in connection with problems concerning planetary motion, but a complete analytic solution has not been obtained in spite of considerable efforts by many writers over a long period. Not only are solutions of this problem needed for astronomical purposes, but also for satellite orbits and in many such cases the solution must be immediately available for control purposes.

8.2. NUMERICAL SOLUTIONS

The numerical solution of a differential equation can consist of a set of tabulated values of the dependent variable y to the required number of significant figures or, more particularly in some real time applications, the solution can be produced directly on a video screen in the form of a graph. However, it is important to realise that the numerical solution of a differential equation is essentially a set of numbers which are approximations to the true solution at the corresponding values $x_0, x_1, x_2, \ldots$ of the independent variable x; in general these points are equidistant, so that $x_r = x_0 + rh$. The distance between any two successive points is

$$x_{r+1} - x_r = x_0 + (r + 1)h - x_0 - rh$$

$$= h,$$

and the value of h is referred to as the step width. The numerical methods for obtaining these approximations involve the formulation of the equation in a discrete form and the use of truncated Taylor series [see IV, §§ 2.10 and 3.6] and finite difference formulae [see § 1.4].

In this chapter two general categories of methods will be described: one-step methods and multi-step methods. In a one-step method, only the current value

y_r of the solution is used in calculating the next value y_{r+1}. Examples of one-step methods are Euler's method and the Runge–Kutta methods, both of which are described in section 8.2. In a multi-step method considerably more information is used in calculating the next value y_{r+1} of the solution; not only are some of the previously calculated values of the solution $y_r, y_{r-1}, y_{r-2}, \ldots$ employed, but also previously calculated values of the derivatives $y'_r, y'_{r-1}, y'_{r-2}, \ldots$ where y'_r is written for $dy/dx|_{y=y_r, x=x_r}$. Some multi-step methods are described in section 8.3.

In examining different numerical methods, problems of accuracy, stability and convergence have to be considered, since these factors have to be taken into account when choosing a particular method and a suitable step width.

Accuracy

In common with all other numerical computations (excepting a few specialized tasks involving only integers), the arithmetic operations involved in the solution of differential equations can never be carried out with complete accuracy because of the requirement that numbers are stored in a computer with a given number of digits. Thus in general, numbers have to be 'chopped off', or rounded, and the error due to this limitation is termed rounding error [see also § 1.3]. Rounding errors occur at each stage of the calculation and this can be regarded as a sort of 'noise' which is always present in a computation. Some idea of the errors caused by this can be found by repeating the calculation using double length working, but such an operation can be extremely time consuming.

The other major cause of loss of accuracy is truncation error. This generally arises through the use of a truncated Taylor series to represent a function at a particular point, and can be illustrated by the following example. Suppose that the value of a function $f(x)$ and its derivatives $f'(x), f''(x), \ldots$ are known at a particular point $x = x_0$. Then the value of the function at the point $x_1 = x_0 + h$ is given by the Taylor series

$$f(x_1) = f(x_0 + h) = f(x_0) + hf'(x_0) + \frac{h^2}{2!} f''(x_0) + \ldots .$$

This can be written as

$$f(x_1) = f(x_0) + hf'(x_0) + \frac{h^2}{2!} f''(x_0) + \ldots + \frac{h^p}{p!} f^{(p)}(x_0) + R_{p+1}(x_0)$$

where $R_{p+1}(x_0)$, the remainder term, can be expressed in the form

$$R_{p+1}(x_0) = \frac{h^{p+1}}{(p+1)!} f^{(p+1)}(x_0 + \theta h)$$

where $0 < \theta < 1$. A full account of this result can be found in Volume IV, § 3.6.

If the remainder term is now omitted, so that $f(x_1)$ is given by the approximation

$$f(x_1) = f(x_0) + hf'(x_0) + \ldots + \frac{h^p}{p!} f^{(p)}(x_0)$$

the truncation error is then

$$\frac{h^{p+1}}{(p+1)!} f^{(p+1)}(x_0 + \theta h)$$

and the approximation is said to be of *order of accuracy p*. In the numerical solution of a differential equation, this type of approximation occurs at each step of the calculation and accordingly truncation errors are introduced at each step. At first sight the errors could be minimized by decreasing h, that is taking smaller steps, or increasing p, the order of accuracy of this method. However, we shall see in later sections (e.g. § 8.2.7) that such measures do not always work. The cumulative effect of the truncation errors in any method determines the stability properties and this will now be explained in more detail.

Stability

There are essentially two types of instability which can occur in the numerical solution of ordinary differential equations and these are termed *inherent* instability and *induced* instability.

Inherent instability arises when a differential equation has both an increasing and a decreasing solution—for example the solution could be of the form

$$y = A e^x + B e^{-x},$$

but because of the boundary and/or initial conditions imposed, we only require the decreasing solution $y = B e^{-x}$. Because of the truncation errors introduced at each step of the numerical solution a component of the unwanted increasing solution $y = A e^x$ can creep in and will quickly overwhelm the solution we are trying to find. A worked example of this type of instability is given in section 8.2.7 where it is shown that a way of avoiding this difficulty is to reverse the direction of integration.

Induced stability can be defined as instability which arises through the method of solution. The truncation errors propagate through the solution with increasing effect, and as in the case of inherent instability they can swamp the required solution. Induced instability can occur with both one-step methods and multi-step methods and an example of induced stability of a particular one-step method (the Runge–Kutta method) is given in section 8.2.6. In this case it is possible to avoid the instability by decreasing the step width h, and several of the methods described in this chapter have built in procedures to make the step width h as large as possible, to achieve efficiency, while at the same time maintaining stability. The instability of multi-step methods is more complicated but essentially can occur when the difference equation used in the solution (see equation (8.3.1)) possesses extra solutions, referred to as *parasitic solutions*, which grow faster than the required solution as the calculation proceeds, again swamping the true solution. A worked example of this type of instability is shown in Table 8.3.2 and a general account of this topic is contained in the corresponding section 8.3.2. Some multi-step methods always exhibit this form of instability and are referred to as *strongly unstable*; in other cases instability can be avoided by choosing a sufficiently small step width h and such methods are termed *conditionally unstable*. Again, an example of this is given in section 8.3.2. Finally the method adopted must be *convergent*, in the sense that the sequence of approximate solutions, obtained by using successively smaller values of stepwidth h, must converge to the true solution of the original differential equation.

8.2.1.   Euler's Method

The simplest 1-step method available for the integration of the first-order differential equation $dy/dx = f(x, y)$ uses the relation

$$y_{r+1} = y_r + hf(x_r, y_r). \tag{8.2.1}$$

In this formulation $x_r = x_0 + rh$ $(r = 1, 2, \ldots)$, where x_0 is an initial given value of x and h is the step width, while $\{y_n\}$ constitutes the set of numbers which approximates the set $\{y(x_n)\}$, which represents the value of the exact solution to the initial value problem defined by

$$\frac{dy}{dx} = f(x, y), \qquad y(x_0) = y_0.$$

The relation (8.2.1) can be seen to be derived from the Taylor expansion for the exact solution for the corresponding values of x:

$$y(x_{r+1}) = y(x_r) + hy'(x_r) + \frac{h^2}{2!} y''(x_r) + \frac{h^3}{3!} y'''(x_r) + \ldots$$

where

$$y'(x_r) = \frac{dy}{dx}\bigg|_{\substack{x=x_r \\ y=y_r}} = f(x_r, y_r)$$

from the differential equation. If all terms involving h^2 and higher powers are ignored then

$$y(x_{r+1}) = y(x_r) + hf(x_r, y_r).$$

Although the relation (8.2.1) is a very simple one, and easy to program, it suffers from the disadvantage that very small values of h are needed to obtained an accurate answer, although the rapidly decreasing cost of computation has made this objection less significant than previously. The method has the property that the error in the approximate solution tends to zero at a particular point x_r as the step length h tends to zero, and thus the method is convergent.

As an illustration of Euler's method, we consider the example

$$y' = 2y, \qquad y(0) = 1 \tag{8.2.2}$$

which has the exact solution $y = e^{2x}$ [see IV, § 7.2].

The following table shows the result of applying Euler's method to this example, with the step widths indicated; the final column gives the exact solution.

x	$h = 0 \cdot 2$	$h = 0 \cdot 1$	$h = 0 \cdot 05$	$h = 0 \cdot 01$	Exact value
0·2	1·400	1·440	1·464	1·486	1·492
0·4	1·960	2·074	2·144	2·208	2·226
0·6	2·774	2·986	3·138	3·281	3·320
0·8	3·842	4·300	4·595	4·876	4·953
1·0	5·378	6·192	6·727	7·245	7·389

Table 8.2.1: Solutions to the initial-value problem $y' = 2y$, $y(0) = 1$, using Euler's method with the step widths indicated.

Inspection of this table shows that the accuracy of the entries in each row increases as the step width decreases, and this suggests the use of Richardson's extrapolation technique [see § 2.4.1] as a way of getting better accuracy. The tenchnique has a wide applicability wherever a sequence of approximate results is produced by a corresponding sequence of values of a parameter and the required result corresponds to the vanishing of the parameter. In the present case, we assume that the tabulated value y_n (for any h) is related to the exact value Y_n by an approximation of the form

$$y_n = Y_n + \alpha h + \beta h^2 + \dots,$$

where α and β are constants. By using the results given in Table 8.2.1 for any particular value of x for three values of h (say 0·01, 0·05, 0·1) and ignoring terms of order h^3, 3 equations for the unknowns Y_n, α and β are obtained, from which it is simple to eliminate α and β, and hence derive an expression for Y_n. In fact, for the values of h stated it is easily seen that

$$Y_n = \tfrac{1}{18}\{2y_{n,0\cdot1} - 9y_{n,0\cdot05} + 25y_{n,0\cdot01}\}$$

where $y_{n,0\cdot1}$ is the value of y_n calculated using a step width $h = 0 \cdot 1$ and $y_{n,0\cdot05}$, $y_{n,0\cdot01}$ are obtained by using step widths of 0·05 and 0·01 respectively.

Table 8.2.2 shows the results of this calculation for the indicated values of x, and again the exact solution is displayed alongside for the purposes of comparison.

x	Exact value	Y_n
0·2	1·492	1·492
0·4	2·226	2·225
0·6	3·320	3·320
0·8	4·953	4·953
1·0	7·389	7·387

Table 8.2.2: Application of Richardson extrapolation to the results obtained in Table 8.2.1, using values $h = 0·01, 0·05, 0·1$.

The accuracy produced by the technique is good, but it has to be remembered that a considerable number of evaluations of the function $f(x, y)$ are needed at each step; in this particular example a total of $20 + 4 + 2 = 26$ calculations of f are used. This point will be discussed later when various methods are compared.

The convergence of Euler's method can be demonstrated by substituting the expression $2y$ for f in (8.2.1) so that

$$y_{r+1} = y_r + 2hy_r = (1 + 2h)y_r.$$

The solution to this difference equation which satisfies the initial condition $y_0 = 1$ is $y_r = (1 + 2h)^r$, or using the relation $x_r = rh$,

$$y_r = \left(1 + \frac{2x_r}{r}\right)^r \tag{8.2.3}$$

[see I, § 14.13]. If the step width h now tends to zero such that x_r remains fixed, and the result

$$\lim_{n \to \infty} \left(1 + \frac{x}{n}\right)^n = e^x$$

is used, [see IV (2.11.4)] the limiting value of y_r in (8.2.3) is e^{2x_r}. Thus the method is convergent in the sense described above.

8.2.2. Runge–Kutta Methods

Euler's method has a simple geometrical interpretation, since the move from y_r to y_{r+1} occurs in the direction of the gradient calculated at (x_r, y_r). The general class of *Runge–Kutta methods* attempt to improve upon this by using a weighted mean of the gradients at a set of points (X_i, Y_i) in the neighbourhood of the points (x_r, y_r). A particular feature of these methods is that the Y_i's themselves involve the evaluation of the function f and thus considerably more calculation is involved per step than in Euler's method. But these methods are used extensively in numerical applications mainly because they need no special starting arrangements, the step width h can be changed easily and storage requirements are minimal. The most commonly used Runge–Kutta formula is a fourth-order scheme, but the derivation is somewhat complicated and will not be given here. The principle can however be illustrated by the somewhat simpler second-order scheme which will now be derived. The idea is to express y_{r+1} as a linear combination of the form

$$y(x_{r+1}) = y(x_r) + ak_1 + bk_2 + e \tag{8.2.4}$$

where

$$k_1 = hf(x_r, y(x_r))$$

$$k_2 = hf(x_r + \theta h, y(x_r) + \phi k_1)$$

and a, b, θ and ϕ are parameters whose values are to be calculated in order that the error e in (8.2.4) be of order h^3.

The calculation is done by expanding both sides in a Taylor series [see IV, § 3.6] and making use of the differential equation

$$y(x_{r+1}) = y(x_r) + hy'(x_r) + \frac{h^2}{2!} y''(x_r) + \ldots$$

$$= y(x_r) + hf(x_r, y(x_r)) + \frac{h^2}{2!} \left\{ \frac{\partial f}{\partial x} + f \frac{\partial f}{\partial y} \right\}_{x_r} + \ldots$$

$$k_1 = hf(x_r, y(x_r)) = hf_r; \quad \text{say}$$

$$k_2 = hf(x_r + \theta h, y(x_r) + \phi k_1) = hf(x_r, y(x_r)) + \theta h^2 \frac{\partial f}{\partial x} + \phi k_1 h \frac{\partial f}{\partial y} + O(h^3).$$

If the coefficients of powers of h are now equated in [see IV, § 2.3] (8.2.4)

$$h: \quad 1 = a + b$$

$$h^2: \quad \tfrac{1}{2} = b\theta; \qquad \tfrac{1}{2} = b\phi$$

three of the parameters can be expressed in terms of the fourth, thus giving the resulting formula:

$$y_{r+1} = y_r + \left(1 - \frac{1}{2\theta}\right) hf_r + \frac{h}{2\theta} f(x_r + \theta h, y_r + \theta hf_r), \tag{8.2.5}$$

where $y_r = y(x_r)$ and $f_r = f(x_r, y(x_r))$. Two particular values of θ lead to well known cases:

$$\theta = \tfrac{1}{2}: \quad y_{r+1} = y_r + hf(x_r + \tfrac{1}{2}h, y + \tfrac{1}{2}hf_r) \tag{8.2.6}$$

which is the *modified Euler method*, and

$$\theta = 1: \quad y_{r+1} = y_r + \tfrac{1}{2}hf_r + \tfrac{1}{2}hf(x_{r+1}, y_r + hf_r). \tag{8.2.7}$$

which is *Heun's method*. Note that in these schemes, it is necessary to evaluate the function $f(x, y)$ twice for each step, compared with one for Euler's method described earlier. Both these cases have simple geometrical interpretations and the modified Euler method is shown in Figure 8.2.1.

Both results (8.2.6) and (8.2.7) are closely related to the simple Euler's

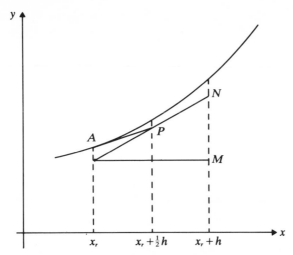

Figure 8.2.1: Geometrical interpretation of (8.2.6); the increment in y is represented by the line MN, PA is the tangent to the curve at A, and the slope of PN is obtained from the original differential equation.

method described above. If Euler's method is employed with a step width $h/2$, two successive applications give:

$$y_{r+\frac{1}{2}} = y_r + \frac{h}{2} f_r$$

$$y_{r+1} = y_{r+\frac{1}{2}} + \frac{h}{2} f(x_r + \tfrac{1}{2}h, y_{r+\frac{1}{2}}) \qquad (8.2.8)$$

$$= y_r + \tfrac{1}{2}hf_r + \tfrac{1}{2}hf(x_r + \tfrac{1}{2}h, y_r + \tfrac{1}{2}hf_r)$$

while a single application with a step width of h gives

$$y_{r+1} = y_r + hf_r.$$

If it is now assumed that the true value Y of y at x_{r+1} satisfies a relation of the form

$$y = Y + \alpha h + \beta h^2 + \dots$$

the equations (8.2.8) can be used in this relation, assuming terms of $O(h^2)$ are neglected, to give the result

$$Y = y_r + hf(x_r + \tfrac{1}{2}h, y_r + \tfrac{1}{2}hf_r)$$

which is seen to correspond to the case $\theta = \frac{1}{2}$ given in equation (8.2.6). Returning to the simple example considered above:

$$y' = 2y, \qquad y(0) = 1$$

the results of using the general second order Runge–Kutta formula with a step width of 0·01 are shown in the following table, which should be compared

with the results in Tables 8.2.1 and 8.2.2. These show that the accuracy of
the solution is considerably better than that produced by Euler's method for
the same step-width, but at the expense of twice as many evaluations of the
function $f(x, y)$

x	y_r	Exact solution
0·2	1·492	1·492
0·4	2·226	2·226
0·6	3·320	3·320
0·8	4·952	4·953
1·0	7·388	7·389

Table 8.2.3: Second order Runge–
Kutta method applied to $y' = 2y$.

In practice the most commonly used Runge–Kutta scheme is a fourth-order
formula. As in the second-order scheme the essence of the method is to express
the value of the increment in the dependent variable y as a weighted mean of
the derivatives at a set of points in the (x, y) plane in the neighbourhood of
(x_r, y_r).

It is assumed that $y(x_{r+1})$ can be written as

$$y(x_{r+1}) = y(y_r + ak_1 + bk_2 + ck_3 + dk_4)$$

where the constants a,b,c,d are termed *weighting coefficients* and

$$k_1 = hf(x_r, y_r)$$
$$k_2 = hf(x_r + \theta h, y_r + \phi k_1)$$
$$k_3 = hf(x_r + \xi h, y_r + \eta k_2)$$ (8.2.9)
$$k_4 = hf(x_r + \zeta h, y_r + \psi k_3).$$

The ten parameters are chosen in order to make the Taylor series expansions
agree up to and including the term in h^4. The solution gives considerable
freedom in selecting values of the ten parameters, but the most commonly used
set gives the following formulae

$$y_{r+1} = y_r + \tfrac{1}{6}(k_1 + 2k_2 + 2k_3 + k_4)$$

where

$$k_1 = hf_r(x_r, y_r)$$
$$k_2 = hf(x_r + \tfrac{1}{2}h, y_r + \tfrac{1}{2}k_1)$$
$$k_3 = hf(x_r + \tfrac{1}{2}h, y_r + \tfrac{1}{2}k_2)$$ (8.2.10)
$$k_4 = hf(x_r + h, y_r + k_3).$$

One interesting aspect of this result is in the special cases where $f(x, y)$ is
independent of y, so that $k_2 = k_3$ and the formula reduces to Simpson's rule for

the integration of a function of one variable [see § 7.1.5]. Notice that this scheme now requires the evaluation of the function $f(x, y)$ four times for each integration step, so that in cases where the function is complicated the fourth-order Runge–Kutta method involves a considerable amount of computation, in fact considerably more than some of the finite difference methods described later in this chapter.

For the nth-order equation, written as

$$y_i' = f_i(x, y_1, y_2, \ldots y_n) \qquad 1 \le i \le n$$

(which is a generalization of (8.1.10)), the corresponding formula for advancing the solution one step is

$$y_{i,r+1} = y_{i,r} + \tfrac{1}{6}(k_{i,1} + 2k_{i,2} + 2k_{i,3} + k_{i,4}) \qquad 1 \le i \le n$$

where

$$
\begin{aligned}
y_{i,r+1} &= y_i(x_{r+1}) = y_i(x_0 + (r+1)h) \\
k_{i1} &= hf_i(x_r, y_{1,r}, y_{2,r}, \ldots, y_{n,r}) \\
k_{i2} &= hf_i(x_{r+\frac{1}{2}}, y_{1,r} + \tfrac{1}{2}k_{1,1}, \ldots, y_{n,r} + \tfrac{1}{2}k_{n,1}) \\
k_{i3} &= hf_i(x_{r+\frac{1}{2}}, y_{1,r} + \tfrac{1}{2}k_{i,2}, \ldots, y_{n,r} + \tfrac{1}{2}k_{n,2}) \\
k_{i4} &= hf_i(x_{r+1}, y_{1,r} + k_{1,3}, \ldots, y_{n,r} + k_{n,3}).
\end{aligned}
\qquad (8.2.11)
$$

In performing this computation the quantities $k_{i,1}$ $(1 \le i \le n)$ must first be found, then $k_{i,2}$, $k_{i,3}$ and lastly $k_{i,4}$.

8.2.3.   Merson's Method

One of the main disadvantages of Runge–Kutta methods is the lack of knowledge of the size of the truncation error at each step. Even for the second-order scheme given by equation (8.2.4) and (8.2.5) the error term is

$$\frac{h^3}{6}(y''' - 3\theta^2(f_{xx} + 2ff_{xy} + f^2f_y)) \qquad (8.2.12)$$

while that for the fourth-order scheme (8.2.10) has a considerably more complex form. In order that the calculation proceed as quickly as possible the step width must be made as large as possible, provided that the truncation error at each step does not exceed some preassigned value. A method of achieving this to some extent is due to Merson (1957) and it is this variation of the Runge–Kutta scheme which is most commonly used in practice, and indeed is found in program libraries, e.g. the NAG Library (1978).

Essentially, Merson's modification involves an additional evaluation of the function $f(x, y)$ in the differential equation at each step. This enables an estimate of the truncation error to be made and the algorithm can then leave the step width unchanged, or double or halve it as required.

The formulae used by Merson are:

$$k_1 = hf(x_r, y(x_r))$$

$$k_2 = hf(x_r + \tfrac{1}{3}h, y(x_r) + \tfrac{1}{3}k_1)$$

$$k_3 = hf(x_r + \tfrac{1}{3}h, y(x_r) + \tfrac{1}{6}k_1 + \tfrac{1}{6}k_2) \qquad (8.2.13)$$

$$k_4 = hf(x_r + \tfrac{1}{2}h, y(x_r) + \tfrac{1}{8}k_1 + \tfrac{3}{8}k_3)$$

$$k_5 = hf(x_r + h, y(x_r) + \tfrac{1}{2}k_1 - \tfrac{3}{2}k_3 + 2k_4)$$

which are used to advance the integration by one step width

$$y(x_{r+1}) = y(x_r) + \tfrac{1}{6}k_1 + \tfrac{2}{3}k_4 + \tfrac{1}{6}k_5.$$

The truncation error in this formula is approximated by

$$e_{r+1} = \tfrac{1}{15}k_1 - \tfrac{3}{10}k_3 + \tfrac{4}{15}k_4 - \tfrac{1}{30}k_5 \qquad (8.2.14)$$

which clearly tends to zero with h. The actual interval changing can then be decided by a 'dipstick' procedure: if e_{r+1} lies outside given limits (say one eighth and twice the preassigned value) then the step width is either doubled or halved; otherwise it is left unchanged. The extension to an nth-order equation is straightforward, with formulae analogous to those in (8.2.11). In this case the set $\{e_{i,r+1}\}$ $(1 \leq i \leq n)$ is evaluated, and the maximum element is used to decide whether the step width is to be altered.

Merson's method has the following advantages:

(i) The method is self starting, requiring only the initial values of the dependent variables $\{y_i\}$. No special procedures are necessary as in the case of the multistep methods discussed below.

(ii) The evaluation of the terms $\{e_{i,r+1}\}$ enables the step width to be adjusted automatically as the solution proceeds, with the aim of keeping the truncation error within a prescribed bound at each step.

However, a major disadvantage of the method is that the estimate e_{r+1} of the error is asymptotically correct for linear equations only. For non-linear equations, no justification for its validity exists and in such cases the method can lead to highly inefficient calculations.

8.2.4.   Runge–Kutta–Fehlberg Methods

The difficulty of providing a satisfactory step width control procedure when using Runge–Kutta methods has been largely overcome by a recent development due to Fehlberg (1968) who has presented results for fourth-, fifth-, sixth-, seventh- and eighth-order formulae (see also Bulirsch and Stoer (1966)). The technique adopted is to estimate the leading truncation error term at each step by computing two approximations to the solution y_1. For the fifth-order method [cf. (8.2.5) and (8.2.9)]

$$y_1 = y_0 + h \sum_{k=0}^{5} c_k f_k + O(h^6) \qquad (8.2.15)$$

and also

$$\hat{y}_1 = y_0 + h \sum_{k=0}^{7} \hat{c}_k f_k + O(h^7)$$

where $c_k, \hat{c}_k$ are the usual weighting coefficients corresponding to the coefficients a,b,c,d in equation (8.2.9) and

$$f_k = f\left(x_0 + \alpha_k h, \ y_0 + h \sum_{\lambda=0}^{k-1} \beta_{k\lambda} f_\lambda \right)$$

where $\alpha_k, \beta_{k\lambda}$ are the same in the expression for y_1 and $\hat{y}_1$, so that both can be calculated with eight evaluations of the function f_k. The difference between these values gives the leading term in the local truncation error as

$$E = h \sum_{k=0}^{5} (\hat{c}_k - c_k) f_k + h\{\hat{c}_6 f_6 + \hat{c}_7 f_7\}. \tag{8.2.16}$$

By performing the usual Taylor series expansions and equating coefficients of powers of h, a set of 37 equations is obtained for the coefficients $\alpha_r, \beta_{rs}, c_r$ in case of the sixth-order formula and a set of 17 equations for the corresponding coefficients in the fifth-order formula. The expression for the leading truncation error term of the fifth-order Runge–Kutta formula is of the form

$$h^6 \sum_{\nu=1}^{20} T_\nu F_\nu \tag{8.2.17}$$

where the F_ν are complicated expressions involving the function f and its partial derivatives [see IV, § 5.2] and the coefficients T_ν are constants which can be expressed in terms of the coefficients $\alpha_r, \beta_{rs}, c_r$. A variety of solutions is possible, and the choice is made in such a way as to minimize the truncation error, *consistent with being able to use it for stepwidth control. This has the advantage that a larger stepwidth can be used, and thus contributes to the overall efficiency of this method.* The particular solution adopted by Fehlberg for the fifth-order scheme (8.2.15) is given in the following tableau:

k	λ / 0	1	2	3	4	5	6	α_k	c_k	$\hat{c}_k$
0								0	$\frac{31}{380}$	$\frac{7}{1408}$
1	$\frac{1}{6}$							$\frac{1}{6}$	0	0
2	$\frac{4}{75}$	$\frac{16}{75}$						$\frac{4}{15}$	$\frac{1125}{2816}$	$\frac{1125}{2816}$
3	$\frac{5}{6}$	$-\frac{8}{3}$	$\frac{5}{2}$					$\frac{2}{3}$	$\frac{9}{32}$	$\frac{9}{32}$
4	$-\frac{8}{5}$	$\frac{144}{25}$	-4	$\frac{16}{25}$				$\frac{4}{5}$	$\frac{125}{768}$	$\frac{125}{768}$
5	$\frac{361}{320}$	$-\frac{18}{5}$	$\frac{407}{128}$	$-\frac{11}{80}$	$\frac{55}{128}$			1	$\frac{5}{66}$	0
6	$-\frac{11}{640}$	0	$\frac{11}{256}$	$-\frac{11}{160}$	$\frac{11}{256}$	0		0	—	$\frac{5}{66}$
7	$\frac{93}{640}$	$-\frac{18}{5}$	$\frac{803}{256}$	$-\frac{11}{160}$	$\frac{99}{256}$	0	1	1	—	$\frac{5}{66}$

Table header spanning columns 0–6: $\beta_{k\lambda}$

Runge–Kutta–Fehlberg method of order five.

With these values of the coefficients, the leading truncation error is considerably reduced, since all the coefficients T_ν are zero with the exception of

$$T_3 = -3T_{13}, \qquad T_4 = -T_{13}, \qquad T_{10} = -1/2160,$$

$$T_{12} = 3T_{13}, \qquad T_{13} = 1/32400.$$

Furthermore, the expression for the local truncation error is, by (8.2.16) easily seen to take the following simple form:

$$E = \frac{5h}{66}(f_0 + f_5 - f_6 - f_7).$$

One consequence of this expression, however, is that this method (and the higher-order ones derived by Fehlberg) cannot be used for quadrature. For, since $\alpha_6 = 0$, and $\alpha_5 = \alpha_7$ then, if the function f is independent of y, $f_0 = f_6$, $f_5 = f_7$, so that E is identically zero and cannot be used for error estimation. The derivation of the schemes for the higher order formulae follows similar lines to those described although the algebra is somewhat complicated; for example the eighth order formula involves a set of 287 equations.

Runge–Kutta–Fehlberg Sixth-Order Formulae

The sixth order solution is given [cf. (8.2.5) and (8.2.9)] by the equations

$$y_1 = y_0 + h \sum_{k=0}^{7} c_k f_k + O(h^7)$$

$$\hat{y}_1 = y_0 + h \sum_{k=0}^{7} \hat{c}_k f_k + O(h^8). \tag{8.2.18}$$

where the form of the definition of f_k is the same as that for the fifth order case. The seventh-order equation leads to a set of 85 equations for the unknown α_r, β_{rs}, c_r and these give the leading truncation error term in the expression for y_1 as a sum of the form $h^7 \sum_{\nu=1}^{48} T_\nu F_\nu$. Again a solution is selected which

| k | $\beta_{k\lambda}$ | | | | | | | | | α_k | c_k | $\hat{c}_k$ |
	0	1	2	3	4	5	6	7	8			
0	0									0	$\frac{77}{1440}$	$\frac{11}{864}$
1	$\frac{2}{33}$									$\frac{2}{33}$	0	0
2	0	$\frac{4}{33}$								$\frac{4}{33}$	0	0
3	$\frac{1}{22}$	0	$\frac{3}{22}$							$\frac{2}{11}$	$\frac{1771561}{6289920}$	$\frac{1771561}{6289920}$
4	$\frac{43}{64}$	0	$-\frac{165}{64}$	$\frac{77}{32}$						$\frac{1}{2}$	$\frac{32}{105}$	$\frac{32}{105}$
5	$-\frac{2383}{486}$	0	$\frac{1067}{54}$	$-\frac{26312}{1701}$	$\frac{2176}{1701}$					$\frac{2}{3}$	$\frac{243}{2560}$	$\frac{243}{2560}$
6	$\frac{10077}{4802}$	0	$-\frac{5643}{686}$	$\frac{116259}{16807}$	$-\frac{6240}{16807}$	$\frac{1053}{2401}$				$\frac{6}{7}$	$\frac{16807}{74880}$	$\frac{16807}{74880}$
7	$-\frac{733}{176}$	0	$\frac{141}{8}$	$-\frac{335763}{23296}$	$\frac{216}{77}$	$-\frac{4617}{2816}$	$\frac{7203}{9152}$			1	$\frac{11}{270}$	0
8	$\frac{15}{352}$	0	0	$-\frac{5445}{46592}$	$\frac{18}{77}$	$-\frac{1215}{5632}$	$\frac{1029}{18304}$	0		0	—	$\frac{11}{270}$
9	$-\frac{1833}{352}$	0	$\frac{141}{8}$	$-\frac{51237}{3584}$	$\frac{18}{7}$	$-\frac{729}{512}$	$\frac{1029}{1408}$	0	1	1	—	$\frac{11}{270}$

minimizes the truncation error, and the results can be conveniently expressed in the previous tableau.

The leading truncation error term is

$$y_1 - \hat{y}_1 = h\{(c_0 - \hat{c}_0)f_0 + \tfrac{11}{270}f_7 - \tfrac{11}{270}f_8 - \tfrac{11}{270}f_9\}$$

$$= \tfrac{11h}{270}(f_0 + f_7 - f_8 - f_9) \tag{8.2.19}$$

and this vanishes identically when f does not depend on y.

Runge–Kutta–Fehlberg Seventh-Order Formulae

The seventh-order solution is given [cf. (8.2.5) and (8.2.6)] by the equation

$$y_1 = y_0 + \sum_{k=0}^{10} c_k f_k + O(h^3) \tag{8.2.20}$$

$$\hat{y}_1 = y_0 + \sum_{k=0}^{12} \hat{c}_k f_k + O(h^9)$$

and these now give a total of 200 equations for the coefficients. The algebra is very complicated, and in order to select an appropriate solution which makes the error term T_ν reasonably small, a variety of cases were tested numerically. The particular solution adopted, which is displayed in the tableau below produces values of $O(10^{-5})$ for the leading error terms.

For this scheme the truncation error takes the form

$$\frac{41h}{840}(f_0 + f_{10} - f_{11} - f_{12}) \tag{8.2.21}$$

which again vanishes when $f = f(x)$.

k	0	1	2	3	4	5	6	7	8	9	10	11	α_k	c_k	$\hat{c}_k$
						$\beta_{k\lambda}$									
0	0												0	$\frac{41}{840}$	0
1	$\frac{2}{27}$												$\frac{2}{27}$	0	0
2	$\frac{1}{36}$	$\frac{1}{12}$											$\frac{1}{9}$	0	0
3	$\frac{1}{24}$	0	$\frac{1}{8}$										$\frac{1}{6}$	0	0
4	$\frac{5}{12}$	0	$-\frac{25}{16}$										$\frac{5}{12}$	0	0
5	$\frac{1}{20}$	0	0	$\frac{1}{4}$	$\frac{1}{5}$								$\frac{1}{2}$	$\frac{34}{105}$	$\frac{34}{105}$
6	$-\frac{25}{108}$	0	0	$\frac{125}{108}$	$-\frac{65}{27}$	$\frac{125}{54}$							$\frac{5}{6}$	$\frac{9}{35}$	$\frac{9}{35}$
7	$\frac{31}{300}$	0	0	0	$\frac{61}{225}$	$-\frac{2}{9}$	$\frac{13}{900}$						$\frac{1}{6}$	$\frac{9}{35}$	$\frac{9}{35}$
8	2	0	0	$-\frac{53}{6}$	$\frac{704}{45}$	$-\frac{107}{9}$	$\frac{67}{90}$	3					$\frac{2}{3}$	$\frac{9}{280}$	$\frac{9}{280}$
9	$-\frac{91}{108}$	0	0	$\frac{23}{108}$	$-\frac{976}{135}$	$\frac{311}{54}$	$-\frac{19}{60}$	$\frac{17}{6}$	$-\frac{1}{12}$				$\frac{1}{3}$	$\frac{9}{280}$	$\frac{9}{280}$
10	$\frac{2383}{4100}$	0	0	$-\frac{341}{164}$	$\frac{4496}{1025}$	$-\frac{301}{82}$	$\frac{2133}{4100}$	$\frac{45}{82}$	$\frac{45}{164}$	$\frac{18}{41}$			1	$\frac{41}{840}$	0
11	$\frac{3}{205}$	0	0	0	0	$-\frac{6}{41}$	$-\frac{3}{205}$	$-\frac{3}{41}$	$\frac{3}{41}$	$\frac{6}{41}$	0	0	0	—	$\frac{41}{840}$
12	$-\frac{1777}{4100}$	0	0	$-\frac{341}{164}$	$\frac{4496}{1025}$	$-\frac{289}{82}$	$\frac{2193}{4100}$	$\frac{51}{82}$	$\frac{33}{164}$	$\frac{12}{41}$	0	1	1	—	$\frac{41}{840}$

The eighth-order method starts with the equations

$$y_i = y_0 + \sum_{k=0}^{14} c_k f_k + O(h^9)$$

$$\hat{y}_1 = y_0 + \sum_{k=0}^{16} \hat{c}_k f_k + O(h^{10})$$

(8.2.22)

and the ninth-order formula for y_1 involves 286 equations. For this case, Fehlberg finds that in general the coefficients are not readily expressible in rational form and accordingly gives a table of values of the coefficients to 32 significant figures (not reproduced here), which have been produced using a 40-digit arithmetic package (see Fehlberg (1968)).

The number of function evaluations required per step increases with the order of the Runge–Kutta–Fehlberg formula, and is summarized in the following table:

Order of RKF Method	Number of function evaluations
5	8
6	10
7	13
8	17

The choice of order for a particular problem must take account of a number of factors including the cost of evaluating the function and the accuracy required; it is not possible to select any one of the above schemes as one which should be adopted in every case. In principle there is no reason why higher order schemes should not be considered, and indeed Fehlberg (loc. cit.) has tried a ninth-order formula, using a tenth-order formula for calculation of the truncation error. However, the increased cost of function evaluation appears to more than compensate for the gain in accuracy, so that there seems little point in trying such schemes. This will be further considered in the section dealing with comparisons of methods.

8.2.5. Convergence of Runge–Kutta Methods

All the Runge–Kutta methods described in this section are convergent in the sense that the error in the approximate solution tends to zero at a particular point x, as the step length h tends to zero. For the particular example

$$y' = \lambda y, \qquad y(0) = 1$$

which has the exact solution $y = e^{\lambda x}$, the general second order Runge–Kutta formula (8.2.5) gives the following recurrence relation

$$y_{r+1} = y_r + \left(1 - \frac{1}{2\theta}\right) h\lambda y_r + \frac{h}{2\theta}\lambda(y_r + \theta h\lambda y_r)$$

$$= \left(1 + \lambda h + \frac{\lambda^2 h^2}{2}\right) y_r. \tag{8.2.23}$$

Thus the solution which satisfies the initial condition $y_0 = 1$ is

$$y_r = \left(1 + \lambda h + \frac{\lambda^2 h^2}{2}\right)^r, \tag{8.2.24}$$

[see I, § 14.13] and in order to examine the convergence of this at, say, x_r it is necessary to study the behaviour of this function as h tends to zero in such a manner that x_r remains fixed.

Now

$$y_r = \left(1 + \lambda h + \frac{\lambda^2 h^2}{2}\right)^{x_r/h}$$

so that

$$\log y_r = \frac{x_r}{h}\log\left(1 + \lambda h + \frac{\lambda^2 h^2}{2}\right).$$

Then by de l'Hospital's rule [see IV, Corollarys 3.4.3 and 3.4.4]

$$\lim_{h\to 0}\frac{1}{h}\log\left(1 + \lambda h + \frac{\lambda^2 h^2}{2}\right) = \lim_{h\to 0}\frac{\lambda + \lambda^2 h}{1 + \lambda h + (\lambda^2 h^2/2)} = \lambda.$$

Hence $\lim_{h\to 0}\log y_r = \lambda x_r$ and thus $\lim_{h\to 0} y_r = e^{\lambda x_r}$.

The convergence of the fourth-order Runge–Kutta method can be demonstrated in a similar manner, for the same example,

$$y' = \lambda y, \qquad y(0) = 1.$$

The application of the fourth-order formulae (8.2.10) gives the recurrence relation

$$y_{r+1} = \left(1 + \lambda h + \tfrac{1}{2}\lambda^2 h^2 + \frac{1}{3!}\lambda^3 h^3 + \frac{1}{4!}\lambda^4 h^4\right) y_r \tag{8.2.25}$$

with the corresponding solution

$$y_r = \left(1 + \lambda h + \frac{1}{2!}\lambda^2 h^2 + \frac{1}{3!}\lambda^3 h^3 + \frac{1}{4!}\lambda^4 h^4\right)^r \tag{8.2.26}$$

when the initial condition $y(0) = 1$ is applied. As before, it can be shown that for a fixed $x_r = rh$, y_r tends to $e^{\lambda x_r}$ as h tends to zero. Finally, Merson's method gives the result

$$y_{r+1} = \left(1 + \lambda h + \frac{\lambda^2 h^2}{2!} + \frac{1}{3!}\lambda^3 h^3 + \frac{1}{4!}\lambda^4 h^4 + \tfrac{1}{144}\lambda^5 h^5\right) y_r \tag{8.2.27}$$

which can also be proved convergent.

8.2.6. Stability of Runge–Kutta Methods

Although the Runge–Kutta methods are convergent as shown above, they suffer from the disadvantage that instability can arise if too large a step-width h is used, when the solution is exponentially decreasing. Returning again to the simple example

$$y' = \lambda y, \qquad y(0) = 1$$

the Runge–Kutta (and Merson's) methods all give relations of the general form

$$y_{r+1} = P(\lambda h) y_r$$

where the polynomial $P(\lambda h)$ depends on the order of the particular method employed; examples of $P(\lambda h)$ are given in equations (8.2.23) and (8.2.25).

Now the exact solution is $y(x) = e^{\lambda x}$ [see IV, § 7.2] so that

$$y(x_{r+1}) = e^{\lambda(r+1)h} = e^{\lambda h} e^{\lambda rh} = e^{\lambda h} y(x_r)$$

Thus the Runge–Kutta methods can be expected to succeed in cases where $e^{\lambda h}$ is approximated by the polynomial $P(\lambda h)$, but not otherwise.

Figure 8.2.2 shows how the polynomials $P(\lambda h)$ compare with $e^{\lambda h}$ in the case of the second order method, given by equation (8.2.23), the fourth order (equation (8.2.25)) and also for Merson's method, equation (8.2.27). For values of λh in the range $(0, 1)$ agreement is close; indeed the graphs of $e^{\lambda h}$ and the Merson polynomial are undistinguishable. But when λh is negative, a different situation arises, and in particular when $\lambda h < -1$ large differences occur. Indeed the polynomial appropriate to Merson's method is negative when $\lambda h < -2.5$.

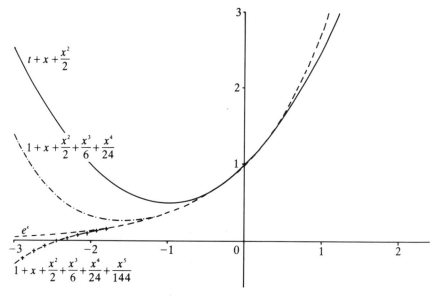

Figure 8.2.2

Since for negative values of λ the true solution is a decreasing function, the relative error is large and quickly swamps the true solution. It can be seen from Figure 8.2.2 that the polynomial $P(\lambda h)$ corresponding to the second-order method satisfies $P > 1$ when $\lambda h < -2$ and that for the fourth-order method when $\lambda h < -2.8$. For values of $|\lambda h|$ greater than this, Runge–Kutta methods are unstable, but even for values slightly less than these limits, they can be expected to be of little value. This is also illustrated in Table 8.2.4, where the results of integrating the equation $y' = -20y$, using the fourth-order method, are given for values of λh ranging from -0.1 to -2. The exact solution, e^{-20x}, is given in column 2 of the table and it can be seen that values of λh in the range -0.1 to -0.5 lead to reasonably accurate results, but that the remaining values displayed give increasingly large errors.

x	e^{-20x}	-0.1	-0.2	-0.5	-1	-2
0	1	1	1	1	1	1
0.2	0.01832	0.01832	0.01832	0.01837	0.01978	0.11111
0.4	0.00034	0.00035	0.00035	0.00033	0.00039	0.01235
0.6	6.144×10^{-6}	6.144×10^{-6}	6.145×10^{-6}	6.203×10^{-6}	7.733×10^{-6}	0.00137
0.8	1.125×10^{-7}	1.125×10^{-7}	1.126×10^{-7}	1.140×10^{-7}	1.529×10^{-7}	0.00015
1.0	2.061×10^{-9}	2.061×10^{-9}	2.062×10^{-9}	2.094×10^{-9}	3.024×10^{-9}	0.00002

Table 8.2.4: Solution of $y' = -20y$, using the fourth-order Runge–Kutta method, showing the effect of different values of λh. The exact solution is given in column 2; the particular value of λh is given at the head of each succeeding column.

8.2.7. Inherent Instability of Runge–Kutta Methods

Inherent instability has already been discussed in a general way on page 318; it arises in situations where the general solution of an equation includes a rapidly increasing term, but where the boundary or initial conditions in a particular problem cause the component to be absent. As an example of this let us consider the first-order differential equation

$$y' = 10y + 10x - 1.$$

The general solution of this equation is

$$y = A e^{10x} - x$$

where A is an arbitrary constant which can be determined by an initial condition [see IV, § 7.2]. For example, we might require the solution to satisfy the initial condition

$$y(0) = \alpha.$$

It then follows that

$$\alpha = A \cdot e^{10 \times 0} - 0 = A$$

so that the solution is

$$y = \alpha e^{10x} - x.$$

If now, however, we require the particular solution of this equation which passes through the origin $x = 0$, $y = 0$ then $\alpha = 0$ and the solution is

$$y = -x.$$

But as explained on page 318 a component of the unwanted solution αe^{10x} is introduced into the numerical solution and for integration methods proceeding in the direction of increasing value of x this component grows rapidly and the true solution is swamped. The effect of this is shown in Table 8.2.5, which gives the results of applying the fourth-order Runge–Kutta method to this example, using step widths of $0 \cdot 01$, $0 \cdot 001$ and $0 \cdot 0001$.

x	$-x$	$h = 0 \cdot 0001$	$h = 0 \cdot 001$	$h = 0 \cdot 01$
0·2	−0·2	−0·19986	−0·19998	−0·20000
0·4	−0·4	−0·39920	−0·39988	−0·39998
0·6	−0·6	−0·59483	−0·59918	−0·59986
0·8	−0·8	−0·76311	−0·79394	−0·79896
1·0	−1·0	−0·72925	−0·95529	−0·99232
1·2	−1·2	0·79465	−0·86994	−1·1434
1·4	−1·4	13·316	1·0378	−0·98318
1·6	−1·6	107·03	16·410	1·4702
1·8	−1·8	800·28	131·26	20·815
2·0	−2·0	5921·3	981·07	164·59

Table 8.2.5: Solution of $y' = 10y + 10x - 1$, $y(0) = 0$, using the fourth-order Runge–Kutta method with the step widths indicated; the exact solution, $y = -x$, is given in column 2.

The results show that this problem cannot be avoided by taking smaller step widths, as was the case in the instabilities considered previously. The effect of the smaller step width is to introduce additional 'noise' into the calculation, so that the solution for $h = 0 \cdot 0001$ is in fact worse than that for $h = 0 \cdot 01$. In this type of problem a simple and often effective method of avoiding such difficulties is to reverse the direction of integration. For the particular example considered above, the constant A can be expressed in terms of the value of the solution $y(2)$ at $x = 2$ as

$$A = e^{-20}(2 + y(2)).$$

Any solution starting at $x = 2$ with values of $y(2)$ in the vicinity of -2 will give results which agree closely with the required solution, as shown in Table 8.2.6, where values of -1, -2 and -3 have been used as starting values for the fourth-order Runge–Kutta method.

x	$y(2) = -1$	$y(2) = -2$	$y(2) = -3$
2	−1	−2	−3
1·8	−1·56	−1·80	−1·94
1·6	−1·58	−1·60	−1·62
1·4	−1·40	−1·40	−1·40
1·2	−1·20	−1·20	−1·20
1·0	−1·00	−1·00	−1·00
0·8	−0·80	−0·80	−0·80
0·6	−0·60	−0·60	−0·60
0·4	−0·40	−0·40	−0·40
0·2	−0·20	−0·20	−0·20
0	0	0	0

Table 8.2.6: Solution of $y' = 10x + 10y - 1$, starting
at $x = 2$ with the values of $y(2)$ indicated; the step
length used in each case is $h = 0·01$

8.3.   MULTI-STEP METHODS

So far, all the methods described have been one step methods; they have the
property that the value of y_{n+1} is found at each step by using only the previous
value of the dependent variable y_n and of course the differential equation
$y' = f(x, y)$. Such methods involve the evaluation of the function $f(x, y)$ at
several points within the interval $(x + nh, x + (n + 1)h)$; for example Merson's
method (equation (8.2.13)) requires no fewer than 5 evaluations of the function
$f(x, y)$ per step and thus, especially for high-order systems, involves substantial
amounts of computation. These methods do not make use of other, previously
calculated, values of the dependent variable $y_{n-1}, y_{n-2}, \ldots$ nor of the derivatives $f_{n-1}, f_{n-2}, \ldots$. Since none of this information is used (indeed, all previous
values are discarded at each step) such methods can intuitively be expected to
involve more computation at each step when compared with methods which
use this information. Such methods are termed multistep methods and in
general they can be represented by an equation of the form

$$\sum_{s=0}^{k} \alpha_s y_{n+s} = h \sum_{s=0}^{k} \beta_s f_{n+s} \tag{8.3.1}$$

where y_{n+s} and f_{n+s} have the conventional meaning defined earlier, i.e.

$$y_{n+s} \equiv y(x_{n+s}) \quad \text{and} \quad f_{n+s} \equiv f(x_{n+s}, y_{n+s}).$$

α_s and β_s, are suitably chosen constants and k is defined as the *order* of the
particular method employed.

Such equations are conveniently divided into two classes depending upon the
value of the coefficient β_k. When $\beta_k = 0$, the right-hand side of (8.3.1) does not
involve f_{n+k} and hence the equation can be solved explicitly for y_{n+k}. Such cases
are referred to as *predictor* or *open formulae*. When β_k is non zero then y_{n+k} is
present in the right-hand side of (8.3.1), through the function $f(x_{n+k}, y_{n+k})$ and

in general, since f is a non-linear function of y, the equation cannot then be solved explicitly for y_{n+k}. In these cases the solution requires use of an iterative procedure and such formulae are termed *corrector* or *closed formulae*.

Multi-step methods, particularly of high order, lead to efficient methods of solving differential equations, but they also possess several disadvantages. Since the methods require, at each step, information concerning the previous history of the solution, special methods are needed to start the integration. The Runge–Kutta methods which have already been described are commonly used, but it is important to ensure that the degree of accuracy of the results obtained from these initial calculations matches that of the subsequent multistep method. In practice this usually means a considerably smaller stepwidth over this initial stage of the integration. Multistep methods can also be unstable and this is discussed more fully below. The requirements to use some past information about the solution implies a need for more storage in the computer than for single step methods. Finally the general method of approach makes the process of changing the step width both awkward and time consuming. This disadvantage is particularly apparent when the solution enters a region of rapid change and several changes of stepwidth are necessary.

8.3.1. Predictor–Corrector Methods

Predictor–corrector methods are defined as those which use a pair of multi-step formulae, of the form displayed in (8.3.1), in conjunction with each other. The general principle is to employ a fairly coarse method of extrapolation [see § 2.4] to obtain an approximation (the predicted) value of y_{n+1}^p, using an open or predictor formulae. The value thus obtained is substituted into the differential equation so that a value for the derivative f_{n+1} can be calculated. This value is then inserted into a closed or corrector formula in order to obtain a more accurate value y_{n+1}^c (the corrected value).

Finally, this corrected value is then substituted into the differential equation so that a more accurate value of f_{n+1} can be obtained and the integration can then be advanced one step, unless a further iteration is performed using the corrector formula. But the value of repeated iterations at this stage is questionable since each iteration involves one calculation of the function $f(x, y)$ appearing in the differential equation and in any case the corrector formula contains a truncation error [see § 1.3]. There is a danger that repeated iterations at this stage will merely consume substantial amounts of machine time to produce convergence to a wrong answer.

The method can be explained in detail in terms of a particular case, the *Milne–Simpson* method.

Step 1.   Evaluate the predicted value y_{n+4}^p from the formula

$$y_{n+4}^p = y_n + \frac{4h}{3}(2y_{n+1}' - y_{n+2}' + 2y_{n+3}') + \tfrac{14}{45}h^5 y^{(v)}(\xi_1) \qquad (8.3.2)$$

where the truncation error term $\tfrac{14}{45}h^5 y^{(v)}(\xi_1)$ is neglected.

Step 2.   Use this value of y in the original differential equation

$$y' = f(x, y)$$

to obtain a value of y'_{n+4}.

Step 3.   Calculate an improved value of y from the corrector formula, where the value of y'_{n+4} found at Step 2 is used on the right-hand side

$$y^c_{n+4} = y_{n+2} + \frac{h}{3}(y'_{n+2} + 4y'_{n+3} + y'_{n+4}) - \frac{h^5}{90} y^{(v)}(\xi_2) \qquad (8.3.3)$$

where again the truncation error term $h\frac{5}{90}y^{(v)}(\xi_2)$ is neglected.

Step 4.   Substitute the value of y^c_{n+1} into the differential equation to obtain an improved value of y'_{n+1}

$$y'_{n+1} = f(x_{n+1}, y^c_{n+1}).$$

Step 5.   Provided that the changes between steps 2 and 4 and steps 1 and 3 are *consistent with the order of the truncation error terms* [*cf.* § 8.2] then the integration can be advanced one step; otherwise repeat steps 3 and 4.

The estimation of the truncation error at each step, and hence its elimination, is possible if it is assumed that the fifth derivative $y^{(v)}(\xi)$ remains effectively constant over the range in question. With this assumption, which is not too bad if the stepwidth h is sufficiently small and the solution is well behaved, it then follows from equations (8.3.2) and (8.3.3) that

$$y^c_{n+1} - y^p_{n+1} = \tfrac{29}{90}h^5 y^{(v)}(\xi),$$

giving

$$y_{n+1} = \frac{28y^c_{n+1} + y^p_{n+1}}{29}. \qquad (8.3.4)$$

A comparison of equations (8.3.2) and (8.3.3) also displays a common feature of predictor–corrector methods, namely that the truncation error of the corrector formula is considerably less than that of the corresponding predictor.

Other predictor–corrector methods can be derived by appropriate choices of the coefficient α_i, β_i in the equation (8.3.1) and the truncation error evaluated by use of Taylor's series [see IV, § 3.6]. For the case of methods of order $k = 3$ predictor formulae can be found by setting $\beta_{n+3} = 0$ so that

$$y_{n+3} = -\alpha_2 y_{n+2} - a_1 y_{n+1} - \alpha_0 y_n + h(\beta_0 f_n + \beta_1 f_{n+1} + \beta_2 f_{n+2})$$

and then requiring that this is exact for polynomials of degree ≤ 4. Corrector formulae can then be derived in a similar fashion, except that the coefficient β_2 is taken to be non-zero, thus giving a closed formula. A wide choice is available, but two commonly used systems are the *Adams–Bashforth method*, given by

$$y_{n+4}^p = y_{n+3} + \frac{h}{24}(55f_{n+3} - 59f_{n+2} + 37f_{n+1} - 9f_n) + \tfrac{251}{720}h^5 y^{(v)}(\xi_1)$$

$$y_{n+4}^c = y_{n+3} + \frac{h}{24}(9f_{n+4} + 19f_{n+3} - 5f_{n+2} + f_{n+1}) - \tfrac{19}{720}h^5 y^{(v)}(\xi_2)$$

(8.3.5)

and a method due to Hamming (1962), given by

$$y_{n+4}^p = \tfrac{1}{3}(2y_{n+2} + y_{n+1}) + \frac{h}{72}(191f_{n+3} - 107f_{n+2} + 109f_{n+1} - 25f_n) + \tfrac{707}{2160}h^5 y^{(v)}(\xi_1)$$

(8.3.6)

$$y_{n+4}^c = \tfrac{1}{3}(2y_{n+2} + y_{n+1}) + \frac{h}{72}(25f_{n+4} + 91f_{n+3} + 43f_{n+2} + 9f_{n+1}) - \tfrac{43}{2160}h^5 y^{(v)}(\xi_2).$$

Both these methods exhibit satisfactory convergence and stability properties, and these are now examined in detail.

8.3.2. Accuracy and Stability of Multi-Step Methods

The fundamental results described in this section are due to Dahlquist (1956) and the interested reader is referred to this important paper for a full discussion of the problem. Returning to equation (8.3.1), the accuracy of this formula can be defined by substituting the values for the true solution and then expanding each term in powers of h as a Taylor series [see IV, § 3.6]. The *degree of accuracy, p,* of the method for any particular choice of the coefficients is then given by the highest power of h having a zero coefficient. In fact the error in (8.3.1) is $O(h^{p+1})$ if and only if the following relations hold:

$$\sum_{s=0}^{k} \alpha_s = 0 \tag{8.3.7}$$

$$\sum_{s=0}^{k} \alpha_s s^r = r \sum_{s=0}^{k} \beta_s s^{r-1} \qquad (r = 1, 2, \ldots, p). \tag{8.3.8}$$

These equations can also be obtained directly by assuming y to be of the form x^r $(r = 0, 1, 2, \ldots, p)$, and it immediately follows that (8.3.1) is satisfied exactly whenever the relations (8.3.7) and (8.3.8) hold and y is a polynomial whose degree is less than or equal to p. These relations are often referred to as the *consistency conditions* and they can also be expressed in terms of the two related polynomials

$$\rho(Z) = \alpha_k Z^k + \alpha_{k-1} Z^{k-1} + \ldots + \alpha_1 Z + \alpha_0$$

$$\sigma(Z) = \beta_k Z^k + \beta_{k-1} Z^{k-1} + \ldots + \beta_1 Z + \beta_0$$

(8.3.9)

so that the first two consistency conditions are

$$\rho(1) = 0, \qquad \rho'(1) = \sigma(1). \tag{8.3.10}$$

Now equations (8.3.7) and (8.3.8) are homogeneous linear relations involving $2k+2$ coefficients. It can be expected, therefore, solutions for the coefficients exist for values of p up to and including $2k$, and Dahlquist (loc. cit.) has proved this and obtained an explicit solution for the case $p = 2k$. Thus the accuracy of the fundamental equation (8.3.1) is limited by the order k, satisfying the constraint

$$p \leq 2k.$$

Since the error is $O(h^{p+1})$ it would appear at first sight that the coefficients α_i, β_i should be chosen in order to ensure that p attains its permitted maximum value of $2k$. However, methods based on values of p satisfying $p > k+2$ where k is even, and $p > k+1$ (k odd) have been proved by Dahlquist to be *strongly unstable*, in the sense that parasitic solutions arise which swamp the true solution, thus making the computation useless. Such instabilities cannot be eliminated by taking smaller values of h, and indeed this exacerbates the problem [cf. Table 8.2.5]. In addition, Dahlquist has shown that when $p = k+2$ instabilities can still arise, but that in certain cases it is possible to control these by choosing a sufficiently small value of h. This phenomenon is an example of *conditional stability* in contrast to the strong instability which occurs when $p > k+2$ [see also § 8.2].

8.3.3. Conditional Stability

As an example of the occurrence of conditional stability, the case $p = 2$, $k = 4$ will be examined in detail.

EXAMPLE 8.3.3. On using the consistency conditions (8.3.7) and (8.3.8) it follows that the appropriate form of (8.3.1) is

$$y_{n+2} - y_n = h(\tfrac{1}{3}f_n + \tfrac{4}{3}f_{n+1} + \tfrac{1}{3}f_{n+2}) \tag{8.3.11}$$

which is the Milne–Simpson corrector formula, equation (8.3.3).

If this is applied to the solution of the differential equation

$$y' = Ay, \qquad y(0) = 1$$

(with the solution $y = e^{Ax}$), equation (8.3.11) takes the form

$$(1 - \theta)y_{n+2} - 4\theta y_{n+1} - (1 + \theta)y_n = 0 \tag{8.3.12}$$

where $\theta \equiv Ah/3$ and $y_1 = 1$. This is a *linear difference equation* [see I, §§ 14.12 and 14.13] whose solution is given by

$$y_n = C_1 \xi_1^n + C_2 \xi_2^n$$

where C_1 and C_2 are constants determined by initial conditions, and ξ_1 and ξ_2 are the (distinct) roots of the quadratic equation

$$(1 - \theta)\xi^2 - 4\theta\xi - (1 + \theta) = 0.$$

For small values of θ, the roots can be written as

$$\xi_1 = 1 + 3\theta + O(\theta^2)$$

$$\xi_2 = -1 + \theta + O(\theta^2)$$

and on using the initial condition $y_0 = C_1 + C_2 = 1$, the solution for y_n takes the form

$$y_n = (1 - C_2)\left(1 + \frac{Ax_n}{n} + \ldots\right)^n + (-1)^n C_2 \left(1 - \frac{Ax_n}{3n} + \ldots\right)^n.$$

If now $n \to \infty$ in such a manner that $x_n = nh$ remains fixed, then

$$\xi_1 \to e^{Ax_n}, \qquad \xi_2 \to \pm e^{-Ax_n/3}.$$

and

$$y_n \to (1 - C_2) \, e^{Ax_n} + (-1)^n C_2 \, e^{-Ax_n/3}. \tag{8.3.13}$$

The correct solution to the differential equation corresponds to the case $C_2 = 0$, but in practice a component of the parasitic solution will occur, either through a slight inaccuracy in the starting values or through the effect of rounding errors as the calculation proceeds. If A is positive then the true solution is increasing exponentially and the component is unimportant. This is illustrated in Table 8.3.1, where the results of using (8.3.12) with $h = 0 \cdot 1$ and

x_n	y_n	$\exp(3x_n)$
0	1	1
0·1	1·34986	1·34986
0·2	1·82216	1·82212
0·3	2·45967	2·4596
0·4	3·32027	3·32012
0·5	4·48194	4·48169
0·6	6·05008	6·04965
0·7	8·16685	8·16617
0·8	11·0242	11·0232
0·9	14·8814	14·8797
1·0	20·088	20·0855
1·1	27·1163	27·1126
1·2	36·6036	36·5982
1·3	49·4104	49·4025
1·4	66·6979	66·6862
1·5	90·0339	90·0171
1·6	121·535	121·511
1·7	164·057	164·022
1·8	221·456	221·407
1·9	298·939	298·867
2·0	403·53	403·429

Table 8.3.1: Solution of the equation $y' = 3y$, with $y(0) = 1$ using the finite difference relation (8.3.12).

$A = 3$ are given, together with the correct solution. The exact values of y_0 and y_1 were used, in order to start the calculation.

However, when A is negative the parasitic solution eventually swamps the correct solution and an example of this is given in Table 8.3.2 where $A = -3$ and the other conditions are the same as those shown in Table 8.3.1; again, the exact values of y_0 and y_1 were used to start the calculation.

x_n	$y_n \cdot 10^5$	$\exp(-3x_n) \cdot 10^5$
0	100000	100000
0·5	22312·5	22313
1·0	4976·65	4978·71
1·5	1113·01	1110·9
2·0	243·981	247·875
2·5	61·5824	55·3084
3·0	1·99625	12·3410
3·1	20·5585	9·14241
3·2	−5·84252	6·77286
3·3	18·9451	5·01747
3·4	−11·6694	3·71702
3·5	19·7440	2·75364
3·6	−16·7273	2·03995
3·7	22·2368	1·51123
3·8	−21·7721	1·11955
3·9	26·1109	0·82938
4·0	−27·3084	0·61442
4·1	31·2938	4·55174
4·2	−33·7228	3·37201
4·3	37·8668	2·49805
4·4	−41·3611	1·85060
4·5	46·0223	1·37096
4·6	−50·5763	1·01563
4·7	56·0460	7·52397
4·8	−61·7610	5·57388
4·9	68·3144	0·0412924
5·0	−75·3733	0·0305901

Table 8.3.2: An example of conditional stability: the solution of the equation $y' = -3y$, with $y(0) = 1$, using the finite difference relation (8.3.12).

Conditionally stable methods can be used for some problems, where the solution is exponentially increasing, but in other cases such methods will fail. In the particular example illustrated the choice of a smaller stepwidth will not avert the failure, and care is therefore needed in applying these methods.

A detailed analysis of the particular equations to be solved should be performed, and a partial discussion of this problem is given at the end of this section.

8.3.4. Strong Instability

As an example of strong instability satisfying Dahlquist's criterion $p > k + 1$, the case $p = 6$, $k = 3$ is examined.

EXAMPLE 8.3.4.   It follows from the consistency equations (8.3.7) and (8.3.8) that the multistep formula (8.3.1) has the form

$$-y_n - \tfrac{27}{11}y_{n+1} + \tfrac{27}{11}y_{n+2} + y_{n+3} = h\left(\tfrac{3}{11}f_n + \tfrac{27}{11}f_{n+1} + \tfrac{27}{11}f_{n+2} + \tfrac{3}{11}f_{n+3}\right). \quad (8.3.14)$$

If this is applied to the differential equation

$$\frac{dy}{dr} = 0, \qquad y(0) = 1$$

(having the solution $y = 1$), equation (8.3.14) reduces to

$$y_{n+3} + \tfrac{27}{11}y_{n+2} - \tfrac{27}{11}y_{n+1} - y_n = 0. \qquad (8.3.15)$$

As above, the solution of this linear difference equation is obtained by assuming a solution of the form $y_n = K\xi^n$ [see I, § 14.13], and this leads to the following equation for ξ:

$$\xi^3 + \tfrac{27}{11}\xi^2 - \tfrac{27}{11}\xi - 1 = 0 \qquad (8.3.16)$$

which has the roots

$$\xi = 1, \qquad \tfrac{1}{11}\{-19 \pm 4\sqrt{15}\}$$

$$= 1, \, -3{\cdot}14, \, -0{\cdot}32 \quad \text{(to 2 decimal places)}.$$

The first of these three roots corresponds to the desired solution of the differential equation, while the remaining two roots have appeared as a result of the order of the difference equation (8.3.15) which in turn is due to imposing the high-order accuracy defined by $p = 6$.

The solution to equation (8.3.15) can thus be written as

$$y_n = C_1(1)^n + C_2(-0{\cdot}32)^n + C_3(-3{\cdot}14)^n.$$

As before, the occurrence of errors in the initial values found for y_1 and y_2 together with rounding errors introduce components corresponding to non-zero values of C_2 and C_3. The root $\xi = -0{\cdot}32$ will produce a decaying contribution, but the root $\xi = -3{\cdot}14$ gives a rapidly growing oscillatory component which swamps the true solution and renders the method useless. A similar analysis can be performed for the differential equation of Example 8.3.3;

$$y' = Ay, \qquad y(0) = 1.$$

For small values of h, the roots of the difference equation differ only slightly from those of (8.3.16) and the method is unstable for all values of A. A numerical example of this is illustrated in Tables 8.3.3 and 8.3.4, which show the results obtained from applying (8.3.14) to the differential equation $y' = Ay$ in the cases $A = \pm 3$, with $h = 0{\cdot}1$.

x_n	y_n		exp $(3x_n)$
0	1		1
0·2	1·82212		1·82212
0·4	3·32012		3·32012
0·6	6·04965		6·04965
0·8	11·0232		11·0232
1·0	20·0863		20·0855
1·2	36·6048		36·5982
1·4	66·7434		66·6862
1·6	122		121·511
1·8	225·592		221·407
2·0	439·181		403·429
2·2	1040·47		735·095
2·4	3947·66		1339·43
2·6	24717·2		2440·6
2·8	194709		4447·07
3·0	1·63311	E − 06	8103·08

Table 8.3.3: An example of strong instability: the solution of the equation $y' = 3y$, with $y(0) = 1$, using the finite difference scheme (8.3.14) and exact values for y_0, y_1 and y_2.

x_n	y_n		exp $(-3x_n)$	
0	1		1	
0·2	0·548812		0·548812	
0·4	0·301191		0·301194	
0·6	0·165267		0·165299	
0·8	9·03545	E − 02	9·07179	E − 02
1·0	4·57110	E − 02	4·97871	E − 02
1·2	−1·83726	E − 02	2·73237	E − 02
1·4	−0·497287		1·49956	E − 02
1·6	−5·73476		8·22974	E − 03
1·8	−64·3777		4·51658	E − 03
2·0	−721·76		2·47875	E − 03

Table 8.3.4: An example of strong instability: the solution of the equation $y' = -3y$, $y(0) = 1$, using the finite difference scheme (8.3.14) and exact starting values for y_0, y_1 and y_2.

8.3.5. Summary of Stability Properties of Linear Multi-Step Methods

The stability properties of multi-step methods are closely connected with properties of the polynomials $\rho(Z)$ and $\sigma(Z)$ defined in (8.3.9) and the main results obtained by Dahlquist can be summarized as follows:

(i) A linear multi-step scheme is stable if the zeros of the polynomial $\rho(Z)$ lie within or on the unit circle and those roots which lie on the unit circle are distinct [see I, § 2.7.2 and V, (1.2.8)].

(ii) If k is even, the degree p of a stable scheme of order k cannot exceed $k+2$, and if k is odd, the degree p cannot exceed $k+1$.

(iii) If a method of even order k is stable, then the conditions

$$\alpha_s = -\alpha_{k-s}$$

$$\beta_s = \beta_{k-s} \qquad 0 \le s \le \frac{k}{2}$$

are both necessary and sufficient in order that it be of maximum degree $k+2$. In this case all the roots of $\rho(Z)$ have unit modulus.

(iv) If $p>k$ for a stable method, then the method is closed and, further, $\beta_k/\alpha_k > 0$.

Some of the limitations on the degree p of linear multi-step methods of order k which are implied by these results, are illustrated in Figure 8.3.1, for small values of p and k. The region of possible combinations is bounded by $p \le 2k$ and in this region those combinations of p and k which lead to strongly unstable and conditionally stable schemes are indicated. It must be borne in mind, however, that the remaining combinations do not necessarily represent

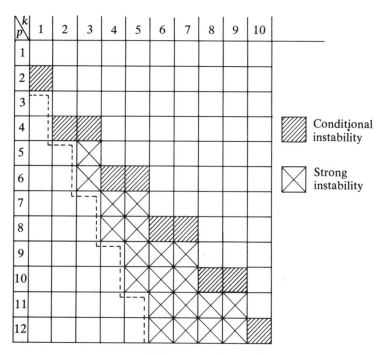

Figure 8.3.1: Diagram showing the limitations implied by Dahlquist's theory on the degree p of linear multistep methods of order k. All squares marked with crosses indicate strongly unstable combinations of p and k and shaded ones correspond to conditionally stable combinations.

strongly stable methods; as will be shown below, it is perfectly possible for a particular combination of p and k to give a range of schemes which exhibit all types of stability characteristics.

8.3.6.   Stability of Multi-Step Methods Applied to the Differential Equation $y' = Ay$

Because of its simple form, this differential equation can be treated directly and the results studied in the light of the general theory already described. For this case, equation (8.3.1) takes the form

$$\alpha_0 y_n + \alpha_1 y_{n+1} + \ldots + \alpha_k y_{n+k} = Ah\{\beta_0 y_n + \beta_1 y_{n+1} + \ldots + \beta_k y_{n+k}\} \quad (8.3.17)$$

which can be written as

$$a_k y_{n+k} + a_{k-1} y_{n+k-1} + \ldots + a_1 y_{n+1} + a_0 = 0 \quad (8.3.18)$$

where

$$a_r = \alpha_r - Ah\beta_r, \qquad 0 \le r \le k.$$

The solution of this linear difference equation is

$$y_n = C_1 Z_1^n + C_2 Z_2^n + \ldots + C_k Z_k^n \quad (8.3.19)$$

where Z_i $(1 \le i \le k)$ are the roots (assumed distinct) of the equation

$$a_k Z^k + a_{k-1} Z^{k-1} + \ldots + a_1 Z + a_0 = 0 \quad (8.3.20)$$

and c_i are constants determined by initial conditions [see I, § 14.13].

If one of the roots, say Z_1, has multiplicity m then the solution of (8.3.18) is

$$y_n = (d_{m-1} n^{m-1} + d_{m-2} n^{m-2} + \ldots + d_1 n + a_0) Z_1^n + C_{m+1} Z_{m+1}^n + \ldots + C_k Z_k^n.$$

In view of the definition of a_r, equation (8.3.19) can be written in the form

$$\rho(Z) - Ah\sigma(Z) = 0$$

where $\rho(Z)$ and $\sigma(Z)$ are defined in (8.3.9) and when $h \to 0$ this equation becomes simply

$$\rho(Z) = 0. \quad (8.3.21)$$

If the multi-step scheme (8.3.1) satisfies the first of the consistency conditions (8.3.10) then $\rho(1) = 0$, so that one of the roots of (8.3.21) is unity. Now the zeros of a polynomial are continuous functions of the coefficients $\{a_r\}$ (see for example Isaacson and Keller (1966), p. 24). Furthermore, from the theory of functions of a complex variable we have the following:

If $Z = Z_0$ is a simple zero of the polynomial $P(Z)$, then for $|\varepsilon|$ sufficiently small, the polynomial $P(Z) - \varepsilon Q(Z)$ (where $P(Z)$ and $Q(Z)$ are both of degree k) has a zero $Z_0(\varepsilon)$, such that

$$\left| Z_0(\varepsilon) - Z_0 + \frac{\varepsilon Q(Z_0)}{P'(Z_0)} \right| = O(\varepsilon^2).$$

For the particular case under consideration, application of this result gives

$$\left| Z_1 - 1 + hA\frac{\sigma(1)}{\rho'(1)} \right| = O(h^2)$$

where Z_1 is a solution of equation (8.3.20). On using the consistency condition (8.3.10) it follows that

$$Z_1 = 1 + Ah + O(h^2)$$

and applying a limiting process similar to that used in equation (8.2.24)–(8.2.26) it can be seen that this is the solution of the difference equation which corresponds to the solution of the differential equation $y' = Ay$. If this solution dominates the remaining roots of (8.3.20), then provided h is sufficiently small the corresponding multi-step method is either strongly or conditionally stable. But if one or more of the extraneous solutions $Z_2, \ldots, Z_k$, (which occur because of the order of the difference equation (8.3.18)) dominate the root Z_1, then the method will be unstable. Since the lemma quoted above shows the manner in which the zeros of the polynomial $\rho(Z) - Ah\sigma(Z)$ approach those of $\rho(Z)$ as $h \to 0$, it is possible in a large class of problems to define the stability properties of multi-step methods in terms of zeros of the polynomial $\rho(Z)$. Only in the cases when $\rho(Z)$ has more than one distinct zero on the unit circle is it necessary to examine the zeros of the polynomial $\rho(Z) - Ah\sigma(Z)$.

We summarize as follows:

If $\xi_1, \xi_2, \ldots, \xi_k$ are the zeros of the polynomial $\rho(\xi)$, so that

$$Z_r \to \xi_r \quad \text{as } h \to 0$$

and so $\xi_1 = 1$, then:

   (i) If $|\xi_r| < 1$ for $r = 2, 3, \ldots, k$ the linear multi-step method is strongly stable.
   (ii) If $|\xi_r| > 1$ for at least one r, $(2 \le r \le k)$, the method is strongly unstable.
   (iii) If $\rho(\xi)$ possess a multiple root of modulus unity the method is strongly unstable.
   (iv) If $|\xi_r| = 1$ for at least one r, $(2 \le r \le k)$, then it is necessary to examine the zeros of the polynomial $\rho(Z) - Ah\sigma(Z)$ to see whether
      (a) All zeros Z_r $(2 \le r \le k)$ satisfy $|Z_r| \le |Z_1|$ for h sufficiently small in which case the method is *weakly stable*

   or

      (b) at least one zero Z_r $(2 \le r \le k)$ exists such that $|Z_r| > |Z_1|$ for h sufficiently small, in which case the method is *weakly unstable*.

As a final illustration the case $p = 3$, $k = 2$ is now examined. With these values of p and k, equation (8.3.7) and (8.3.8) give the following system of equations for the coefficient $\{\alpha_r, \beta_r\}$ $(0 \le r \le 2)$:

$$\alpha_0 + \alpha_1 + \alpha_2 = 0$$
$$\alpha_1 + 2\alpha_2 = (\beta_0 + \beta_1 + \beta_2)$$
$$\alpha_1 + 4\alpha_2 = 2(\beta_1 + 2\beta_2)$$
$$\alpha_1 + 8\alpha_2 = 3(\beta_1 + 4\beta_2).$$

Setting $\alpha_2 = 1$, and using $\beta_2 \equiv \theta$ as a parameter the corresponding multi-step scheme is

$$y_{n+2} + 4(1 - 3\theta)y_{n+1} + (12\theta - 5)y_n = Ah\{\theta y_{n+2} + 4(1 - 2\theta)y_{n+1} + (2 - 5\theta)y_n\}$$

so that

$$\rho(\xi) \equiv \xi^2 + 4(1 - 3\theta)\xi + 12\theta - 5$$

and

$$\sigma(\xi) \equiv \theta\xi^2 + 4(1 - 2\theta)\xi + 2 - 5\theta.$$

The zeros ξ_i of $\rho(\xi)$ are

$$\xi_1 = 1$$
$$\xi_2 = 12\theta - 5.$$

The method is strongly stable if $|\xi_2| < 1$, i.e. $\frac{1}{3} < \theta < \frac{1}{2}$ and strongly unstable if $\theta < \frac{1}{3}$ or $\theta \ge \frac{1}{2}$. Note that the value $\theta = \frac{1}{2}$ gives $\xi_2 = 1$, so that $\rho(\xi)$ has a double root and (iii) applies. In particular the method is strongly unstable when $\theta = 0$ which is an example of result (iv) of Dahlquist [see § 8.3.5], since the value of θ leads to an open formula. When $\theta = \frac{1}{3}$, $\xi_2 = -1$ and conditional stability occurs. Reference to equation (8.3.8), however, shows that this value of θ in fact corresponds to the case $p = 4$, $k = 2$ since the coefficients also satisfy the equation

$$\alpha_1 + 16\alpha_2 = 4\{\beta_1 + 8\beta_2\}$$

and the resulting formula is the Milne–Simpson corrector formula, which has already been discussed [see (8.3.11)]. Setting $Ah/3 \equiv \phi$, and assuming h small, the zeros of $\rho(Z) - Ah\sigma(Z)$ can be shown to be

$$Z_1 = 1 + 3\phi + O(\phi^2)$$
$$Z_1 = -1 + \phi + O(\phi^2).$$

If $\phi > 0$ then $|Z_2| < 1 < |Z_1|$ and the method is weakly stable. If $\phi < 0$ then $|Z_2| > 1 > |Z_1|$ and the method is weakly unstable and the results of a numerical example where $A = \pm 3$ have already been given in Tables 8.3.1 and 8.3.2.

8.3.7. Variable Order Variable Step Multi-Step Methods

The theory of multi-step methods described has been restricted to those with constant stepsize and constant order. However, a good deal of effort is currently being expended into developing methods which permit both the stepsize and order to vary as the integration proceeds. These methods give very good results when applied to a wide class of problems but a comprehensive theoretical study of such methods has not yet been achieved. The rules for selecting the order have largely been determined experimentally, although Krogh (1968) has found that the efficiency of the method is relatively insensitive to the choice of rule.

Krogh's method is based on the use of Adams predictor–corrector formulae [see (8.3.5)] with orders varying from 1 to 13 for first order systems. The formulae are given in terms of backward differences [see § 1.4] and at each stage the order of the corrector is taken to be one higher than that of the predictor. After each step the local truncation error [see § 1.3] is estimated and this is used to decide whether the stepsize or order is changed. The aim is to adjust the order so that the stepsize is always as large as possible. Initially the method starts with a predictor of order one, and this builds up as additional points are calculated; one effect of this is to impose a very small stepwidth initially in order that the error tolerance criterion is satisfied. It is significant that Krogh also found that multi-step methods which directly integrate equations of the form $y'' = f(x, y)$ are almost free from the instability problems which are encountered when multi-step methods are applied to the equivalent system of first order equations. In the particular example $y'' = -y$, $y(0) = 0$, $y'(0) = 1$, it was found possible to go to values of 15 or 16 for the order k and in consequence, a larger stepwidth was permitted. Special methods for particular equations will be discussed in more detail below.

A similar approach has been tried by Gear (1971) (page 158) who used a Taylor series representation of Adams predictor–corrector formulae [see § 8.3.1]; in general this approach appears to be slightly less successful than methods based on either divided differences [see § 1.5] or backward differences, although giving better results than the fixed order fixed step size methods described earlier.

Finally, a recent book by Shampine and Gordon (1975) gives an extensive treatment of a variable order variable step method, using divided differences to represent Adams predictor–corrector formulae. Problems of stability convergence, the detection of stiffness and criteria for changing step size and order are discussed at length, and the interested reader should refer to the book for full details.

8.4. EXTRAPOLATION METHODS

A different approach has been adopted by Bulirsch and Stoer (1966) based on the idea of Richardson's extrapolation to the limit [see § 2.4.1 and Richardson (1927)]. An elementary example of the use of this technique has already been given in connection with the solution to a simple initial value problem, using Euler's method with a set of different step widths $\{h_i\}$ [see § 8.2.1]. The basic idea in Bulirsch and Stoer's method is to integrate the differential equation over an interval several times, using a succession of step widths $\{h_i\}$, a strictly decreasing sequence tending to zero, and apply extrapolation over that interval before proceeding further with the integration.

Extrapolation techniques can be used to accelerate the convergence of a sequence of approximations $\{T(h_i)\}$, say, which have been computed using a sequence $\{h_i\}$ of values for the discretization interval. Assuming that for small values of h the approximation $T(h)$ can be expressed as

$$T(h) = T_0 + T_1 h^{p_1} + T_2 h^{p_2} + \ldots$$

where the coefficients T_r are independent of h, then two calculations using h_0 and h_1, giving $T(h_0)$ and $T(h_1)$ respectively can be used to eliminate T_1, and each successive value of h_i allows the elimination of succeeding coefficients T_i. In particular, if $h_1 = \frac{1}{2}h_0$ and $p_1 = 2$,

$$T_0 = \tfrac{1}{3}\{4T(h_1) - T(h_0)\}$$

and the leading error term is $T_2 h^{p_2}$. Effectively, the method involves fitting a polynomial in h to the successive approximations [see Chapter 6] and then extrapolating it to the value zero for h. The method of Bulirsch and Stoer uses a rational function rather than a polynomial, for extrapolation and this is found to lead to a somewhat more efficient algorithm; a complete Algol program is published in their paper and a Fortran implementation is given by Fox (1971), Chapter 9 pp. 477–507.

The particular method of integration adopted, the modified mid-point method [cf. § 7.1.4], is well suited to this approach and Bulirsch and Stoer use the sequence of step widths $h, h/2, h/4, h/8, \ldots$. The integration from x_0 to $x_0 + h$ proceeds as follows:

Letting $N \equiv 2^n$, $h_n = h/N$ and $x_r = x_0 + rh_n$, the following values are calculated:

$$y_1 = y_0 + h_n f_0$$

$$y_{r+1} = y_{r-1} + 2h_n f_r, \qquad 1 \le r \le N-1$$

$$Y_N = y_{N-1} + h_n f_N$$

$$T(x, h_n) = \tfrac{1}{2}(y_N + Y_N).$$

where $f_r = f(x_r, y_r)$.

The first step is necessary to start the process, and a similar step at the end gives a second estimate Y_N of the value of the dependent variable at the end of the interval; the value used by the extrapolation process is the mean of these two end values y_N, Y_N. With suitable differentiability assumptions, the asymptotic expansion of $T(x, h_n)$ has the form

$$T(x, h_n) = y(x) + t_1(x)h^2 + t_2(x)h^4 + \ldots$$

where $y(x)$ is the required solution. In the implementation of Fox (1971) a rational extrapolation of order 6 is used, so that at each step up to 6 applications of the midpoint rule are computed for successively smaller values of h and extrapolation to $h = 0$; Bulirsch and Stoer have adopted a limit of 9 extrapolation steps.

8.5. METHODS FOR SPECIAL EQUATIONS

Although all the methods described so far can be immediately generalized to a system of first order equations

$$y_i' = f_i(x; y_1, y_2, \ldots, y_r) \qquad 1 \le i \le n,$$

[see IV, §§ 7.9 and 7.11] in practice more efficient methods are available for equations, or systems of equations of the form

$$y_i^{(p)} = f_i(x; y_1, y_2, \ldots, y_r) \quad \text{where } p > 1$$

which avoid the necessity for rewriting them as a system of first order equations as in (8.1.10). Such equations occur widely in a range of applications, and in particular the differential equation

$$y'' = f(x, y)$$

arises in orbit theory, satellite trajectory calculations, the problem of primary cosmic rays and solutions to the Schroedinger equation. A special method for this equation has been devised independently by several authors, and it is usually referred to as the *Cowell–Numerov method*, following Numerov's papers in the 1920's. However the method had already been used by Carl Störmer in 1906 to analyse the aurora borealis, and Cowell and Crommelin used a variant of the method to predict the return of Halley's comet in 1910. The most straightforward way to produce the scheme is to use Taylor's series [see IV, § 3.6]

$$y_{n+1} = y_n + hy_n' + \frac{h^2}{2} y_n'' + \frac{h^3}{3!} y_n''' + \ldots$$

$$y_{n-1} = y_n - hy_n' + \frac{h^2}{2} y_n'' - \frac{h^3}{3!} y_n''' + \ldots$$

so that

$$y_{n+1} - 2y_n + y_{n-1} = h^2 y_n'' + \frac{h^4}{12} y_n^{(iv)} + \frac{h^6}{360} y_n^{(vi)} + O(h^8). \qquad (8.5.1)$$

Now from the differential equation we have $y_n'' = f(x_n, y_n) \equiv f_n$ and it follows that

$$h^2 f_{n+1} = h^2 y_n'' + h^3 y_n''' + \frac{h^4}{2} y_n^{(iv)} + \ldots$$

$$h^2 f_{n-1} = h^2 y_n'' - h^3 y_n''' + \frac{h^4}{2} y_n^{(iv)} - \ldots$$

giving

$$h^2(\alpha f_{n+1} + \beta f_n + \alpha f_{n-1}) = h^2(2\alpha + \beta)y_n'' + \alpha h^4 y^{(iv)} + \frac{\alpha h^6}{12} y^{(vi)} + O(h^8)$$

On setting $\alpha = 1, \beta = 10$,

$$h^2 y_n'' = \frac{h^2}{12}(f_{n+1} + 10f_n + f_{n-1}) - \frac{h^4}{12} y_n^{(iv)} - \frac{h^6}{144} y_n^{(vi)} + O(h^8)$$

and equation (8.5.1) becomes

$$y_{n+1} = 2y_n - y_{n-1} + \frac{h^2}{12}(f_{n+1} + 10f_n + f_{n-1}) \tag{8.5.2}$$

with a local truncation error term of $-h^6 y_n^{(vi)}/240$.

In general (8.5.2) is an implicit scheme, since the term f_{n+1} on the right-hand side of the equation involves the unknown y_{n+1}. However, for an important sub-class of equations of the form

$$y'' = g(x)y$$

the Cowell–Numerov method takes the particularly simple form

$$y_{n+1} = \frac{1}{1 - (h^2 g_{n+1}/12)}\left[2\left(1 + \frac{5h^2}{12} g_n\right)y_n - \left(1 - \frac{h^2}{12} g_{n-1}\right)y_{n-1}\right] \tag{8.5.3}$$

with the leading truncation error term of

$$\frac{-h^6 y_n^{(vi)}}{240(1 - (h^2/12)g_{n+1})}.$$

In this case, each step in the integration consists of the following sequence:
Calculate

(i) $g_{n+1} \equiv g(x_{n+1})$
(ii) $A_{n+1} = 1 - \theta g_{n+1}$ (where $\theta = h^2/12$)
(iii) $B_{n+1} = 2 + \phi g_{n+1}$ (where $\phi = 10h^2/12$)
(iv) $A_{n+1}y_{n+1} = B_n y_n - A_{n-1}y_{n-1}$
(v) y_{n+1}
(vi) $B_{n+1}y_{n+1}$.

The quantities $A_{n+1}y_{n+1}, B_{n+1}y_{n+1}$ are stored in readiness for the next cycle, so that each integration step requires only three additions, three subtractions, three multiplications and one division in addition to one evaluation of the function $g(x)$. The Cowell–Numerov method is thus extremely fast except in cases where a change in the step width is required. As in the case of linear multi-step methods for equations of the first order, a comprehensive theory of stability and convergence exists, and a full account is given in Henrici (1962).

A similar scheme to the Cowell–Numerov method can be derived for the fourth order differential equation

$$\frac{d^4 y}{dx^4} = f(x, y).$$

Again, the use of Taylor's series provides a straightforward way of obtaining the appropriate formula and as in the second-order equation, the method of solution is considerably simplified if $f(x, y) = g(x)y$.

M.H.R.

REFERENCES

Bulirsch, R. and Stoer, J. (1966). Numerical Treatment of Ordinary Differential Equations by Extrapolation Methods, *Numerische Mathematik*. **8**, 1–13.

Dahlquist, G. (1956). *Mathematica Scandanavicia*, Vol. 4, pp. 33–53.

Fehlberg, H. (1968). Classical Fifth-, Sixth-, Seventh-, and Eighth-Order Runge–Kutta Formulas with Stepsize Control, *NASA TR R-287*, Huntsville, Alabama.

Fox, P. A. (1971). *Mathematical Software*, Edited J. R. Rice, Academic Press, New York.

Gear, C. W. (1971). *Numerical Initial-Value Problems in Ordinary Differential Equations*, Prentice-Hall, Englewood Cliffs, N.J.

Hamming, R. W. (1962). *Numerical Methods for Engineers and Scientists*, McGraw Hill, New York.

Henrici, P. (1962). *Discrete Variable Methods in Ordinary Differential Equations*, J. Wiley and Sons, New York.

Isaacson, E. and Keller, H. B. (1966). *Analysis of Numerical Methods*, J. Wiley.

Krogh, F. T. A Variable Stop Variable Order Multistep Method for the Numerical Solution of Ordinary Differential Equations, *Proc. Information Processing 68*, Vol. 1 (A. J. H. Morrell, Editor) North Holland Publishing Co., Amsterdam, The Netherlands. pp. 194–199.

Merson, R. H. (1957). An Operational Method for the Study of Integration Processes, *Proc. Symp. Data processing Weapons Research Establishment*, Salisbury, S. Australia.

NAG Library (1978). NAG Library Manual Mk. 6. NAG Ltd., Oxford.

Phillips, G. M. and Taylor, P. J. (1972). *Theory and Applications of Numerical Analysis*, Academic Press, New York.

Richardson, L. F. and Gaunt, J. (1927). The Deferred Approach to the Limit, *Phil. Trans. A*. **226**, 299–361.

Shampine, L. F. and Gordon, M. K. (1975). *Computer Solution of Ordinary Differential Equations: The Initial Value Problem*, W. H. Freeman, San Francisco, Calif.

Shampine, L. F. (1977). Stiffness and Nonstiff Differential Equation Solvers, II: Detecting Stiffness with Runge–Kutta Methods, *ACM Transactions on Mathematical Software*, **3**, 1, 44–53.

Stoker, J. J. (1950). *Non Linear Vibrations*, Interscience Publishers Inc., New York.

Wasow, W. (1965). *Asymptotic Expansions for Ordinary Differential Equations*, Interscience, New York.

PROGRAMS

In the Appendix there are programs for the use of the Runge–Kutta Method: (14) for a first-order differential equation, (15) for a second-order differential equation. There is also code for the Runge–Kutta–Fehlberg method of order 5.

CHAPTER 9

Partial Differential Equations

9.1. INTRODUCTION

The purpose of this chapter is to present some of the popular methods for the numerical solution of partial differential equations. In this section it is intended to introduce the reader to some of the basic properties which will be used later. It is not intended to give rigorous mathematical proofs but merely to indicate some results that will enable the reader to follow the methods outlined in the ensuing sections. Further details can be found in texts such as Smith (1965) and Ames (1977). For the analytical solution of these equations see Volume IV, Chapter 8.

Consider the function $u(x, y)$ satisfying the quasi-linear first-order partial differential equation

$$a\frac{\partial u}{\partial x}+b\frac{\partial u}{\partial y}=c, \tag{9.1.1}$$

where a, b and c are, in general, functions of x, y and u. For a solution to be possible we require that the values of u be specified along some initial curve C (in many cases C will be a part of the x-axis and then u will be given as a function of x on $y = 0$).

It is anticipated that (9.1.1) will have to be solved by numerical integration and it would be convenient to integrate in one direction only with no disturbances from derivatives in other directions. Thus we seek a curve Γ called the *characteristic* in the x, y-plane along which this can be achieved. Suppose that such a curve is given 'parametrically' by $x = x(s)$, $y = y(s)$; thus u reduces to a function of s only, i.e. $u(s) = u(x(s), y(s))$. It follows that [see IV, § 5.4]

$$\frac{du}{ds}=\frac{\partial u}{\partial x}\frac{dx}{ds}+\frac{\partial u}{\partial y}\frac{dy}{ds}. \tag{9.1.2}$$

Eliminating $\partial u/\partial x$ from (9.1.1) and (9.1.2) we obtain that

$$c\frac{dx}{ds}-a\frac{du}{ds}=\left(b\frac{dx}{ds}-a\frac{dy}{ds}\right)\frac{\partial u}{\partial y}. \tag{9.1.3}$$

The effect of $\partial u/\partial y$ can be eliminated provided that we choose Γ such that

$$b\frac{dx}{ds} = a\frac{dy}{ds} = \lambda, \quad \text{say.} \tag{9.1.4}$$

Equations (9.1.3) and (9.1.4) give

$$\frac{dx}{a} = \frac{dy}{b} = \frac{du}{c} \quad (=\lambda\, ds). \tag{9.1.5}$$

Those familiar with matrix theory will recognize conditions (9.1.5) as equivalent to stating that the matrix

$$\begin{pmatrix} a & b & c \\ dx & dy & du \end{pmatrix}$$

must be of rank unity [see I, § 5.6]. Thus all 2×2 determinants obtained from the matrix must be zero [see I, Prop. 6.13.1]. For example we have

$$\begin{vmatrix} a & b \\ dx & dy \end{vmatrix} = 0,$$

that is $a\,dy - b\,dx = 0$; or

$$\frac{dx}{a} = \frac{dy}{b}.$$

Relations (9.1.5) will be used in section 9.2 to obtain a solution to equation (9.1.1) provided that u is given on some initial curve C in the x, y-plane.

Many of the problems encountered in engineering and science are of second order and take the general quasi-linear form

$$a\frac{\partial^2 u}{\partial x^2} + b\frac{\partial^2 u}{\partial x\,\partial y} + c\frac{\partial^2 u}{\partial y^2} = e, \tag{9.1.6}$$

where a, b, c and e are functions of $x, y, u, \partial u/\partial x$ and $\partial u/\partial y$. For a solution to be possible we again require some initial condition specified. The analysis of the first-order equation can be extended to second-order equations. That is along some curve Γ in the x, y-plane, given parametrically by $x = x(s)$ and $y = y(s)$, we have that

$$\frac{du}{ds} = \frac{\partial u}{\partial x}\frac{dx}{ds} + \frac{\partial u}{\partial y}\frac{dy}{ds}, \tag{9.1.7}$$

together with

$$\frac{d}{ds}\left(\frac{\partial u}{\partial x}\right) = \frac{\partial^2 u}{\partial x^2}\frac{dx}{ds} + \frac{\partial^2 u}{\partial y\,\partial x}\frac{dy}{ds}, \tag{9.1.8}$$

and

$$\frac{d}{ds}\left(\frac{\partial u}{\partial y}\right) = \frac{\partial^2 u}{\partial x\,\partial y}\frac{dx}{ds} + \frac{\partial^2 u}{\partial y^2}\frac{dy}{ds}. \tag{9.1.9}$$

Eliminating $(\partial^2 u/\partial x^2)$ and $(\partial^2 u/\partial y^2)$ from (9.1.6), (9.1.8) and (9.1.9) we find that

$$\left[a\left(\frac{dy}{ds}\right)^2 - b\left(\frac{dx}{ds}\right)\left(\frac{dy}{ds}\right) + c\left(\frac{dx}{ds}\right)^2 \right] \frac{\partial^2 u}{\partial x \, \partial y}$$

$$= a\left(\frac{dy}{ds}\right)\frac{d}{ds}\left(\frac{\partial u}{\partial x}\right) - e\left(\frac{dy}{ds}\right)\left(\frac{dx}{ds}\right) + c\left(\frac{dx}{ds}\right)\frac{d}{ds}\left(\frac{\partial u}{\partial y}\right). \qquad (9.1.10)$$

Thus we can eliminate the effect of $(\partial^2 u/\partial x \partial y)$ provided that Γ is such that

$$a\left(\frac{dy}{ds}\right)^2 - b\left(\frac{dx}{ds}\right)\left(\frac{dy}{ds}\right) + c\left(\frac{dx}{ds}\right)^2 = 0.$$

Unless $dx/ds = 0$ this can be written as a quadratic equation in dy/dx

$$a\left(\frac{dy}{dx}\right)^2 - b\left(\frac{dy}{dx}\right) + c = 0 \qquad (9.1.11)$$

and defines the characteristics [see I, Prop. 14.5.4]. The right-hand side of (9.1.10) must now be zero, that is

$$e \, dy - a\frac{dy}{dx} d\left(\frac{\partial u}{\partial x}\right) - cd\left(\frac{\partial u}{\partial y}\right) = 0. \qquad (9.1.12)$$

For a point (x, y) associated with given values of u and $\partial u/\partial x$, there will be two directions for which (9.1.11) is satisfied provided that $b^2 > 4ac$; in this case the partial differential equation is classified as 'hyperbolic' [see IV, § 8.5]. The simplest and best known example of a hyperbolic partial differential equation is the wave equation [see IV, § 8.1] which, in non-dimensional form, is given by

$$\frac{\partial^2 u}{\partial x^2} - \frac{\partial^2 u}{\partial y^2} = 0.$$

If $b^2 = 4ac$, then the characteristics are coincident and the partial differential equation is termed 'parabolic' [see IV, § 8.5], for example the non-dimensional diffusion equation [see IV, § 8.1]

$$\frac{\partial^2 u}{\partial x^2} = \frac{\partial u}{\partial y}.$$

The condition $b^2 < 4ac$ gives complex roots to the quadratic equation (9.1.11) and the partial differential equation is classed as 'elliptic' [see IV, § 8.5], the classic example being Laplace's equation [see IV, § 8.1]

$$\frac{\partial^2 u}{\partial x^2} + \frac{\partial^2 u}{\partial y^2} = 0.$$

It should be noted that the class to which the equation belongs may depend on the solution or on the region in which the solution is being sought.

For example

$$u\frac{\partial^2 u}{\partial x^2}+\frac{\partial^2 u}{\partial y^2}=0,$$

where $a = u$, $b = 0$, $c = 1$, is elliptic in any region over which the solution u is positive and hyperbolic in any region over which u is negative. On the other hand

$$\frac{\partial^2 u}{\partial x^2}+2x\frac{\partial^2 y}{\partial x\,\partial y}+(1-y^2)\frac{\partial^2 u}{\partial y^2}=0,$$

has $a = 1$, $b = 2x$, $c = 1 - y^2$ and hence $b^2 - 4ac$ becomes $4x^2 - 4(1 - y^2)$ or $4(x^2 + y^2 - 1)$. Thus it is hyperbolic outside the unit circle (i.e. $x^2 + y^2 > 1$) and elliptic inside the unit circle.

Hyperbolic equations can be solved by the method of characteristics using equations (9.1.11) and (9.1.12) together with (9.1.7) as detailed in section 9.2. Parabolic and elliptic equations, however, cannot be solved by this method and other techniques are required.

A popular numerical method for solving all types of partial differential equations is to replace the differential coefficients by finite-difference approximations [see § 1.4]. If $(x_j, y_k) = (x_0 + j\,\Delta x, y_0 + k\,\Delta y)$ is a given point with reference to some origin (x_0, y_0) and Δx and Δy are constant increments in the x and y directions respectively, then we denote the solution $u(x_j, y_k)$ at this point by u_j^k and terms such as $\partial u/\partial x$ evaluated at (x_j, y_k) by $\partial u_j^k/\partial x$. Assuming that $u(x, y)$ and its derivatives of sufficiently high order are single-valued, finite and continuous functions of x and y [see IV, Definition 5.1.1], we have by Taylor's expansion [see IV, Theorem 5.8.1]

$$u_{j\pm 1}^k = \left\{1 \pm \Delta x\frac{\partial}{\partial x}+\frac{(\Delta x)^2}{2}\frac{\partial^2}{\partial x^2}\pm\frac{(\Delta x)^3}{6}\frac{\partial^3}{\partial x^3}+\ldots\right\}u_j^k. \qquad (9.1.13)$$

Adding and rearranging these two expansions we have

$$(\Delta x)^2\frac{\partial^2 u}{\partial x^2}\bigg|_{x_j,y_k} = u_{j-1}^k - 2u_j^k + u_{j+1}^k + O[(\Delta x)^4], \qquad (9.1.14)$$

Similarly we have

$$(\Delta y)^2\frac{\partial^2 u}{\partial y^2}\bigg|_{x_j,y_k} = u_j^{k-1} - 2u_j^k + u_j^{k+1} + O[(\Delta y)^4]. \qquad (9.1.15)$$

Expansion (9.1.13) also gives

$$u_{j\pm 1}^k = u_j^k \pm \Delta x\frac{\partial u}{\partial x}\bigg|_{x_j,y_k} + O[(\Delta x)^2],$$

that is the forward-difference approximation

$$\Delta x\frac{\partial u}{\partial x}\bigg|_{x_j,y_k} = u_{j+1}^k - u_j^k + O[(\Delta x)^2], \qquad (9.1.16)$$

and the backward-difference approximation

$$\Delta x \frac{\partial u}{\partial x}\bigg|_{x_j, y_k} = u_j^k - u_{j-1}^k + O[(\Delta x)^2].$$

(9.1.17)

The more accurate central-difference approximation

$$2(\Delta x) \frac{\partial u}{\partial x}\bigg|_{x_j, y_k} = u_{j+1}^k - u_{j-1}^k + O[(\Delta x)^3],$$

(9.1.18)

can be derived by subtracting the two expansions in expression (9.1.13).

Obviously similar expansions can be obtained for $\Delta y(\partial u/\partial y)$ at x_j, y_k.

By replacing the differential coefficients in the partial differential equation under consideration we can obtain a difference equation which can be solved. For example taking the diffusion equation

$$\frac{\partial^2 u}{\partial x^2} = \frac{\partial u}{\partial y},$$

we replace $\partial u/\partial y|_{x_j, y_k}$ by $(u_j^{k+1} - u_j^k)/\Delta y + O(\Delta y)$ (cf. equation 9.1.16) and replace $\partial^2 u/\partial x^2|_{x_j, y_k}$ by $(u_{j-1}^k - 2u_j^k + u_{j+1}^k)/\Delta x^2 + O(\Delta x)^2$ to give

$$u_j^{k+1} = \frac{\Delta y}{(\Delta x)^2}(u_{j-1}^k + u_{j+1}^k) + \left(1 - \frac{2\Delta y}{(\Delta x)^2}\right)u_j^k + O[\Delta y(\Delta x)^2, (\Delta y)^2].$$

(9.1.19)

If we know the three values u_{j-1}^k, u_j^k and u_{j+1}^k then we have a direct or explicit method for evaluating u_j^{k+1} (see § 9.4).

It is our aim to illustrate the use of the above methods by solving a number of differential equations that are of common occurrence in physical situations.

9.2.   HYPERBOLIC EQUATIONS I—METHOD OF CHARACTERISTICS

To familiarize ourselves with the idea of characteristics we proceed by considering the first-order differential equation

$$\frac{\partial u}{\partial x} + 2\frac{\partial u}{\partial y} = 1,$$

(9.2.1)

with initial values of u specified along some curve C.

The characteristic equation is $dy/dx = 2$ and the characteristics are the family of straight lines $y = 2x + A$, where A is a constant, and are illustrated in Figure 9.2.1.

Along any characteristic we have from equation (9.1.5)

$$du = dx$$

that is

$$u = x + B.$$

(9.2.2)

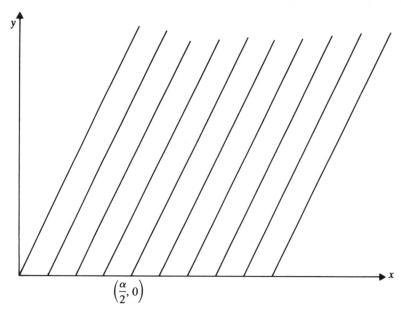

$$\left(\frac{\alpha}{2}, 0\right)$$

Figure 9.2.1

Let us now assume some initial conditions along various initial curves.

(i) Suppose $u = x(x-1)$ on $y = 0$, $0 \le x \le 1$. That is, C is that part of the x-axis in the range $[0, 1]$.

The characteristic $y = 2x - \alpha$ cuts the initial curve when $x = \alpha/2$, and at this point u is given as $\alpha(\alpha - 2)/4$.

Substituting into (9.2.2) we find $B = \alpha(\alpha - 4)/4$, giving the solution at all points along the characteristic $y = 2x - \alpha$ as

$$u = x + \alpha(\alpha - 4)/4.$$

By varying the parameter α we have the solution along all characteristics. We note, however, that if $\alpha < 0$ or $\alpha > 2$, then the corresponding characteristic cuts the initial line (the x-axis) at a point where the initial solution is not given and we cannot obtain the solution along this characteristic. That is, we cannot obtain a solution outside the bounding characteristics $y = 2x$ and $y = 2(x-1)$.

(ii) Suppose that u is given on the line $y = 2x$. That is, C, the initial curve, is the characteristic through the origin. In this case the bounding characteristics are coincident and the only solution is along $y = 2x$.

(iii) Suppose that the initial condition is given by

$$u = \begin{cases} x, & 0 \le x < \tfrac{1}{2}, \\ 0, & \tfrac{1}{2} < x \le 1, \end{cases} \qquad y = 0.$$

In this case C is the x-axis in the range $[0, 1]$; but there is a discontinuity in u at $x = \tfrac{1}{2}$ which will persist along the characteristic $y = 2x - 1$. We can obtain the solution to the left and to the right of this characteristic but not along the characteristic.

In the above illustration we were able to solve $dy/dx = b/a$ and $du/c = dx/a$, analytically. This is not always so and in such cases we have to resort to numerical integration of the ordinary differential equations arising. To illustrate this point we take the example given in Smith ((1965), Chapter 4, Exercise 5b) where we attempt to evaluate $u(1{\cdot}1, 0{\cdot}1)$ given that

$$x\frac{\partial u}{\partial x} + u\frac{\partial u}{\partial y} = x + y,$$

with

$$u = 1 \quad \text{on } y = 0.$$

It follows from equation (9.1.5) that

$$\frac{dx}{x} = \frac{dy}{u} = \frac{du}{x+y}.$$

Hence the characteristics are given by

$$u\,dx = x\,dy,$$

and the solution by

$$(x + y)\,dx = x\,du.$$

Neither expression is readily integrable.

We know the point $R(1{\cdot}1, 0{\cdot}1)$ at which the solution is required but we do not know where the characteristic through R cuts the x-axis (the initial curve). Let this unknown point be $P(x_p, 0)$. Thus integrating from P to R along the characteristic yields

$$\int_{x_P}^{x_R} u\,dx = \int_{y_P}^{y_R} x\,dy,$$

and

$$\int_{x_P}^{x_R} (x+y)\,dx = \int_{u_P}^{u_R} x\,du.$$

We resort to numerical quadrature to evaluate these integrals. In this case the trapezium rule [§ 7.1.3] seems adequate so long as $|x_R - x_P|$, $|y_R - y_P|$ and $|u_R - u_P|$ do not become too large and produce an unacceptable truncation error [see (7.1.7)].

Thus we have

$$\tfrac{1}{2}(u_R + u_P)(x_R - x_P) = \tfrac{1}{2}(x_R + x_P)(y_R - y_P),$$

and

$$\tfrac{1}{2}(x_R + y_R + x_P + y_P)(x_R - x_P) = \tfrac{1}{2}(x_R + x_P)(u_R - u_P).$$

The initial condition gives $u_P = 1$ and on substituting the known values into the above expressions we have that

$$(u_R + 1)(1 \cdot 1 - x_P) = (1 \cdot 1 + x_P)(0 \cdot 1),$$

$$(1 \cdot 2 + x_P)(1 \cdot 1 - x_P) = (1 \cdot 1 + x_P)(u_R - 1),$$

or

$$(1 \cdot 1 + u_R)x_P = 1 \cdot 1 u_R + 0 \cdot 99, \tag{9.2.3}$$

and

$$(1 \cdot 1 + x_P)u_R = (1 \cdot 2 + x_P)(1 \cdot 1 - x_P) + (1 \cdot 1 + x_P). \tag{9.2.4}$$

The two non-linear equations (9.2.3) and (9.2.4) can be solved iteratively [see § 5.2]. To start the iteration we assume $u_R = u_P$, that is $u_R = 1$. Then (9.2.3) gives $x_P = 0 \cdot 9952$, and substituting this value in (9.2.4) gives $u_R = 1 \cdot 1098$.

Repeating the iteration yields $x_P = 1 \cdot 0004$ and $u_R = 1 \cdot 1043$. Further iterations lead to

$$x_P = 1 \cdot 0002, \qquad u_R = 1 \cdot 1046,$$

$$x_P = 1 \cdot 0002, \qquad u_R = 1 \cdot 1045.$$

Thus to three decimal places we have that $u(1 \cdot 1, 0 \cdot 1) = 1 \cdot 105$.

A problem that arises, and which we shall leave unanswered at this stage, is how do we obtain a solution at a point such as $(1 \cdot 1, 1 \cdot 0)$ where the interval $(y_R - y_P)$ is too large to allow a single step of the trapezium rule?

The method of characteristics can be successfully applied to systems of first-order equations. When solving the pair of equations

$$a\frac{\partial u}{\partial x} + b\frac{\partial u}{\partial y} + c\frac{\partial v}{\partial x} + d\frac{\partial v}{\partial y} = e,$$

$$\alpha\frac{\partial u}{\partial x} + \beta\frac{\partial u}{\partial y} + \gamma\frac{\partial v}{\partial x} + \delta\frac{\partial v}{\partial y} = \varepsilon,$$

the analysis carried out previously and suitably extended indicates that the characteristics (in this case there are two) and the solutions (u and v) are given by demanding that the matrix

$$\begin{bmatrix} a & b & c & d & e \\ \alpha & \beta & \gamma & \delta & \varepsilon \\ dx & dy & 0 & 0 & du \\ 0 & 0 & dx & dy & dv \end{bmatrix}$$

be of rank 3.

To illustrate the technique we solve the equations

$$\frac{\partial u}{\partial y} + \frac{\partial v}{\partial x} = -u, \qquad \frac{\partial u}{\partial x} + \frac{\partial v}{\partial y} = 0, \tag{9.2.5}$$

with $u = x$, $v = 0$ on $y = 0$, $0 \le x \le 1$.

The characteristics are given by

$$\begin{vmatrix} 0 & 1 & 1 & 0 \\ 1 & 0 & 0 & 1 \\ dx & dy & 0 & 0 \\ 0 & 0 & dx & dy \end{vmatrix} = 0,$$

that is

$$-dx^2 + dy^2 = 0,$$

or

$$\frac{dy}{dx} = \pm 1.$$

One set of characteristics is the family of straight lines having slope 1, and the other set is the family of straight lines having slope -1.

The solutions are given by equations of the form

$$\begin{vmatrix} 0 & 1 & 1 & -u \\ 1 & 0 & 0 & 0 \\ dx & dy & 0 & du \\ 0 & 0 & dx & dv \end{vmatrix} = 0,$$

that is

$$du + \left(\frac{dy}{dx}\right) dv + u\, dy = 0.$$

We now replace (dy/dx) by ± 1 to give the two equations

$$du + dv + u\, dy = 0, \qquad du - dv + u\, dy = 0, \tag{9.2.6}$$

which we solve numerically as follows:

Suppose we require the solution at some given point R. Then we must find where the characteristics through R cut the initial curve, that is find the coordinates of two points P and Q on the initial curve. Alternatively, if we specify P and Q we can find the position of R. For the purpose of this example we shall assume that $x_Q - x_P = 0 \cdot 1$ and working along the x-axis we shall find the solution at all points such as R (see Figure 9.2.2).

The geometry is well defined for this example and we immediately have $x_R = x_P + 0 \cdot 05$, $y_R = 0 \cdot 05$. Integrating the first of (9.2.6) from P to R and the second of (9.2.6) from Q to R using the trapezium rule gives

$$(u_R - u_P) + (v_R - v_P) + 0 \cdot 5(u_R + u_P)(y_R - y_P) = 0$$

$$(u_R - u_Q) - (v_R - v_Q) + 0 \cdot 5(u_R + u_Q)(y_R - y_Q) = 0.$$

Taking x_P to be $0 \cdot 6$ and using the given conditions, that is $u_P = 0 \cdot 6$, $u_Q = 0 \cdot 7$, $v_P = v_Q = 0$ we have that

$$u_R = 0 \cdot 6183, \qquad v_R = -0 \cdot 04875.$$

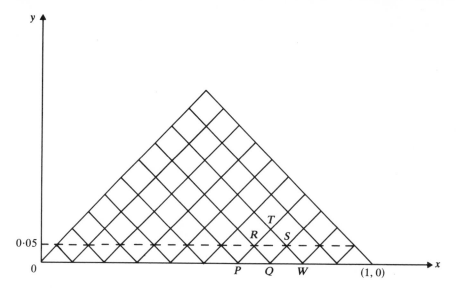

Figure 9.2.2

Proceeding in this way we can also obtain the solution at $S(x_S = 0.75,$ $y_S = 0.05)$, $u_S = 0.7134$, $v_S = -0.04875$ and all such points on the line $y = 0.05$.

Having computed the solution at R and S we can estimate the solution at the next level of intersection T. From the geometry $x_T = x_Q = 0.7$ and $y_T = 2y_R = 0.1$. We now integrate from R to T and from S to T (in this way we keep the step length reasonably small) that is, we consider the initial curve as the line through all points such as R and S ($y = 0.05$).

Once we obtain the solution at all points such as T along the line $y = 0.1$, we can take these as our 'initial condition' for computing the solution at the next level of intersections, and so on. The solution to equation (9.2.6) at the points of intersection of the characteristics shown in Figure 9.2.2 are given in Table 9.2.1.

We note once again that the solution can only be obtained within the triangle bounded by the initial curve ($y = 0$) and the bounding characteristics ($y = x$ and $y = 1 - x$). The solution outside this triangle can be obtained provided that we have boundary conditions specified at $x = 0$ and $x = 1$ as well as the necessary initial conditions.

Let us consider the effect of the boundary conditions $v = \sin(5\pi/6)y$ at $x = 0$, $y > 0$ and $v = 0$ at $x = 1$, $y > 0$.

Figure 9.2.3 shows some of the points outside the 'bounding characteristics' at which the solution can be obtained when boundary conditions are specified.

The known solution at C on the bounding characteristic $y = x$ (given in Table 9.2.1) can be used to give the value of u at A by integrating from C to A along

y		x = 0·05	0·1	0·15	0·2	0·25	0·3	0·35	0·4	0·45	0·5	0·55	0·6	0·65	0·7	0·75	0·8	0·85	0·9	0·95
0·05	u	0·0476		0·1427		0·2378		0·3329		0·4280		0·5232		0·6183		0·7134		0·8085		0·9037
	v	−0·0488		−0·0488		−0·0488		−0·0487		−0·0487		−0·0488		−0·0487		−0·0488		−0·0488		−0·0487
0·10	u		0·0905		0·1810		0·2714		0·3619		0·4524		0·5429		0·6334		0·7239		0·8143	
	v		−0·0951		−0·0951		−0·0951		−0·0951		−0·0951		−0·0951		−0·0951		−0·0951		−0·0951	
0·15	u			0·1291		0·2152		0·3012		0·3873		0·4734		0·5594		0·6455		0·7316		
	v			−0·1392		−0·1392		−0·1392		−0·1392		−0·1392		−0·1392		−0·1392		−0·1392		
0·20	u				0·1637		0·2456		0·3275		0·4093		0·4912		0·5731		0·6550			
	v				−0·1812		−0·1812		−0·1812		−0·1812		−0·1812		−0·1812		−0·1812			
0·25	u					0·1947		0·2726		0·3504		0·4283		0·5062		0·5841				
	v					−0·2211		−0·2211		−0·2211		−0·2211		−0·2211		−0·2211				
0·30	u						0·2222		0·2963		0·3704		0·4445		0·5185					
	v						−0·2591		−0·2591		−0·2591		−0·2591		−0·2591					
0·35	u							0·2466		0·3171		0·3876		0·4580						
	v							−0·2952		−0·2952		−0·2952		−0·2952						
0·40	u								0·2681		0·3351		0·4022							
	v								−0·3295		−0·3295		−0·3295							
0·45	u									0·2869		0·3507								
	v									−0·3622		−0·3622								
0·50	u										0·3032									
	v										−0·3933									

Table 9.2.1

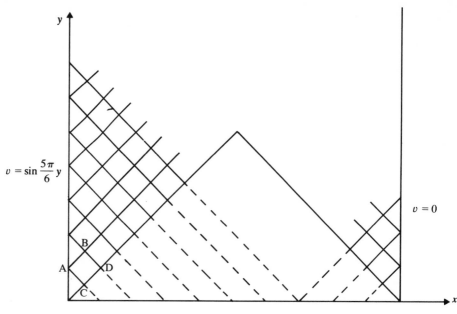

Figure 9.2.3

the characteristic $y = 0 \cdot 1 - x$. Thus

$$(u_A - u_C) - (v_A - v_C) + \tfrac{1}{2}(u_A + u_C)(y_A - y_C) = 0,$$

giving $u_A = 0 \cdot 3453$.

Similarly the solutions at A and D are used to compute u_B and v_B. Some results are given in Table 9.2.2.

The pair of equations (9.2.5) can be reduced to a single second-order equation. By differentiation and elimination of cross derivative terms we obtain

$$\frac{\partial^2 u}{\partial x^2} - \frac{\partial^2 u}{\partial y^2} - \frac{\partial u}{\partial y} = 0, \qquad (9.2.7)$$

where

$$u = x, \quad \frac{\partial u}{\partial y} = -u \quad \text{on } y = 0, \quad 0 \le x \le 1,$$

or

$$\frac{\partial^2 v}{\partial x^2} - \frac{\partial^2 v}{\partial y^2} - \frac{\partial v}{\partial y} = 0, \qquad (9.2.8)$$

where

$$v = 0, \quad \frac{\partial v}{\partial y} = -1 \quad \text{on } y = 0, \quad 0 \le x \le 1,$$

y \ x		0·0	0·05	0·1	0·15	0·2	0·25	0·3	0·35	0·4	0·45	0·5	0·55	0·6	0·65	0·7	0·75	0·8	0·85	0·9	0·95	1·0
0·05	u		0·0476		0·1427		0·2378		0·3329		0·4280		0·5232		0·6183		0·7134		0·8085		0·9037	
	v		−0·0488		−0·0488		−0·0488		−0·0487		−0·0487		−0·0488		−0·0487		−0·0488		−0·0488		−0·0487	
0·10	u	0·3453		0·0905		0·1810		0·2714		0·3619		0·4524		0·5429		0·6334		0·7239		0·8143		0·8120
	v	0·2588		−0·0951		−0·0951		−0·9051		−0·0951		−0·0951		−0·0951		−0·0951		−0·0951		−0·0951		0·0000
0·15	u		0·3799		0·1291		0·2152		0·3012		0·3873		0·4734		0·5594		0·6455		0·7316		0·7271	
	v		0·2061		−0·1392		−0·1392		−0·1392		−0·1392		−0·1392		−0·1392		−0·1392		−0·1392		−0·0464	
0·20	u	0·6481		0·4105		0·1637		0·2456		0·3275		0·4093		0·4912		0·5731		0·6550		0·6485		0·6463
	v	0·5000		0·1557		−0·1812		−0·1812		−0·1812		−0·1812		−0·1812		−0·1812		−0·1812		−0·0907		0·0000
0·25	u		0·6715		0·4375		0·1947		0·2726		0·3504		0·4283		0·5062		0·5841		0·5758		0·5716	
	v		0·4437		0·1076		−0·2211		−0·2211		−0·2211		−0·2211		−0·2211		−0·2211		−0·1328		−0·0443	
0·30	u	0·8957		0·6914		0·4610		0·2222		0·2963		0·3704		0·4445		0·5185		0·5085		0·5025		0·5005
	v	0·7071		0·3897		0·0616		−0·2591		−0·2591		−0·2591		−0·2591		−0·2591		−0·1729		−0·0865		0·0000
0·35	u		0·9097		0·7081		0·4814		0·2466		0·3171		0·3876		0·4580		0·4465		0·4387		0·4349	
	v		0·6480		0·3380		0·0177		−0·2952		−0·2952		−0·2952		−0·2952		−0·2111		−0·1268		−0·0423	
0·40	u	1·0780		0·9207		0·7220		0·4988		0·2681		0·3351		0·4022		0·3892		0·3799		0·3743		0·3724
	v	0·8660		0·5912		0·2884		−0·0243		−0·3295		−0·3295		−0·3295		−0·2475		−0·1652		−0·0826		0·0000
0·45	u		1·0846		0·9290		0·7332		0·5136		0·2869		0·3507		0·3363		0·3256		0·3184		0·3148	
	v		0·8053		0·5367		0·2408		−0·0644		−0·3622		−0·3622		−0·2822		−0·2018		−0·1212		−0·0404	
0·50	u	1·1884		1·0888		0·9349		0·7419		0·5260		0·3032		0·2877		0·2756		0·2670		0·2618		0·2600
	v	0·9659		0·7469		0·4842		0·1952		−0·1028		−0·3933		−0·3152		−0·2368		−0·1580		−0·0790		0·0000
0·55	u		1·1899		1·0906		0·9385		0·7484		0·5361		0·2430		0·2297		0·2196		0·2130		0·2096	
	v		0·9050		0·6906		0·4338		0·1514		−0·1394		−0·3467		−0·2701		−0·1932		−0·1160		−0·0387	

Table 9.2.2

which are classified as hyperbolic and can be solved by the method of characteristics.

Some of the theory has been outlined in section 9.1 and we now develop this further with the view to obtaining numerical results. To simplify the notation we define

$$p = \frac{\partial u}{\partial x} \quad \text{and} \quad q = \frac{\partial u}{\partial y}.$$

Then (9.1.7) becomes

$$du = p\, dx + q\, dy, \tag{9.2.9}$$

and (9.1.12) becomes

$$e\, dy - a\left(\frac{dy}{dx}\right) dp - c\, dq = 0. \tag{9.2.10}$$

The characteristic equation (9.1.11) gives 2 curves illustrated in Figure 9.2.4 and defined by

$$\frac{dy}{dx} = f \quad \text{and} \quad \frac{dy}{dx} = g,$$

where in general f and g will be functions of x, y, u, p and q.

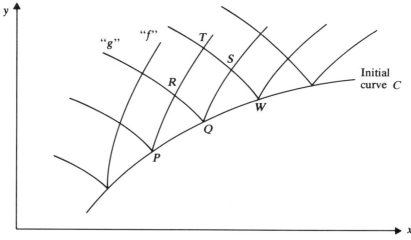

Figure 9.2.4

By integrating from P to R along the f-characteristic and from Q to R along the g-characteristic we can obtain an approximation to the solution at R.

When these differential equations cannot be solved analytically we use the trapezium rule [see § 7.1.3]. The characteristics give

$$\int_P^R dy = \int_P^R f\, dx, \qquad \int_Q^R dy = \int_Q^R g\, dx,$$

hence

$$y_R - y_P = \tfrac{1}{2}(f_R + f_P)(x_R - x_P), \tag{9.2.11}$$

and

$$y_R - y_Q = \tfrac{1}{2}(g_R + g_Q)(x_R - x_Q). \tag{9.2.12}$$

Similarly from equation (9.1.12)

$$(e_R + e_P)(y_R - y_P) - (a_R f_R + a_P f_P)(p_R - p_P) - (c_R + c_P)(q_R - q_P) = 0, \tag{9.2.13}$$

$$(e_R + e_Q)(y_R - y_Q) - (a_R g_R + a_Q g_Q)(p_R - p_Q) - (c_R + c_Q)(q_R - q_Q) = 0. \tag{9.2.14}$$

The solution at R is given by (integrating (9.2.9)) either

$$u_R - u_P = \tfrac{1}{2}(p_R + p_P)(x_R - x_P) + \tfrac{1}{2}(q_R + q_P)(y_R - y_P), \tag{9.2.15}$$

or

$$u_R - u_Q = \tfrac{1}{2}(p_R + p_Q)(x_R - x_Q) + \tfrac{1}{2}(q_R + q_Q)(y_R - y_Q). \tag{9.2.16}$$

After making a choice of equations for u_R we have five equations for the unknowns x_R, y_R, p_R, q_R and u_R. Equations (9.2.11) to (9.2.15) are non-linear and are best solved iteratively [see § 5.6]. If we assume that f_R and g_R are known—they can be taken as f_P and g_Q as an initial approximation—then (9.2.11) and (9.2.12) are a pair of linear equations for x_R and y_R. Having obtained an approximation for the coordinates of R we can proceed to estimate p_R, q_R from the linearized equations (9.2.13) and (9.2.14) and finally, u_R from (9.2.15). Using the approximations to the unknown quantities at R we can evaluate f_R and g_R and repeat the computation until convergence takes place. The iterations will usually converge rapidly provided that the interval PQ is not too large, a condition which is also necessary to minimize the truncation error in the trapezium rule.

Having found the solution at all points such as R and S at the first level of intersection of the characteristics we can consider these as a new set of 'initial conditions' and proceed to estimate the solution at points such as T. By this process of moving our 'initial' curve we can obtain the solution at discrete points within the bounding characteristics. To obtain solutions at points outside the bounding characteristics we must have boundary values specified as well as the necessary initial condition.

As an illustration of the method we solve equation (9.2.8)

$$\frac{\partial^2 v}{\partial x^2} - \frac{\partial^2 v}{\partial y^2} - \frac{\partial v}{\partial y} = 0, \qquad 0 \le x \le 1, \qquad y > 0,$$

where

$$v = 0, \qquad \frac{\partial v}{\partial y} = -1, \qquad y = 0, \qquad 0 \le x \le 1.$$

We shall also impose the boundary conditions

$$v = \sin\frac{5\pi y}{6}, \qquad x = 0 \atop v = 0, \qquad\qquad x = 1 \right\} y > 0.$$

The characteristic equation is

$$\left(\frac{dy}{dx}\right)^2 = 1$$

giving the characteristics as the family of straight lines with slope $+1$ and the family of straight lines of slope -1 and we can make use of Figure 9.2.2 and Figure 9.2.3.

Taking x_P and x_Q to be $0\cdot6$ and $0\cdot7$ respectively we find $(x_R, y_R) = (0\cdot65, 0\cdot05)$. For this differential equation we have that

$$a = 1, \qquad b = 0, \qquad c = -1, \qquad e = q$$

(cf. equation (9.1.6)). If we take f as $+1$ and g as -1 then equations (9.2.13) and (9.2.14) reduce to

$$(q_R + q_P)0\cdot05 - 2(p_R - p_P) + 2(q_R - q_P) = 0 \tag{9.2.17}$$

$$(q_R + q_Q)0\cdot05 + 2(p_R - p_Q) + 2(q_R - q_Q) = 0. \tag{9.2.18}$$

From the initial conditions

$$p = \frac{\partial v}{\partial x} = 0 \atop q = \frac{\partial v}{\partial y} = -1 \right\} \text{on } y = 0, \tag{9.2.19}$$

it follows that

$$p_P = 0 = p_Q, \qquad q_P = -1 = q_Q.$$

Hence (9.2.17) and (9.2.18) imply that

$$p_R = 0, \qquad q_R = -0\cdot9512.$$

Finally (9.2.15) gives

$$v_R = \tfrac{1}{2}(-0\cdot9512 - 1)(0\cdot05), \qquad v_R = -0\cdot04878.$$

The solution can be obtained at other points of intersection in a similar manner use being made of the boundary conditions to obtain solutions outside the bounding characteristics. This can be illustrated by seeking to obtain a solution at the point $(0\cdot8, 1\cdot4)$.

Because the geometry of the characteristics is well defined we can trace the f-characteristic $(dy/dx = 1)$ and the g-characteristic $(dy/dx = -1)$ through R to the points P and Q on the initial line. With reference to Figure 9.2.5 the characteristics through $R(0\cdot8, 1\cdot4)$ are $x + y = 2\cdot2$ and $y - x = 0\cdot6$. These cut the boundaries $x = 1$ and $x = 0$ at M and L whose coordinates we find to be

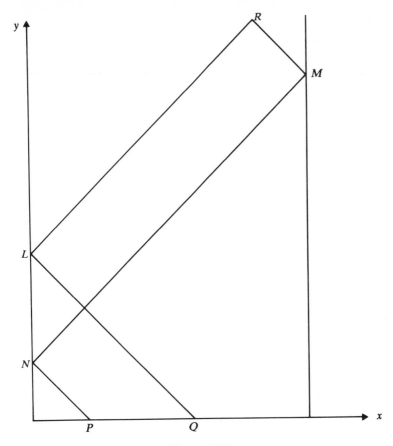

Figure 9.2.5

(1, 1·2) and (0, 0·6) respectively. The characteristic $y - x = 0·2$ through M cuts the boundary $x = 0$ at $N(0, 0·2)$ and the characteristic $y + x = 0·2$ cuts the x-axis (the initial line) at $P(0·2, 0)$. Similarly we find that the characteristic $y + x = 0·6$ through L cuts the x-axis at Q (0·6, 0).

Equation (9.1.12) reduces to

$$q\, dy - \frac{dy}{dx}\, dp + dq = 0, \tag{9.2.20}$$

and integrating from P to N along $dy/dx = -1$ gives (cf. equation (9.2.14))

$$(q_N + q_P)(y_N - y_P) + 2(p_N - p_P) + 2(q_N - q_P) = 0. \tag{9.2.21}$$

Since $q = \partial v/\partial y$ and $v = \sin(5\pi y/6)$ on $x = 0$, it follows that

$$q_N = \frac{5\pi}{6}\cos\frac{5\pi}{6} y_N = \frac{5\pi}{6} \cdot \frac{\sqrt{3}}{2}.$$

The initial condition (see equation (9.2.19)) gives $p_P = 0$, $q_P = -1$ and hence equation (9.2.21) gives

$$p_N = -0.9 - 1.1\left(\frac{5\pi}{6}\sqrt{\frac{3}{2}}\right).$$

Similarly integrating (9.2.20) from N to M along $dy/dx = 1$ gives

$$(q_M + q_N)(y_M - y_N) - 2(p_M - p_N) + 2(q_M - q_N) = 0.$$

The boundary condition at $x = 1$ implies that $q_M = 0$. Hence

$$p_M = -\frac{q_N}{2} + p_N = -4.5276.$$

On integrating from M to R we obtain that

$$(q_R + q_M)(y_R - y_M) + 2(p_R - p_M) + 2(q_R - q_M) = 0,$$

or

$$2.2q_R + 2p_R = 2p_M = 1.8 - \pi\sqrt{\tfrac{3}{2}}. \tag{9.2.22}$$

The other equation is obtained by integrating (9.2.20) from Q to L and L to P; we find that

$$p_L = -0.7,$$

and

$$2.8q_R - 2p_R = -2p_L = 1.4. \tag{9.2.23}$$

Equations (9.2.22) and (9.2.23) give

$$p_R = -2.8434, \qquad q_R = -1.5310.$$

Finally integrating equation (9.1.7) (cf. equation (9.2.15)) from M to R gives

$$v_R - v_M = \tfrac{1}{2}(p_R + p_M)(x_R - x_M) + \tfrac{1}{2}(q_R + q_M)(y_R - y_M), \qquad v_R = 0.0584.$$

Not all differential equations have such a well-defined geometry. In such cases it is necessary to proceed step by step from one level of inter-sections to the next. To add to the complication the quadrature formulae yield non-linear equations which have to be solved iteratively.

To provide an example with a more complicated geometry for the characteristics we solve the equation given in Smith ((1965), Chapter 4, exercise 4.1)

$$\frac{\partial^2 u}{\partial x^2} - u^2 \frac{\partial^2 u}{\partial y^2} = 0, \qquad 0 \le x \le 1,$$

$$\left.\begin{array}{l} u = 0.2 + 5x^2, \\[2mm] \dfrac{\partial u}{\partial y} = 3x, \end{array}\right\} \quad y = 0, \qquad 0 \le x \le 1.$$

The initial curve is that part of the x-axis between $x = 0$ and $x = 1$ and for convenience we take points equally spaced along the axis with $x_Q - x_P = 0.1$. To illustrate the computation we choose $x_P = 0.3$ (see Figure 9.2.6).

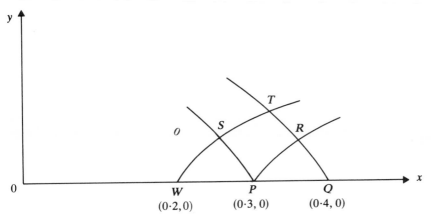

Figure 9.2.6

The characteristic equation is

$$\left(\frac{dy}{dx}\right)^2 - u^2 = 0,$$

or

$$\frac{dy}{dx} = \pm u.$$

Let $f = u$ and $g = -u$. Since

$$u = 0.2 + 5x^2 \quad \text{on } y = 0, \qquad 0 \le x \le 1,$$

we have that $p = \partial u/\partial x = 10x$ on the initial curve and we are given that $q = \partial u/\partial y = 3x$. Then

$$u_P = 0.65, \qquad u_Q = 1,$$
$$p_P = 3, \qquad p_Q = 4,$$
$$q_P = 0.9, \qquad q_Q = 1.2,$$
$$f_P = 0.65, \qquad g_Q = -1.$$

Taking $f_R = f_P$, $g_R = g_Q$ as a first approximation, equations (9.2.11) and (9.2.12) give

$$y_R = 0.65(x_R - 0.3)$$
$$y_R = -1(x_R - 0.4),$$

whence

$$x_R = 0.3606, \qquad y_R = 0.03939.$$

Next we use equations (9.2.13) and (9.2.14) to estimate p_R and q_R. We note that for this problem $a = 1$, $c = -u^2$ and $e = 0$ (cf. equation (9.1.6)); in particular we take

$$c_R = c_P = -0 \cdot 4225, \quad \text{or} \quad \dot{c}_R = c_Q = -1.$$

Thus we have that

$$0 \cdot 65(p_R - 3) - 0 \cdot 4225(q_R - 0 \cdot 9) = 0,$$

$$-(p_R - 4) - (q_R - 1 \cdot 2) = 0.$$

Hence $p_R = 3 \cdot 5121$, $q_R = 1 \cdot 6879$. Equation (9.2.15) now gives $u_R = 0 \cdot 8983$. From this we are able to estimate $f_R = -g_R = 0 \cdot 8983$. We repeat the calculations using equations (9.2.11) to (9.2.15) with the improved values for f_R and g_R and with $c_R = -u_R^2 = -0 \cdot 8069$. We obtain

$$x_R = 0 \cdot 3551, \quad y_R = 0 \cdot 04264,$$

$$p_R = 3 \cdot 5847, \quad q_R = 1 \cdot 6363,$$

$$u_R = 0 \cdot 8854.$$

One more iteration gives

$$x_R = 0 \cdot 3551, \quad y_R = 0 \cdot 04231,$$

$$p_R = 3 \cdot 5825, \quad q_R = 1 \cdot 6513,$$

$$u_R = 0 \cdot 8852.$$

Thus to 3 decimal places we have $u_R = 0 \cdot 885$ (in fact each value given above is correct to 4 significant figures).

By following the same procedure we find the solution at the point S shown in Figure 9.2.6 to be (see Smith (1965), Chapter 4, example 4.1)

$$x_S = 0 \cdot 2558, \quad y_S = 0 \cdot 02668,$$

$$p_S = 2 \cdot 5288, \quad q_S = 1 \cdot 6764,$$

$$u_S = 0 \cdot 5567.$$

Using the computed solutions at S and R as our 'initial solution' we can now obtain estimates for the solution at T on the next level of intersection.

The first iteration with $f_T = f_S$ and $g_T = g_R$ gives

$$y_T - 0 \cdot 02668 = 0 \cdot 5567(x_T - 0 \cdot 2558),$$

$$y_T - 0 \cdot 04231 = -0 \cdot 8852(x_T - 0 \cdot 3551),$$

whence

$$x_T = 0 \cdot 3276, \quad y_T = 0 \cdot 06665,$$

$$0 \cdot 5567(p_T - 2 \cdot 5288) - 0 \cdot 3099(q_T - 1 \cdot 6764) = 0,$$

$$-0 \cdot 8852(p_T - 3 \cdot 5825) - 0 \cdot 7836(q_T - 1 \cdot 6413) = 0,$$

whence

$$p_T = 2 \cdot 9238, \qquad q_T = 2 \cdot 3856.$$

Hence $u_T = 0 \cdot 8336$. We must now continue with the iteration until the process converges.

The truncation error in the trapezium rule which was used to derive equations (9.2.11) to (9.2.15) is proportional to the square of the step length [see (7.1.7)]. In equations (9.2.11) and (9.2.12) the step length is $|x_R - x_P|$ and $|x_R - x_Q|$ each of which is of order $0 \cdot 05$ in the above example, and we can expect a reasonably small error. In equation (9.2.13) the step length is $|p_R - p_P|$ and $|q_R - q_P|$ and are of order $0 \cdot 5$ and $0 \cdot 75$ respectively (with similar step lengths occurring in equation (9.2.14). These are rather large, and it would seem desirable to decrease the interval. This can be achieved by decreasing the interval PQ, which will result in additional computation and the estimation of the solutions at a completely different set of points R, S, W.

As has been shown, the coordinates of the point of intersection of the characteristics are not known in advance except for the simplest of cases, and we have to accept the solution at the points determined by the computation. Frequently we require the solution at a set of prespecified points and this can cause problems when the points are too far from the given initial curve to allow the use of the trapezium rule. In particular we often require the solution at the points of intersection of a rectangular grid.

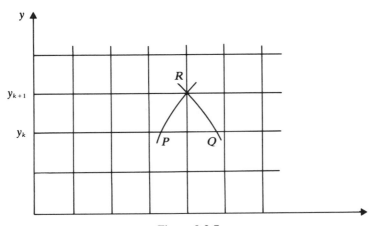

Figure 9.2.7

With reference to Figure 9.2.7, the coordinates of R are obviously known but in this case the x coordinates of P and Q have to be determined. Thus equations (9.2.11) to (9.2.15) can be used but with x_P and x_Q as unknown quantities rather than x_R and y_R. The complication, however, is that the solution (u, p, q) are not known at P and Q (unless P and Q are on the initial curve when an analytic initial condition would, in general, be available). Thus having estimated the coordinates x_P and x_Q we have to find (u, p, q) at these points by linear interpolation using the known solution at the nodes along the line $y = y_k$ [see § 2.3].

When the hyperbolic differential equation is relatively simple with uncomplicated families of characteristics then a method of finite-differences can be used to obtain estimates of the solution at the nodes of a rectangular mesh [see § 1.4]. We discuss such methods in the next section.

9.3. HYPERBOLIC EQUATIONS II—FINITE DIFFERENCE METHODS

The methods developed in this section can only be applied to relatively simple hyperbolic equations, but they do provide an efficient and uncompli-cated technique for such partial differential equations. We will restrict our attention to the simple non-dimensional wave equation

$$\frac{\partial^2 u}{\partial x^2} = \frac{\partial^2 u}{\partial y^2} \tag{9.3.1}$$

with initial conditions

$$\left. \begin{array}{l} u = \tfrac{1}{2}x(1-x), \\[2mm] \dfrac{\partial u}{\partial y} = 0, \end{array} \right\} \ 0 \le x \le 1, \ y = 0,$$

and boundary conditions $u = 0$ on $x = 0$ and $x = 1$, $y > 0$.

Identities (9.1.14) and (9.1.15) are used to replace the differential coefficients in (9.3.1) to give

$$\frac{u^k_{j+1} - 2u^k_j + u^k_{j-1}}{(\Delta x)^2} = \frac{u^{k+1}_j - 2u^k_j + u^{k-1}_j}{(\Delta y)^2}$$

or

$$u^{k+1}_j = \rho^2(u^k_{j+1} + u^k_{j-1}) + 2(1-\rho^2)u^k_j - u^{k-1}_j, \tag{9.3.2}$$

where $\rho = (\Delta y)/(\Delta x)$.

Relation (9.3.2) means, with reference to Figure 9.3.1, that the value of u at P is calculated directly or explicitly from the known values at the points marked Q, R, S, W. The weighting given to the values of u at these points can be represented by the *molecule*

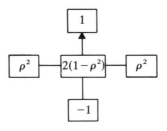

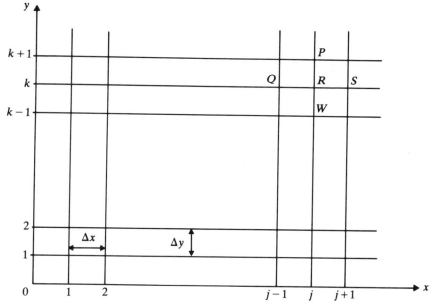

Figure 9.3.1

The computation starts with $k = 0$ using the given initial conditions to find the solution at the level $k = 1$. When $k = 0$ (9.3.2) introduces the 'fictitious' value u_j^{-1} and this value has to be eliminated. This can be done by replacing $\partial u / \partial y$ by the central difference approximation (9.1.18), i.e.

$$\left. \frac{\partial u}{\partial y} \right|_{j,0} = \frac{u_j^1 - u_j^{-1}}{2(\Delta y)}. \tag{9.3.3}$$

Thus the initial condition for the derivative gives $u_j^{-1} = u_j^1$, and the first step of computation becomes

$$u_j^1 = \rho^2(u_{j+1}^0 + u_{j-1}^0) + 2(1 - \rho^2)u_j^0 - u_j^1,$$

and since $u_j^0 = \frac{1}{2}j\,\Delta x(1 - j\,\Delta x)$, $j = 0, 1, 2, \ldots$, we have that

$$u_j^1 = \frac{1}{2}\rho^2\,\Delta x[j - (j^2 + 1)\Delta x] + \frac{1}{2}(1 - \rho^2)j\,\Delta x[1 - j\,\Delta x], \qquad j = 1, 2, \ldots. \tag{9.3.4}$$

Having computed u_j^1 for all j we use (9.3.2) to compute u_j^2 for all j and hence u_j^3 etc. Some results for the computation are given in Table 9.3.1, where we have taken $\Delta x = 0 \cdot 1$ and $\rho = 1$ (hence $\Delta y = 0 \cdot 1$).

Clearly $\rho = 1$ seems convenient but we must consider the maximum value of ρ that can be used. A graph for the case $\rho = 1$ is given in Figure 9.3.2 which confirms that the computed values look reasonable.

If we now take $\rho = 2$ and keep $\Delta x = 0 \cdot 1$ then the results become those of Table 9.3.2.

y \ x	0·1	0·2	0·3	0·4	0·5	0·6	0·7	0·8	0·9
0	0·045	0·080	0·105	0·120	0·125	0·120	0·105	0·080	0·045
0·1	0·040	0·075	0·100	0·115	0·120	0·115	0·100	0·075	0·040
0·2	0·030	0·060	0·085	0·100	0·105				
0·3	0·020	0·040	0·060	0·075	0·080				
0·4	0·010	0·020	0·030	0·040	0·045		Solution is		
0·5	0·000	0·000	0·000	0·000	0·000		symmetric		
0·6	−0·010	−0·020	−0·030	−0·040	−0·045		about $x = 0.5$		
0·7	−0·020	−0·040	−0·060	−0·075	−0·080				
0·8	−0·030	−0·060	−0·085	−0·100	−0·105				

Table 9.3.1

We note that we obtain $u(0·1, 0·2)$ as $0·025$ and not $0·03$ as for the case $\rho = 1$.

At the level $y = 0·4$ there are further discrepancies (illustrated in Figure 9.3.2) when x is $0·1$, $0·2$ and $0·3$, and it can be seen that these discrepancies are going to propogate through all values as the computation proceeds and y increases. Clearly the value $\rho = 2$ is unacceptable and does not yield a correct

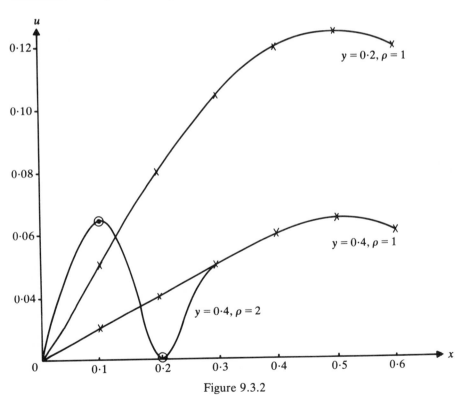

Figure 9.3.2

y \ x	0·1	0·2	0·3	0·4	0·5
0	0·045	0·080	0·105	0·120	0·125
0·2	0·025	0·060	0·085	0·100	0·105
0·4	0·045	0·000	0·025	0·040	0·045
0·6	−0·295	0·220	−0·075	−0·060	−0·055
0·8	2·605	−2·800	1·065	−0·200	−0·195

Table 9.3.2: Solution symmetric about $x = 0.5$.

approximation to the given differential equation. We say that $\rho = 2$ leads to *instability*.

A finite difference scheme is said to be *stable* if small errors introduced into the computation (due to round off, etc.) decrease as the value of y increases. A full discussion of stability is beyond the scope of this book and we refer the interested reader to such texts as Smith ((1965), Chapter 3), Ames ((1977), § 2.2, § 2.4–2.7; Chapter 4, § 4.7), Cohen ((1973), Chapter 12, § 12.5). We state here, without proof, that the finite difference scheme (9.3.2) for solving the wave equation (9.3.1) is stable for $0 < \rho \le 1$.

If the condition $\rho \le 1$ proves to be restrictive and it becomes desirable to use larger values of ρ then we can use the finite difference scheme (see Ames (1977), Chapter 4, § 4.8)

$$-\rho^2 u_{j+1}^{k+1} + 2(1+\rho^2)u_j^{k+1} - \rho^2 u_{j-1}^{k+1} = 4u_j^k + \rho^2 u_{j-1}^{k-1} - 2(1+\rho^2)u_j^{k-1} + \rho^2 u_{j-1}^{k-1},$$
(9.3.5)

which can be represented by the molecule

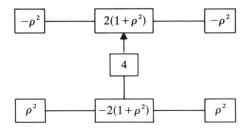

This scheme is stable for all ρ and the penalty for this extension in the region of stability is that we no longer have an explicit scheme. The value of u_j^{k+1} cannot be obtained directly from (9.3.5) but rather the value is implied. Finite difference schemes such as (9.3.5) are called *implicit* schemes and in general have a wider range of stability for the mesh ratio than explicit schemes.

We can illustrate the use of the implicit scheme (9.3.5) by solving again equation (9.3.1) with the specified conditions. Note that u_j^0 is given and that u_j^1 cannot be computed from (9.3.5) without first eliminating the fictitious value u_j^{-1}.

We know from (9.3.3) that $u_j^{-1} = u_j^1$, so the first step of the computation becomes after simplification

$$-\rho^2 u_{j+1}^1 + 2(1+\rho^2)u_j^1 - \rho^2 u_{j-1}^1 = j\,\Delta x(1-j\,\Delta x).$$

Taking increments $\Delta x = 0{\cdot}1$ we have the system of equations

$$\begin{bmatrix} 2(1+\rho^2) & -\rho^2 & & & \\ -\rho^2 & 2(1+\rho^2) & -\rho^2 & & \\ \ddots & \ddots & \ddots & \\ & -\rho^2 & 2(1+\rho^2) & -\rho^2 \\ & & -\rho^2 & 2(1+\rho^2) \end{bmatrix} \begin{bmatrix} u_1^1 \\ u_2^1 \\ \vdots \\ u_8^1 \\ u_9^1 \end{bmatrix} = \begin{bmatrix} 0{\cdot}09 \\ 0{\cdot}16 \\ \vdots \\ 0{\cdot}16 \\ 0{\cdot}09 \end{bmatrix}.$$

Note that in order to have a complete set of equations we must have two boundary points—in this example at $x = 0$ and $x = 1$—and that implicit schemes cannot cope with an infinite or semi-infinite x region.

Since the required solution is symmetric, in particular $u_6^1 = u_4^1$, the equations to be solved reduce to

$$\begin{bmatrix} 2(1+\rho^2) & -\rho^2 & & & \\ -\rho^2 & 2(1+\rho^2) & -\rho^2 & & \\ & -\rho^2 & 2(1+\rho^2) & -\rho^2 & \\ & & -\rho^2 & 2(1+\rho^2) & -\rho^2 \\ & & & -2\rho^2 & 2(1+\rho^2) \end{bmatrix} \begin{bmatrix} u_1^1 \\ u_2^1 \\ u_3^1 \\ u_4^1 \\ u_5^1 \end{bmatrix} = \begin{bmatrix} 0{\cdot}09 \\ 0{\cdot}16 \\ 0{\cdot}21 \\ 0{\cdot}24 \\ 0{\cdot}25 \end{bmatrix}, \quad (9.3.6)$$

which can be solved by the techniques of Chapter 3.

The system of equations (9.3.6) can be expressed in matrix form

$$\mathbf{A}\mathbf{u}_k = \mathbf{b}$$

where $\mathbf{A}$ is the matrix involving ρ, $\mathbf{u}_k$ is the solution vector at the level k and $\mathbf{b}$ is the known right hand side [see also I, § 5.7–5.10].

To obtain the solution $\mathbf{u}_2$ we use the implicit scheme (9.3.5) to give the algebraic system

$$\mathbf{A}\mathbf{u}_2 = \mathbf{B}\mathbf{u}_1 + 4\mathbf{I}\mathbf{u}_0$$

where $\mathbf{B} = -\mathbf{A}$ and $\mathbf{I}$ is the unit matrix [see I (6.2.7)]. In general we have

$$\mathbf{A}\mathbf{u}_{k+1} = -\mathbf{A}\mathbf{u}_k + 4\mathbf{I}\mathbf{u}_{k-1}, \qquad k = 1, 2, 3, \ldots.$$

Taking $\rho = 1$ we have the results given in Table 9.3.3 and when $\rho = 2$ the results obtained are those shown in Table 9.3.4.

The agreement in the figures again leaves much to be desired but there are no outstanding differences. The explanation for loss of accuracy lies in a discussion of the truncation error, which for both explicit and implicit schemes (9.3.2) and (9.3.5) is $O((\Delta x)^2 + (\Delta y)^2)$ which means that with Δx and Δy of order $0{\cdot}1$ we cannot expect the second decimal place to be accurate. This problem can be surmounted by decreasing Δx and keeping ρ relatively small $O(1)$ say. If in the

x / y	0·1	0·2	0·3	0·4	0·5
0	0·0450	0·0800	0·1050	0·1200	0·1250
0·1	0·0413	0·0754	0·1001	0·1150	0·1200
0·2	0·0319	0·0623	0·0857	0·1002	0·1051
0·3	0·0202	0·0431	0·0630	0·0762	0·0807
0·4	0·0091	0·0210	0·0342	0·0442	0·0478
0·5	−0·0005	−0·0007	0·0025	0·0067	0·0085
0·6	−0·0094	−0·0205	−0·0287	−0·0325	−0·0335
0·7	−0·0186	−0·0380	−0·0564	−0·0689	−0·0730
0·8	−0·0277	−0·0537	−0·0788	−0·0977	−0·1047

Table 9.3.3: Solution symmetric about $x = 0·5$.

x / y	0·1	0·2	0·3	0·4	0·5
0	0·0450	0·0800	0·1050	0·1200	0·1250
0·2	0·0350	0·0651	0·0877	0·1015	0·1062
0·4	0·0119	0·0273	0·0413	0·0508	0·0541
0·6	−0·0110	−0·0169	−0·0185	−0·0183	−0·0180
0·8	−0·0261	−0·0519	−0·0717	−0·0836	−0·0876

Table 9.3.4: Solutions symmetric about $x = 0·5$.

above example we reduce Δx to 0·01, then we will have 50 equations to solve at each y level and the aid of an electronic computer will be required.

Other implicit schemes are discussed briefly in Ames ((1977), Chapter 4, § 4.8). But since these step by step methods can only be applied to relatively simple hyperbolic equations we terminate this discussion with the comment that integration along characteristics (§ 9.2) is usually more convenient and accurate.

Further information on finite difference methods, some of the difficulties that can arise, the treatment of derivative boundary conditions, etc., can be found in § 9.4 and § 9.5. The differential equations are different, but the general finite-difference approach will be the same.

9.4. PARABOLIC EQUATIONS—EXPLICIT AND IMPLICIT METHODS

Partial differential equations are classified as parabolic when the characteristics are coincident (§ 9.1) thus the method of integration along characteristics cannot be applied to this type of equation. We can, however, use finite difference methods where we approximate to the differential coefficients by suitable finite difference expansions.

Consider the non-dimensional diffusion equation

$$\frac{\partial^2 u}{\partial x^2} = \frac{\partial u}{\partial y}. \tag{9.4.1}$$

Thus, as shown in § 9.1, we have the explicit scheme

$$u_j^{k+1} = \rho u_{j-1}^k + (1 - 2\rho)u_j^k + \rho u_{j+1}^k, \tag{9.4.2}$$

where

$$\rho = (\Delta y)/(\Delta x)^2.$$

This scheme can be represented by the molecule

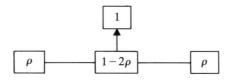

The direct evaluation of u_j^{k+1} from the known values at the level $y = k(\Delta y)$ is straightforward and presents no difficulties (cf. § 9.3). No special first step is required other than the initial specification of u_j^0 ($j = 0, 1, 2, \ldots$) and (9.4.2) holds for $k = 0, 1, 2, \ldots$. Unfortunately the explicit scheme (9.4.2) suffers from the stability restriction $0 < \rho \le \frac{1}{2}$ (see Cohen (1973), Chapter 12, § 12.5). This means that when the x-increment, Δx, is $0 \cdot 1$, the y-increment Δy has to be less than or equal to $0 \cdot 005$ resulting in a substantial increase in computation.

The instability of the explicit scheme (9.4.2) can be demonstrated by solving the diffusion equation (9.4.1) subject to the conditions

$$u = 0 \quad \text{on } x = 0 \qquad \text{and} \qquad x = 1, \quad y \ge 0,$$

$$u = 0 \cdot 5x(1 - x) \quad \text{on } y = 0, \quad 0 \le x \le 1,$$

and taking $\rho = 0 \cdot 25, \ 0 \cdot 5, \ 1$.

Some results are given in Table 9.4.1 and illustrated in Figure 9.4.1.

We must have some idea of the accuracy that we can hope to achieve, and for this purpose it is necessary to discuss the convergence of the finite difference solution to the required solution of the differential equation. A detailed discussion is beyond the scope of this book and the interested reader is referred to Smith ((1965), Chapter 3) and Ames ((1977), Chapter 2, § 2.7, § 2.11). We merely make some superficial investigations.

y		x 0·1	0·2	0·3	0·4	0·5
0		0·0450	0·0800	0·1050	0·1200	0·1250
0·0025	(a)	0·0425	0·0775	0·1025	0·1175	0·1225
0·005	(a)	0·0406	0·0750	0·1000	0·1150	0·1200
	(b)	0·0400	0·0750	0·1000	0·1150	0·1200
0·0075	(a)	0·0391	0·0727	0·0975	0·1125	0·1175
0·01	(a)	0·0377	0·0705	0·0950	0·1100	0·1150
	(b)	0·0375	0·0700	0·0950	0·1100	0·1150
	(c)	0·0350	0·0700	0·0950	0·1100	0·1150
0·02	(a)	0·0333	0·0629	0·0858	0·1002	0·1051
	(b)	0·0331	0·0625	0·0856	0·1000	0·1050
	(c)	0·0350	0·0600	0·0850	0·1000	0·1050
0·03	(a)	0·0299	0·0566	0·0776	0·0910	0·0955
	(b)	0·0297	0·0562	0·0773	0·0906	0·0953
	(c)	0·0250	0·0600	0·0750	0·0900	0·0950
0·04	(a)	0·0269	0·0511	0·0702	0·0825	0·0867
	(b)	0·0268	0·0508	0·0699	0·0820	0·0863
	(c)	0·0350	0·0400	0·0750	0·0800	0·0850
0·05	(a)	0·0243	0·0462	0·0636	0·0747	0·0785
	(b)	0·0242	0·0459	0·0632	0·0742	0·0781
	(c)	0·0050	0·0700	0·0450	0·0800	0·0750
0·06	(a)	0·0220	0·0419	0·0576	0·0677	0·0712
	(b)	0·0219	0·0415	0·0572	0·0671	0·0707
	(c)	0·0650	−0·0200	0·1050	0·0400	0·0850

Table 9.4.1: (a) $\rho = 0{\cdot}25$, (b) $\rho = 0{\cdot}5$, (c) $\rho = 1$.

The *local truncation error* is defined as the difference between the finite difference equation and the differential equation and should tend to zero as the mesh size tends to zero. For the explicit scheme (9.4.2) the local truncation error is $O[\Delta y + (\Delta x)^2]$ unless $\rho = \frac{1}{6}$ when it reduces to $O[(\Delta y)^2 + (\Delta x)^4]$. The increase in accuracy with $\rho = \frac{1}{6}$ may well justify the small y increment and the extra computation involved.

An explicit scheme that is stable for all positive ρ is due to Du Fort and Frankel (1953) and takes the form

$$\frac{u_j^{k+1} - u_j^{k-1}}{2(\Delta y)} = \frac{u_{j+1}^k - u_j^{k+1} - u_j^{k-1} + u_{j-1}^k}{(\Delta x)^2},$$

or

$$(1 + 2\rho)u_j^{k+1} = 2\rho(u_{j+1}^k + u_{j-1}^k) + (1 - 2\rho)u_j^{k-1}, \tag{9.4.3}$$

where $\rho = \Delta y/(\Delta x)^2$, giving the molecule

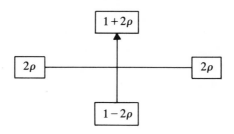

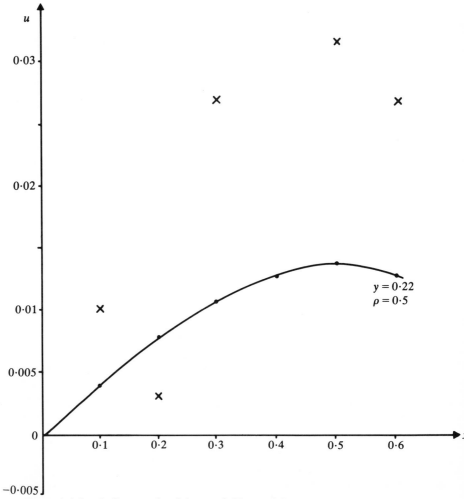

Figure 9.4.1: • indicates $u(x, y)$ for $y = 0.22$, $\rho = 0.5$; × indicates $u(x, y)$ for $y = 0.22$, $\rho = 0.55$.

This is a three level problem $(k+1, k, k-1)$ and when $k=0$ we have to eliminate the fictitious term u_j^{-1} (see (9.3.2) and the corresponding discussion). It is not usual for parabolic partial differential equations to have derivative initial conditions and therefore the process discussed in section 9.3 cannot be used to remove u_j^{-1} from the difference equation (9.4.3). We can, however, use (9.4.2) to start the process and then continue with (9.4.3).

For illustrative purposes we again solve

$$\frac{\partial^2 u}{\partial x^2} = \frac{\partial u}{\partial y}, \qquad 0 \le x \le 1, \quad y > 0,$$

$$u = 0.5x(1-x), \qquad 0 \le x \le 1, \quad y = 0,$$

$$u = 0, \quad \text{on } x = 0 \quad \text{and} \quad x = 1, \quad y > 0,$$

this time using the Du Fort and Frankel scheme (9.4.3). Taking $\rho = 1$ and $\Delta x = 0.1$ we require the solution at the level $y = 0.01$ and we obtain this from Table 9.4.1(a). Using the explicit scheme (9.4.3) we find for $\Delta y = 0.02, 0.03$, the results given in Table 9.4.2.

y \ x	0·1	0·2	0·3	0·4	0·5
0	0·0450	0·0800	0·1050	0·1200	0·1250
0·01	0·0377	0·0705	0·0950	0·1100	0·1150
0·02	0·0320	0·0618	0·0853	0·1000	0·1050
0·03	0·0286	0·0547	0·0762	0·0902	0·0950
0·04	0·0259	0·0493	0·0682	0·0808	0·0853
0·05	0·0233	0·0444	0·0613	0·0722	0·0761
0·06	0·0210	0·0400	0·0550	0·0647	0·0679

Table 9.4.2: Solution symmetric about $x = 0.5$.

The local truncation error in the Du Fort and Frankel scheme is

$$\left(\frac{\Delta y}{\Delta x}\right)^2 \frac{\partial^2 u}{\partial y^2} + O[(\Delta x)^2 + (\Delta y)^2].$$

This will tend to zero as Δx and Δy tend to zero provided that Δy tends to zero faster than Δx. If Δx and Δy tend to zero at the same rate, i.e. $(\Delta y)/(\Delta x) = \alpha$, then we are solving the hyperbolic equation

$$\frac{\partial u}{\partial y} - \frac{\partial^2 u}{\partial x^2} + \alpha^2 \frac{\partial^2 u}{\partial y^2} = 0.$$

We can also obtain unconditionally stable methods by constructing implicit schemes. For example we can use the backward difference approximation

$$\Delta y \frac{\partial u}{\partial y}\bigg|_{x_j, y_{k+1}} = u_j^{k+1} - u_j^k, \tag{9.4.4}$$

(see equation (9.1.17)) and the central difference approximation

$$(\Delta x)^2 \left. \frac{\partial^2 u}{\partial x^2} \right|_{x_j, y_{k+1}} = u_{j+1}^{k+1} - 2u_j^{k+1} + u_{j-1}^{k+1}, \tag{9.4.5}$$

to obtain the implicit scheme

$$-\rho u_{j+1}^{k+1} + (1+2\rho)u_j^{k+1} - \rho u_{j-1}^{k+1} = u_j^k, \tag{9.4.6}$$

which is stable for all positive $\rho = (\Delta y)/(\Delta x)^2$, or in molecule form

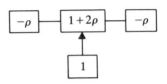

A more popular implicit method is a combination of the explicit scheme (9.4.2) and the implicit scheme (9.4.6) due to Crank and Nicolson (1947), which is

$$u_j^{k+1} - u_j^k = \tfrac{1}{2}\rho[u_{j+1}^{k+1} - 2u_j^{k+1} + u_{j-1}^{k+1} + u_{j+1}^k - 2u_j^k + u_{j-1}^k],$$

or

$$-\tfrac{1}{2}\rho u_{j+1}^{k+1} + (1+\rho)u_j^{k+1} - \tfrac{1}{2}\rho u_{j-1}^{k+1} = \tfrac{1}{2}\rho u_{j+1}^k + (1-\rho)u_j^k + \tfrac{1}{2}\rho u_{j-1}^k. \tag{9.4.7}$$

This is again stable for all positive ρ; see for example Ames ((1977), Chapter 2, § 2.5), Cohen ((1973), Chapter 12, § 12.5).

The local truncation error can be deduced to be $O[(\Delta x)^2 + (\Delta y)^2]$ and tends to zero as the mesh decreases.

Implicit methods cannot be solved in a step-wise manner and are unsuitable for an infinite x-region. Problems in a finite range $x_0 \leq x \leq x_n$ reduce to a system of linear equations. If we are given values of u on the boundaries $x = x_0$ and $x = x_n = x_0 + n\,\Delta x$ for all y, then at each level $y_{k+1} = y_0 + (k+1)\,\Delta y$ we will have $(n-1)$ equations to solve for $u_1^{k+1}, u_2^{k+1}, \ldots, u_{n-1}^{k+1}$.

Suppose we wish to use the Crank–Nicolson method (9.4.7) to solve the diffusion equation (9.4.1) subject to the initial condition that u is known for $x_0 \leq x \leq x_n$ on $y = y_0$, and the boundary conditions that u is specified on $x = x_0$ and on $x = x_n$ for all $y > y_0$. The equations to be solved take the form

$$\begin{bmatrix} (1+\rho) & -\tfrac{1}{2}\rho & & & & \\ -\tfrac{1}{2}\rho & (1+\rho) & -\tfrac{1}{2}\rho & & & \\ & -\tfrac{1}{2}\rho & (1+\rho) & -\tfrac{1}{2}\rho & & \\ & & \cdot & \cdot & \cdot & \\ & & & -\tfrac{1}{2}\rho & (1+\rho) & -\tfrac{1}{2}\rho \\ & & & & -\tfrac{1}{2}\rho & (1+\rho) \end{bmatrix} \begin{bmatrix} u_1^{k+1} \\ u_2^{k+1} \\ u_3^{k+1} \\ \vdots \\ u_{n-2}^{k+1} \\ u_{n-1}^{k+1} \end{bmatrix} =$$

$$
\begin{bmatrix}
(1-\rho) & \tfrac{1}{2}\rho \\
\tfrac{1}{2}\rho & (1-\rho) & \tfrac{1}{2}\rho \\
 & \tfrac{1}{2}\rho & (1-\rho) & \tfrac{1}{2}\rho \\
 & & \ddots & \ddots & \ddots \\
 & & & \tfrac{1}{2}\rho & (1-\rho) & \tfrac{1}{2}\rho \\
 & & & & \tfrac{1}{2}\rho & (1-\rho)
\end{bmatrix}
\begin{bmatrix}
u_1^k \\ u_2^k \\ u_3^k \\ \vdots \\ u_{n-2}^k \\ u_{n-1}^k
\end{bmatrix}
+ \tfrac{1}{2}\rho
\begin{bmatrix}
u_0^{k+1} + u_0^k \\ 0 \\ 0 \\ \vdots \\ 0 \\ u_n^{k+1} + u_n^k
\end{bmatrix}.
$$

The right-hand side reduces to a vector of known terms giving us a diagonally dominant set of tridiagonal equations [see (3.4.3)] that can be solved by the usual methods.

The solution to the problem

$$
\frac{\partial u}{\partial y} = \frac{\partial^2 u}{\partial x^2}, \qquad 0 \leq x \leq 1, \quad y > 0,
$$

$$
u = 0.5x(1-x), \qquad 0 \leq x \leq 1, \quad y = 0,
$$

$$
u = 0 \quad \text{on } x = 0 \quad \text{and} \quad x = 1, \quad y > 0, \qquad \cdot
$$

using the Crank–Nicolson method with $\Delta x = 0.1$ for the cases $\rho = 1$ and $\rho = \tfrac{1}{2}$ is summarized in Table 9.4.3. A comparison can be made between Tables 9.4.1, 9.4.2 and 9.4.3.

The explicit and implicit formulae given above are for the diffusion equation (9.4.1). If the differential equation has a different form or contains extra terms, then the difference scheme has to be modified. The basis of the Crank–Nicolson

y \ x		0·1	0·2	0·3	0·4	0·5
0		0·0450	0·0800	0·1050	0·1200	0·1250
0·005	(a)	0·0409	0·0751	0·1000	0·1150	0·1200
0·01	(a)	0·0379	0·0707	0·0952	0·1100	0·1150
	(b)	0·0377	0·0707	0·0952	0·1101	0·1150
0·02	(a)	0·0335	0·0631	0·0861	0·1003	0·1052
	(b)	0·0335	0·0631	0·0861	0·1004	0·1052
0·03	(a)	0·0300	0·0569	0·0779	0·0912	0·0958
	(b)	0·0300	0·0568	0·0779	0·0912	0·0959
0·04	(a)	0·0271	0·0514	0·0707	0·0828	0·0870
	(b)	0·0271	0·0514	0·0706	0·0828	0·0870
0·05	(a)	0·0245	0·0465	0·0640	0·0752	0·0790
	(b)	0·0245	0·0465	0·0640	0·0751	0·0790

Table 9.4.3: (a) $\rho = 0.5$, (b) $\rho = 1$.

method is to take the mean of the forward difference explicit scheme (cf. (9.4.2)) and the backward difference implicit scheme (cf. (9.4.6)). Thus when solving a parabolic equation such as

$$\frac{\partial u}{\partial y} = \frac{\partial^2 u}{\partial x^2} + au, \qquad 0 \le x \le 1, \quad y > 0,$$

where a is a known constant, we have that

$$\frac{u_j^{k+1} - u_j^k}{(\Delta y)} = \frac{1}{2}\left[\frac{u_{j+1}^{k+1} - 2u_j^{k+1} + u_{j-1}^{k+1}}{(\Delta x)^2} + au_j^{k+1} + \frac{u_{j+1}^k - 2u_j^k + u_{j-1}^k}{(\Delta x)^2} + au_j^k\right],$$

or

$$-\tfrac{1}{2}\rho u_{j+1}^{k+1} + \left(1 + \rho - \frac{a\,\Delta y}{2}\right)u_j^{k+1} - \tfrac{1}{2}\rho u_{j-1}^{k+1} = \tfrac{1}{2}\rho u_{j+1}^k + \left(1 - \rho + \frac{a\,\Delta y}{2}\right)u_j^k + \tfrac{1}{2}\rho u_{j-1}^k.$$

The presence of the extra term au does not present any difficulties. Indeed a could be a function of x and y and would still not disrupt the linearity of the equations.

If, however, we have an equation of the form

$$\frac{\partial u}{\partial y} = \frac{\partial^2 u}{\partial x^2} + u^4, \qquad 0 \le x \le 1, \quad y > 0,$$

then the Crank–Nicolson scheme becomes

$$\frac{u_j^{k+1} - u_j^k}{(\Delta y)} = \frac{1}{2}\left\{\frac{u_{j+1}^{k+1} - 2u_j^{k+1} + u_{j-1}^{k+1}}{(\Delta x)^2} + (u_j^{k+1})^4 + \frac{u_{j+1}^k - 2u_j^k + u_{j-1}^k}{(\Delta x)^2} + (u_j^k)^4\right\},$$

$$(9.4.8)$$

giving rise to non-linear algebraic equations owing to the presence of the term $(u_j^{k+1})^4$. The term $(u_j^k)^4$ presents no problems because it can be computed from the known value u_j^k. A possible approach to this problem is to linearize (9.4.8). Since

$$\frac{u_j^{k+1} - u_j^k}{\Delta y} \simeq \left.\frac{\partial u}{\partial y}\right|_{x_j, y_k},$$

it follows that (retaining only first order terms)

$$(u_j^{k+1})^4 \simeq \left[u_j^k + \Delta y\left(\frac{\partial u}{\partial y}\right)_{x_j, y_k}\right]^4$$

$$\simeq (u_j^k)^4 + 4(\Delta y)(u_j^k)^3\left(\frac{\partial u}{\partial y}\right)_{x_j, y_k}$$

$$\simeq (u_j^k)^4 + 4(u_j^k)^3(u_j^{k+1} - u_j^k).$$

Using this expression for $(u_j^{k+1})^4$ in (9.4.8) we have the linear difference equation

$$-\tfrac{1}{2}\rho u_{j+1}^{k+1} + [1+\rho - 2(\Delta y)(u_j^k)^3]u_j^{k+1} - \tfrac{1}{2}\rho u_{j-1}^{k+1}$$
$$= \tfrac{1}{2}\rho u_{j+1}^k + [1-\rho]u_j^k + \tfrac{1}{2}\rho u_{j-1}^k - \Delta y(u_j^k)^4,$$

which can be solved in the usual way, provided that we have the requisite initial and boundary conditions [see I, § 14.13].

The equation

$$\frac{\partial u}{\partial y} = \frac{\partial^2 u}{\partial x^2} + \frac{1}{x}\frac{\partial u}{\partial x}, \tag{9.4.9}$$

is a parabolic equation which frequently arises. We now solve this problem using the Crank–Nicolson scheme and replacing $(\partial u/\partial x)_{j,k}$ by the central difference approximation $(u_{j+1}^k - u_{j-1}^k)/(2\Delta x)$.

The differential equation reduces to

$$\frac{u_j^{k+1} - u_j^k}{\Delta y} = \frac{1}{2}\left\{ \frac{u_{j+1}^{k+1} - 2u_j^{k+1} + u_{j-1}^{k+1}}{(\Delta x)^2} + \frac{1}{j\,\Delta x}\left(\frac{u_{j+1}^{k+1} - u_{j-1}^{k+1}}{2\,\Delta x} \right) \right.$$
$$\left. + \frac{u_{j+1}^k - 2u_j^k + u_{j-1}^k}{(\Delta x)^2} + \frac{1}{j\,\Delta x}\left(\frac{u_{j+1}^k - u_{j-1}^k}{2\,\Delta x} \right) \right\},$$

or

$$-\left(1 - \frac{1}{2j}\right)\frac{\rho}{2}u_{j-1}^{k+1} + (1+\rho)u_j^{k+1} - \left(1 + \frac{1}{2j}\right)\frac{\rho}{2}u_{j+1}^{k+1}$$
$$= \left(1 - \frac{1}{2j}\right)\frac{\rho}{2}u_{j-1}^k + (1-\rho)u_j^k + \left(1 + \frac{1}{2j}\right)\frac{\rho}{2}u_{j+1}^k, \tag{9.4.10}$$

where $\rho = \Delta y/(\Delta x)^2$. This can be solved provided that we have the appropriate initial and boundary conditions.

One typical boundary condition for equations of the type (9.4.9) is

$$\frac{\partial u}{\partial x} = 0 \quad \text{on } x = 0, \quad y > 0. \tag{9.4.11}$$

Thus u_0^k is not specified explicitly and there is obviously going to be some difficulty with the term $(1/x)\,\partial u/\partial x$ occurring in the differential equation (9.4.9). A possible way around this problem is to write down the Taylor expansion for $\partial u(x, y)/\partial x$ about the point $(0, y)$ [see IV, Theorem 5.8.1]

$$\frac{\partial u}{\partial x}(x, y) = \frac{\partial u}{\partial x}(0, y) + x\frac{\partial^2 u}{\partial x^2}(0, y) + \frac{x^2}{2}\frac{\partial^3 u}{\partial x^3}(0, y) + \ldots,$$

and we see that

$$\lim_{x \to 0} \frac{1}{x}\frac{\partial u}{\partial x}(x, y) = \frac{\partial^2 u}{\partial x^2}(0, y).$$

This means that, when $x = 0$, the differential equation (9.4.9) reduces to

$$2\frac{\partial^2 u}{\partial x^2} = \frac{\partial u}{\partial y}. \tag{9.4.12}$$

This can be approximated by the Crank–Nicolson scheme

$$-\rho u_{-1}^{k+1} + (1+2\rho)u_0^{k+1} - \rho u_1^{k+1} = \rho u_{-1}^k + (1-2\rho)u_0^k + \rho u_1^k, \tag{9.4.13}$$

which introduces the fictitious term u_{-1}^k. The boundary condition (9.4.11) can be represented by the central difference approximation

$$\frac{u_1^k - u_{-1}^k}{2(\Delta x)} = 0,$$

that is $u_{-1}^k = u_1^k$ for all k. Hence the first equation, obtained from (9.4.13), is

$$(1+2\rho)u_0^{k+1} - 2\rho u_1^{k+1} = (1-2\rho)u_0^k + 2\rho u_1^k.$$

The implicit scheme (9.4.10) can now be used with $j = 1, 2, \ldots, n-1$ and, assuming that the second boundary condition is

$$u(x_n, y) = u_n^k = 0,$$

we have the system

$$
\begin{bmatrix}
(1+2\rho) & -2\rho & & & & \\
-\dfrac{\rho}{4} & (1+\rho) & -\dfrac{3\rho}{4} & & & \\
& -\dfrac{3\rho}{8} & (1+\rho) & -\dfrac{5\rho}{8} & & \\
& & \ddots & \ddots & \ddots & \\
& & & -\left(1-\dfrac{1}{2(n-1)}\right)\dfrac{\rho}{2} & (1+\rho)
\end{bmatrix}
\begin{bmatrix}
u_0^{k+1} \\
u_1^{k+1} \\
u_2^{k+1} \\
\vdots \\
u_{n-1}^{k+1}
\end{bmatrix}
$$

$$
=
\begin{bmatrix}
(1-2\rho) & 2\rho & & & & \\
\dfrac{\rho}{4} & (1-\rho) & \dfrac{3\rho}{4} & & & \\
& \dfrac{3\rho}{8} & (1-\rho) & \dfrac{5\rho}{8} & & \\
& & \ddots & \ddots & \ddots & \\
& & & \left(1-\dfrac{1}{2(n-1)}\right)\dfrac{\rho}{2} & (1-\rho)
\end{bmatrix}
\begin{bmatrix}
u_0^{k} \\
u_1^{k} \\
u_2^{k} \\
\vdots \\
u_{n-1}^{k}
\end{bmatrix},
$$

which can be solved by the usual method.

In this section we have tried to illustrate how different parabolic equations can be approximated by finite difference schemes. The initial and boundary conditions have been kept as simple as possible so that we can concentrate on the replacement of the differential equation by the finite difference approximation. The same replacement technique is used in the next section where we also take the opportunity of considering other types of boundary conditions.

9.5. ELLIPTIC EQUATIONS

Elliptic equations frequently arise from steady state considerations and give rise to boundary-value problems, that is, the solution is required inside some region bounded by a closed surface. For the two-dimensional problem the region R, say, will be a plane area bounded by a closed curve Γ. To obtain a unique solution some condition of the solution must be known at each point of the boundary Γ. We require either a Dirichlet condition, that is, the solution u has to be specified at each point on the boundary or a Neumann condition where the normal derivative $\partial u/\partial n$ of u, must be specified on Γ [see also IV, § 8.2]. Mixed problems can also arise when u and $\partial u/\partial n$ are specified on different parts of the boundary. The Dirichlet condition is the easier to deal with, and we shall commence our discussion by examining problems with this type of boundary condition.

Consider Poisson's equation

$$\frac{\partial^2 u}{\partial x^2} + \frac{\partial^2 u}{\partial y^2} = -2, \tag{9.5.1}$$

inside a unit square $0 \le x \le 1$, $0 \le y \le 1$, subject to the boundary conditions $u = 0$ on $x = 0$ and $x = 1$ and

$$u = \begin{cases} x, & 0 \le x \le \frac{1}{2} \\ 1-x, & \frac{1}{2} \le x \le 1 \end{cases} \quad \text{on } y = 0 \quad \text{and} \quad y = 1.$$

The region of integration is now covered by a rectangular mesh, and when the differential coefficients are replaced by the finite-difference approximations (9.1.14) and (9.1.15) at each node we obtain that

$$\frac{u_{j+1}^k - 2u_j^k + u_{j-1}^k}{(\Delta x)^2} + \frac{u_j^{k+1} - 2u_j^k + u_j^{k-1}}{(\Delta y)^2} = -2.$$

With equations of this type it is convenient to use a square mesh, that is, $\Delta x = \Delta y = h$. Thus we have the 5-point molecule

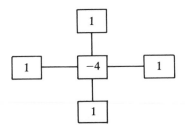

to represent $h^2 \nabla^2$ where

$$\nabla^2 \equiv \frac{\partial^2}{\partial x^2} + \frac{\partial^2}{\partial y^2}.$$

A further simplification is to make a slight change in the notation and number the internal mesh points sequentially. Using the notation of Figure 9.5.1 we see that

$$h^2 \nabla^2 u_P = u_N + u_W + u_S + u_E - 4u_P. \tag{9.5.2}$$

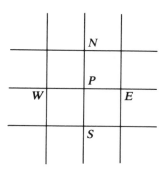

Figure 9.5.1

Taking $h = 0.25$ and referring to Figure 9.5.2 the finite-difference approximation to Poisson's equation (9.5.1) at point 1 becomes

$$-4u_1 + u_A + u_B + u_4 + u_2 = -0.125.$$

Points A and B lie on the boundary so that u_A and u_B are given by the boundary condition, that is $u_A = 0.25$ and $u_B = 0$. Hence

$$-4u_1 + u_2 + u_4 = 0.375.$$

Working sequentially through the other points we find at point 2

$$u_1 - 4u_2 + u_3 + u_5 = -0.625,$$

at point 3

$$u_2 - 4u_3 + u_6 = -0.375,$$

at point 4

$$u_1 - 4u_4 + u_5 + u_7 = -0.125,$$

at point 5

$$u_2 + u_4 - 4u_5 + u_6 + u_8 = -0.125,$$

at point 6

$$u_3 + u_5 - 4u_6 + u_9 = -0.125,$$

at point 7

$$u_4 - 4u_7 + u_8 = -0 \cdot 375,$$

at point 8

$$u_5 + u_7 - 4u_8 + u_9 = -0 \cdot 625,$$

at point 9

$$u_6 + u_8 - 4u_9 = -0 \cdot 375.$$

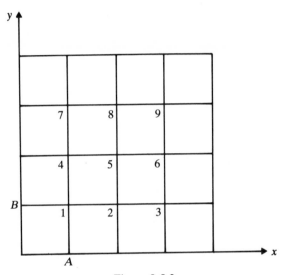

Figure 9.5.2

We have nine equations in nine unknowns, which can be written in the matrix form:

$$\mathbf{M u} = \mathbf{b}$$

where **u** is the vector of unknowns, **b** the 'given' right-hand side and **M** the sparse matrix

$$
\begin{bmatrix}
-4 & 1 & 0 & 1 & 0 & 0 & 0 & 0 & 0 \\
1 & -4 & 1 & 0 & 1 & 0 & 0 & 0 & 0 \\
0 & 1 & -4 & 0 & 0 & 1 & 0 & 0 & 0 \\
1 & 0 & 0 & -4 & 1 & 0 & 1 & 0 & 0 \\
0 & 1 & 0 & 1 & -4 & 1 & 0 & 1 & 0 \\
0 & 0 & 1 & 0 & 1 & -4 & 0 & 0 & 1 \\
0 & 0 & 0 & 1 & 0 & 0 & -4 & 1 & 0 \\
0 & 0 & 0 & 0 & 1 & 0 & 1 & -4 & 1 \\
0 & 0 & 0 & 0 & 0 & 1 & 0 & 1 & -4
\end{bmatrix}
$$

The dotted lines have been drawn in to show the natural partitioning of the matrix **M** into the block tridiagonal form [see I, § 6.6]

$$\begin{bmatrix} \mathbf{A} & \mathbf{I} & \mathbf{0} \\ \mathbf{I} & \mathbf{A} & \mathbf{I} \\ \mathbf{0} & \mathbf{I} & \mathbf{A} \end{bmatrix}$$

where

$$\mathbf{A} = \begin{bmatrix} -4 & 1 & 0 \\ 1 & -4 & 1 \\ 0 & 1 & -4 \end{bmatrix}$$

and **I** is the 3×3 unit matrix and **0** the 3×3 null matrix.

The solution to the elliptic partial differential equation defined by (9.5.1) is given in Table 9.5.1. Nine equations in nine unknowns is not too difficult to handle, but if we halve the interval to $\Delta x = \Delta y = 0 \cdot 125$ then we have 49 equations or if we take $\Delta x = \Delta y = 0 \cdot 01$ we have 9801 equations to solve. An interval h generates $(N-1)^2$ equations where $N = 1/h$ for this problem. Clearly one of the major problems of solving elliptic partial differential equations is handling the large matrices that arise. A full discussion of this problem is beyond the scope of this chapter and the reader is referred to Chapter 3 and (Smith (1965) Chapter 5, pp. 143 ff and Ames (1977) Chapter 3, § 3.2–§ 3.11).

$u_1 = 0 \cdot 2422, \quad u_2 = 0 \cdot 3594, \quad u_3 = 0 \cdot 2422,$
$u_4 = 0 \cdot 2344, \quad u_5 = 0 \cdot 3281, \quad u_6 = 0 \cdot 2344,$
$u_7 = 0 \cdot 2422, \quad u_8 = 0 \cdot 3594, \quad u_9 = 0 \cdot 2422.$

Table 9.5.1

We could have anticipated the symmetry of the required solution for Poisson's equation (9.5.1) and saved some effort in solving the linear equations. The usual technique is to label all points with the same solution with the same number. Thus we have the numbering system shown in Figure 9.5.3.

The equations to be solved now become

$$\begin{bmatrix} -4 & 1 & 1 & 0 \\ 2 & -4 & 0 & 1 \\ 2 & 0 & -4 & 1 \\ 0 & 2 & 2 & -4 \end{bmatrix} \begin{bmatrix} u_1 \\ u_2 \\ u_3 \\ u_4 \end{bmatrix} = - \begin{bmatrix} 0 \cdot 375 \\ 0 \cdot 625 \\ 0 \cdot 125 \\ 0 \cdot 125 \end{bmatrix}.$$

Decreasing the interval decreases the magnitude of the truncation error thus producing more accurate results at the expense of increasing the amount of work involved in solving the equations. The five-point formula for $h^2 \nabla^2$ has a truncation error $O(h^4)$ hence a truncation error $O(h^2)$ in the finite-difference

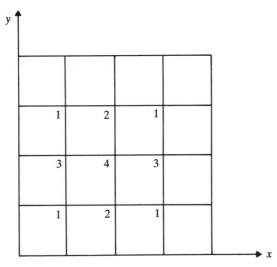

Figure 9.5.3

approximation to Poisson's equation (see Smith (1965), Chapter 5, exercise 9). In an attempt to reduce the truncation error a 9-point formula given by the molecule

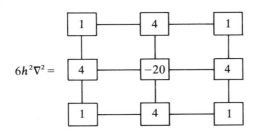

is often used to represent the Laplacian ∇^2. It should be noted that when solving Poisson's equation $\nabla^2 u = f(x, y)$ the truncation error will again be $O(h^2)$ unless $f(x, y)$ is harmonic, that is, $\nabla^2 f(x, y) = 0$ when it reduces to $O(h^4)$.

The 9-point molecule can gainfully be applied to equation (9.5.1), since $\nabla^2(-2) = 0$. Using Figure 9.5.3 we have that

$$
\begin{bmatrix}
-20 & 4 & 4 & 1 \\
4 & -10 & 1 & 2 \\
4 & 1 & -10 & 2 \\
1 & 2 & 2 & -5
\end{bmatrix}
\begin{bmatrix}
u_1 \\ u_2 \\ u_3 \\ u_4
\end{bmatrix}
= -
\begin{bmatrix}
2 \cdot 25 \\
1 \cdot 615 \\
0 \cdot 375 \\
0 \cdot 1875
\end{bmatrix}
$$

giving

$$u_1 = 0 \cdot 2448, \qquad u_2 = 0 \cdot 3477$$
$$u_3 = 0 \cdot 2340, \qquad u_4 = 0 \cdot 3191.$$

Note that the results given in Table 9.5.1 agree with these figures to two decimal places.

The number of equations to be solved can be reduced if we use a method first developed by Peaceman and Rachford (1955). Consider the equation $\nabla^2 u = f(x, y)$ in a square region R with the Dirichlet condition $u = 0$ on the boundary. The 5-point molecule gives

$$u_N + u_W + u_S + u_E - 4u_P = h^2 f_P,$$

where

$$f_P = f(x_P, y_P),$$

or

$$u_E - 2u_P + u_W = -u_N + 2u_P - u_S + h^2 f_P.$$

Introducing $-\rho u_P$ to both sides, where ρ is a scalar parameter we have that

$$u_E - (2 + \rho)u_P + u_W = -u_N + (2 - \rho)u_P - u_S + h^2 f_P,$$

which forms the basis of the iterative formula

$$u_E^{(i+1)} - (2 - \rho)u_P^{(i+1)} + u_W^{(i+1)} = -u_N^{(i)} + (2 - \rho)u_P^{(i)} - u_S^{(i)} + h^2 f_P,$$

$$i = 0, 1, 2, \ldots, \tag{9.5.3}$$

where $u_P^{(i)}$ is the ith approximation to u_P, $u_P^{(0)}$ being the initial estimate for u_P.

Using Figure 9.5.2 and working along successive rows we have for the first row

$$-(2 + \rho)u_1^{(i+1)} + u_2^{(i+1)} = (2 - \rho)u_1^{(i)} - u_4^{(i)} + h^2 f_1,$$

$$u_1^{(i+1)} - (2 + \rho)u_2^{(i+1)} + u_3^{(i+1)} = (2 - \rho)u_2^{(i)} - u_5^{(i)} + h^2 f_2,$$

$$u_2^{(i+1)} - (2 + \rho)u_3^{(i+1)} = (2 - \rho)u_3^{(i)} - u_6^{(i)} + h^2 f_3.$$

three equations which can be solved for $u_1^{(i+1)}$, $u_2^{(i+1)}$, $u_3^{(i+1)}$.

Working along the second row yields another three equations for $u_4^{(i+1)}$, $u_5^{(i+1)}$ and $u_6^{(i+1)}$ and the third row enables us to find $u_7^{(i+1)}$, $u_8^{(i+1)}$ and $u_9^{(i+1)}$.

Using the 5-point formula directly resulted in nine equations in nine unknowns. Using (9.5.3) we have three equations to be solved three times. In general (9.5.3) produces $(N - 1)$ equations to be solved $(N - 1)$ times compared with the 5-point molecule which produces $(N - 1)^2$ equations. Thus some of the serious difficulties associated with the storage of large matrices inside a computer are greatly reduced.

Unfortunately using (9.5.3) for each successive iteration leads to instability. However, the 5-point molecule can also be used to produce

$$u_N^{(i+1)} - (2 + \rho)u_P^{(i+1)} + u_S^{(i+1)} = -u_E^{(i)} + (2 - \rho)u_P^{(i)} - u_W^{(i)} + h^2 f_P,$$

$$i = 0, 1, 2, \ldots, \tag{9.5.4}$$

implying that we now work along columns. Unilateral use of (9.5.4) would again prove unstable, but a combination of (9.5.3) and (9.5.4), where we alternate between solving along rows and then along columns, yields a stable method.

The alternating direction implicit method described here consists of two steps per iteration defined by

$$u_W^{(i+\frac{1}{2})} - (2+\rho)u_P^{(i+\frac{1}{2})} + u_E^{(i+\frac{1}{2})} = -u_S^{(i)} + (2-\rho)i_P^{(i)} - u_N^{(i)} + h^2 f_P$$

$$u_S^{(i+1)} - (2+\rho)u_P^{(i+1)} + u_N^{(i+1)} = -u_E^{(i+\frac{1}{2})} + (2-\rho)u_P^{(i+\frac{1}{2})} - u_W^{(i+\frac{1}{2})} + h^2 f_P$$

$$i = 0, 1, 2, \ldots.$$

The parameter ρ must be kept constant for the 2 steps involved in going from $u^{(i)}$ to $u^{(i+1)}$ but can be changed for the next and subsequent iterations. If ρ is kept constant then its optimum value for best rate of convergence for a square region using a square mesh with $(N-1)^2$ internal nodes is $\rho = 2 \sin \pi/N$. A discussion on the rate of convergence and references to other alternating direction implicit methods can be found in Ames ((1965), Chapter 3, § 3.12 and § 3.13).

Throughout the remainder of this section we shall, for simplicity, use the 5-point molecule to approximate $h^2 \nabla^2$.

Whenever we have rectangular boundaries it is generally possible to choose a mesh system such that the boundary lines are themselves a part of the mesh lines and that mesh lines intersect on the boundary. This is not always possible with curved boundaries. Consider Figure 9.5.4.

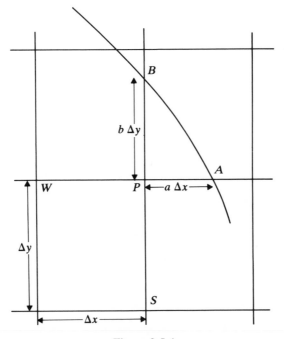

Figure 9.5.4

The 5-point molecule can be applied to all internal nodes such as W which are entirely within the closed region R and at least a full mesh length away from the boundary Γ. The point P, however, is not a full mesh length from the boundary and in the illustration is a distance $a\,\Delta x$ from the boundary at A and a distance $b\,\Delta y$ from the boundary at B where $0 < a,\ b < 1$. A special formula is therefore required to approximate the partial differential equation at P. Using Taylor's expansion and the ideas outlined in section 9.1 (cf. equation (9.1.13)) we have that

$$u_A = u_P + a\,\Delta x\,\frac{\partial u_P}{\partial x} + \tfrac{1}{2}(a\,\Delta x)^2\frac{\partial^2 u}{\partial x^2} + O((\Delta x)^3)$$

$$u_W = u_P - \Delta x\,\frac{\partial u_P}{\partial x} + \tfrac{1}{2}(\Delta x)^2\frac{\partial^2 u_P}{\partial x^2} + O((\Delta x)^3).$$

(9.5.5)

Eliminating $(\partial u/\partial x)_P$ and rearranging gives

$$\frac{\partial^2 u_P}{\partial x^2} = \frac{2}{(\Delta x)^2}\left[\frac{u_A}{a(1+a)} + \frac{u_W}{(1+a)} - \frac{u_P}{a}\right].$$

(9.5.6a)

Similarly we can obtain

$$\frac{\partial^2 u_P}{\partial y^2} = \frac{2}{(\Delta y)^2}\left[\frac{u_B}{b(1+b)} + \frac{u_S}{(1+b)} - \frac{u_P}{b}\right].$$

(9.5.6b)

If $(\partial u/\partial x)_P$ is required, then we must eliminate the second order terms from (9.5.5) to give

$$\frac{\partial u_P}{\partial x} = \frac{1}{\Delta x}\left[\frac{u_A}{a(1+a)} - \frac{au_W}{(1+a)} - \frac{(1-a)u_P}{a}\right].$$

(9.5.6c)

To illustrate the use of these approximations we solve

$$\frac{\partial^2 u}{\partial x^2} + (1+x^2)\frac{\partial^2 u}{\partial y^2} = 0,$$

inside the region shown in Figure 9.5.5, where the curve BCD is a semi-circle with diameter BD, AB is the line $y = 0\cdot5$ and ED is $y = -0\cdot5$ with $0 \le x \le 0\cdot5$, BD is $x = 0\cdot5$, subject to the mixed boundary conditions

$$\frac{\partial u}{\partial x} = 0 \qquad \text{on } OA,$$

$$u = 0\cdot5 \qquad \text{on } AB,$$

$$u = 1 - y \qquad \text{on } BC,$$

and there is symmetry about the x-axis.

Taking for convenience a square mesh with $\Delta x = \Delta y = 0\cdot25$ then we require the solution at the points labelled $1, 2, \ldots, 7, 8$. A modification of the 5-point formula can be used at nodes 2, 3, 4, 6 and 7 (note that the 5-point molecule

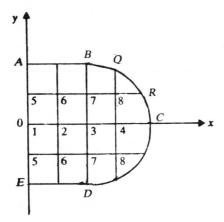

Figure 9.5.5

given earlier approximates $h^2\nabla^2$ only) that is with the notation of Figure 9.5.1.

$$\frac{u_W - 2u_P + u_E}{(0\cdot 25)^2} + (1 + x_P^2)\frac{(u_S - 2u_P + u_N)}{(0\cdot 25)^2} = 0. \qquad (9.5.7)$$

Node 8 is adjacent to the boundary and less than a full mesh length away and we have to use approximations (9.5.6) (a) and (b). First, however, we have to determine how far node 8 is from the boundary. Using the geometry of the circle (of radius $0\cdot 5$ and centred at point 3) we find that $a = b = \sqrt{3} - 1 = 0\cdot 7321$. Thus at point 8 we must use

$$\frac{2}{(0\cdot 25)^2}\left\{\frac{u_R}{(1\cdot 7321)(0\cdot 7321)} + \frac{u_7}{1\cdot 7321} - \frac{u_8}{0\cdot 7321}\right\} +$$

$$(1 + x_8^2)\frac{2}{(0\cdot 25)^2}\left\{\frac{u_Q}{(1\cdot 7321)(0\cdot 7321)} + \frac{u_4}{1\cdot 7321} - \frac{u_8}{0\cdot 7321}\right\} = 0. \qquad (9.5.8)$$

The value of y at R is $0\cdot 25$ hence $u_R = 0\cdot 75$. Similarly we find $u_Q = 0\cdot 5670$.

Nodes 1 and 5 are on the boundary. Thus using (9.5.7) introduces 'fictitious' values outside the region (cf. § 9.2). For example, at point 1 equation (9.5.7) gives after simplification

$$u_W - 4u_1 + u_2 + 2u_5 = 0,$$

where u_W is the value of u at the point with coordinates $(0, -0\cdot 25)$. The derivative boundary condition $\partial u/\partial x = 0$ implies $u_W - u_2 = 0$, that is $u_W = u_2$. Therefore at point 1 we have that

$$-4u_1 + 2u_2 + 2u_5 = 0.$$

Similarly at point 5 we have that

$$u_1 - 4u_5 + 2u_6 = -0.5.$$

The complete set of equations to be solved becomes:

$$\begin{bmatrix} -2 & 0 & 0 & 0 & 1 & 0 & 0 & 0 \\ 1 & -4\cdot125 & 1 & 0 & 0 & 2\cdot125 & 0 & 0 \\ 0 & 1 & -4\cdot5 & 0 & 0 & 0 & 2\cdot5 & 0 \\ 0 & 0 & 1 & -5\cdot125 & 0 & 0 & 0 & 3\cdot125 \\ 1 & 0 & 0 & 0 & -4 & 2 & 0 & 0 \\ 0 & 1\cdot0625 & 0 & 0 & 1 & -4\cdot125 & 1 & 0 \\ 0 & 0 & 1\cdot25 & 0 & 0 & 1 & -4\cdot5 & 1 \\ 0 & 0 & 0 & 0\cdot9021 & 0 & 0\cdot5774 & 0 & -2\cdot7683 \end{bmatrix} \times \begin{bmatrix} u_1 \\ u_2 \\ u_3 \\ u_4 \\ u_5 \\ u_6 \\ u_7 \\ u_8 \end{bmatrix}$$

$$= \begin{bmatrix} 0 \\ 0 \\ 0 \\ 0 \\ 0\cdot5 \\ 0\cdot53125 \\ 0\cdot625 \\ 0\cdot9944 \end{bmatrix}.$$

Special care must be taken also with rectangular boundaries involving re-entrant corners, that is, boundaries with internal angles greater than π. Two typical regions are shown in Figure 9.5.6 where the re-entrant corners have been marked.

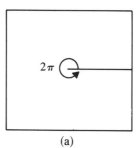

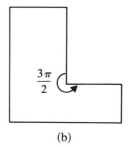

(a)   (b)

Figure 9.5.6

The re-entrant corner causes a disturbance or singularity, not in the solution, but in its derivative and special care is required with the use of finite-differences in the neighbourhood of the corner. A full discussion is beyond our scope and the interested reader is referred to the work of Motz (1946) and Woods (1953); see also Ames ((1965), Chapter 5, § 5.1) for a discussion of the 'cracked plate', Figure 9.5.6(a) and Crank ((1975), Chapter 8, § 8.10.2) for a discussion on the *L*-shaped region of Figure 9.5.6(b). The basis of the method is to use a finite-difference approximation at mesh points far enough away from the re-entrant corner to be unaffected by the singularity and a Fourier series solution near the corner [see IV, Chapter 20].

Throughout this section we have concentrated on second order equations. We now turn our attention to higher order partial differential equations in particular the fourth order biharmonic equation.

$$\nabla^4 u = f(x, y),$$

where

$$\nabla^4 \equiv \frac{\partial^4}{\partial x^4} + \frac{2 \,\partial^4}{\partial x^2 \,\partial y^2} + \frac{\partial^4}{\partial y^4} \equiv (\nabla^2)^2,$$

is the biharmonic operator.

This equation is elliptic in the sense that for a well-posed problem we required a closed region R with two conditions specified at every point of the closed boundary Γ.

The operator ∇^4 can be represented by the thirteen-point molecule; see Smith ((1965), Chapter 5) and Ames ((1977), Chapter 5, § 5.3)

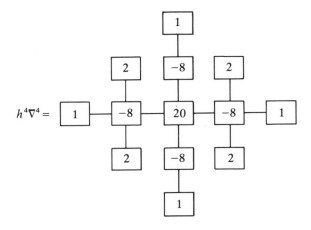

and is used in the same way as all other molecules given in this chapter.

Suppose we require the solution to $\nabla^4 u = x^2$, inside a square of side 2 with centre at the origin and sides parallel to the cartesian axes, subject to the boundary conditions

$$u = \frac{\partial u}{\partial x} = 0 \quad \text{on } |x| = 1,$$

$$u = \nabla^2 u = 0 \quad \text{on } |y| = 1.$$

For simplicity we choose a square mesh of side 0·4 and number the internal nodes as shown in Figure 9.5.7 due account having been taken of the symmetry relative to the axes.

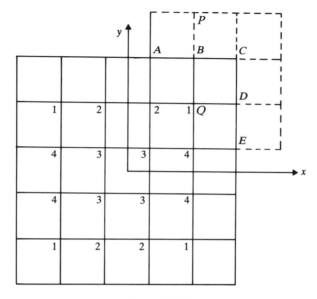

Figure 9.5.7

The dotted lines show an extension of the mesh system that is necessary to deal with the fictitious values that are introduced by the 13-point molecule.

At point 1 we have that

$$20u_1 - 7u_2 + 2u_3 - 7u_4 + 2u_A - 8u_B + 2u_C - 8u_D$$
$$+ 2u_E + u_P + u_Q = (0{\cdot}4)^4 \times (0{\cdot}6)^2.$$

The boundary conditions give

$$u_A = u_B = u_C = u_D = u_E = 0,$$

and

$$\frac{u_P - 2u_B + u_1}{(0{\cdot}4)^2} = 0,$$

that is

$$u_P = -u_1$$

and

$$\frac{u_Q - u_1}{2(0{\cdot}4)} = 0.$$

Hence

$$u_Q = u_1.$$

Thus at point 1

$$20u_1 - 7u_2 + 2u_3 - 7u_4 = (0{\cdot}4)^4(0{\cdot}6)^2.$$

Similarly at point 2

$$-7u_1+11u_2-5u_3+2u_4=1{\cdot}024\times10^{-3},$$

at point 3

$$2u_1-5u_2+6u_3-5u_4=1{\cdot}024\times10^{-3},$$

at point 4

$$-7u_1+2u_2-5u_3+13u_4=9{\cdot}216\times10^{-3}.$$

Giving

$$u_1=2{\cdot}143\times10^{-3},\qquad u_2=2{\cdot}949\times10^{-3},$$
$$u_3=4{\cdot}545\times10^{-3},\qquad u_4=3{\cdot}158\times10^{-3}.$$

A discussion on boundary value problems would not be complete without considering eigenvalue problems. In its simplest form the problem is to find one (often the smallest) or more of the constants λ, called the eigenvalue, and the corresponding eigenfunction $u(x, y)$, satisfying the partial differential equation

$$\nabla^2 u+\lambda u=0 \quad \text{in } R,$$

with $u=0$ on Γ (the closed boundary of R).

Using the 5-point molecule and the notation of Figure 9.5.1, the finite-difference approximation becomes

$$u_N+u_W+u_S+u_E+(\Lambda-4)u_P=0,$$

where $\Lambda=\lambda h^2$.

Taking a rectangular region and the mesh shown in Figure 9.5.8 we have the algebraic eigenvalue problem [see Chapter 4 and I, (7.2.1)]

$$\begin{bmatrix} \Lambda-4 & 1 & 0 & 1 & 0 & 0 \\ 1 & \Lambda-4 & 1 & 0 & 1 & 0 \\ 0 & 1 & \Lambda-4 & 0 & 0 & 1 \\ 1 & 0 & 0 & \Lambda-4 & 1 & 0 \\ 0 & 1 & 0 & 1 & \Lambda-4 & 1 \\ 0 & 0 & 1 & 0 & 1 & \Lambda-4 \end{bmatrix}\begin{bmatrix} u_1 \\ u_2 \\ u_3 \\ u_4 \\ u_5 \\ u_6 \end{bmatrix}=\mathbf{0}.$$

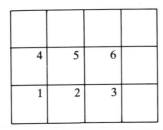

Figure 9.5.8

We note that the differential eigenvalue problem has an infinite number of eigenvalues, whereas the algebraic eigenvalue problem has a finite-number (equal to the number of mesh points). If $\Lambda_1, \Lambda_2, \ldots, \Lambda_n$ are the eigenvalues of the algebraic system and $\lambda_1, \lambda_2, \ldots$ are the eigenvalues of the differential system both taken in increasing order of magnitude then

$$\Lambda_i \simeq \lambda_i h^2, \qquad i = 1, 2, \ldots, n,$$

where $|\Lambda_i - \lambda_i h^2|$ increases as i increases. That is, with a small mesh, we can expect Λ_1 / h^2 to be a good approximation to λ_1 but that the goodness of the approximation decreases as we progress to $\Lambda_2, \Lambda_3, \ldots$. Further information on this problem can be found in Ames ((1965), Chapter 5, § 5.3) and Forsythe and Wasow ((1900), Chapter 3, § 24).

Throughout this section (and most of the chapter) we have concentrated on standard and well tested finite-difference approximations. A method gaining in popularity is the finite-element method but as this involves ideas beyond the scope of this text the interested reader is referred to an introduction to the subject given in Ames ((1965), Chapter 6) and the more detailed study of Mitchell and Wait (1977).

D.E.J.

REFERENCES

Ames, W. F. (1977). *Numerical Methods for Partial Differential Equations* (2nd Ed.), Nelson.

Cohen, A. M. (1973). *Numerical Analysis*, McGraw-Hill.

Crank, J. (1975). *Mathematics of Diffusion*, 2nd Ed., O.U.P.

Crank, J. and Nicolson, P. (1947). *Proc. Camb. Phil. Soc.* **43**, 50–67.

Du Fort, E. C. and Frankel, S. P. (1953). *Mathl. Tabl. Natn. Res. Coun., Wash.* **7**, 135.

Forsythe, G. E. and Wasow, W. R. (1969). *Finite-Difference Methods for Partial Differential Equations*, Wiley, New York.

Fox, L. (1962). *Numerical Solution of Ordinary and Partial Differential Equations*, O.U.P.

Mitchell, A. R. and Wait, R. (1977). *The Finite-Element Method in Partial Differential Equations*, Wiley.

Motz, H. (1946). *Quart. Appl. Math.* **4**, 371–377.

Peaceman, D. N. and Rachford, H. H. Jr. (1955). *J. Soc. Ind. Appl. Math.* **3**, 28.

Smith, G. D. (1965). *Numerical Solution of Partial Differential Equations*, O.U.P.

Woods, L. C. (1953). *Quart. J. Mech.* **6**, 163–185.

CHAPTER 10

Integral Equations

10.1. LINEAR INTEGRAL EQUATIONS

10.1.1. Classification of Linear Integral Equations

An *integral equation* is an equation in which a function $\phi(x)$, which we desire to determine, appears in the integrand of the equation [see IV, § 4.1]. Examples of linear integral equations are (10.1.1)–(10.1.4) below. The equation

$$\int_a^b K(x, y)\phi(y)\, dy = f(x), \qquad c < x < d \qquad (10.1.1)$$

where $f(x)$ and $K(x, y)$ are given functions and a and b are given fixed numbers is known as a *Fredholm equation of the first kind*. $K(x, y)$ is referred to as the *kernel* of the integral equation. When $f(x) = 0$ in (10.1.1) the equation is called a *homogeneous* equation of the first kind.

The equation

$$\phi(x) = f(x) + \lambda \int_a^b K(x, y)\phi(y)\, dy, \qquad a < x < b \qquad (10.1.2)$$

where $f(x)$ and $K(x, y)$ are given functions and a, b and λ are fixed numbers is called a *Fredholm equation of the second kind*. λ can be a real or complex-valued parameter but we shall assume for the most part that λ is real. When $f(x) = 0$ in (10.1.2) the equation is referred to as a homogeneous equation of the second kind.

In another type of integral equation, which also arises in practice, the upper limit of integration is variable. A typical equation is

$$\phi(x) = f(x) + \lambda \int_a^x K(x, y)\phi(y)\, dy, \qquad x > a \qquad (10.1.3)$$

and this is known as a *Volterra equation of the second kind*. An equation of the form

$$\int_a^x K(x, y)\phi(y)\, dy = f(x), \qquad x > a \qquad (10.1.4)$$

405

is called a *Volterra equation of the first kind*. A Volterra equation may be thought of as a Fredholm equation for which $K(x, y) = 0$ when $y > x$. If the kernel $K(x, y)$ in the above equations satisfies $K(x, y) = K(y, x)$ then it is termed *symmetric*.

All the equations given above can be thought of as particular examples of the general linear integral equation

$$\sigma(x)\phi(x) = f(x) + \lambda \int_a^{\rho(x)} K(x, y)\phi(y)\, dy$$

where $\rho(x) = x$ for Volterra equations, $\rho(x) = b$ for Fredholm equations. For equations of the first kind $\sigma(x) = 0$ whilst $\sigma(x) = 1$ for equations of the second kind.

10.1.2.   Connection with Ordinary Differential Equations. Green's Functions

If $p(x)$ and $q(x)$ are given functions the second order ordinary differential equation system [see IV, § 7.6]

$$\phi''(x) + p(x)\phi(x) = q(x), \qquad \phi(0) = \phi(1) = 0 \qquad (10.1.5)$$

is equivalent to the Fredholm equation of the second kind

$$\phi(x) = f(x) + \int_0^1 K(x, y)p(y)\phi(y)\, dy \qquad (10.1.6)$$

where

$$K(x, y) = \begin{cases} y(1-x) & y \le x \\ x(1-y) & x \le y \end{cases}$$

and

$$f(x) = -\int_0^1 K(x, y)q(y)\, dy.$$

This can be established as follows. Since $K(0, y) = 0 = K(1, y)$ it follows from (10.1.6) that

$$\phi(0) = \phi(1) = 0.$$

Further, since (10.1.6) can be expressed as

$$\phi(x) = -\int_0^x K(x, y)q(y)\, dy - \int_x^1 K(x, y)q(y)\, dy$$

$$+ \int_0^x K(x, y)p(y)\phi(y)\, dy + \int_x^1 K(x, y)p(y)\phi(y)\, dy$$

$$= -(1-x)\int_0^x yq(y)\, dy - x\int_x^1 (1-y)q(y)\, dy$$

$$+ (1-x)\int_0^x yp(y)\phi(y)\, dy + x\int_x^1 (1-y)p(y)\phi(y)\, dy$$

it follows that [see IV, Theorem 4.7.3]

$$\phi'(x) = \int_0^x yq(y)\,dy - (1-x)xq(x) + x(1-x)q(x) - \int_x^1 (1-y)q(y)\,dy$$

$$- \int_0^x yp(y)\phi(y)\,dy + (1-x)xp(x)\phi(x)$$

$$- x(1-x)p(x)\phi(x) + \int_x^1 (1-y)p(y)\phi(y)\,dy$$

$$= \int_0^1 yq(y)\,dy - \int_x^1 q(y)\,dy - \int_0^1 yp(y)\phi(y)\,dy$$

$$+ \int_x^1 p(y)\phi(y)\,dy.$$

A further differentiation produces

$$\phi''(x) = q(x) - p(x)\phi(x)$$

i.e. the solution of the integral equation (10.1.6) satisfies the differential equation system (10.1.5).

In a similar fashion we can establish the equivalence of the Volterra equation of the second kind

$$\phi(x) = \int_0^x (x-y)q(y)\,dy - \int_0^x (x-y)p(y)\phi(y)\,dy + \alpha + \beta x \quad (10.1.7)$$

with the initial value problem for ordinary differential equations

$$\phi''(x) + p(x)\phi(x) = q(x), \qquad \phi(0) = \alpha, \qquad \phi'(0) = \beta. \quad (10.1.8)$$

We have established the equivalence of the integral equations (10.1.6) and (10.1.7) with the ordinary differential equation systems (10.1.5) and (10.1.8) respectively by showing that the integral equation satisfies the differential system. However, it is instructive to find the integral equation which corresponds to a given differential equation system. Suppose we wish to solve the Sturm–Liouville problem [see IV, § 7.8]

$$L\phi = \frac{d}{dx}\left\{ p(x)\frac{d\phi}{dx} \right\} - q(x)\phi = f(x), \qquad a \le x \le b \quad (10.1.9)$$

with the associated boundary conditions

$$\alpha_1\phi(a) + \beta_1\phi'(a) = 0$$

$$\alpha_2\phi(b) + \beta_2\phi'(b) = 0.$$

Assume $\phi_1(x)$ and $\phi_2(x)$ are independent solutions of the homogeneous equation $L\phi = 0$ such that $\phi_1(x)$ and $\phi_2(x)$ satisfies the boundary conditions at *a* and *b* respectively (but not both points), i.e.

$$\alpha_1\phi_1(a)+\beta_1\phi_1'(a)=0$$

$$\alpha_2\phi_2(b)+\beta_2\phi_2'(b)=0.$$

We look for a solution of (10.1.9) of the form

$$\phi(x)=z_1(x)\phi_1(x)+z_2(x)\phi_2(x),\qquad(10.1.10)$$

where $z_1(x)$ and $z_2(x)$ are to be determined, by the method of variation of parameters [see IV, § 7.4.2]. Differentiation of (10.1.10) produces [see IV, (3.2.6)]

$$\phi'=z_1'\phi_1+z_2'\phi_2+z_1\phi_1'+z_2\phi_2'$$

and we choose z_1 and z_2 such that

$$z_1'\phi_1+z_2'\phi_2=0.$$

It then follows that

$$L\phi=\frac{d}{dx}[p\{z_1\phi_1'+z_2\phi_2'\}]-q\{z_1\phi_1+z_2\phi_2\}$$

$$=p(z_1'\phi_1'+z_2'\phi_2'),$$

on using the results $L\phi_1=L\phi_2=0$. Thus z_1' and z_2' satisfy the linear equations

$$z_1'\phi_1+z_2'\phi_2=0$$

$$p(z_1'\phi_1'+z_2'\phi_2')=f$$

and hence

$$z_1'=f\phi_2/p(\phi_2\phi_1'-\phi_1\phi_2'),$$

$$z_2'=-f\phi_1/p(\phi_2\phi_1'-\phi_1\phi_2').$$

The denominator in these 2 expressions is a constant since ϕ_1, ϕ_2 satisfy

$$0=\phi_2 L\phi_1-\phi_1 L\phi_2=\frac{d}{dx}\{p(\phi_2\phi_1'-\phi_1\phi_2')\},$$

so that by suitable scaling

$$z_1'=-f\phi_2,\qquad z_2'=f\phi_1$$

i.e.,

$$z_1(x)=\int_x \phi_2(y)f(y)\,dy,\qquad z_2(x)=\int^x \phi_1(y)f(y)\,dy$$

and all we have to do is to specify the limits of integration in both integrals. These limits are determined from the requirement that $\phi(x)$ satisfies the boundary conditions. For instance, since

$$\alpha_1\phi(x)+\beta_1\phi'(x)=\alpha_1(z_1\phi_1+z_2\phi_2)+\beta_1(z_1\phi_1'+z_2\phi_2')$$

$$=z_1(\alpha_1\phi_1(x)+\beta_1\phi_1'(x))+z_2(\alpha_1\phi_2(x)+\beta_1\phi_2'(x))$$

we have at $x = a$, by our choice of $\phi_1(x)$,

$$\alpha_1\phi(a) + \beta_1\phi'(a) = z_2(a)[\alpha_1\phi_2(a) + \beta_1\phi_2'(a)].$$

It follows that for the boundary condition at $x = a$ to be satisfied by $\phi(x)$ we must have $z_2(a) = 0$. This indicates that

$$z_2(x) = \int_a^x \phi_1(y)f(y)\,dy.$$

Similarly

$$z_1(x) = \int_x^b \phi_2(y)f(y)\,dy.$$

Hence, from (10.1.10)

$$\phi(x) = z_1(x)\phi_1(x) + z_2(x)\phi_2(x)$$

i.e.

$$\phi(x) = \int_a^b G(x, y)f(y)\,dy$$

where   (10.1.11)

$$G(x, y) = \begin{cases} \phi_1(y)\phi_2(x), & a \le y \le x \\ \phi_1(x)\phi_2(y), & x \le y \le b. \end{cases}$$

Equation (10.1.11) is the required integral equation formulation of the differential system (10.1.9). The kernel $G(x, y)$ is known as the *Green's function* associated with the operator L and the specified boundary conditions. The reader should consult Chambers (1976), regarding the case where $\phi_1(x)$, for example, satisfies both boundary conditions of the Sturm–Liouville problem simultaneously.

10.1.3. The Liouville–Neumann Series

If the kernel of a Fredholm integral equation of the second kind is small enough then the integral equation can be approximately solved by iteration. Write

$$\phi^{(0)}(x) = f(x)$$

and   (10.1.12)

$$\phi^{(n)}(x) = f(x) + \lambda \int_a^b K(x, y)\phi^{(n-1)}(y)\,dy$$

where a and b are assumed finite. Then the solution of the equation

$$\phi(x) = f(x) + \lambda \int_a^b K(x, y)\phi(y)\,dy \qquad\qquad (10.1.2)$$

is given by the series, called the *Liouville–Neumann series*,

$$\phi(x) = \lim_{n \to \infty} \phi^{(n)}(x) = f(x) + \lambda Kf + \lambda^2 K^2 f + \ldots + \lambda^n K^n f + \ldots \qquad (10.1.13)$$

where we define

$$Kf = \int_a^b K(x, y)f(y)\, dy$$

$$K^n f = \int_a^b K(x, y)K^{n-1} f\, dy, \quad n = 2, 3, \ldots \qquad (10.1.14)$$

The series converges [see IV, § 1.7] when $f(x)$ and $K(x, y)$ are continuous for $a \le x,\, y \le b$ [see IV, §§ 2.1 and 5.1] and

$$|\lambda|(b - a) \max_{a \le x, y \le b} |K(x, y)| < 1 \qquad (10.1.15)$$

this condition being sufficient but not necessary for convergence. The solution given by (10.1.13) is unique.

EXAMPLE 10.1.1.   Solve the Fredholm integral equation

$$\phi(x) = x + \tfrac{1}{5} \int_0^1 xy\phi(y)\, dy$$

by iteration.

  We have $\phi^{(0)}(x) = x$,

$$\phi^{(1)}(x) = x + \tfrac{1}{5} \int_0^1 xy^2\, dy = (1 + \tfrac{1}{15})x,$$

$$\phi^{(2)}(x) = x + \tfrac{1}{5} \int_0^1 x(1 + \tfrac{1}{15})y^2\, dy = \left(1 + \frac{1}{15} + \frac{1}{15^2}\right)x,$$

and so on. It is clear that we are converging to

$$\phi(x) = \left(1 + \frac{1}{15} + \frac{1}{15^2} + \ldots\right)x = \tfrac{15}{14}x,$$

which is the unique solution of the integral equation.

  Since we saw in section 10.1.1 that the Volterra equation is a particular case of the Fredholm equation it follows that we can use iteration to solve the Volterra equation of the second kind. But, whereas the kernel for the Fredholm equation has, in general, to satisfy a bound such as (10.1.15) no such restriction on the kernel is necessary for the Volterra equation. This is because

$$|Kf| = \left| \int_a^x K(x, y)f(y)\, dy \right|$$

$$\le |x - a| \max_{a \le y \le x} |K(x, y)f(y)| = MF(x - a), \quad \text{say,}$$

where M and F are, respectively, the maximum values in modulus of $K(x, y)$ and $f(y)$.

$$|K^2 f| = \left| \int_a^x K(x, y)Kf \, dy \right|$$

$$\leq \int_a^x |K(x, y)||Kf| \, dy$$

$$\leq \int_a^x M^2 F(y-a) \, dy = M^2 F(x-a)^2/2!,$$

and so on. The Liouville–Neumann series is thus smaller in magnitude than

$$F + F|\lambda| M(x-a) + F|\lambda|^2 M^2 \frac{(x-a)^2}{2!} + \dots,$$

which we know to be convergent as it is the exponential series [see IV (2.11.1)] for

$$F e^{|\lambda| M(x-a)}.$$

EXAMPLE 10.1.2.   Solve the Volterra integral equation

$$\phi(x) = x + \tfrac{1}{5} \int_0^x xy\phi(y) \, dy, \qquad x > 0.$$

We take $\phi^{(0)}(x) = x$ and compute the sequence

$$\phi^{(1)}(x) = x + \tfrac{1}{5} \int_0^x xy^2 \, dy = x + \tfrac{1}{15}x^4$$

$$\phi^{(2)}(x) = x + \tfrac{1}{5} \int_0^x xy(y + \tfrac{1}{15}y^4) \, dy = x + \tfrac{1}{15}x^4 + \tfrac{1}{2}(\tfrac{1}{15})^2 x^7$$

$$= x\left[1 + \tfrac{1}{15}x^3 + \frac{1}{2!}(\tfrac{1}{15}x^3)^2\right]$$

and, in general,

$$\phi^{(n)}(x) = x\left[1 + (\tfrac{1}{15}x^3) + \frac{1}{2!}(\tfrac{1}{15}x^3)^2 + \dots + \frac{1}{n!}(\tfrac{1}{15}x^3)^n\right]$$

so that

$$\phi(x) = \lim_{n\to\infty} \phi^{(n)}(x) = x \exp(x^3/15).$$

It should be noted that there is no Liouville–Neumann series for equations of the first kind. Some theory relating to integral equations of the first kind will be given when we deal with this type of equation (§ 10.4).

10.1.4. Fredholm Alternative Theory

It has been assumed previously that the kernel $K(x, y)$ is small enough. Even when $K(x, y)$ is not small an existence theory has been developed for Fredholm equations. Assume, as before, that the known functions $f(x)$ and $K(x, y)$ are continuous and a and b are finite. For convenience, we shall consider only real λ, although the results extend easily to complex λ. We will start by assuming that the kernel $K(x, y)$ is *degenerate*, that is, $K(x, y)$ can be written in the form

$$K(x, y) = \sum_{k=1}^{n} A_k(x)B_k(y) \tag{10.1.16}$$

where the functions $A_k(x)$ are linearly independent [cf. I, Definition 5.3.2] and so are the functions $B_k(y)$. Then, substituting the series (10.1.16) into the Fredholm equation (10.1.2), which we recall is

$$\phi(x) = f(x) + \lambda \int_a^b K(x, y)\phi(y) \, dy, \tag{10.1.2}$$

gives

$$\phi(x) = f(x) + \lambda \sum_{k=1}^{n} \left(\int_a^b B_k(y)\phi(y) \, dy \right) A_k(x) \tag{10.1.17}$$

$$= f(x) + \sum_{k=1}^{n} \alpha_k A_k(x), \tag{10.1.18}$$

where

$$\alpha_k = \lambda \int_a^b B_k(y)\phi(y) \, dy.$$

The solution of (10.1.2) and hence (10.1.17) is completely equivalent to determining the appropriate α_k in (10.1.18). Formally substituting the ϕ given by (10.1.18) in equation (10.1.17) we have

$$f(x) + \sum_{k=1}^{n} \alpha_k A_k(x) = f(x) + \lambda \sum_{k=1}^{n} \left(\int_a^b B_k(y)\left\{ f(y) + \sum_{m=1}^{n} \alpha_m A_m(y) \right\} dy \right) A_k(x).$$

As the $A_k(x)$ are linearly independent, we have for each k $(k = 1, \ldots, n)$

$$\alpha_k = \lambda \int_a^b B_k(y)f(y) \, dy + \lambda \sum_{m=1}^{n} \alpha_m \left(\int_a^b B_k(y)A_m(y) \, dy \right). \tag{10.1.19}$$

Let

$$\beta_k = \int_a^b B_k(y)f(y) \, dy, \qquad \gamma_{km} = \int_a^b B_k(y)A_m(y) \, dy$$

and [see I, § 6.5]

$$\boldsymbol{\beta} = (\beta_1, \ldots, \beta_n)', \qquad \boldsymbol{\alpha} = (\alpha_1, \ldots, \alpha_n)', \qquad \boldsymbol{\Gamma} = (\gamma_{km}), \quad 1 \leq k, m \leq n$$

Then equation (10.1.19) leads to the matrix equation [see I, § 5.7]

$$(\mathbf{I} - \lambda \boldsymbol{\Gamma})\boldsymbol{\alpha} = \lambda \boldsymbol{\beta} \tag{10.1.20}$$

where $\mathbf{I}$ is the unit $n \times n$ matrix [see I (6.2.7)].

The integral equation (10.1.2) has thus been expressed in the equivalent algebraic form (10.1.20). It follows from the theory of simultaneous linear equations that the equation (10.1.20) has a unique solution $\boldsymbol{\alpha}$ provided that

$$\det (\mathbf{I} - \lambda \boldsymbol{\Gamma}) \neq 0 \tag{10.1.21}$$

[see Theorem 3.2.1 and I, § 5.8]. Values of λ for which $\det (\mathbf{I} - \lambda \boldsymbol{\Gamma}) \neq 0$ are termed *regular* values. For all regular values of λ, $\boldsymbol{\alpha}$ is uniquely determined by

$$\boldsymbol{\alpha} = \lambda (\mathbf{I} - \lambda \boldsymbol{\Gamma})^{-1} \boldsymbol{\beta},$$

and this leads to the unique solution

$$\phi(x) = f(x) + (\lambda/\det \mathbf{C}) \sum_{m=1}^{n} \sum_{k=1}^{n} C_{mk} \beta_m A_k(x) \tag{10.1.22}$$

where

$$\mathbf{C} = \mathbf{I} - \lambda \boldsymbol{\Gamma} = (c_{mk})$$

and C_{mk} is the cofactor of the element c_{mk} [see I, § 6.11].

We shall illustrate the above by a simple example.

EXAMPLE 10.1.3.   Solve the integral equation

$$\phi(x) = 1 - 3x + \lambda \int_0^1 (1 - 3xy)\phi(y) \, dy$$

for regular values of λ.

In this example the kernel $K(x, y) = 1 - 3xy$ is degenerate. We can take $A_1(x) = 1$, $A_2(x) = x$ (and consequently $B_1(y) = 1$, $B_2(y) = -3y$) and our solution $\phi(x)$ must then, from (10.1.18), be of the form

$$\phi(x) = 1 - 3x + \alpha_1 + \alpha_2 x.$$

Substituting this value in the given integral equation we obtain, after cancellation of the common term $(1 - 3x)$ on both sides

$$\alpha_1 + \alpha_2 x = \lambda \int_0^1 (1 - 3xy)(1 - 3y + \alpha_1 + \alpha_2 y) \, dy$$

$$= \lambda(-\tfrac{1}{2} + \tfrac{3}{2}x + \alpha_1 + \tfrac{1}{2}\alpha_2) - \lambda x(\tfrac{3}{2}\alpha_1 + \alpha_2)$$

comparison of the coefficients of $A_1(x)$ and $A_2(x)$ gives

$$\alpha_1(1-\lambda)-\tfrac{1}{2}\lambda\alpha_2 = -\tfrac{1}{2}\lambda$$
$$\tfrac{3}{2}\lambda\alpha_1+\alpha_2(1+\lambda) = \tfrac{3}{2}\lambda. \tag{10.1.23}$$

It follows that

$$\mathbf{I}-\lambda\mathbf{\Gamma} = \begin{pmatrix} 1-\lambda & -\tfrac{1}{2}\lambda \\ \tfrac{3}{2}\lambda & 1+\lambda \end{pmatrix}$$

and

$$\det(\mathbf{I}-\lambda\mathbf{\Gamma}) = 1-\tfrac{1}{4}\lambda^2.$$

Note that $\mathbf{\Gamma}$ can be obtained directly by working out each element of the matrix using the result

$$\gamma_{km} = \int_a^b B_k(y)A_m(y)\,dy.$$

Since $\det(\mathbf{I}-\lambda\mathbf{\Gamma}) = 0$ when $\lambda = \pm 2$ it is clear that all values other than $\lambda = \pm 2$ are regular values of the integral equation. For a regular value of λ we have, solving the simultaneous equations (10.1.23),

$$\alpha_1 = -\lambda/(2+\lambda), \qquad \alpha_2 = 3\lambda/(2+\lambda)$$

and thus

$$\phi(x) = 2(1-3x)/(2+\lambda)$$

for regular values of λ.

We now examine what happens when $\det(\mathbf{I}-\lambda\mathbf{\Gamma}) = 0$. There are at most n distinct values of λ for which this can happen and these values are called *characteristic values* or *eigenvalues* [see I, § 7.2]. For these values of λ the homogeneous equation

$$(\mathbf{I}-\lambda\mathbf{\Gamma})\boldsymbol{\alpha} = \mathbf{0} \tag{10.1.24}$$

has non-trivial solutions and so also has the homogeneous integral equation

$$\phi(x) = \lambda\int_a^b K(x,y)\phi(y)\,dy \tag{10.1.25}$$

These non-trivial solutions are called *characteristic functions* or *eigen-functions* of the kernel $K(x,y)$ and we shall denote by $\Phi_1(x),\dots,\Phi_r(x)$, $r \leq n-1$ the independent eigenfunctions corresponding to the eigenvalue λ. Associated with the equation (10.1.25) is the homogeneous integral equation

$$\Psi(x) = \lambda\int_a^b K(y,x)\Psi(y)\,dy$$

which has independent solutions $\Psi_1(x), \ldots, \Psi_r(x)$, $r \leq n-1$. When λ is a characteristic value a solution of the Fredholm equation (10.1.2) exists provided that

$$\int_a^b f(x)\Psi_k(x)\,dx = 0, \qquad k = 1, \ldots, r \tag{10.1.26}$$

and the solution of (10.1.2) has the form

$$\phi(x) = g(x) + a_1\Phi_1(x) + \ldots + a_r\Phi_r(x) \tag{10.1.27}$$

where $g(x)$ is a particular solution of the integral equation and $a_1, \ldots, a_r$ are arbitrary. Note that when $K(x, y) = K(y, x)$ the functions $\Psi_k(x)$ and $\Phi_k(x)$ are identical.

We illustrate the above by means of some examples.

EXAMPLE 10.1.4. Solve the integral equation

$$\phi(x) = 1 - 3x + \lambda \int_0^1 (1 - 3xy)\phi(y)\,dy$$

when λ is a characteristic value of the integral equation.

The characteristic values were shown in Example 10.1.3 to be $\lambda = \pm 2$. For either value of λ the characteristic solution can be found from the equations

$$(1 - \lambda)\alpha_1 - \tfrac{1}{2}\lambda\alpha_2 = 0$$

$$\tfrac{3}{2}\lambda\alpha_1 + (1 + \lambda)\alpha_2 = 0$$

or $\alpha_1 = \tfrac{1}{2}\lambda\alpha_2/(1 - \lambda)$, as one of the equations is redundant. When $\lambda = 2$, the characteristic function $\Phi_1(x)$ is a multiple of

$$\alpha_1 + \alpha_2 x = \alpha_2(-1 + x) = A(1 - x), \quad \text{say.} \quad (A \neq 0, \text{ arbitrary}).$$

As $K(x, y) = 1 - 3xy$ is a symmetric kernel $\Phi_1(x) = \Psi_1(x)$ and thus there is a solution of the integral equation if

$$\int_0^1 (1 - x)(1 - 3x)\,dx = 0.$$

This is the case and we find the general solution of the integral equation is given by

$$\phi(x) = \frac{2}{2 + \lambda}(1 - 3x) + A(1 - x)$$

$$= \tfrac{1}{2}(1 - 3x) + A(1 - x).$$

When $\lambda = -2$,

$$\alpha_1 = -\tfrac{1}{3}\alpha_2$$

so that

$$\Phi_1(x) = \Psi_1(x) = \alpha_1 + \alpha_2 x = A(1 - 3x), \qquad (A \neq 0, \text{ arbitrary}).$$

It follows that

$$\int_0^1 f(x)\Psi_1(x)\, dx = A \int_0^1 (1 - 3x)^2\, dx \neq 0$$

and thus no solution of the integral equation exists for $\lambda = -2$.

It is also instructive to consider an example where the kernel $K(x, y)$ is not symmetric:

EXAMPLE 10.1.5.   Solve the integral equation

$$\phi(x) = 4 - 7x + \lambda \int_0^1 (x + 2y - 2xy)\phi(y)\, dy.$$

The solution is of the form

$$\phi(x) = 4 - 7x + \alpha_1 + \alpha_2 x.$$

Substitution in the integral equation yields

$$\alpha_1 + \alpha_2 x = \lambda \int_0^1 (x + 2y - 2xy)(4 - 7y + \alpha_1 + \alpha_2 y)\, dy$$

$$= \lambda(-\tfrac{2}{3} + \alpha_1 + \tfrac{2}{3}\alpha_2) + \lambda x(\tfrac{7}{6} - \tfrac{1}{6}\alpha_2)$$

Comparison of powers of x on both sides produces the system of equations

$$(1 - \lambda)\alpha_1 - \tfrac{2}{3}\lambda\alpha_2 = -\tfrac{2}{3}\lambda$$

$$(1 + \tfrac{1}{6}\lambda)\alpha_2 = \tfrac{7}{6}\lambda.$$

It is clear from these equations that λ is a regular value of the integral equation provided that

$$(1 - \lambda)(1 + \tfrac{1}{6}\lambda) \neq 0$$

i.e.

$$\lambda \neq 1 \quad \text{or} \quad \lambda \neq -6.$$

It follows that for regular values of λ

$$\alpha_1 = -\frac{4\lambda}{6 + \lambda}, \qquad \alpha_2 = \frac{7\lambda}{6 + \lambda}$$

and hence the solution of the integral equation for regular values of λ is

$$\phi(x) = [24 - 42x]/(6 + \lambda). \qquad (10.1.28)$$

Now consider what happens when λ is an eigenvalue, say $\lambda = 1$. We have to investigate in this case the solutions of the homogeneous integral equations

$$\phi(x) = \lambda \int_0^1 (x + 2y - 2xy)\phi(y)\, dy, \qquad \lambda = 1$$

and

$$\psi(x) = \lambda \int_0^1 (y + 2x - 2xy)\psi(y)\, dy, \qquad \lambda = 1.$$

The solutions of these equations are $\Phi_1(x) = A$ (A arbitrary) and $\Psi_1(x) = B(2 + 3x)$ (B arbitrary). Since

$$\int_0^1 (4 - 7x)(2 + 3x)\, dx = 0$$

the conditions necessary for a solution apply and the general solution is thus, from (10.1.27) and (10.1.28),

$$\phi(x) = A + \tfrac{24}{7} - 6x.$$

When $\lambda = -6$, $\Psi_1(x) = C(1 - 2x)$, C arbitrary and since

$$\int_0^1 f(x)\Psi_1(x)\, dx = C \int_0^1 (4 - 7x)(1 - 2x)\, dx \neq 0$$

we do not have a solution in this case.

We conclude this section by giving some general results about Fredholm integral equations. Firstly, when the kernel is symmetric, i.e. $K(x, y) = K(y, x)$, then it is known that the characteristic values λ_n are all real and at least one characteristic value exists. Further, the characteristic functions $\Phi_m(x)$, $\Phi_n(x)$ belonging to distinct characteristic values λ_m, λ_n are orthogonal, i.e.

$$\int_a^b \Phi_m(x)\Phi_n(x)\, dx = 0. \qquad (10.1.29)$$

These results extend also to Hermitian kernels but, in general, we shall only be concerned with real kernels. We give an example which shows that a real symmetric kernel need not have more than one characteristic value.

EXAMPLE 10.1.6. The symmetric kernel $K(x, y) = \sin x \sin y$, $0 \le x, y \le \pi$ has only one characteristic value.

For, consider the homogeneous equation

$$\phi(x) = \lambda \int_0^\pi \sin x \sin y\, \phi(y)\, dy$$

$$= \lambda \sin x \int_0^\pi \sin y\, \phi(y)\, dy.$$

From examination of this equation it is clear that any eigensolution $\Phi(x)$ is a multiple of $\sin x$. Now multiply both sides of the above equation by $\sin x$ and integrate between the limits 0 and π. We have

$$\int_0^\pi \phi(x) \sin x \, dx = \lambda \left(\int_0^\pi \sin^2 x \, dx \right) \left(\int_0^\pi \phi(y) \sin y \, dy \right)$$

$$= \tfrac{1}{2}\pi\lambda \int_0^\pi \phi(y) \sin y \, dy.$$

As the integral cannot be zero because the eigensolution is a multiple of $\sin x$ it follows by cancellation that

$$1 = \tfrac{1}{2}\pi\lambda$$

or $\lambda = 2/\pi$, indicating that there is just 1 characteristic value.

Finally, we remark that even when $K(x, y)$ is not degenerate it can be shown that the Fredholm alternative theory applies. That is,

either λ is a regular value in which case the equation

$$\phi(x) = f(x) + \lambda \int_a^b K(x, y)\phi(y) \, dy \qquad (10.1.2)$$

has a unique solution for any given function $f(x)$;

or λ is a characteristic value of the homogeneous equation

$$\phi(x) = \lambda \int_a^b K(x, y)\phi(y) \, dy.$$

The homogeneous equation has a finite number of linearly independent solutions $\Phi_1(x), \ldots, \Phi_m(x)$ and the *transposed homogeneous* equation

$$\psi(x) = \lambda \int_a^b K(y, x)\psi(y) \, dy$$

also has m solutions $\Psi_1(x), \ldots, \Psi_m(x)$. The equation (10.1.2) has a solution if, and only if,

$$\int_a^b f(x)\Psi_j(x) \, dx = 0, \qquad j = 1, \ldots, m.$$

This solution is not unique as it contains any linear combination of the $\Phi_j(x)$, $j = 1, \ldots, m$.

For more background information on the mathematical theory of integral equations the reader should consult Cochran (1972), or the extensive list of references given in Delves and Walsh (1974). For a comprehensive treatment of the numerical solution of integral equations Delves and Walsh (1974) and Baker (1978) should be consulted.

10.2. FREDHOLM EQUATIONS OF THE SECOND KIND

10.2.1. Approximate Solution

In this section we shall be concerned with the approximate solution of the equation

$$\phi(x) = f(x) + \lambda \int_0^1 K(x, y)\phi(y)\, dy \tag{10.2.1}$$

where $f(x)$ and $K(x, y)$ are given functions which are continuous [see IV, §§ 2.1 and 5.1], and all of whose derivatives exist [see IV, §§ 2.9 and 5.2] and are continuous, and λ is a regular value of the integral equation. If the integral equation is given in the form of equation (10.1.2) then it can be expressed in the form (10.2.1) by means of the transformation $y = (b - a)u + a$ [see IV, § 4.3] which produces a change in the interval of integration from $[a, b]$ to $[0, 1]$ [cf. (7.3.3)].

To find the approximate solution to (10.2.1) we select a quadrature formula (see Chapter 7). Assume that this quadrature formula is

$$\int_0^1 g(x)\, dx = \sum_{k=1}^n \alpha_k g(x_k) + R_n(x). \tag{10.2.2}$$

Then, if we evaluate the integral in (10.2.1) by this formula, we have for all x

$$\phi(x) = f(x) + \lambda \sum_{k=1}^n \alpha_k K(x, x_k)\phi(x_k) + R_n^*(x) \tag{10.2.3}$$

where $R_n^*(x)$ is the appropriate remainder when the integrand is $g(y) = K(x, y)\phi(y)$. In particular, this equation is satisfied by $x = x_r$, $r = 1, \ldots, n$, so that

$$\phi(x_r) = f(x_r) + \lambda \sum_{k=1}^n \alpha_k K(x_r, x_k)\phi(x_k) + R_n^*(x_r), \qquad r = 1, \ldots, n. \tag{10.2.4}$$

The approximate solution of the integral equation (10.2.1) for $x = x_r$, $r = 1, \ldots, n$ is $\phi_r^* = \phi^*(x_r)$ where ϕ_r^* satisfies the system of linear equations

$$\phi_r^* = f_r + \lambda \sum_{k=1}^n \alpha_k K(x_r, x_k)\phi_k^*, \qquad r = 1, \ldots, n \tag{10.2.5}$$

which can be written in matrix form [see I, § 5.7] as

$$\mathbf{A}\boldsymbol{\phi}^* = \mathbf{f} \tag{10.2.6}$$

where

$$\mathbf{A} = \begin{pmatrix} 1 - \lambda\alpha_1 K(x_1, x_1) & -\lambda\alpha_2 K(x_1, x_2) & \cdots & -\lambda\alpha_n K(x_1, x_n) \\ -\lambda\alpha_1 K(x_2, x_1) & 1 - \lambda\alpha_2 K(x_2, x_2) & \cdots & -\lambda\alpha_n K(x_2, x_n) \\ \cdots & \cdots & \cdots & \cdots \\ -\lambda\alpha_1 K(x_n, x_1) & -\lambda\alpha_2 K(x_n, x_2) & \cdots & 1 - \lambda\alpha_n K(x_n, x_n) \end{pmatrix}$$

$$\boldsymbol{\phi}^* = \begin{pmatrix} \phi_1^* \\ \phi_2^* \\ \vdots \\ \phi_n^* \end{pmatrix} \qquad \mathbf{f} = \begin{pmatrix} f_1 \\ f_2 \\ \vdots \\ f_n \end{pmatrix} = \begin{pmatrix} f(x_1) \\ f(x_2) \\ \vdots \\ f(x_n) \end{pmatrix}.$$

The system of linear equations can be solved by standard techniques (see Chapter 3) to give $\phi_1^*, \ldots, \phi_n^*$. Once these quantities are known then $\phi^*(x)$ is determined for any x from

$$\phi^*(x) = f(x) + \lambda \sum_{k=1}^{n} K(x, x_n)\phi_k^* \qquad (10.2.7)$$

We shall illustrate the method by means of an example.

EXAMPLE 10.2.1.   Solve the integral equation

$$\phi(x) = 1 - 3x + \int_0^1 (1 - 3xy)\phi(y)\, dy \qquad (10.2.8)$$

and estimate $\phi(x)$ for $x = 0(\frac{1}{4})1$ (the notation $x = a(b)c$ denotes that x takes the values $a, a+b, a+2b, \ldots, c$ for suitable values of a, b and c).
    We shall use two integration formulae to obtain our answers.

(a) the basic Simpson rule [see § 7.1.5]
(b) the Gauss–Legendre 3-point rule [see § 7.3.1]

The exact answer for $\phi(x)$ can be obtained from Example 10.1.3.

(a) The basic Simpson rule is

$$\int_0^1 g(x)\, dx = \tfrac{1}{6}[g(0) + 4g(\tfrac{1}{2}) + g(1)] + \text{Remainder term.}$$

Applying this rule to estimate the integral in (10.2.8) we get

$$\phi(x) = 1 - 3x + \tfrac{1}{6}[\phi(0) + 4(1 - \tfrac{3}{2}x)\phi(\tfrac{1}{2}) + (1 - 3x)\phi(1)] + \text{Remainder term.}$$

If the remainder term is neglected the values of $\phi(x)$ at $x = 0, \frac{1}{2}, 1$ are given approximately by $\phi^*(0), \phi^*(\frac{1}{2}), \phi^*(1)$ where

$$\phi^*(0) = 1 + \tfrac{1}{6}[\phi^*(0) + 4\phi^*(\tfrac{1}{2}) + \phi^*(1)]$$
$$\phi^*(\tfrac{1}{2}) = -\tfrac{1}{2} + \tfrac{1}{6}[\phi^*(0) + \phi^*(\tfrac{1}{2}) - \tfrac{1}{2}\phi^*(1)]$$
$$\phi^*(1) = -2 + \tfrac{1}{6}[\phi^*(0) - 2\phi^*(\tfrac{1}{2}) - 2\phi^*(1)]$$

The exact solution of this system of linear equations is [see § 3.3.1],

$$\phi^*(0) = \tfrac{2}{3}, \qquad \phi^*(\tfrac{1}{2}) = -\tfrac{1}{3}, \qquad \phi^*(1) = \frac{-4}{3}.$$

The estimates for $\phi(\tfrac{1}{4})$ and $\phi(\tfrac{3}{4})$ are

$$\phi^*(\tfrac{1}{4}) = 1 - 3 \times \tfrac{1}{4} + \tfrac{1}{6}[\phi^*(0) + 4(1 - \tfrac{3}{2} \times \tfrac{1}{4})\phi^*(\tfrac{1}{2}) + (1 - 3 \times \tfrac{1}{4})\phi^*(1)]$$

$$= \tfrac{1}{6}$$

and

$$\phi^*(\tfrac{3}{4}) = 1 - 3 \times \tfrac{3}{4} + \tfrac{1}{6}[\phi^*(0) + 4(1 - \tfrac{3}{2} \times \tfrac{3}{4})\phi^*(\tfrac{1}{2}) + (1 - 3 \times \tfrac{3}{4})\phi^*(1)]$$

$$= -\tfrac{5}{6}.$$

These results are exact because we have used exact arithmetic and the remainder term in the integration formula is, fortuitously, zero. This latter fact follows because $\phi(y)$ is a polynomial of degree 1 in y (see Example 10.1.3) and this makes $K(x, y)\phi(y) = (1 - 3xy)\phi(y)$ a polynomial of degree 2. We know that Simpson's rule is exact for polynomials of degree 3 or less and consequently the remainder term is zero.

(b) The Gauss–Legendre 3-point rule is

$$\int_0^1 g(x)\,dx = \tfrac{1}{18}[5g(\alpha) + 8g(\tfrac{1}{2}) + 5g(\beta)] + \text{Remainder term}$$

where

$$\alpha = \tfrac{1}{2}[1 - \sqrt{\tfrac{3}{5}}] = 0{\cdot}112702, \qquad \beta = \tfrac{1}{2}[1 + \sqrt{\tfrac{3}{5}}] = 0{\cdot}887298.$$

Applying this rule to estimate the integral in (10.2.8) we get

$$\phi(x) = 1 - 3x + \tfrac{1}{18}[5(1 - 3x\alpha)\phi(\alpha) + 8(1 - \tfrac{3}{2}x)\phi(\tfrac{1}{2})$$

$$+ 5(1 - 3x\beta)\phi(\beta)] + \text{Remainder term}.$$

We wish to determine $\phi(\alpha)$, $\phi(\tfrac{1}{2})$ and $\phi(\beta)$ and these are approximately $\phi^*(\alpha)$, $\phi^*(\tfrac{1}{2})$, $\phi^*(\beta)$ where

$$\phi^*(\alpha) = 0{\cdot}661895 + 0.267193\phi^*(\alpha) + 0.369310\phi^*(\tfrac{1}{2}) + 0{\cdot}194444\phi^*(\beta)$$

$$\phi^*(\tfrac{1}{2}) = -0{\cdot}5 + 0{\cdot}230819\phi^*(\alpha) + 0{\cdot}111111\phi^*(\tfrac{1}{2}) - 0{\cdot}091930\phi^*(\beta)$$

$$\phi^*(\beta) = -1{\cdot}661894 + 0{\cdot}194444\phi^*(\alpha) - 0{\cdot}147088\phi^*(\tfrac{1}{2}) - 0{\cdot}378304\phi^*(\beta)$$

or

$$0{\cdot}732807\phi^*(\alpha) - 0{\cdot}369310\phi^*(\tfrac{1}{2}) - 0{\cdot}194444\phi^*(\beta) = 0{\cdot}661895$$

$$-0{\cdot}230819\phi^*(\alpha) + 0{\cdot}888889\phi^*(\tfrac{1}{2}) + 0{\cdot}091930\phi^*(\beta) = -0{\cdot}5$$

$$-0{\cdot}194444\phi^*(\alpha) + 0{\cdot}147088\phi^*(\tfrac{1}{2}) + 1{\cdot}378304\phi^*(\beta) = -1{\cdot}661894.$$

These equations were solved by Gaussian elimination [see § 3.3.1] and it was found that

$$\phi^*(\alpha) = 0.441264, \qquad \phi^*(\tfrac{1}{2}) = -0.333333, \qquad \phi^*(\beta) = -1.107929.$$

We now have to determine approximations for $\phi(0)$, $\phi(\tfrac{1}{4})$, $\phi(\tfrac{3}{4})$ and $\phi(1)$ and these are obtained from

$$\phi^*(x) = 1 - 3x + \tfrac{1}{18}[5(1 - 3x\alpha)\phi^*(\alpha) + 8(1 - \tfrac{3}{2}x)\phi^*(\tfrac{1}{2}) + 5(1 - 3x\beta)\phi^*(\beta)].$$

Thus, for example,

$$\phi^*(0) = 1 + \tfrac{5}{18}\phi^*(\alpha) + \tfrac{4}{9}\phi^*(\tfrac{1}{2}) + \tfrac{5}{18}\phi^*(\beta) = 0.666667$$

and we can similarly deduce that

$$\phi^*(\tfrac{1}{4}) = 0.166667, \qquad \phi^*(\tfrac{3}{4}) = -0.833334, \qquad \phi^*(1) = -1.333334.$$

These answers are correct to within 1 unit in the sixth decimal place and the discrepancy between these results and the ones obtained by Simpson's rule can be accounted for by rounding errors incurred during the calculation. If exact arithmetic had been used throughout we would have obtained exact answers for the same reasons that were given above for Simpson's rule.

10.2.2. Error Considerations

We shall now consider a more difficult example and see how we can improve on the answers obtained.

EXAMPLE 10.2.2. $\phi(x)$ satisfies the integral equation

$$\phi(x) = e^x + \int_0^1 (1 + xy)\phi(y)\, dy, \qquad 0 \le x \le 1. \qquad (10.2.9)$$

Determine $\phi(0)$, $\phi(\tfrac{1}{2})$ and $\phi(1)$.

For convenience we shall use the basic Simpson rule [see § 7.1.5] and we find that, if the remainder term is neglected, the value of $\phi(x)$ for any given x in $[0, 1]$ is approximately $\phi^*(x)$ where

$$\phi^*(x) = e^x + \tfrac{1}{6}[\phi^*(0) + 4(1 + \tfrac{1}{2}x)\phi^*(\tfrac{1}{2}) + (1 + x)\phi^*(1)]. \qquad (10.2.10)$$

In particular, when the values of $x = 0, \tfrac{1}{2}, 1$ are substituted in this equation we obtain the system of simultaneous linear equations

$$\tfrac{5}{6}\phi^*(0) - \tfrac{2}{3}\phi^*(\tfrac{1}{2}) - \tfrac{1}{6}\phi^*(1) = 1$$

$$-\tfrac{1}{6}\phi^*(0) + \tfrac{1}{6}\phi^*(\tfrac{1}{2}) - \tfrac{1}{4}\phi^*(1) = e^{1/2}$$

$$-\tfrac{1}{6}\phi^*(0) - \phi^*(\tfrac{1}{2}) + \tfrac{2}{3}\phi^*(1) = e.$$

The solution of this system of equations is obtained by Gaussian elimination in Table 10.2.1 and is $\phi^*(0) = -5.588903$, $\phi^*(\tfrac{1}{2}) = -6.659033$, $\phi^*(1) = -7.308339$. The question we now have to ask ourselves is 'How well do these answers approximate to the exact values of $\phi(0)$, $\phi(\frac{1}{2})$ and $\phi(1)$ given by the integral equation (10.2.9)?' In this particular problem we are fortunate that we can determine the exact answer of the integral equation which is

$$\phi(x) = e^x - \tfrac{2}{3}(4e - 1) - 2(e - 1)x, \qquad 0 \le x \le 1.$$

It follows that the correctly rounded values of $\phi(0)$, $\phi(\frac{1}{2})$ and $\phi(1)$ are (to 5 decimal places)

$$\phi(0) = -5 \cdot 58208, \qquad \phi(\tfrac{1}{2}) = -6 \cdot 65165, \qquad \phi(1) = -7 \cdot 30037$$

and thus the computed values are in error in the third decimal place.

In general, we do not know the exact solution but we can estimate the error if we have some idea of the magnitude of the remainder term in formula (10.2.4). For $\phi(x)$ at $x_1, x_2, \ldots, x_n$ is obtained exactly from the equation

$$\mathbf{A}\boldsymbol{\phi} = \mathbf{f} + \mathbf{R}$$

where

$$\mathbf{R} = (R_n^*(x_1), \ldots, R_n^*(x_n))'.$$

Consequently

$$\boldsymbol{\phi} = \mathbf{A}^{-1}\mathbf{f} + \mathbf{A}^{-1}\mathbf{R}$$

implying that

$$\boldsymbol{\phi} - \boldsymbol{\phi}^* = \mathbf{A}^{-1}\mathbf{R},$$

and, as we can determine the elements in $\mathbf{A}^{-1}$ exactly [see I, § 6.4] a bound for the error in each computed ϕ^* is given by

$$|\phi(x_r) - \phi_r^*| \le n \times \max_i |A_{ri}| \times \max_i |R_n^*(x_i)|,$$

where A_{ij} denotes the (i, j) element of $\mathbf{A}^{-1}$.

EXAMPLE 10.2.3.   Estimate the error in the computed values $\phi^*(0)$, $\phi^*(\frac{1}{2})$, $\phi^*(1)$ of Example 10.2.2.

From Example 10.2.2 the exact value of $\mathbf{A}$ is given by

$$\mathbf{A} = \begin{pmatrix} \frac{5}{6} & -\frac{2}{3} & -\frac{1}{6} \\ -\frac{1}{6} & \frac{1}{6} & -\frac{1}{4} \\ -\frac{1}{6} & -1 & \frac{2}{3} \end{pmatrix}$$

and the exact value of $\mathbf{A}^{-1}$ can be shown to be

$$\mathbf{A}^{-1} = \begin{pmatrix} \frac{5}{9} & -\frac{22}{9} & -\frac{7}{9} \\ -\frac{11}{18} & -\frac{19}{9} & -\frac{17}{18} \\ -\frac{7}{9} & -\frac{34}{9} & -\frac{1}{9} \end{pmatrix}.$$

It is clear that the maximum element in modulus in $\mathbf{A}^{-1}$ is $\frac{34}{9}$. We can get some idea of the magnitude of the remainder term by setting up a difference table for the function $g(x, y) = K(x, y)\phi^*(y)$ and noting the results [see § 1.4]

$$h^{2k}g_0^{(2k)} \approx \delta_0^{2k}$$

$$h^{2k+1}g_0^{(2k+1)} \approx \mu\delta_0^{2k+1}.$$

If we take $h = \frac{1}{4}$ then, to determine the fourth differences at $y = 0$ and $y = 1$, we need to know the values of $\phi^*(-\frac{1}{2})$, $\phi^*(-\frac{1}{4})$, $\phi^*(\frac{1}{4})$, $\phi^*(\frac{3}{4})$, $\phi^*(\frac{5}{4})$, $\phi^*(\frac{3}{2})$. These quantities can be estimated from (10.2.10) even though $\phi^*(-\frac{1}{2})$ and $\phi^*(\frac{5}{4})$, for example, are values of the function at $y = -\frac{1}{2}$ and $y = \frac{5}{4}$ which are outside the range of integration. (The dangers of using function values outside the range of integration can be seen from section 10.2.3 and it is a practice which should be avoided if possible.) The differences for $g(\frac{1}{2}, y)$ are given in Table 10.2.2. From the table

$$\max_{0 \le x \le 1} h^4 f^{(\text{iv})}(x) \approx 0.037743$$

and as $h = \frac{1}{4}$ it follows that

$$\max_{0 \le x \le 1} f^{(\text{iv})}(x) \approx 256 \times 0.037743.$$

Consequently the maximum error in the computed $\phi^*(\frac{1}{2})$ satisfies

$$|\text{Error}| < 3 \times \tfrac{34}{9} \times \tfrac{1}{2880} \times 256 \times 0.037743 < 0.04.$$

Because of the way in which the error bound is derived this usually over-estimates the error in each $\phi^*(x)$ and we can be confident, therefore, that our answer for $\phi(\frac{1}{2})$ is in error by at most 1%. Similar results can be established for the errors in $\phi^*(0)$ and $\phi^*(1)$ by constructing differences tables for $K(0, y)\phi^*(y)$ and $K(1, y)\phi^*(y)$.

In some problems an error of 1% might not be considered satisfactory and we would wish for an error of, say, at most 0.01% in the computed values of $\phi(x)$. How can this be achieved? We list below a number of possible ways.

(i) *Use of Higher Order Formulae*

In this method we initially solve the integral equation using some specified formula, say a Gauss–Legendre 3-point formula [see § 7.3.1] and we then re-solve the equation using higher order formulae of the same kind (in this instance Gauss–Legendre 4- and 5-point formulae). If the answers for $\phi^*(x)$, for given x, exhibit some trend towards a particular value, i.e. there is agreement in 3 or 4 significant figures then these figures may be regarded as being correct. However, as an example in section 7.3.1 indicates, this approach might, on occasion, yield spurious results but it is, in general, reliable.

(ii) *Extrapolation of Results from Composite Formulae*

In example 10.2.2, the basic Simpson rule was employed to evaluate $\phi(0)$, $\phi(\frac{1}{2})$, $\phi(1)$. An alternative procedure is to apply the basic Simpson rule to subintervals of $(0, 1)$. For instance we could subdivide $(0, 1)$ into the subintervals $(0, \frac{1}{2})$, $(\frac{1}{2}, 1)$ and we would then use the approximation

$$\int_0^1 g(x)\, dx \approx \tfrac{1}{12}[g(0)+4g(\tfrac{1}{4})+2g(\tfrac{1}{2})+4g(\tfrac{3}{4})+g(1)].   \qquad (10.2.11)$$

We illustrate this by means of the following example.

EXAMPLE 10.2.4.   Use the rule (10.2.11) to solve the integral equation

$$\phi(x) = e^x + \int_0^1 (1+xy)\phi(y)\, dy$$

The approximate formulation of the integral equation is now

$$\phi^*(x) = e^x + \tfrac{1}{12}[\phi^*(0)+4(1+\tfrac{1}{4}x)\phi^*(\tfrac{1}{4})+2(1+\tfrac{1}{2}x)\phi^*(\tfrac{1}{2})$$
$$+4(1+\tfrac{3}{4}x)\phi^*(\tfrac{3}{4})+(1+x)\phi^*(1)].$$

Substituting $x = 0, \frac{1}{4}, \frac{1}{2}, \frac{3}{4}, 1$ into this equation we obtain, after a little rearrangement, the system of linear simultaneous equations

$$\tfrac{11}{12}\phi^*(0)-\tfrac{1}{3}\phi^*(\tfrac{1}{4})-\tfrac{1}{6}\phi^*(\tfrac{1}{2})-\tfrac{1}{3}\phi^*(\tfrac{3}{4})-\tfrac{1}{12}\phi^*(1) = 1$$
$$-\tfrac{1}{12}\phi^*(0)+\tfrac{31}{48}\phi^*(\tfrac{1}{4})-\tfrac{3}{16}\phi^*(\tfrac{1}{2})-\tfrac{19}{48}\phi^*(\tfrac{3}{4})-\tfrac{5}{48}\phi^*(1) = e^{1/4} = 1{\cdot}284025$$
$$-\tfrac{1}{12}\phi^*(0)-\tfrac{3}{8}\phi^*(\tfrac{1}{4})+\tfrac{19}{24}\phi^*(\tfrac{1}{2})-\tfrac{11}{24}\phi^*(\tfrac{3}{4})-\tfrac{1}{8}\phi^*(1) = e^{1/2} = 1{\cdot}648721$$
$$-\tfrac{1}{12}\phi^*(0)-\tfrac{19}{48}\phi^*(\tfrac{1}{4})-\tfrac{11}{48}\phi^*(\tfrac{1}{2})+\tfrac{23}{48}\phi^*(\tfrac{3}{4})-\tfrac{7}{48}\phi^*(1) = e^{3/4} = 2{\cdot}117000$$
$$-\tfrac{1}{12}\phi^*(0)-\tfrac{5}{12}\phi^*(\tfrac{1}{4})-\tfrac{1}{4}\phi^*(\tfrac{1}{2})-\tfrac{7}{12}\phi^*(\tfrac{3}{4})+\tfrac{5}{6}\phi^*(1) = e = 2{\cdot}718282.$$

These equations were solved by Gaussian elimination [see § 3.3.1] and the following values were obtained:

$$\phi_2^*(0) = -5{\cdot}582527, \qquad \phi_2^*(\tfrac{1}{2}) = -6{\cdot}652130, \qquad \phi_2^*(1) = -7{\cdot}300900$$

$$\phi^*(\tfrac{1}{4}) = -6{\cdot}157672, \qquad \phi^*(\tfrac{3}{4}) = -7{\cdot}043015,$$

where the suffix 2 is used to distinguish these estimates from those obtained earlier in Example 10.2.2. Since the integrand has no singularities we can use the Romberg technique [see § 7.2] to improve on the answers obtained for $\phi^*(0)$, $\phi^*(\frac{1}{2})$, $\phi^*(1)$ in the two examples. For instance, we have

$$\phi(0) \approx \frac{16\phi_2^*(0)-\phi^*(0)}{15} = -5{\cdot}582102$$

and this is a much improved approximation to either $\phi^*(0)$ or $\phi_2^*(0)$.

(iii)   *Deferred Correction*

An alternative procedure which could be adopted is to solve the approximate equations, as we have done previously, on the assumption that the remainder term, R, is zero and follow this up by estimating the remainder term from the calculated results by means of a finite difference approximation. We then iteratively correct the estimates proceeding until correction terms are sufficiently small. We illustrate the method below in Example 10.2.5. A disadvantage of the technique is that a finite difference table needs to be established for each individual x value so that even in the case of the basic Simpson rule we have to construct three finite difference tables corresponding to the cases $x = 0, \frac{1}{2}, 1$. It is worth remarking here that the method of deferred correction has achieved considerable success in the solution of boundary value problems for second order ordinary differential equations (see § 8.4).

EXAMPLE 10.2.5.   Improve on the results obtained in Example 10.2.2 by application of the method of deferred correction.

Before we can start we have to be able to estimate remainder term. For this purpose we can use a formula analogous to Gregory's formula [see § 7.7.1] namely

$$\int_0^1 g(x)\, dx = \tfrac{1}{6}[g(0) + 4g(\tfrac{1}{2}) + g(1)]$$

$$-\frac{1}{2880h^3}[(\nabla_n^3 - \Delta_0^3) + \tfrac{3}{2}(\nabla_n^4 + \Delta_0^4)$$

$$+ \tfrac{7}{4}(\nabla_n^5 - \Delta_0^5) + \tfrac{15}{8}(\nabla_n^6 + \Delta_0^6) + \ldots]$$

$$+ \frac{1}{96768h^5}[(\nabla_n^5 - \Delta_0^5) + \tfrac{5}{2}(\nabla_n^6 + \Delta_0^6) + \ldots] + \ldots \quad (10.2.12)$$

where h denotes the step size used in setting up the difference table, Δ_0^r denotes the rth forward difference measured from the value $\phi^*(0)$ and ∇_n^r denotes the rth backward difference measured from the value $\phi^*(1)$. The only point we have to decide is where to terminate the differences in the above formula. If we only use those differences up to where ∇_n^r and Δ_0^r coincide the correction terms are found from Tables 10.2.2, 10.2.3 and 10.2.4 and the modified equations to be solved are

$$0\cdot833333\phi_1^*(0) - 0\cdot666667\phi_1^*(\tfrac{1}{2}) - 0\cdot166667\phi_1^*(1) = -0\cdot000583$$

$$-0\cdot166667\phi_1^*(0) + 0\cdot166667\phi_1^*(\tfrac{1}{2}) - 0\cdot25\phi_1^*(1) \quad\;\; = -0\cdot001892$$

$$-0\cdot166667\phi_1^*(0) - \phi_1^*(\tfrac{1}{2}) \qquad\qquad + 0\cdot666667\phi_1^*(1) = -0\cdot003200.$$

The solution of these equations is given in Table 10.2.1. We find the corrected value of $\phi^*(\tfrac{1}{2})$ is $\phi_c^*(\tfrac{1}{2})$

$$\phi_c^*(\tfrac{1}{2}) = -6\cdot659033 + 0\cdot007375 = -6\cdot65166.$$

This answer, rounded to 5 decimal places, is only in error by 1 unit in the fifth decimal place. Similarly, the corrected values of $\phi^*(0)$ and $\phi^*(1)$ are

$$\phi_c^*(0) = -5 \cdot 58211, \qquad \phi_c^*(1) = -7 \cdot 30038$$

and these answers are only in error by at most 2 units in the fifth decimal place. We can improve these answers, if required, by performing another difference correction. It should be noted, however, that this necessitates the correcting of the values of $\phi^*(\frac{1}{4})$ and $\phi^*(\frac{3}{4})$ and for this we have to set up an additional two difference tables.

While we have used Simpson's rule with correction terms to establish our results any formula with correction terms involving forward and backward differences could be used and, in the original demonstration of this method by Fox and Goodwin (1953), Gregory's formula was employed [see (7.7.4)].

(iv) Chebyshev Methods

In this section we shall consider the integral equation in the form

$$\phi(x) = f(x) + \lambda \int_{-1}^{1} K(x, y)\phi(y)\, dy \tag{10.2.13}$$

and approximate to its solution by means of a *Chebyshev series* of the form

$$\phi(x) = \sum_{n=0}^{N}{}' a_n T_n(x) \tag{10.2.14}$$

[see Definition 6.3.3], where the prime after the summation sign indicates that the first term in the series is to be $\frac{1}{2}a_0 T_0(x)$. To determine the coefficients, a_n, we select $(N+1)$ points of collocation, x_i, where

$$x_i = \cos(\pi i/N) \qquad i = 0, 1, \ldots, N$$

and we thus obtain, by substitution in (10.2.13), the $N+1$ equations

$$\phi(x_i) = f(x_i) + \lambda \int_{-1}^{1} K(x_i, y)\phi(y)\, dy. \tag{10.2.15}$$

For each x_i, the kernel $K(x_i, y)$ is approximated by a polynomial of degree M of the form

$$K(x_i, y) = \sum_{n=0}^{M}{}' b_n(x_i) T_n(y). \tag{10.2.16}$$

If we now write

$$I(x_i) = \int_{-1}^{1} K(x_i, y)\phi(y)\, dy$$

then, using the results [see § 6.3.4]

$$T_n(x) T_m(x) = \tfrac{1}{2}(T_{n+m}(x) + T_{|n-m|}(x))$$

and

$$\int_{-1}^{1} T_n(x)\, dx = \begin{cases} 2 & \text{if } n = 0 \\ 0 & \text{if } n = 1 \\ \dfrac{-1}{n^2 - 1}\{1 - (-1)^{n+1}\} & \text{if } n > 1 \end{cases}$$

we find

$$I(x_i) = \sum_{n=0}^{N}{}' a_n \beta_n(x_i)$$

where

$$\beta_n(x_i) = b_n(x_i) - \sum_{r=1}^{|n-2r|\le\max(M,N)} \left(\frac{b_{|n-2r|}(x_i) + b_{n+2r}(x_i)}{4r^2 - 1} \right).$$

In this equation $b_{n+2r}(x_i)$ is taken to be zero when $n + 2r > M$. Equation (10.2.15) can now be written in the matrix form

$$(\mathbf{T} - \lambda \mathbf{B})\mathbf{a} = \mathbf{f} \tag{10.2.17}$$

where

$$\mathbf{T} = \begin{pmatrix} \frac{1}{2} & T_1(x_0) & T_2(x_0) & \dots & T_N(x_0) \\ \frac{1}{2} & T_1(x_1) & T_2(x_1) & \dots & T_N(x_1) \\ \cdot & \cdot & \cdot & \dots & \cdot \\ \frac{1}{2} & T_1(x_N) & T_2(x_N) & \dots & T_N(x_N) \end{pmatrix}, \quad \mathbf{B} = \begin{pmatrix} \frac{1}{2}\beta_0(x_0) & \beta_1(x_0) & \dots & \beta_N(x_0) \\ \frac{1}{2}\beta_0(x_1) & \beta_1(x_1) & \dots & \beta_N(x_1) \\ \cdot & \cdot & \dots & \cdot \\ \frac{1}{2}\beta_0(x_N) & \beta_1(x_N) & \dots & \beta_N(x_N) \end{pmatrix}$$

$$\mathbf{a} = (a_0, a_1, \dots, a_N)', \qquad \mathbf{f} = (f(x_0), f(x_1), \dots, f(x_N))'.$$

The matrix equation (10.2.17) can be solved by standard methods to determine **a**. Note that the same method can be used for any finite range of integration by using the appropriate shifted Chebyshev polynomials [see IV, § 10.5.3].

We have omitted a few points of detail in the above outline of the method. Firstly we need to know M and N. For computational purposes it is probably most convenient to start with $M = N = 8$ and to repeat the computations with larger values of M and N say $M = N = 16$. Secondly we have to compute for each x, the coefficients $b_n(x_i)$. These are found from

$$b_n(x_i) = \begin{cases} \dfrac{2}{M} \sum_{r=0}^{M}{}'' K(x_i, y_r) T_r(y_n), & n = 0, 1, \dots, M-1 \\ \dfrac{1}{M} \sum_{r=0}^{M}{}'' K(x_i, y_r) T_r(y_n), & n = M \end{cases}$$

where $y_r = \cos(\pi r / M)$ and the double prime indicates that the first and last terms of the sum have to be multiplied by a factor of $\frac{1}{2}$. Finally we mention that we can get some idea of the error in the computed series for $f(x)$ by observing the convergence of the computed coefficients a_n (see Elliott, 1963).

We illustrate the method by means of a simple example where, for convenience, we have taken $M = N = 2$.

EXAMPLE 10.2.6. Solve approximately, by the Chebyshev series method, the integral equation

$$\phi(x) = e^x + \int_0^1 (1 + xy)\phi(y)\, dy.$$

This integral equation was solved by other methods in Example 10.2.2. As the interval of integration is $(0, 1)$ it is most convenient to use shifted Chebyshev polynomials and to express $\phi(x)$ in the form

$$\phi(x) = \tfrac{1}{2}a_0 T_0^*(x) + a_1 T_1^*(x) + a_2 T_2^*(x),$$

as we have taken N to be 2. The collocation points for the shifted Chebyshev polynomials are

$$x_i = \tfrac{1}{2}[1 + \cos(\pi i / N)], \qquad i = 0, 1, \ldots, N$$

and as $N = 2$ in this case we have

$$x_0 = 1, \qquad x_1 = \tfrac{1}{2}, \qquad x_2 = 0$$

and the values of $f(x)$ at the collocation points are

$$f(x_0) = e \approx 2{\cdot}71828, \qquad f(x_1) = e^{1/2} \approx 1{\cdot}64872, \qquad f(x_2) = e^0 \approx 1.$$

As $K(x, y)$ is a polynomial of degree 1 it is exactly represented in terms of shifted Chebyshev polynomials in the form

$$K(x, y) = 1 + \tfrac{1}{2}x + \tfrac{1}{2}x(2y - 1) = 1 + \tfrac{1}{2}x + \tfrac{1}{2}x T_1^*(y).$$

In particular, when $x = x_0$ we have

$$K(x_0, y) = 1{\cdot}5 + 0{\cdot}5 T_1^*(y).$$

It follows that

$$K(x_0, y)\phi(y) = (1{\cdot}5 + 0{\cdot}5 T_1^*(y))(\tfrac{1}{2}a_0 T_0^*(y) + a_1 T_1^*(y) + a_2 T_2^*(y))$$

$$= (0{\cdot}75a_0 + 0{\cdot}25a_1) T_0^*(y) + (1{\cdot}5a_1 + 0{\cdot}25a_0 + 0{\cdot}25a_2) T_1^*(y)$$

$$+ (1{\cdot}5a_2 + 0{\cdot}25a_1) T_2^*(y) + 0{\cdot}25a_2 T_3^*(y)$$

using the result $T_r^*(x) T_s^*(x) = \tfrac{1}{2}[T_{r+s}^*(x) + T_{|r-s|}^*(x)]$. Since [see IV, § 10.5.3]

$$\int_0^1 T_r^*(x)\, dx = \begin{cases} 0, & \text{if } r = 2m + 1,\ m = 0, 1, \ldots \\ \dfrac{-1}{r^2 - 1}, & \text{if } r = 2m,\ m = 0, 1, \ldots \end{cases}$$

it follows that

$$I(x_0) = \int_0^1 K(x_0, y)\phi(y)\, dy \approx 0 \cdot 75 a_0 + 0 \cdot 25 a_1 - \tfrac{1}{3}(1 \cdot 5 a_2 + 0 \cdot 25 a_1)$$

and the approximation to the integral equation at the point x_0 is

$$\phi(x_0) = f(x_0) + I(x_0).$$

Since

$$\phi(x_0) = \tfrac{1}{2} a_0 + a_1 T_1^*(1) + a_2 T_2^*(1) = \tfrac{1}{2} a_0 + a_1 + a_2$$

this equation simplifies to

$$-0 \cdot 25 a_0 + 0 \cdot 83333 a_1 + 1 \cdot 5 a_2 = 2 \cdot 71828.$$

Similarly at the collocation points x_1 and x_2 we obtain

$$-0 \cdot 125 a_0 - 0 \cdot 08333 a_1 - 0 \cdot 58333 a_2 = 1 \cdot 64872$$

and

$$-a_1 + 1 \cdot 33333 a_2 = 1.$$

The solution of these equations is

$$a_0 = -13 \cdot 10758, \qquad a_1 = -0 \cdot 85972, \qquad a_2 = 0 \cdot 10521$$

so that

$$\phi(x) \approx -6 \cdot 55379 T_0^*(x) - 0 \cdot 85972 T_1^*(x) + 0 \cdot 10521 T_2^*(x).$$

In particular, at the collocation points we find

$$\phi(0) = -5 \cdot 58886, \qquad \phi(\tfrac{1}{2}) = -6 \cdot 65900, \qquad \phi(1) = -7 \cdot 30830.$$

(Compare Example 10.2.2.)

10.2.3. Singular Kernels

Unfortunately a large number of integral equations which occur in practical problems are *singular*. By this we mean that *either* the kernel $K(x, y)$ *or* one of its derivatives is infinite at some point in the interval of integration, *or* the function $\phi(x)$ which we wish to determine has some derivative infinite or the interval of integration is infinite. For instance, in two dimensional potential problems we encounter the kernel

$$\log|x - y|$$

which is infinite when $y = x$ and the quadrature methods of section 10.2.1 break down completely. This is clearly demonstrated when we use Simpson's rule over $(0, 1)$ with the three points $0, \tfrac{1}{2}, 1$. If we substitute $x = \tfrac{1}{2}$ then the value of the integrand at $y = \tfrac{1}{2}$ is infinite and Simpson's rule will give an infinite result.

This difficulty can be overcome in a number of ways.

(i) *Weakening the Singularity.*

We write

$$\int_0^1 K(x, y)\phi(y)\, dy = \int_0^1 K(x, y)[\phi(y) - \phi(x)]\, dy + \int_0^1 K(x, y)\phi(x)\, dy$$

$$= \int_0^1 K(x, y)[\phi(y) - \phi(x)]\, dy + \phi(x)\xi(x)$$

where

$$\xi(x) = \int_0^1 K(x, y)\, dy$$

is assumed to be a known function. Although $K(x, x)$ is infinite nevertheless, because the factor $\phi(y) - \phi(x)$ vanishes when $y = x$, the product must also vanish otherwise the singularity would be so strong that the integral would not even exist. It should be noticed that this device can be usefully employed to weaken any singular kernel even though that kernel might not be infinite at any point in $(0, 1)$. This is advantageous as the closer the integrand is to an analytic function [see IV, Definition 2.10.3] the more accurate (in general) will be the approximate quadrature formula.

EXAMPLE 10.2.7. Solve approximately the integral equation

$$\phi(x) = x + \int_0^1 \log_e |x - y|\phi(y)\, dy.$$

We first weaken the singularity by using the above device so that our integral equation is

$$\phi(x) = x + \int_0^1 \log_e |x - y|[\phi(y) - \phi(x)]\, dy + \xi(x)\phi(x)$$

where

$$\xi(x) = \int_0^1 \log_e |x - y|\, dy = \int_0^x \log_e (x - y)\, dy + \int_x^1 \log_e (y - x)\, dx$$

$$= [(y - x)\log_e (x - y) - y]_0^x + [(y - x)\log_e (y - x) - y]_x^1$$

by integration by parts,

$$= x \log_e x + (1 - x)\log_e (1 - x) - 1.$$

Thus

$$\phi(x) = x + \int_0^1 \log_e |x - y|[\phi(y) - \phi(x)]\, dy$$

$$+ [x \log_e x + (1 - x)\log_e (1 - x) - 1]\phi(x).$$

Now the basic Simpson rule can be applied, as in section 10.2.1, to obtain approximate values for $\phi(0)$, $\phi(\frac{1}{2})$ and $\phi(1)$. We have, with the usual notation,

$$\phi^*(0) = 0 + \tfrac{1}{6}[4 \log_e \tfrac{1}{2} (\phi^*(\tfrac{1}{2}) - \phi^*(0))] - \phi^*(0)$$

$$\phi^*(\tfrac{1}{2}) = 0 \cdot 5 + \tfrac{1}{6}[\log_e \tfrac{1}{2} (\phi^*(0) - \phi^*(\tfrac{1}{2})) + \log_e \tfrac{1}{2} (\phi^*(1) - \phi^*(\tfrac{1}{2}))] + (\log_e \tfrac{1}{2} - 1)\phi^*(\tfrac{1}{2})$$

$$\phi^*(1) = 1 + \tfrac{1}{6}[4 \log_e \tfrac{1}{2} (\phi^*(\tfrac{1}{2}) - \phi^*(1))] - \phi^*(1)$$

which simplifies to the system of equations

$$1 \cdot 537902\phi^*(0) + 0 \cdot 462098\phi^*(\tfrac{1}{2}) = 0$$

$$0 \cdot 115525\phi^*(0) + 2 \cdot 462098\phi^*(\tfrac{1}{2}) + 0 \cdot 115525\phi^*(1) = 0 \cdot 5$$

$$462098\phi^*(\tfrac{1}{2}) + 1 \cdot 537902\phi^*(1) = 1.$$

The solution of these equations was found to be

$$\phi^*(0) = -0 \cdot 053357, \qquad \phi^*(\tfrac{1}{2}) = 0 \cdot 177576, \qquad \phi^*(1) = 0 \cdot 596880.$$

The difficulty we are faced with now is to decide how much confidence we should attach to the above results. One cannot glean anything from examination of the differences of $K(x, y)[\phi(y) - \phi(x)]$, as we did in section 10.2.2, for now the derivative with respect to y, at $y = x$, is infinite and consequently the differences in the region of $y = x$, are very large. In this type of situation we are forced to do some additional computing in order to attach some confidence to the numerical results. For instance, we can compare the results obtained above with those found by using Simpson's rule applied to the intervals $(0, \frac{1}{2})$, $(\frac{1}{2}, 1)$, and if these do not agree sufficiently well we can make further subdivisions of the interval $(0, 1)$ to get a better approximation to the integral.

(ii) *Infinite Intervals*

Previously we have studied Fredholm integral equations of the second kind for which the range of integration is finite. We sometimes encounter problems in which the range of integration is infinite i.e. we desire to solve equations such as

$$\phi(x) = f(x) + \lambda \int_0^\infty K(x, y)\phi(y)\, dy.$$

One method of doing this is to take a value Y which is so large that

$$\left| \int_Y^\infty K(x, y)\phi(y)\, dy \right| < \varepsilon, \qquad \varepsilon \text{ arbitrarily small}$$

and then to solve the integral equation

$$\phi(x) = f(x) + \lambda \int_0^Y K(x, y)\phi(y)\, dy.$$

The drawback of this approach is that we will require a large number of pivotal points to adequately represent the integral and the amount of work required to carry out any finite difference analysis will be prohibitive.

However, if we possess some knowledge of the asymptotic behaviour of the integrand then it may be possible to employ an 'accurate' quadrature formula. For instance if $K(x, y)\phi(y) \sim e^{-\lambda y}\eta(x, y)$ where $\eta(x, y)$ is a slowly varying function we can employ a Gauss–Laguerre n-point formula [see § 7.3.3]. We illustrate this by means of the following example.

EXAMPLE 10.2.8.   Solve approximately the integral equation

$$\phi(x) = 1 + \int_0^\infty e^{-(x+y)}\phi(y)\,dy.$$

If we use the Gauss–Laguerre 4-point formula we have approximately

$$\phi^*(x) = 1 + e^{-x}[0 \cdot 603154\phi^*(0 \cdot 322548) + 0 \cdot 357419\phi^*(1 \cdot 745761)$$

$$+ 0 \cdot 038888\phi^*(4 \cdot 536620) + 0 \cdot 000539\phi^*(9 \cdot 395071)]. \quad (10.2.18)$$

Substituting the pivotal values $x = 0 \cdot 322548,\ 1 \cdot 745761,\ 4 \cdot 536620,\ 9 \cdot 395071$, in the above equation we get a system of equations which enable us to determine the approximations to ϕ at the pivotal values. These approximations are

$$\phi^*(0 \cdot 322548) = 2 \cdot 447606, \qquad \phi^*(1 \cdot 745761) = 1 \cdot 348785$$

$$\phi^*(4 \cdot 536620) = 1 \cdot 021406, \qquad \phi^*(9 \cdot 395071) = 1 \cdot 000166.$$

To check the validity of these answers we could determine $\phi^*(1)$, say, from formula (10.2.18) and then approximate the integral, equations using the Gauss–Laguerre 5-point formula and again make an estimate for $\phi^*(1)$. If the two estimates for $\phi^*(1)$ agree to 3 significant digits, say, then we can be reasonably confident that we have achieved 3 significant digit accuracy in our result for $\phi^*(1)$. The exact solution in this example is $\phi(x) = 1 + 2\,e^{-x}$.

We can in a similar manner solve integral equations in which the range of integration is $(-\infty, \infty)$ and $K(x, y)\phi(y) \sim e^{-\lambda y^2}\eta(x, y)$ with $\eta(x, y)$ a slowly varying function. In this case the appropriate quadrature formulae needed to enable us to approximate to ϕ, the solution of the integral equation

$$\phi(x) = f(x) + \lambda \int_{-\infty}^\infty K(x, y)\phi(y)\,dy$$

are Gauss–Hermite formulae [see § 7.3.4].

(iii)   *Discontinuous Kernels*

Another complication which we might encounter in practical problems is that of a discontinuous kernel $K(x, y)$ or a continuous kernel $K(x, y)$ which has a discontinuity in its derivative. With regard to the latter, Mayers (in Fox, 1962) gives the example of the integral equation

$$\phi(x) = \tfrac{1}{6}(-2 + 9x - 2x^3) + \int_0^1 |x - y|\phi(y)\,dy$$

which has the solution $\phi(x) = x$ in $[0, 1]$. If the integral equation is used to compute $\phi(x)$ outside $[0, 1]$ from the formula

$$\phi(x) = \tfrac{1}{6}(-2 + 9x - 2x^3) + \int_0^1 |x - y|y\,dy$$

we find

$$\phi(x) = \begin{cases} x - x^3 & \text{if } x < 0 \\ 2x - \tfrac{2}{3} - \tfrac{1}{3}x^3 & \text{if } x > 1 \end{cases}$$

so that $\phi(x)$ has a discontinuous derivative at each end of the range. As far as numerical work is concerned it means that we have to be careful with differences obtained from function values outside $[0, 1]$ as they can be inaccurate and, as we mentioned earlier, this is the reason why it is advisable to use only differences which are generated from points internal to $[0, 1]$.

We now illustrate a technique for dealing with discontinuous kernels and, for this purpose, we consider an example given in Stroud and Secrest (1966).

EXAMPLE 10.2.9.   Determine $\phi(x)$ when $\phi(x)$ satisfies the integral equation

$$\phi(x) = f(x) + \int_0^1 K(x, y)\phi(y)\,dy$$

where, for $0 \le x < \tfrac{1}{2}$,

$$K(x, y) = \begin{cases} \tfrac{1}{2}, & \text{if } 0 \le y < \tfrac{1}{2} \\ 1, & \text{if } \tfrac{1}{2} < y \le 1 \end{cases}$$

$$f(x) = \log_e (1 + x) - 0 \cdot 3321955302$$

and, for $\tfrac{1}{2} \le x \le 1$,

$$K(x, y) = \begin{cases} 2, & \text{if } 0 \le y < \tfrac{1}{2} \\ \tfrac{1}{2}, & \text{if } \tfrac{1}{2} < y \le 1 \end{cases}$$

$$f(x) = \log_e (1 + x) - 0 \cdot 3554436737.$$

In this problem besides the kernel being discontinuous the function $f(x)$ is also discontinuous. For given x, whether it be in $(0, \tfrac{1}{2})$ or $(\tfrac{1}{2}, 1)$, the integrand has a discontinuity at $y = \tfrac{1}{2}$. We therefore employ the device suggested in section 7.6.5, and apply a quadrature formula to evaluate $\int_0^{1/2} K(x, y)\phi(y)\,dy$ and the same quadrature formula to evaluate $\int_{1/2}^1 K(x, y)\phi(y)\,dy$. Thus, if the quadrature formula used is a Gauss 2-point formula, we have

$$\phi(x) = f(x) + \int_0^{1/2} K(x, y)\phi(y)\, dy + \int_{1/2}^1 K(x, y)\phi(y)\, dy, \qquad 0 \le x < \tfrac{1}{2}$$

$$= \log_e (1 + x) - 0{\cdot}3321955302 + \int_0^{1/2} \tfrac{1}{2}\phi(y)\, dy + \int_{1/2}^1 \phi(y)\, dy$$

so that if $\phi^*(x)$ is our approximation to $\phi(x)$ for any x in $[0, \tfrac{1}{2})$ we have

$$\phi^*(x) = \log_e (1 + x) - 0{\cdot}3321955302 + \tfrac{1}{8}(\phi_1^* + \phi_2^*) + \tfrac{1}{4}(\phi_3^* + \phi_4^*) \quad (10.2.19)$$

where $\phi_i^* = \phi^*(x_i)$, $i = 1, 2, 3, 4$ and $x_1, x_2 = \tfrac{1}{4}(1 \mp 1/\sqrt{3})$, $x_3, x_4 = \tfrac{1}{4}(3 \mp 1/\sqrt{3})$. Similarly for x in $[\tfrac{1}{2}, 1]$ we get

$$\phi^*(x) = \log_e (1 + x) - 0{\cdot}3554436737 + \tfrac{1}{2}(\phi_1^* + \phi_2^*) + \tfrac{1}{8}(\phi_3^* + \phi_4^*). \quad (10.2.20)$$

In particular substituting the values x_1, x_2 in (10.2.19) and x_3, x_4 in (10.2.20) we obtain, after simplification, the system of linear equations

$$\tfrac{7}{8}\phi_1^* - \tfrac{1}{8}\phi_2^* - \tfrac{1}{4}\phi_3^* - \tfrac{1}{4}\phi_4^* = -0{\cdot}231751$$

$$-\tfrac{1}{8}\phi_1^* + \tfrac{7}{8}\phi_2^* - \tfrac{1}{4}\phi_3^* - \tfrac{1}{4}\phi_4^* = 0{\cdot}000224$$

$$-\tfrac{1}{2}\phi_1^* - \tfrac{1}{2}\phi_2^* + \tfrac{7}{8}\phi_3^* - \tfrac{1}{8}\phi_4^* = 0{\cdot}118093$$

$$-\tfrac{1}{2}\phi_1^* - \tfrac{1}{2}\phi_2^* - \tfrac{1}{8}\phi_3^* + \tfrac{7}{8}\phi_4^* = 0{\cdot}283426$$

giving

$$\phi_1^* = 0{\cdot}100924$$

$$\phi_2^* = 0{\cdot}332899$$

$$\phi_3^* = 0{\cdot}474227$$

$$\phi_4^* = 0{\cdot}639560.$$

It can be verified by direct substitution that $\phi(x) = \log_e (1 + x)$ is the solution of the integral equation so that the maximum error in the computed values ϕ_i^* is $0{\cdot}000691$. This result, obtained using only four function values, compares very favourably with the results obtained using a 16-point Gauss–Legendre quadrature formula in Stroud and Secrest (1966). Better results can be obtained by using a 3-point formula to approximate the integrals in $[0, \tfrac{1}{2})$ and $(\tfrac{1}{2}, 1]$ and the comparison of the results obtained from using the 2- and 3-point formulae serves as a check on their numerical accuracy.

10.3.   VOLTERRA EQUATIONS OF THE SECOND KIND

Because of the correspondence [see § 10.1.2] between the second-order differential equation

$$f''(x) = \psi[x, f(x)], \qquad x > 0,$$

where $f(0), f'(0)$ are given and the integral equation

$$f(x) = f(0) + xf'(0) + \int_0^x (x-y)\psi[y, f(y)]\, dy, \qquad x > 0$$

it is perhaps not surprising that techniques for solving Volterra integral equations of the second kind bear some resemblance to techniques for the solution of initial-value problems for differential equations. We shall illustrate the application of these techniques in the coming sections.

10.3.1. Multi-step Methods

Assume that the Volterra integral equation of the second kind is given in the form

$$\phi(x) = f(x) + \int_0^x K(x, y)\phi(y)\, dy, \qquad x > 0. \tag{10.3.1}$$

The basic multi-step method is as follows. Assume $x = nh$ and we use an equal interval quadrature formula to evaluate the integral, namely

$$\int_0^{nh} g(x)\, dx = h \sum_{k=0}^{n} \alpha_{nk} g(kh) + R_n, \tag{10.3.2}$$

where R_n is the remainder term for the formula used. Then, neglecting the remainder term, application of (10.3.2) gives the following approximation for $\phi(nh)$,

$$\phi(nh)[1 - \alpha_{nn} hK(nh, nh)] = f(nh) + h \sum_{k=0}^{n-1} \alpha_{nk} K(nh, kh)\phi(kh). \tag{10.3.3}$$

For instance, if the quadrature rule used was the repeated trapezium rule [see § 7.1.3], formula (10.3.3) becomes

$$\phi(nh)[1 - \tfrac{1}{2} hK(nh, nh)] = f(nh) + h \sum_{k=1}^{n-1} K(nh, kh)\phi(kh) + \tfrac{1}{2} hK(nh, 0)\phi(0). \tag{10.3.4}$$

This formula is self-starting since $\phi(0) = f(0)$ from the integral equation (10.3.1) and $\phi(h), \phi(2h), \ldots$ can be obtained from (10.3.4) recursively.

EXAMPLE 10.3.1. Find $\phi(x)$ for $x = 0(0 \cdot 1)2$ when $\phi(x)$ satisfies the Volterra equation

$$\phi(x) = x + \tfrac{1}{5} \int_0^x xy\phi(y)\, dy, \qquad x > 0.$$

The analytic solution of this equation was found in Example 10.1.2 to be

$$\phi(x) = x \exp(x^3/15).$$

From the integral equation we have

$$\phi(0) = 0.$$

Formula (10.3.4) yields

$$\phi(0.1)[1 - 0.05 \times \tfrac{1}{5}(0.1)^2] = 0.1 + 0.05 \times \tfrac{1}{5}(0.1)(0)\phi(0).$$

Hence

$$\phi(0.1) = 0.10001.$$

Likewise

$$\phi(0.2)[1 - 0.05 \times \tfrac{1}{5}(0.2)(0.2)] = 0.2 + 0.05 \times \tfrac{1}{5}(0.2)(0)\phi(0)$$
$$+ 0.1 \times \tfrac{1}{5}(0.2)(0.1)\phi(0.1)$$

from which we find

$$\phi(0.2) = 0.20012.$$

Similarly we can establish the results in Table 10.3.1.

For more accurate computation we must use a higher order quadrature formula in place of the trapezium rule. In the case of the nth order formula (10.3.2) referred to above we require a knowledge of $\phi(h), \ldots, \phi(\overline{n-1}h)$. Exactly this situation occurs in the solution of ordinary differential equations [see § 8.3], and a special starting procedure is needed to compute these quantities. For suitable kernels it might be possible to find a power series expansion for $\phi(kh)$, k small, from which the starting values can be determined.

It is more convenient, however, to use a set of purpose built formulae which are due to Day (1968). These enable one to compute $\phi(h)$, $\phi(2h)$, $\phi(3h)$. Day's formulae are

$$\phi(h) = f(h) + \tfrac{1}{6}h[K(h, 0)f(0) + 4K(h, \tfrac{1}{2}h)\phi_{13} + K(h, h)\phi_{12}] \quad (10.3.4)$$

where

$$\phi_{11} = f(h) + hK(h, 0)f(0)$$

$$\phi_{12} = f(h) + \tfrac{1}{2}h[K(h, 0)f(0) + K(h, h)\phi_{11}]$$

$$\phi_{13} = f(\tfrac{1}{2}h) + \tfrac{1}{4}h[K(\tfrac{1}{2}h, 0)f(0) + K(\tfrac{1}{2}h, \tfrac{1}{2}h)(\tfrac{1}{2}f(0) + \tfrac{1}{2}\phi_{12})]$$

$$\phi(2h) = f(2h) + \tfrac{1}{3}h[K(2h, 0)f(0) + 4K(2h, h)\phi(h) + K(2h, 2h)\phi_{21}] \quad (10.3.5)$$

where

$$\phi_{21} = f(2h) + 2hK(2h, h)\phi(h)$$

$$\phi(3h) = f(3h) + \tfrac{3}{8}h[K(3h, 0)f(0) + 3K(3h, h)\phi(h)$$
$$+ 3K(3h, 2h)\phi(2h) + K(3h, 3h)\phi_{31}] \quad (10.3.6)$$

where

$$\phi_{31} = f(3h) + \tfrac{3}{2}h[K(3h, h)\phi(h) + K(3h, 2h)\phi(2h)].$$

We shall illustrate the use of Day's starting procedure in conjunction with the composite Simpson rule. We immediately run into difficulties as soon as we have to apply the quadrature rule to an even number of points, as the composite Simpson rule requires an odd number of points for its application [see § 7.1.5]. This can be overcome by using a different rule at the upper end of the range of integration. An appropriate rule for this purpose is the $\frac{3}{8}$th rule [see Appendix 7.1] as it has the same local truncation error as the Simpson rule (namely $O(h^5)$).

EXAMPLE 10.3.2.   Find $\phi(x)$ for $x = 0(0\cdot1)2$ when $\phi(x)$ satisfies the Volterra equation

$$\phi(x) = x + \tfrac{1}{5}\int_0^x xy\phi(y)\,dy, \qquad x > 0.$$

We have $\phi(0) = 0$. With $h = 0\cdot1$, $K(x, y) = \tfrac{1}{5}xy$, we have, using Day's starting procedure to determine $\phi(0\cdot1)$,

$\phi_{11} = 0\cdot1$

$\phi_{12} = 0\cdot1 + 0\cdot05[0 + \tfrac{1}{5}(0\cdot1)^3] \approx 0\cdot10001$

$\phi_{13} = 0\cdot05 + 0\cdot025[0 + \tfrac{1}{5}(0\cdot05)^2(0 + 0\cdot05)] \approx 0\cdot05$

and

$\phi(0\cdot1) = 0\cdot1 + \tfrac{1}{6}(0\cdot1)[0 + 4 \times \tfrac{1}{5}(0\cdot1)(0\cdot05)(0\cdot05) + \tfrac{1}{5}(0\cdot1)(0\cdot1)(0\cdot10001)]$

$\qquad\qquad = 0\cdot10001$ (to 5 decimal places).

Simpson's rule gives $\phi(0\cdot2)$ from

$\phi(0\cdot2) = 0\cdot2 + \tfrac{1}{3}(0\cdot1)[\tfrac{1}{5}(0\cdot2)(0)\phi(0) + 4 \times \tfrac{1}{5}(0\cdot2)(0\cdot1)\phi(0\cdot1) + \tfrac{1}{5}(0\cdot2)^2\phi(0\cdot2)]$

i.e.

$$\phi(0\cdot2) = 0\cdot20011 \quad \text{(to 5 decimal places)}.$$

To compute $\phi(0\cdot3)$ we use the $\frac{3}{8}$ths rule to obtain

$\phi(0\cdot3) = 0\cdot3 + \tfrac{3}{8}(0\cdot1)[\tfrac{1}{5}(0\cdot3)(0)\phi(0) + 3 \times \tfrac{1}{5}(0\cdot3)(0\cdot1)\phi(0\cdot1)$

$\qquad\qquad + 3 \times \tfrac{1}{5}(0\cdot3)(0\cdot2)\phi(0\cdot2) + \tfrac{1}{5}(0\cdot3)^2\phi(0\cdot3)].$

Hence

$$\phi(0\cdot3) \approx 0\cdot30054.$$

Simpson's rule is now used to determine $\phi(0\cdot4), \phi(0\cdot6), \ldots, \phi(2)$ and a combination of Simpson's rule and the $\frac{3}{8}$th rule is used to compute $\phi(0\cdot5), \phi(0\cdot7), \ldots, \phi(1\cdot9)$. The results are presented in Table 10.3.1. If the $\frac{3}{8}$th rule is used at the lower end the results, also tabulated in Table 10.3.1, are not quite so satisfactory.

10.3.2. Runge–Kutta Type Methods

One drawback of multi-step methods is the need for starting values and a further shortcoming is the lack of flexibility as far as a change of step length is concerned. These drawbacks do not occur with those methods which are similar to the Runge–Kutta methods which are used to solve the ordinary differential equation system

$$y' = f(x, y), \qquad y(x_0) = y_0.$$

Pouzet (1963) gives the following formulae, for estimating $\phi(x_{n+1})$ when $\phi(x)$ satisfies the integral equation

$$\phi(x) = f(x) + \int_0^x K(x, y)\phi(y)\, dy, \qquad x > 0.$$

He defines

$$\begin{aligned}
\phi_n^{(0)} &= \phi_{n-1}^{(4)} \qquad (\phi_0^{(0)} = f(x_0)) \\
\phi_n^{(1)} &= F_n(x_{n+\frac{1}{2}}) + \tfrac{1}{2}hK(x_{n+\frac{1}{2}}, x_n)\phi_n^{(0)} \\
\phi_n^{(2)} &= F_n(x_{n+\frac{1}{2}}) + \tfrac{1}{2}hK(x_{n+\frac{1}{2}}, x_{n+\frac{1}{2}})\phi_n^{(1)}, \\
\phi_n^{(3)} &= F_n(x_{n+1}) + hK(x_{n+1}, x_{n+\frac{1}{2}})\phi_n^{(2)}
\end{aligned} \qquad (10.3.7)$$

and then forms

$$\begin{aligned}
\phi_n^{(4)} = F_n(x_{n+1}) + \tfrac{1}{6}h\{&K(x_{n+1}, x_n)\phi_n^{(0)} + 2K(x_{n+1}, x_{n+\frac{1}{2}})\phi_n^{(1)} \\
&+ 2K(x_{n+1}, x_{n+\frac{1}{2}})\phi_n^{(2)} + K(x_{n+1}, x_{n+1})\phi_n^{(3)}\}
\end{aligned} \qquad (10.3.8)$$

where

$$\begin{aligned}
F_n(x) = f(x) + \tfrac{1}{6}h \sum_{j=0}^{n-1} \{&K(x, x_j)\phi_j^{(0)} + 2K(x, x_{j+\frac{1}{2}})\phi_j^{(1)} \\
&+ 2K(x, x_{j+\frac{1}{2}})\phi_j^{(2)} + K(x, x_{j+1})\phi_j^{(3)}\} \\
F_0(x) = f(x).
\end{aligned}$$

It can be shown that

$$\phi(x_{n+1}) = \phi_n^{(4)} + O(h^4) \qquad (10.3.9)$$

so that $\phi_n^{(4)}$ is an approximation to $\phi(x_{n+1})$. We illustrate the use of these formulae in the following example.

EXAMPLE 10.3.3. Estimate $\phi(x)$ for $x = 0(0\cdot1)2$ where $\phi(x)$ satisfies the Volterra equation

$$\phi(x) = x + \tfrac{1}{5}\int_0^x xy\phi(y)\, dy, \qquad x > 0.$$

The approximate values of $\phi(0\cdot1)$ and $\phi(0\cdot2)$ are determined in full below and the remaining values are given in Table 10.3.1. We have

$$\phi_0^{(0)} = 0$$

$$\phi_0^{(1)} = 0\cdot05 + 0\cdot05(\tfrac{1}{5})(0\cdot05)(0)(0) = 0\cdot05$$

$$\phi_0^{(2)} = 0\cdot05 + 0\cdot05(\tfrac{1}{5})(0\cdot05)(0\cdot05)(0\cdot05) \approx 0\cdot05$$

$$\phi_0^{(3)} = 0\cdot1 + 0\cdot1(\tfrac{1}{5})(0\cdot1)(0\cdot05)(0\cdot05) \approx 0\cdot10001$$

and thus

$$\phi(0\cdot1) \approx \phi_0^{(4)} = 0\cdot1 + \tfrac{1}{6}(0\cdot1)\{\tfrac{1}{5}(0\cdot1)(0)(0) + 2(\tfrac{1}{5})(0\cdot1)(0\cdot05)(0\cdot05)$$

$$+ 2(\tfrac{1}{5})(0\cdot1)(0\cdot05)(0\cdot05) + \tfrac{1}{5}(0\cdot1)(0\cdot1)(0\cdot10001)\}$$

$$= 0\cdot10001.$$

Likewise,

$$\phi_1^{(0)} = 0\cdot10001$$

$$\phi_1^{(1)} = 0\cdot15 + \tfrac{1}{6}(0\cdot1)\{\tfrac{1}{5}(0\cdot15)(0)(0) + 2(\tfrac{1}{5})(0\cdot15)(0\cdot05)(0\cdot05)$$

$$+ 2(\tfrac{1}{5})(0\cdot15)(0\cdot05)(0\cdot05) + \tfrac{1}{5}(0\cdot15)(0\cdot1)(0\cdot10001)\}$$

$$+ 0\cdot05(\tfrac{1}{5})(0\cdot15)(0\cdot1)(0\cdot10001)$$

$$= 0\cdot15002_5$$

$$\phi_1^{(2)} = 0\cdot15 + \tfrac{1}{6}(0\cdot1)\{\tfrac{1}{5}(0\cdot15)(0)(0) + 2(\tfrac{1}{5})(0\cdot15)(0\cdot05)(0\cdot05)$$

$$+ 2(\tfrac{1}{5})(0\cdot15)(0\cdot05)(0\cdot05) + \tfrac{1}{5}(0\cdot15)(0\cdot1)(0\cdot10001)\}$$

$$+ 0\cdot05(\tfrac{1}{5})(0\cdot15)(0\cdot15)(0\cdot15002_5)$$

$$= 0\cdot15004$$

$$\phi_1^{(3)} = 0\cdot2 + \tfrac{1}{6}(0\cdot1)\{\tfrac{1}{5}(0\cdot2)(0)(0) + 2(\tfrac{1}{5})(0\cdot2)(0\cdot05)(0\cdot05)$$

$$+ 2(\tfrac{1}{5})(0\cdot2)(0\cdot05)(0\cdot05) + \tfrac{1}{5}(0\cdot2)(0\cdot1)(0\cdot10001)\}$$

$$+ 0\cdot1(\tfrac{1}{5})(0\cdot2)(0\cdot15)(0\cdot15004)$$

$$= 0\cdot20010.$$

Thus

$$\phi(0\cdot2) \approx \phi_1^{(4)} = 0\cdot20001 + \tfrac{1}{6}(0\cdot1)\{\tfrac{1}{5}(0\cdot2)(0\cdot1)(0\cdot10001)$$

$$+ 2(\tfrac{1}{5})(0\cdot2)(0\cdot15)(0\cdot15002_5)$$

$$+ 2(\tfrac{1}{5})(0\cdot2)(0\cdot15)(0\cdot15004)$$

$$+ \tfrac{1}{5}(0\cdot2)(0\cdot2)(0\cdot20010)\}$$

$$= 0\cdot20010.$$

One disadvantage of this type of method is the number of function evaluations required at each step which is of the order of $4n^2$ whereas, for the
repeated Simpson rule, the number of function evaluations is of order $\frac{1}{2}n^2$.
Pouzet indicates that the number of function evaluations can be reduced
considerably by the use of multi-step methods, such as the Newton–Gregory
(see next section) for the computation of $F_n(x)$.

10.3.3.   Deferred Correction

This method has been used in section 10.2.2 to solve Fredholm equations of
the second kind. The method of application with Volterra equations of the
second kind is slightly different. We must compute the first few values
$\phi(0), \phi(h), \ldots, \phi(rh)$ which satisfy the integral equation

$$\phi(x) = f(x) + \int_0^x K(x, y)\phi(y)\, dy$$

by either finding the Taylor series for $\phi(x)$ [see IV, § 3.6] or using a starting
procedure such as that of Day or Pouzet. Once we have the required function
values to start the deferred correction procedure we compute successive values
of $\phi(x)$ from an equation of the form

$$\{1 - ha_nK(nh, nh)\}\phi(nh) - h \sum_{r=1}^{n} a_{n-r}K(nh, (n-r)h)\phi\{(n-r)h\} = f(nh).$$

$$(10.3.10)$$

In the case where the Gregory formula [see (7.7.4)] is used we have

$$a_r = a_{n-r}$$

and, if the last difference retained is of order p, then the coefficients a_s
corresponding to various p are given in Appendix 10.1. These coefficients were
worked out by Fox and Goodwin (1953). There are complications regarding
the use of the correction formula (10.2.12) of section 10.2.2 for the solution of
Volterra equations and this method will not therefore be given. We illustrate
the method in the following example.

EXAMPLE 10.3.4.   Solve the Volterra equation

$$\phi(x) = x + \tfrac{1}{5} \int_0^x xy\phi(y)\, dy$$

for $x = 0(0\cdot1)2$ by the method of deferred correction.

If no differences are taken into account we have the trapezium rule method
of section 10.3.1. For simplicity we shall use Gregory's formula with second
differences retained. The starting values $\phi(0), \ldots, \phi(0\cdot4)$ have been calculated correct to 5 decimal places by the Taylor series method. To obtain

$\phi(0.5)$, for instance, we have

$$\{1-(0.1)(0.375)\tfrac{1}{5}(0.5)^2\}\phi(0.5) = 0.5 + (0.1)(\tfrac{1}{5})(0.5)\{1.16667 \times 0.4\phi(0.4)$$

$$+0.95833 \times 0.3\phi(0.3)$$

$$+0.95833 \times 0.2\phi(0.2)$$

$$+1.16667 \times 0.1\phi(0.1)\}$$

Thus

$$\phi(0.5) = 0.50418.$$

The relationship which determines $\phi(0.6)$ is

$$\{1-(0.1)(0.375)\tfrac{1}{5}(0.6)^2\}\phi(0.6) = 0.6 + (0.1)(\tfrac{1}{5})(0.6)\{1.16667 \times 0.5\phi(0.5)$$

$$+0.95833 \times 0.4\phi(0.4)+0.3\phi(0.3)$$

$$+0.95833 \times 0.2\phi(0.2)$$

$$+1.16667 \times 0.1\phi(0.1)\}$$

and this leads to the estimate

$$\phi(0.6) = 0.60870.$$

The complete set of results found using the Gregory second-order formula is given in Table 10.3.1. In practice, we keep a check on the differences of $K(x, y)\phi(y)$ and, if the differences which are not retained are not negligible, we switch over to a higher-order Gregory formula which takes these differences into account.

10.3.4. Block-by-block Methods

In the previous sections the value of $\phi(x)$ was determined a step at a time and, in some methods, it was necessary to have a starting procedure. The present method, due to Young (1954), has the advantage of producing a block of values at a time and is self-starting.

The basic idea of the method is as follows. The range $[0, a]$ in which the solution is required is divided into M parts each of which is in turn sub-divided into p sub-intervals. Assume that the values $\phi_0, \phi_1, \ldots, \phi_{pm}$ are known, where $\phi_r = \phi(ra/pM)$. Then, for any integer n in the interval $mp+1, \ldots, m(p+1)$ the integral equation

$$\phi(x_n) = f(x_n) + \int_0^{x_n} K(x_n, y)\phi(y)\, dy$$

can be approximated by

$$\phi(x_n) = f(x_n) + \int_0^{x_{pm}} K(x_n, y)\phi(y)\, dy + \int_{x_{pm}}^{x_n} K(x_n, y)\phi(y)\, dy. \quad (10.3.11)$$

Since $\phi_0, \ldots, \phi_{pm}$ are known the first integral can be determined by standard quadrature methods. The second integral is computed by using a quadrature rule which uses the value of the integrand at $y = x_{pm}, \ldots, x_{p(m+1)}$. This produces a system of p simultaneous equations

$$\phi_n = f(x_n) + h \sum_{r=0}^{mp} w_{nr} K(x_n, x_r) \phi_r$$

$$+ h \sum_{r=0}^{p} w'_{nr} K(x_n, x_{mp+r}) \phi_{mp+r} \qquad n = mp+1, \ldots, p(m+1) \qquad (10.3.12)$$

where w_{nr} denote the weight coefficients of the quadrature formula used for evaluating the first integral in (10.3.11) and w'_{nr} denote the coefficients of the quadrature formula used for evaluating the second integral in (10.3.11). When this system is solved we obtain the block of p values $\phi_{mp+1}, \ldots, \phi_{p(m+1)}$. In the case where $p = 2$, Linz (1967) recommends the use of the formulae (10.3.13) and (10.3.14) in which the repeated Simpson rule [see § 7.1.5] has been used to effect the integration in $(0, x_{2m})$.

$$\phi_{2m+1} = f(x_{2m+1}) + \tfrac{1}{3}h[K(x_{2m+1}, x_0)\phi_0 + 4K(x_{2m+1}, x_1)\phi_1 + \ldots$$

$$+ K(x_{2m+1}, x_{2m})\phi_{2m}]$$

$$+ \tfrac{1}{6}hK(x_{2m+1}, x_{2m})\phi_{2m} + \tfrac{1}{6}hK(x_{2m+1}, x_{2m+1})\phi_{2m+1}$$

$$+ \tfrac{2}{3}hK(x_{2m+1}, x_{2m+\frac{1}{2}})(\tfrac{3}{8}\phi_{2m} + \tfrac{3}{4}\phi_{2m+1} - \tfrac{1}{8}\phi_{2m+2}) \qquad (10.3.13)$$

$$\phi_{2m+2} = f(x_{2m+2}) + \tfrac{1}{3}h[K(x_{2m+2}, x_0)\phi_0 + 4K(x_{2m+2}, x_1)\phi_1 + \ldots$$

$$+ K(x_{2m+2}, x_{2m+2})\phi_{2m+2}]. \qquad (10.3.14)$$

A disadvantage of the block method approach is that numerical difficulties may arise if the kernel is not well-behaved near the edges of the region of integration. However, the formulae (10.3.13), (10.3.14) above have been constructed so that these difficulties are removed.

10.3.5. Stability of Multi-step Methods

The correspondence between Volterra integral equations and initial-value problems in ordinary differential equations, as stated at the beginning of this section, leads us to suspect that the problems associated with ordinary differential equations are encountered also with Volterra integral equations. This turns out to be the case and it is particularly important, therefore, to examine the stability of any numerical method which we may devise. We shall be content to remind the reader of the importance of the order in which formulae are used (see earlier remarks about Simpson and $\frac{3}{8}$th rule) and to refer those interested in details of the analysis of stability to Kershaw (in Delves and Walsh, 1974) or Baker (1977).

10.4. INTEGRAL EQUATIONS OF THE FIRST KIND

10.4.1. Volterra Equations of the First Kind. Theoretical Approach

We give a number of courses of action which we can take to solve such equations.

(i) *Reduction to Volterra Equations of the Second Kind*

The Volterra equation of the first kind

$$\int_a^x K(x, y)\phi(y)\, dy = f(x), \qquad x > a \tag{10.4.1}$$

can be expressed as a Volterra equation of the second kind in the following way. We differentiate (10.4.1) with respect to x [see IV, Theorem 4.7.3] to obtain

$$K(x, x)\phi(x) + \int_a^x \left(\frac{\partial}{\partial x}K(x, y)\right)\phi(y)\, dy = f'(x). \tag{10.4.2}$$

If $K(x, x) \neq 0$, division of (10.4.2) by $K(x, x)$ produces a Volterra equation of the second kind (see § 10.3). If $K(x, x) = 0$ then (10.4.2) can be differentiated to give

$$K'(x, x)\phi(x) + \int_a^x \left(\frac{\partial^2}{\partial x^2}K(x, y)\right)\phi(y)\, dy = f''(x) \tag{10.4.3}$$

so that, if $K'(x, x) \neq 0$ we again arrive at a Volterra equation of the second kind. Even if $K'(x, x) = 0$, if we carry on differentiating (10.4.3) until we reach a derivative of the kernel $K(x, y)$, evaluated at $y = x$, which does not vanish, say $K^{(n)'}(x, x)$, then we arrive at

$$K^{(n)}(x, x)\phi(x) + \int_a^x \left(\frac{\partial^{n+1}}{\partial x^{n+1}}K(x, y)\right)\phi(y)\, dy = f^{(n+1)}(x) \tag{10.4.4}$$

which, apart from a factor, is a Volterra equation of the second kind. This method will fail if there is no derivative $K^{(n)}(x, x)$ which is non-vanishing and this turns out to be the case in several equations arising in practice such as *Abel's equation*

$$\int_a^x \frac{\phi(y)}{(x-y)^\alpha}\, dy = f(x), \qquad 0 < \alpha < 1, \qquad f(0) = 0.$$

An alternative theoretical approach, which also leads to a Volterra equation of the second kind, is to write

$$\psi(x) = \int_a^x \phi(y)\, dy$$

and integrate by parts in (10.4.1) [see IV, § 4.3]. This produces

$$K(x, x)\psi(x) - \int_a^x \left(\frac{\partial}{\partial y} K(x, y)\right)\psi(y) \, dy = f(x)$$

so that if $K(x, x) \neq 0$ we can obtain an equation of the second kind by division by $K(x, x)$.

EXAMPLE 10.4.1.   Transform the Volterra equation

$$\int_0^x \cos(x - y)\phi(y) \, dy = \sin x$$

into a Volterra equation of the second kind.

Differentiation produces

$$\phi(x) - \int_0^x \sin(x - y)\phi(y) \, dy = \cos x. \tag{10.4.5}$$

The numerical solution of (10.4.5) is given in Table 10.4.1 where comparison is made of several methods for the solution of equations of the first kind.
We give another example of the method which illustrates some problems connected with Volterra equations of the first kind.

EXAMPLE 10.4.2.   Solve the Volterra integral equation of the first kind

$$\int_0^x (x - y)\phi(y) \, dy = f(x).$$

By differentiation we obtain

$$\int_0^x \phi(y) \, dy = f'(x)$$

and a further differentiation yields the exact solution

$$\phi(x) = f''(x).$$

Note, however, that we require

$$f(0) = f'(0) = 0$$

for consistency. If the function $f(x)$ does not satisfy these conditions the problem is not well-posed. Even if the problem is well-posed we might be given $f(x)$ in the form of a table and, in this event, our solution can only be as accurate as our numerical determination of the second derivative (see Fox and Goodwin, 1953).

(ii)   *Use of Transforms*

It often happens that the kernel of the integral equation is of *convolution* type, i.e. $K(x, y) = K(x - y)$. If this is so and the Laplace transform of $\phi(x)$ is denoted by $\bar{\phi}(s)$ then the convolution theorem for Laplace transforms [see IV, § 13.4.4] gives

$$\mathscr{L}\left\{\int_0^x K(x-y)\phi(y)\,dy\right\} = \bar{\phi}(s)\bar{K}(s) = \mathscr{L}\{f(x)\} = \bar{f}(s)$$

and thus

$$\bar{\phi}(s) = \bar{f}(s)/\bar{K}(s). \tag{10.4.6}$$

$\phi(x)$ can then be determined by application of the inversion theorem for Laplace transforms [see IV, § 13.4.2].

EXAMPLE 10.4.3.   Find $\phi(x)$ given that

$$\int_0^x \cos(x-y)\phi(y)\,dy = \sin x.$$

We have

$$\bar{\phi}(s) = \frac{1/(s^2+1)}{s/(s^2+1)} = \frac{1}{s}.$$

Hence

$$\phi(x) = 1.$$

EXAMPLE 10.4.4.   Solve Abel's equation

$$\int_0^x \frac{\phi(y)}{(x-y)^\alpha}\,dy = f(x), \qquad 0 \le \alpha < 1, \qquad f(0) = 0.$$

Since [see IV, § 10.2]

$$\mathscr{L}\{x^{-\alpha}\} = \frac{1}{s^{1-\alpha}}\Gamma(1-\alpha)$$

it follows that

$$\bar{\phi}(s) = \frac{s^{1-\alpha}\bar{f}(s)}{\Gamma(1-\alpha)} = \frac{1}{\Gamma(1-\alpha)} \cdot s\{s^{-\alpha}\bar{f}(s)\}.$$

Now, by the convolution theorem, $s^{-\alpha}\bar{f}(s)$ is the transform of

$$\frac{1}{\Gamma(\alpha)}\int_0^x \frac{f(y)\,dy}{(x-y)^{1-\alpha}}$$

and since, in addition,

$$\mathscr{L}\left\{\frac{dg}{dx}\right\} = s\mathscr{L}\{g(x)\} - g(0)$$

it follows that

$$\bar{\phi}(s) = \frac{1}{\Gamma(\alpha)\Gamma(1-\alpha)} \mathcal{L}\left\{\frac{d}{dx}\int_0^x \frac{f(y)\,dy}{(x-y)^{1-\alpha}}\right\}$$

and hence

$$\phi(x) = \frac{1}{\Gamma(\alpha)\Gamma(1-\alpha)}\left\{\frac{d}{dx}\int_0^x \frac{f(y)\,dy}{(x-y)^{1-\alpha}}\right\}.$$

Since $\Gamma(\alpha)\Gamma(1-\alpha) = \pi/\sin \pi\alpha$ the result can be put in the neater form

$$\phi(x) = \frac{\sin \pi\alpha}{\pi}\left\{\frac{d}{dx}\int_0^x \frac{f(y)\,dy}{(x-y)^{1-\alpha}}\right\}.$$

It might not be possible to obtain a simple result for $\phi(x)$ from the knowledge of $\bar{\phi}(s)$ and even if we can determine $\phi(x)$ explicitly it might be more difficult to compute $\phi(x)$ numerically from this formula than to use some other technique. Nevertheless the method is often worth employing for, even if we cannot obtain a useful result for $\phi(x)$, it is possible to glean information about the asymptotic behaviour of $\phi(x)$ from $\bar{\phi}(s)$. For more details of the theory of transforms and their application in the solution of integral equations see Titchmarsh (1948).

10.4.2.   Volterra Equations of the First Kind

(i)   *Simple Numerical Techniques*

The object of our numerical methods is to find an approximation for $\phi(x)$ at $x = rh$, call it $\bar{\phi}(rh)$ such that the difference $\phi(rh) - \bar{\phi}(rh)$ is small and, as $h \to 0$, with rh fixed,

$$\bar{\phi}(rh) \to \phi(rh).$$

Some of the simplest numerical approaches, such as the trapezium rule [see § 7.1.3] and the mid-point rule [see § 7.1.4], turn out to be fairly effective in this respect. We illustrate this by means of an example.

EXAMPLE 10.4.5.   Solve the Volterra equation

$$\int_0^x \cos (x-y)\phi(y)\,dy = \sin x$$

for $x = 0(0\cdot1)2$ by means of the trapezium rule.

With $x = 0\cdot1$ we have

$$\tfrac{1}{2}\times 0\cdot1[(\cos 0\cdot1)\phi(0) + (\cos 0)\phi(0\cdot1)] = \sin 0\cdot1$$

and immediately we have a difficulty as we cannot determine $\phi(0\cdot1)$ without knowledge of $\phi(0)$. However, $\phi(0)$ can be determined from (10.4.2). For, if $K(x, x) \neq 0$, we have

$$\phi(a) = f'(a)/K(a, a). \tag{10.4.7}$$

If $f'(a)$ is not known explicitly we can use the approximation $[f(a+h)-f(a-h)]/2h$ instead. In our example

$$\phi(0) = \cos 0/\cos 0 = 1$$

and thus

$$\phi(0\cdot1) = [(2 \sin 0\cdot1)/0\cdot1 - \cos 0\cdot1] = 1\cdot00166.$$

Similarly with $x = 0\cdot2$ we have

$$\tfrac{1}{2}\times0\cdot1[(\cos 0\cdot2)\phi(0)+2(\cos 0\cdot1)\phi(0\cdot1)+(\cos 0)\phi(0\cdot2)]= \sin 0\cdot2$$

which leads to

$$\phi(0\cdot2) = 1\cdot00000.$$

We carry on in this manner, using the formula

$$\tfrac{1}{2}h \sum_{r=0}^{n-1} [K(nh, rh)\phi(rh)+K(nh, (r+1)h)\phi\{(r+1)h\}]=f(nh), \qquad n=1,2,\ldots$$

which in this example is

$$\tfrac{1}{2}h \sum_{r=0}^{n-1} [\cos\{(n-r)h\}\phi(rh)+\cos\{(n-r-1)h\}\phi\{(r+1)h\}]=\sin nh,$$

$$n=1,2,\ldots$$

$\phi(nh)$ is easily determined as we have a triangular system of equations. The complete set of results is given in column T, Table 10.4.1. As we know from Example 10.4.3 that $\phi(x)\equiv1$ these results are reasonable.

It will be noticed that there are some small oscillations about the true solution and this phenomenon appears to be a characteristic of this method. Linz (1969), justifies a method of smoothing given by Jones (1961), but, from Linz's paper, it would appear to be more favourable to solve problems of this type by means of the mid-point formula. The equations we have now are

$$h \sum_{r=0}^{n-1} K(nh, (r+\tfrac{1}{2})h)\phi\{(r+\tfrac{1}{2})h\}=f(nh)$$

or

$$K(nh, (n-\tfrac{1}{2})h)\phi\{(n-\tfrac{1}{2})h\}=[f(nh)/h]- \sum_{r=0}^{n-2} K(nh, (r+\tfrac{1}{2})h)\phi\{(r+\tfrac{1}{2})h\}.$$

$$(10.4.8)$$

EXAMPLE 10.4.6.   Solve the integral equation $\int_0^x \cos(x-y)\phi(y)\,dy=\sin x$ for $x<2$ by means of the mid-point rule.

For $x = 0\cdot1$ we have

$$(0\cdot1) \cos 0\cdot05\phi(0\cdot05)=\sin 0\cdot1.$$

Thus

$$\phi(0\cdot05) = 0\cdot99958.$$

Application of the mid-point rule when $x = 0\cdot2$ gives

$$0\cdot1[\cos 0\cdot15\phi(0\cdot05) + \cos 0\cdot05\phi(0\cdot15)] = \sin 0\cdot2$$

and consequently

$$\phi(0\cdot15) = 0\cdot99958.$$

Further calculation produces

$$\phi\{(r + \tfrac{1}{2})h\} = 0\cdot99958, \qquad h = 0\cdot1, \qquad r = 0, 1, \ldots, 19.$$

Better results can be obtained by application of Richardson's extrapolation technique [see § 2.4.1]. If we carry out the numerical calculation with $H = 3h$ then we obtain another set of approximations for the computed values of $\phi(\tfrac{1}{2}H), \phi(\tfrac{3}{2}H), \ldots$, call them $\phi_2(\tfrac{1}{2}H), \phi_2(\tfrac{3}{2}H), \ldots$ and the previous approximations $\phi_1(\tfrac{1}{2}H), \phi_1(\tfrac{3}{2}H), \ldots$. It follows from the theory of Richardson's extrapolation that a better approximation is, for each r,

$$\bar\phi\{(r + \tfrac{1}{2})H\} = \tfrac{1}{8}[9\phi_1\{(r + \tfrac{1}{2})H\} - \phi_2\{(r + \tfrac{1}{2})H\}]. \qquad (10.4.9)$$

Thus, in Example 10.4.6

$$\phi_2(0\cdot15) = 0\cdot996255$$

and it follows that

$$\bar\phi(0\cdot15) = \tfrac{1}{8}(9 \times 0\cdot99958 - 0\cdot996255)$$

$$= 0\cdot999995.$$

Similar results are obtained for $\bar\phi(0\cdot45), \bar\phi(0\cdot75), \ldots$.

Linz (1969), points out that higher order formulae are frequently unstable and he cites as examples Simpson's rule [see § 7.1.5] and Gregory's third (and higher order) formulae [see (7.7.4)] (see Table 10.4.1).

(ii) *Product Integration Methods*

In the previous section the integral

$$\int_0^{nh} K(x, y)\phi(y)\, dy = \sum_{r=0}^{n-1} \int_{rh}^{(r+1)h} K(x, y)\phi(y)\, dy, \qquad x = nh$$

was evaluated by employing the trapezium rule or mid-point rule. This is in many ways unsatisfactory as the integrand might only be approximated by a high-order polynomial in $(rh, (r+1)h)$ and the trapezium rule represents the integrand as a polynomial of degree 1. An alternative approach is to approximate $\phi(y)$ by a constant in $(rh, \{r + 1\}h)$, call it $\phi(\{r + \tfrac{1}{2}\}h)$. Then

$$\int_{rh}^{(r+1)h} K(x, y)\phi(y)\, dy \approx \phi\{(r + \tfrac{1}{2})h\} \int_{rh}^{(r+1)h} K(x, y)\, dy$$

and we can use a tailor-made formula to integrate

$$A_r = \int_{rh}^{(r+1)h} K(x, y)\, dy.$$

If we call this approximation $\bar{A}_r$, we obtain as our representation of the integral equation

$$\int_0^x K(x, y)\phi(y)\, dy = f(x), \qquad x = nh$$

the system of linear equations.

$$\sum_{r=0}^{n-1} \bar{A}_r \phi\{(r+\tfrac{1}{2})h\} = f(nh), \qquad n = 1, 2, \ldots \qquad (10.4.10)$$

EXAMPLE 10.4.7. Solve approximately the integral equation

$$\int_0^x \cos(x-y)\phi(y)\, dy = \sin x, \qquad 0 \leq x \leq 2.$$

With $h = 0\cdot 1$ we have

$$\int_0^{0\cdot 1} \cos(0\cdot 1 - y)\phi(y)\, dy = \sin 0\cdot 1$$

and our approximation is

$$\phi(0\cdot 05) \int_0^{0\cdot 1} \cos(0\cdot 1 - y)\, dy = \sin 0\cdot 1.$$

In this particular example the integral can be evaluated exactly. In fact we have

$$\int_0^{0\cdot 1} \cos(0\cdot 1 - y)\, dy = \sin 0\cdot 1$$

so that

$$\phi(0\cdot 05) = 1.$$

Even if we had taken a Gauss 2-point formula [see (7.3.1)] to evaluate the integral we obtain $\phi(0\cdot 05) = 1\cdot 000000$ (correct to 6 decimal places). Similarly, we find

$$\phi(0\cdot 15) = \ldots = \phi(1\cdot 95) = 1.$$

Anderssen and White (1971) give a product rule based on the trapezium rule in addition to the method described above. They assert that product integration rules are particularly appropriate when $K(x, y)$ is badly behaved but the solution $\phi(x)$ is not. These methods have been found to be satisfactory for the numerical solution of Abel's equation.

Baker (in Delves and Walsh, 1974) gives references to other methods which have been used to solve Volterra equations of the first kind (see also Baker, 1977).

10.4.3. Fredholm Equations of the First Kind. Theoretical Aspects

Fredholm equations of the first kind of the form

$$K\Phi = \int_0^1 K(x, y)\phi(y)\, dy = f(x), \qquad c \le x \le d \qquad (10.4.11)$$

present many difficulties. For instance, solutions are only possible, for some types of kernels, for a restricted class of functions $f(x)$. Thus, if $K(x, y)$ satisfies a linear differential equation of the form

$$\left\{ \frac{\partial^n}{\partial x^n} + p_1(x) \frac{\partial^{n-1}}{\partial x^{n-1}} + \ldots + p_n(x) \right\} K(x, y) = 0,$$

it follows, by differentiation of (10.4.11) with respect to x, that a solution of (10.4.11) exists only if $f(x)$ satisfies the same differential equation.

Again, given a kernel of the form

$$K(x, y) = \sum_1^n A_k(x) B_k(y)$$

then no solution of (10.4.11) exists unless $f(x)$ has the form

$$f(x) = \sum_1^n \alpha_k A_k(x)$$

and then any solution is possible for which

$$\int_0^1 B_k(y)\phi(y)\, dy = \alpha_k \qquad (k = 1, 2, \ldots, n).$$

We can, however, obtain a formal solution of (10.4.11) in terms of the eigensolutions of the kernel $K(x, y)$ [see §10.1.4]. We recall that the eigensolutions satisfy

$$\int_0^1 K(x, y)\phi_k(y)\, dy = \lambda_k \phi_k(x).$$

Thus, if we assume

$$\phi(x) = \sum \alpha_k \phi_k(x)$$

and $f(x)$ can be expressed in terms of the eigensolutions in the form

$$f(x) = \sum \beta_k \phi_k(x) \qquad (10.4.12)$$

where the β_k are known, then

$$\alpha_k = \beta_k / \lambda_k$$

and we have the solution. The formal approach conceals the fact that several conditions must be satisfied to ensure the success of this technique. Firstly, we require the existence of an infinity of distinct eigenvalues thereby ensuring a complete set of eigenfunctions. Secondly, none of the eigenvalues must be zero for, if $\lambda_k = 0$, then $\alpha_k = \beta_k / \lambda_k$ does not exist unless β_k is also zero which destroys the generality of $f(x)$. A further important requirement for the existence of the solution $f(x)$ of the form (10.4.12) is that

$$\sum (\beta_k / \lambda_k)^2$$

converges. The solution is then unique (see Tricomi, 1957).

To see the difficulties associated with Fredholm equations of the first kind consider the simple integral equation considered by Fox and Goodwin (1953) namely

$$\int_0^1 (x+y)\phi(y)\,dy = g(x) \qquad (10.4.13)$$

As we have already indicated the right hand side must be of the form $g(x) = a + bx$ for a solution to be possible. Suppose

$$g(x) = x.$$

Then Fox and Goodwin note that a solution of the integral equation is

$$\phi_1(x) = 4 - 6x$$

and, by substitution, this is clearly seen to satisfy (10.4.13). Unfortunately, this solution is not unique. For instance,

$$\phi_2(x) = 3 - 6x^2$$

also satisfies (10.4.13) and so does every linear combination of the form

$$\alpha\phi_1(x) + (1-\alpha)\phi_2(x).$$

Moreover, if $\theta(x)$ is any other function of x which is linearly independent of $\phi_1(x)$ and $\phi_2(x)$ and which satisfies

$$\int_0^1 (x+y)\theta(y)\,dy = 0,$$

and there are an infinity of such functions as all we require is for $\theta(x)$ to be orthogonal [see Definition 6.3.1] to the functions 1 and x over the range $[0, 1]$, then

$$\alpha\phi_1(x) + (1-\alpha)\phi_2(x) + \beta\theta(x)$$

is also a solution of the integral equation (10.4.13).

In attempting to solve Fredholm equations of the first kind it will therefore be necessary to limit enquiries to finding simple functions which are linear combinations of the eigensolutions.

10.4.4.   Fredholm Equations of the First Kind. Numerical Techniques

(i)   *The Method of Regularization*

In this method it is assumed that a solution exists to the ill-posed problem

$$K\Phi = f, \qquad c \leq x \leq d \tag{10.4.11}$$

and one obtains approximations to the solution by minimizing the quadratic functional

$$\|K\Phi - f\| + \alpha\|L\Phi\| \tag{10.4.14}$$

where $L\Phi$ is some linear operator and $\|\ldots\|$ denotes some appropriate norm [see IV, § 19.5]. This minimization problem is a well-posed problem which has a unique solution for a value of α which must be determined. Lewis (1975) found that the best numerical answers were nearly always obtained using the *zero-order regularization method* of Bakushinskii (1965). This consists of solving the equation

$$\alpha\phi(x) + \int_0^1 K(x, y)\phi(y)\, dy = f(x), \tag{10.4.15}$$

when $K(x, y)$ is real and symmetric, i.e. $K(x, y) = K(y, x)$. If $K(x, y)$ is not symmetric then we form the symmetric kernel $K^*(x, y)$ where

$$K^*(x, y) = \int_c^d K(\theta, x)K(\theta, y)\, d\theta \tag{10.4.16}$$

and solve instead the integral equation

$$\alpha\phi(x) + \int_0^1 K^*(x, y)\phi(y)\, dy = F(x) \tag{10.4.17}$$

where

$$F(x) = \int_c^d K(\theta, x)f(\theta)\, d\theta.$$

We illustrate the method by solving the problem of Fox and Goodwin discussed earlier.

EXAMPLE 10.4.8.   Solve the Fredholm equation of the first kind

$$\int_0^1 (x + y)\phi(y)\, dy = x \tag{10.4.13}$$

by the method of zero-order regularization.

In this example $K(x, y) = x + y = K(y, x)$ so that the problem we have to solve is, by (10.4.15)

$$\alpha\phi(x) + \int_0^1 (x+y)\phi(y)\,dy = x \qquad (\alpha \geq 0). \tag{10.4.18}$$

This is a Fredholm equation of the second kind whose solution we can determine by the methods of section 10.2. For simplicity, we shall employ the trapezium rule [see § 7.1.3] to approximate the integral in (10.4.18) so that for given x and α and step length h we have

$$\alpha\phi(x) + \tfrac{1}{2}h \sum_{i=0}^{n-1} \{(x + ih)\phi(ih) + (x + [i+1]h)\phi[(i+1)h]\} = x,$$

where $nh = 1$. Setting $x = 0, h, \ldots, nh$, in turn we obtain a system of $n+1$ simultaneous equations in the $n+1$ unknowns $\phi(0), \phi(h), \ldots, \phi(nh)$ namely

$$\alpha\phi(rh) + \tfrac{1}{2}h^2 \sum_{i=0}^{n-1} \{(r+i)\phi(ih) + (r+i+1)\phi[(i+1)h]\} = rh, \qquad r = 0, 1, \ldots, n.$$

The solution of these equations is given in Table 10.4.2 for various α and $h = 0 \cdot 1$. It can be seen that as α decreases the results tend to $\phi(x) = 4 - 6x$, and for a range of values of α the estimated values of $\phi(x)$ are fairly consistent. When α is decreased further the estimated values of $\phi(x)$ diverge from $4 - 6x$. This behaviour is a characteristic of the method.

(ii)   *Symm's Method in Potential Theory*

Symm (1963) considers the numerical solution of singular integral equations arising in two dimensional potential theory. To illustrate Symm's method, however, we shall consider the simpler one dimensional problem of a straight line of length $2c$ with centre at the origin. The charge density $\sigma(x)$ at a point x on the line satisfies the integral equation

$$\int_{-c}^c \log_e|x - y|\sigma(y)\,dy = 1 \tag{10.4.19}$$

and it is required to determine $\sigma(x)$ so that we can estimate the total charge, S,

$$S = \int_{-c}^c \sigma(x)\,dx. \tag{10.4.20}$$

The exact solution of (10.4.19) is known in this case to be

$$\sigma(x) = \{\pi \log_e (\tfrac{1}{2}c)\sqrt{(c^2 - x^2)}\}^{-1} \tag{10.4.21}$$

from which it follows that

$$S = [\log_e (\tfrac{1}{2}c)]^{-1}.$$

These results provide a check on the numerical answers obtained.

Symm's approach is to divide $(-c, c)$ into n equal intervals of width $2h$. Denote by I_m the interval $[-c+2(m-1)h, -c+2mh]$ and assume that $\sigma(x)$ is constant in I_m and has value σ_m in that interval. With this assumption it follows that

$$\int_{-c}^{c} \log_e |x - y| \sigma(y) \, dy = \sum_{m=1}^{n} \sigma_m \int_{I_m} \log_e |x - y| \, dy.$$

If x lies outside I_m we use the Simpson rule approximation [see § 7.1.5]

$$\int_{I_m} \log_e |x - y| \, dy = \tfrac{1}{3}h\{\log_e |x - y_{m-1}| + 4 \log_e |x - y_{m-\frac{1}{2}}| + \log_e |x - y_m|\}$$

where

$$y_{m-r} = -c + 2(m - r)h, \qquad r = 0, \tfrac{1}{2}, 1,$$

while if $x = y_{m-\frac{1}{2}}$ we use the approximation

$$\int_{I_m} \log_e |y_{m-\frac{1}{2}} - y| \, dy = 2h(\log_e h - 1).$$

By putting $x = y_{1/2}, y_{3/2}, \ldots, y_{n-\frac{1}{2}}$ in turn we obtain a system of n equations which enable us to determine $\sigma_1, \ldots, \sigma_n$. The value of S can then be estimated from

$$S = 2h \sum_{m=1}^{n} \sigma_m$$

which is the mid-point rule approximation [see § 7.1.4] to the integral in (10.4.20). In Table 10.4.3 the charge S has been tabulated for $n = 8, 16, 32, 64$ when $c = \sqrt{3}$ and the results are compared with the exact analytic solution. As n increases it can be seen that the computed values are tending to the analytic value for S.

(iii)   *Other Methods*

Several methods have been devised for solving Fredholm equations of the first kind which have had some measure of success for the problems treated. Examples of such methods are the method of singular function expansion due to Baker *et al.* (1964), the method of least squares and iteration. For a comprehensive account of work carried out on the problem before 1973 see G. F. Miller's article in Delves and Walsh (1974).

10.5.   EIGENVALUE PROBLEMS

In this section we consider numerical methods for solving the equation

$$\int_{a}^{b} K(x, y)\phi(y) \, dy = \lambda \phi(x) \qquad\qquad (10.5.1)$$

in order to determine an eigenvalue λ and a corresponding eigenfunction $\phi(x)$ where a, b are constants and $K(x, y)$ is given [see § 10.1.4]. The simplest approach is to approximate the integral in (10.5.1) using an appropriate quadrature formula, say

$$\int_a^b f(x)\, dx \approx \sum_{k=1}^n \alpha_k f(x_k), \qquad \alpha_k > 0, \quad x_k \in [a, b].$$

We thus have

$$\sum_{k=1}^n \alpha_k K(x, x_k)\phi^*(x_k) = \lambda^* \phi^*(x),$$

where λ^* and $\phi^*(x)$ are approximations to λ and $\phi(x)$ respectively. In particular, when we set $x = x_j$ $(j = 1, \ldots, n)$ we have the algebraic eigenvalue problem

$$\sum_{k=1}^n \alpha_k K(x_j, x_k)\phi^*(x_k) = \lambda^* \phi^*(x_j), \qquad j = 1, \ldots, n \qquad (10.5.2)$$

or, in matrix form [see I, § 5.7 and § 7.1],

$$\mathbf{KD\Phi^*} = \lambda^* \mathbf{\Phi^*} \qquad (10.5.3)$$

where

$$\mathbf{K} = (K(x_j, x_k)), \qquad 1 \le j, k \le n$$

$$\mathbf{D} = \operatorname{diag}(\alpha_1, \ldots, \alpha_n) \qquad [\text{see I (6.7.3)}]$$

$$\mathbf{\Phi^*} = (\phi^*(x_1), \ldots, \phi^*(x_n))'.$$

If the kernel $K(x, y)$ is real and symmetric then there are advantages in recasting (10.5.3) in the form

$$(\mathbf{D}^{1/2}\mathbf{KD}^{1/2})\mathbf{\Psi^*} = \lambda^* \mathbf{\Psi^*} \qquad (10.5.4)$$

where $\mathbf{\Psi^*} = \mathbf{D}^{1/2}\mathbf{\Phi^*}$, as the matrix $\mathbf{D}^{1/2}\mathbf{KD}^{1/2}$ in the algebraic eigenvalue problem (10.5.4) is real and symmetric. The methods developed in Chapter 4 can be employed to find the eigenvalues and eigenvectors of (10.5.4). It should be noted that the number of solutions of (10.5.3) or (10.5.4) do not necessarily correspond to those of (10.5.1). For instance, there might be an infinity of eigenvalues and eigenfunctions satisfying (10.5.1) whereas there can only be a finite number of eigenvalues and eigenvectors satisfying (10.5.3) or (10.5.4). Alternatively, there might be a finite number, m say, of eigenvalues of (10.5.1) while the number of eigenvalues of (10.5.3) or (10.5.4) is n and this varies according to the particular quadrature formula used. This will become clear from the following example.

EXAMPLE 10.5.1. Find the eigenvalues of the equation

$$\int_0^1 (1-3xy)\phi(y)\,dy = \lambda\phi(x) \qquad (10.5.5)$$

and the corresponding eigenfunctions.

If the integral in (10.5.9) is approximated by the basic trapezium rule with $h = 1$ [see § 7.1.3] we obtain the approximate equation for each x,

$$\tfrac{1}{2}\phi^*(0) + \tfrac{1}{2}(1-3x)\phi^*(1) = \lambda^*\phi^*(x). \qquad (10.5.6)$$

In particular, substituting $x = 0$ and $x = 1$ in turn we have

$$\begin{pmatrix} \tfrac{1}{2} & \tfrac{1}{2} \\ \tfrac{1}{2} & -1 \end{pmatrix}\begin{pmatrix} \phi^*(0) \\ \phi^*(1) \end{pmatrix} = \lambda^*\begin{pmatrix} \phi^*(0) \\ \phi^*(1) \end{pmatrix}$$

The eigenvalues of this algebraic eigenvalue problem are easily found to be $-\tfrac{1}{4} \pm \tfrac{\sqrt{13}}{4}$ (0·6514 and 1·1514). Once the eigenvector $(\phi^*(0), \phi^*(1))'$ corresponding to an eigenvalue has been determined the eigenfunction $\phi^*(x)$ can be determined from (10.5.6).

With such a large value of h we cannot really expect very accurate results and we have solved the problem again by taking $h = \tfrac{1}{2}$ and $h = \tfrac{1}{4}$ respectively. With $h = \tfrac{1}{2}$ the repeated trapezium rule approximation to (10.5.5) is

$$\tfrac{1}{4}\phi^*(0) + \tfrac{1}{2}(1-\tfrac{3}{2}x)\phi^*(\tfrac{1}{2}) + \tfrac{1}{4}(1-3x)\phi^*(1) = \lambda^*\phi^*(x)$$

and substituting $x = 0, \tfrac{1}{2}, 1$ in the above equation we obtain the matrix equation

$$\begin{pmatrix} \tfrac{1}{4} & \tfrac{1}{2} & \tfrac{1}{4} \\ \tfrac{1}{4} & \tfrac{1}{8} & -\tfrac{1}{8} \\ \tfrac{1}{4} & -\tfrac{1}{4} & -\tfrac{1}{2} \end{pmatrix}\begin{pmatrix} \phi^*(0) \\ \phi^*(\tfrac{1}{2}) \\ \phi^*(1) \end{pmatrix} = \lambda^*\begin{pmatrix} \phi^*(0) \\ \phi^*(\tfrac{1}{2}) \\ \phi^*(1) \end{pmatrix}$$

or, in symmetric form,

$$\begin{pmatrix} \tfrac{1}{4} & 1/2\sqrt{2} & \tfrac{1}{4} \\ 1/2\sqrt{2} & \tfrac{1}{8} & -(1/4\sqrt{2}) \\ \tfrac{1}{4} & -(1/4\sqrt{2}) & -\tfrac{1}{2} \end{pmatrix}\begin{pmatrix} \psi^*(0) \\ \psi^*(\tfrac{1}{2}) \\ \psi^*(1) \end{pmatrix} = \lambda^*\begin{pmatrix} \psi^*(0) \\ \psi^*(\tfrac{1}{2}) \\ \psi^*(1) \end{pmatrix}$$

where

$$(\psi^*(0), \psi^*(\tfrac{1}{2}), \psi^*(1)) = \left(\tfrac{1}{2}\phi^*(0), \frac{1}{\sqrt{2}}\phi^*(\tfrac{1}{2}), \tfrac{1}{2}\phi^*(1)\right)$$

The eigenvalues satisfying the above equation are easily shown to be $-\tfrac{1}{16} \pm \sqrt{\tfrac{26}{8}}$, 0 (0·5749, −0·6999, 0).

Similarly, with $h = \tfrac{1}{4}$ the approximation to (10.5.5) is

$$\tfrac{1}{8}\phi^*(0) + \tfrac{1}{4}(1-\tfrac{3}{4}x)\phi^*(\tfrac{1}{4}) + \tfrac{1}{4}(1-\tfrac{3}{2}x)\phi^*(\tfrac{1}{2}) + \tfrac{1}{4}(1-\tfrac{9}{4}x)\phi^*(\tfrac{3}{4}) + \tfrac{1}{8}(1-3x)\phi^*(1)$$

$$= \lambda^*\phi^*(x)$$

and writing

$$(\psi^*(0), \psi^*(\tfrac{1}{4}), \psi^*(\tfrac{1}{2}), \psi^*(\tfrac{3}{4}), \psi^*(1))'$$

$$= \left(\frac{1}{\sqrt{8}} \phi^*(0), \tfrac{1}{2}\phi^*(\tfrac{1}{4}), \tfrac{1}{2}\phi^*(\tfrac{1}{2}), \tfrac{1}{2}\phi^*(\tfrac{3}{4}), \frac{1}{\sqrt{8}}\phi^*(1) \right)'$$

we find the eigenvalues of the integral equation (10.5.5) satisfy approximately the algebraic equation

$$\begin{vmatrix} \frac{1}{8} & \alpha & \alpha & \alpha & \frac{1}{8} \\ \alpha & \frac{13}{64} & \frac{5}{32} & \frac{7}{64} & \frac{1}{4}\alpha \\ \alpha & \frac{5}{32} & \frac{1}{16} & -\frac{1}{32} & -\frac{1}{2}\alpha \\ \alpha & \frac{7}{64} & -\frac{1}{32} & -\frac{11}{64} & -\frac{5}{4}\alpha \\ \frac{1}{8} & \frac{1}{4}\alpha & -\frac{1}{2}\alpha & -\frac{5}{4}\alpha & -\frac{1}{8} \end{vmatrix} \begin{pmatrix} \psi^*(0) \\ \psi^*(\tfrac{1}{4}) \\ \psi^*(\tfrac{1}{2}) \\ \psi^*(\tfrac{3}{4}) \\ \psi^*(1) \end{pmatrix} = \lambda^* \begin{pmatrix} \psi^*(0) \\ \psi^*(\tfrac{1}{4}) \\ \psi^*(\tfrac{1}{2}) \\ \psi^*(\tfrac{3}{4}) \\ \psi^*(1) \end{pmatrix}$$

where $\alpha = 1/4\sqrt{2}$. The eigenvalues of the algebraic equation are (to 4 decimal places) 0·5150, −0·4958, 0·0746, 0, 0 so that we are approaching the correct values $\tfrac{1}{2}$, $-\tfrac{1}{2}$ of the eigenvalues of the integral equation (see Example 10.1.3) but due to our discretization of the problem we have also introduced some spurious results.

If the kernel $K(x, y)$ is singular e.g.

$$K(x, y) = H(x, y)/|x - y|^\alpha, \qquad 0 < \alpha < 1$$

where $H(x, y)$ s a well behaved function then it is best to modify the quadrature method using the device of section 10.2.3. If we write

$$K(x) = \int_a^b K(x, y)\, dy \tag{10.5.7}$$

then (10.5.1) can be expressed in the form

$$\int_a^b K(x, y)\{\phi(y) - \phi(x)\}\, dy = \phi(x)\{\lambda - K(x)\}.$$

Application of the quadrature rule produces the approximate equation

$$\sum_{k=1}^{n} \alpha_k K(x, x_k)[\phi^*(x_k) - \phi^*(x)] = \phi^*(x)[\lambda^* - K(x)],$$

so that when $x_1, \ldots, x_n$ are substituted for x we obtain the system of equations

$$\sum_{\substack{k=1 \\ j \neq k}}^{n} \alpha_k K(x_j, x_k)[\phi^*(x_k) - \phi^*(x_j)] = \phi^*(x_j)[\lambda^* - K(x_j)], \qquad j = 1, \ldots, n$$

or, in matrix form,

$$\mathbf{MD\Phi^*} = \lambda^* \mathbf{\Phi^*}$$

where

$$\mathbf{D} = \text{diag} (\alpha_1, \ldots, \alpha_n) \qquad [\text{see I (6.7.2)}]$$

$$\boldsymbol{\Phi}^* = (\phi^*(x_1), \ldots, \phi^*(x_n))'$$

$$\mathbf{M} = (M_{jk}), \qquad 1 \le j, k \le n$$

and

$$M_{jk} = \begin{cases} K(x_j, x_k), & j \ne k \\ \alpha_j^{-1} \left[K(x_j) - \sum_{k \ne j} \alpha_k K(x_j, x_k) \right], & j = k. \end{cases}$$

In the case of a real symmetric kernel we can recast the equation (as we did earlier in (10.5.4)) in the form

$$(\mathbf{D}^{1/2}\mathbf{M}\mathbf{D}^{1/2})\boldsymbol{\Psi}^* = \lambda^*\boldsymbol{\Psi}^*, \qquad \boldsymbol{\Psi}^* = \mathbf{D}^{1/2}\boldsymbol{\Phi}^*$$

and we again have a real symmetric algebraic eigenvalue problem to solve.
   Another method of attack is to apply the method of product integration. We write

$$K(x, y) = \sum_{r=1}^{R} A_r(x, y)B_r(x, y)$$

where the functions $B_r(x, y)$ are chosen so that we have available approximate integration formulae of the form

$$\int_a^b B_r(x, y)\phi(y) \, dy = \sum_{k=1}^{n} \beta_{r,k}(x)\phi(x_k), \qquad r = 1, \ldots, R$$

where $x_k \in [a, b]$ for $k = 1, \ldots, n$. The equation (10.5.1) can now be put in the approximate form

$$\sum_{r=1}^{R} \sum_{k=1}^{n} \beta_{r,k}(x)A_r(x, x_k)\phi^*(x_k) = \lambda^*\phi^*(x)$$

and when x is replaced by $x_1, \ldots, x_n$ in turn we get the approximate eigensystem

$$\sum_{r=1}^{R} \sum_{k=1}^{n} \beta_{r,k}(x_j)A_r(x_j, x_k)\phi^*(x_k) = \lambda^*\phi^*(x_j), \qquad j = 1, \ldots, n.$$

Unfortunately we usually lose symmetry by adopting the above technique although it can be advantageous when $K(x, y)$ is singular and the decomposition is constructed so that the functions $A_r(x, y)$ are continuous and the $B_r(x, y)$ are singular.

10.6. MISCELLANEOUS TOPICS

10.6.1. Non-Linear Integral Equations

We shall give particular examples of non-linear integral equations which occur frequently in practice and give some indication of the methods which have been employed to solve them.

The non-linear Volterra equation has the form

$$\phi(x) = f(x) + \int_0^x g(x, y, \phi(y)) \, dy. \tag{10.6.1}$$

The equation

$$\phi(x) = f(x) + \int_0^1 K(x, y)g(y, \phi(y)) \, dy \tag{10.6.2}$$

is called a *Hammerstein equation* and the equation

$$\phi(x) = f(x) + \int_0^1 g(x, y, \phi(y)) \, dy \tag{10.6.3}$$

is known as a *Urysohn equation*. Clearly the Hammerstein equation is a special form of the Urysohn equation.

When dealing with non-linear equations there is not such a well-developed theory available as there is for linear equations and extra care must be exercised in interpreting results. One problem which can arise with non-linear equations is bifurcation or branching of solutions at certain points. This happens in the case of *Duffing's equation* which arises out of the motion of a pendulum. The equation of motion of the pendulum is

$$\ddot{\phi} + a \sin \phi = \cos t, \qquad \phi(0) = \phi(\pi) = 0.$$

Duffing approximated $\sin \phi$ by $\phi - \tfrac{1}{6}\phi^3$ and obtained the differential equation

$$\ddot{\phi} + a(\phi - \tfrac{1}{6}\phi^3) = \cos t.$$

This differential equation has an equivalent integral equation representation

$$\phi(t) - \int_0^t (t - y)\{a[\phi(y) - \tfrac{1}{6}\phi^3(y)] + \sin y\} \, dy = ct \tag{10.6.4}$$

where c is the unknown value of $\phi'(0)$ which is often called the *shooting parameter*.

A different problem is illustrated by the Volterra equation

$$\phi(x) = x + \int_0^x \phi^2(y) \, dy,$$

given in Delves and Walsh (1974). In this case there is no continuous solution for $x \geq \tfrac{1}{2}\pi$ ($\phi(x) = \tan x$).

We give some examples of the solution of non-linear equations.

EXAMPLE 10.6.1. Solve the equation of radiative transfer

$$H(\mu) = 1 + \tfrac{1}{2}w_0\mu H(\mu) \int_0^1 \frac{H(\mu')}{\mu + \mu'}\, d\mu' \qquad (10.6.5)$$

where $w_0 = 1$ and $0 \le \mu \le 1$.

It follows directly from (10.6.5), by substituting $\mu = 0$, that $H(0) = 1$. To find other values of $H(\mu)$ it is convenient to recast equation (10.6.5) in the form

$$\frac{1}{H(\mu)} = \tfrac{1}{2}w_0 \int_0^1 \frac{\mu' H(\mu')}{\mu + \mu'}\, d\mu' \qquad (10.6.6)$$

which follows from the fact that

$$\int_0^1 H(\mu')\, d\mu' = 1, \quad \text{when } w_0 = 1,$$

(see Chandrasekhar, 1960). If we approximate to the integral by means of the basic Simpson rule [see § 7.1.5] we get

$$\frac{1}{H(\mu)} = \frac{1}{12}\left\{0 + \frac{4 \cdot \tfrac{1}{2}H(\tfrac{1}{2})}{\mu + \tfrac{1}{2}} + \frac{H(1)}{\mu + 1}\right\}, \qquad \mu > 0. \qquad (10.6.7)$$

When the values $\mu = \tfrac{1}{2}, 1$ are substituted in turn we obtain the system of non-linear equations for $H(\tfrac{1}{2})$, $H(1)$,

$$\frac{12}{H(\tfrac{1}{2})} = 2H(\tfrac{1}{2}) + \tfrac{2}{3}H(1)$$

$$\frac{12}{H(1)} = \tfrac{4}{3}H(\tfrac{1}{2}) + \tfrac{1}{2}H(1).$$

These non-linear equations can be solved very quickly by the Newton–Raphson technique [see § 5.4.1] and the solutions we obtain are

$$H(\tfrac{1}{2}) = 2 \cdot 0131, \qquad H(1) = 2 \cdot 9020.$$

$H(\mu)$ can be estimated for any μ in $(0, 1)$ from (10.6.7). The above results compare favourably with those given in Chandrasekhar (1960) which were derived using the more accurate Gauss–Legendre 5-point quadrature formula [see § 7.3.1].

We now give another approach to the solution of non-linear integral equations. Further methods are given in Delves and Walsh (1974).

EXAMPLE 10.6.2. Find the solution of the integral equation

$$F^2(x) = 1 - \tfrac{1}{2}\cos x + \int_0^x F(t)\, dt, \qquad F(0) > 0$$

in $0 \le x \le 1$.

We write

$$F(x) = a_0 + a_1 x + a_2 x^2 + \ldots + a_n x^n + \ldots,$$

so that

$$F^2(x) = a_0^2 + 2a_1 a_0 x + (2a_2 a_0 + a_1^2)x^2 + \ldots + \left(\sum_{r=0}^{n} a_r a_{n-r} \right) x^n + \ldots.$$

The right-hand side of the integral equation is

$$\tfrac{1}{2} + \tfrac{1}{4}x^2 - \tfrac{1}{48}x^4 + \ldots + \frac{(-1)^{k-1}x^{2k}}{2(2k)!} + \ldots + a_0 x + \tfrac{1}{2}a_1 x^2 + \ldots + \frac{1}{n}a_{n-1}x^n + \ldots.$$

Comparison of coefficients of x^0 on both sides gives

$$a_0^2 = \tfrac{1}{2}$$

and hence

$$a_0 = \pm 1/\sqrt{2}.$$

The stipulation that $F(0) > 0$ resolves the ambiguity of sign in a_0 so that $a_0 = 1/\sqrt{2}$. The coefficients $a_1, a_2, \ldots$ can be obtained recursively by comparing powers of $x, x^2, \ldots$ and we find

$$a_1 = \tfrac{1}{2}, \qquad a_2 = \sqrt{2}/8, \qquad a_3 = -\tfrac{1}{12}, \qquad a_4 = 1/96\sqrt{2}, \qquad a_5 = \tfrac{1}{60}, \ldots.$$

Thus

$$F(x) = 0{\cdot}70711 + 0{\cdot}5x + 0{\cdot}17678x^2 - 0{\cdot}08333x^3 + 0{\cdot}00737x^4$$
$$+ 0{\cdot}01667x^5 - 0{\cdot}01608x^6 + 0{\cdot}00645x^7 + 0{\cdot}00195x^8$$
$$- 0{\cdot}00491x^9 + \ldots.$$

An alternative procedure for this problem is to evaluate $a_0 = F(0)$ and solve the ordinary differential equation

$$2F(x)F'(x) = \tfrac{1}{2}\sin x + F(x), \qquad F(0) = a_0$$

by the method of Runge–Kutta [see § 8.2.2] or some other suitable technique.

10.6.2.   Integro-Differential Equations

We shall just content ourselves with a description of one method for the solution of Volterra integro-differential equations. For more details about other methods the reader is referred to the article by Baker in Delves and Walsh (1974) and Baker (1977).

The general form of the Volterra integro-differential equation is

$$\phi'(x) = F(x, \phi(x), V\phi(x)), \qquad \phi(n) = \phi_0 \tag{10.6.8}$$

where V is the *Volterra operator* acting on $\phi(x)$, that is,

$$V\phi(x) = \int_0^x K(x, y, \phi(y))\, dy. \tag{10.6.9}$$

If we write $V\phi(x) = V(x)$ in (10.6.8) we can develop a multi–step method for the solution of the system (10.6.8)–(10.6.9) and this has the form

$$\sum_{r=0}^{k} \alpha_r \phi(x_{n+r}) = h \sum_{r=0}^{k} \beta_r F(x_{n+r}, \phi(x_{n+r}), V(x_{n+r})) \qquad (10.6.10)$$

where $h(>0)$, $x_k = kh$ and $\alpha_0, \alpha_1, \ldots, \alpha_k(\neq 0)$, $\beta_0, \beta_1, \ldots, \beta_k$ are suitable parameters. The unknown values $V(x_{n+r}) = V_{n+r}$, can be obtained from a quadrature rule of the form

$$V_{n+r} = \sum_{j=0}^{n+r} w_{rj} K(x_{n+r}, x_j, \phi_j), \qquad r = 0, 1, \ldots, k. \qquad (10.6.11)$$

Note that a knowledge of $\phi(h), \ldots, \phi((k-1)h)$ is required before (10.6.10) can be applied. These values can be obtained by the method of Taylor series or by means of a starting procedure such as Day's [see § 10.3.1]. Linz (1969) advocates the use of Simpson's rule and quadratic interpolation. Thus for $n = 0, 2, 4, \ldots$ he writes

$$\phi_{n+1} = \phi_n + \tfrac{1}{6} h \{ F(x_n, \phi_n, V_n) + 4F(x_{n+\frac{1}{2}}, \phi_{n+\frac{1}{2}}, V_{n+\frac{1}{2}}) + F(x_{n+1}, \phi_{n+1}, V_{n+1}) \}$$

$$\phi_{n+2} = \phi_n + \tfrac{1}{3} h \{ F(x_n, \phi_n, V_n) + 4F(x_{n+1}, \phi_{n+1}, V_{n+1}) + F(x_{n+2}, \phi_{n+2}, V_{n+2}) \}$$

$$V_{n+1} = \tfrac{1}{3} h \sum_{i=0}^{n} w_i K(x_{n+1}, x_i, \phi_i) + \tfrac{1}{6} h \{ K(x_{n+1}, x_{\mu}, \phi_n) + 4K(x_{n+1}, x_{n+\frac{1}{2}}, \phi_{n+\frac{1}{2}}) $$
$$\qquad\qquad\qquad\qquad\qquad\qquad (10.6.12)$$
$$+ K(x_{n+1}, x_{n+1}, \phi_{n+1}) \}$$

$$V_{n+2} = \tfrac{1}{3} h \sum_{i=0}^{n+2} w_i K(x_{n+2}, x_i, \phi_i)$$

where w_i are the Simpson rule weights $1, 4, 2, 4, \ldots, 2, 4, 1$. The values of $\phi_{n+\frac{1}{2}}$, $V_{n+\frac{1}{2}}$ which are required in the above formulae are found from the quadratic interpolation formulae

$$\left. \begin{array}{c} \phi_{n+\frac{1}{2}} = \tfrac{3}{8} \phi_n + \tfrac{3}{4} \phi_{n+1} - \tfrac{1}{8} \phi_{n+2} \\[2mm] V_{n+\frac{1}{2}} = \tfrac{3}{8} V_n + \tfrac{3}{4} V_{n+1} - \tfrac{1}{8} V_{n+2} \end{array} \right\}. \qquad (10.6.13)$$

EXAMPLE 10.6.3. Given the integro-differential equation

$$\phi'(x) = 1 + 2x - \phi(x) + \int_0^x x(1+2x) e^{y(x-y)} \phi(y) \, dy, \qquad \phi(0) = 1, \quad (10.6.14)$$

determine the approximate values of $\phi(0\cdot1)$, $\phi(0\cdot2)$.

In this example $K(x, y, \phi(y)) = x(1+2x) e^{y(x-y)}\phi(y)$. The calculations required to obtain $\phi_1 = \phi(0\cdot1)$ and $\phi_2 = \phi(0\cdot2)$ are, from (10.6.12), (10.6.13),

$$V_0 = V(0) = 0$$

$$V_1 = \tfrac{1}{6}(0\cdot1)\{(0\cdot1)(1\cdot2)e^0\phi_0 + 4(0\cdot1)(1\cdot2) e^{0\cdot05(0\cdot1-0\cdot05)}\phi_{1/2} + (0\cdot1)(1\cdot2)e^0\phi_1\}$$

i.e.

$$V_1 = 0\cdot002\phi_0 + 0\cdot00802\phi_{1/2} + 0\cdot002\phi_1$$

$$V_2 = \tfrac{1}{3}(0\cdot1)\{(0\cdot2)(1\cdot4)e^0\phi_0 + 4(0\cdot2)(1\cdot4) e^{0\cdot1(0\cdot2-0\cdot1)}\phi_1 + (0\cdot2)(1\cdot4)e^0\phi_2\}$$

i.e.

$$V_2 = 0\cdot009333\phi_0 + 0\cdot037708\phi_1 + 0\cdot009333\phi_2$$

$$V_{1/2} = \tfrac{3}{8}V_0 + \tfrac{3}{4}V_1 - \tfrac{1}{8}V_2$$

while

$$\phi_1 = 1\cdot11 + \tfrac{1}{6}(0\cdot1)[-\phi_0 - \phi_1 - 4\phi_{1/2} + V_0 + V_1 + 4V_{1/2}]$$

$$\phi_2 = 1\cdot24 + \tfrac{1}{3}(0\cdot1)[-\phi_0 - \phi_2 - 4\phi_1 + V_0 + V_2 + 4V_1]$$

$$\phi_{1/2} = \tfrac{3}{8}\phi_0 + \tfrac{3}{4}\phi_1 - \tfrac{1}{8}\phi_2.$$

These equations are solved below in Table 10.6.1 by iteration. The exact solution is $\phi(x) = e^{x^2}$.

10.6.3. Iterative Integral Equations

This type of integral equation arises in the investigation of sequential processes, for example. Downton (1969) considers the equation

$$1 - F_{n+1}(x) = \int_0^\infty \frac{1 - F_n(u+x)}{1 - F_n(u)} \, dF_n(u), \qquad x \geq 0 \qquad (10.6.15)$$

where $F_n(x)$ is a distribution function [see II, § 4.3] i.e.

$$F_n(0) = 0, \qquad \int_0^\infty dF_n(u) = 1 \qquad (10.6.16)$$

and $F_n(u)$ satisfies the monotonicity property

$$F_n(u) \leq F_n(u+x) \leq 1 \quad \text{all } x \geq 0. \qquad (10.6.17)$$

Downton shows that, as $n \to \infty$, a likely outcome of the sequence $\{F_n(x)\}$ is that the function $1 - F_n(x)$ tends to an exponential function [see IV, § 2.11]. A simple argument to support this assertion is now given. We write (10.6.15) in the form

$$\int_0^\infty \frac{1 - F_n(u+x)}{(1 - F_{n+1}(x))(1 - F_n(u))} \, dF_n(u) = 1$$

and subtract (10.6.16) to get

$$\int_0^\infty \left[\frac{1-F_n(u+x)}{(1-F_{n+1}(x))(1-F_n(u))} - 1 \right] dF_n(u) = 0. \qquad (10.6.18)$$

Since the probability density function $dF_n(u) = f_n(u) \geq 0$ one obvious possibility, which is strengthened by the fact that

$$0 < \frac{1-F_n(u+x)}{1-F_n(u)} \leq 1,$$

is that the part of the integrand in square brackets in (10.6.18) is identically zero. If this is the case, letting

$$\phi_n(x) = 1 - F_n(x) \qquad (10.6.19)$$

we have

$$\phi_n(u+x) = \phi_n(u)\phi_{n+1}(x), \qquad \phi_n(0) = 1. \qquad (10.6.20)$$

Differentiation with respect to u gives

$$\phi'_n(u+x) = \phi'_n(u)\phi_{n+1}(x)$$

and hence, by division,

$$\frac{\phi'_n(u+x)}{\phi_n(u+x)} = \frac{\phi'_n(u)}{\phi_n(u)}.$$

Integration produces the result

$$\phi_n(u+x) = K_n(x)\phi_n(u).$$

When $u = 0$ this equation informs us that $K_n(x) = \phi_n(x)$ so that $\phi_n(x)$ satisfies Cauchy's equation

$$\phi_n(u+x) = \phi_n(x)\phi_n(u) \qquad (10.6.21)$$

and the solution of this equation is known to be $\phi_n(x) = e^{-\lambda x}$. Thus, in view of our analysis, it seems reasonable to expect the distribution function $F_n(x)$ to behave like

$$F_n(x) = 1 - e^{-\lambda x} \qquad (10.6.22)$$

as n increases.

To obtain numerical evidence to support the assertion it will be necessary to tabulate $1 - F_n(x)$ for increasing n. If the assertion is valid, $x^{-1}\log_e (1 - F_n(x))$ will be constant $(x > 0)$.

Some idea of the difficulties in computing $F_n(x)$ can be gained from the following simple example.

EXAMPLE 10.6.4.   Given $F_0(x) = (2+x)^2$ and

$$F_{n+1}(x) = \int_{-\infty}^\infty e^{-u^2} F_n(u+x) \, du$$

find the value of $F_2(4)$.

Because of the weight function e^{-u^2} and the range of integration $(-\infty, \infty)$ it is appropriate in this case to use a Gauss–Hermite quadrature rule to effect the integration [see § 7.3.4] and, because the function $F_0(x)$ is quadratic, it will be sufficient to use a 2 point Gauss–Hermite rule. Now if $F_{n+1}(x)$ is to be determined at the value $x = x_1$ we need to know $F_n(x)$ at the values $x = x_1 + 1/\sqrt{2}$ and $x = x_1 - 1/\sqrt{2}$. It follows that if $F_n(x)$ is required at $x = x_1 + 1/\sqrt{2}$ it is necessary to know $F_{n-1}(x)$ at the values $x = x_1$ and $x = x_1 + 2/\sqrt{2}$ and this reasoning shows that for a complete knowledge of $F_{n+1}(x)$ at $x = x_1$ we need to know $F_{n-1}(x)$ at $x = x$, and $x = x_1 \pm 2/\sqrt{2}$. When $n = 1$ and $x_1 = 4$ we have

$$F_1(4 + 1/\sqrt{2}) = \tfrac{1}{2}\sqrt{\pi}[\{2 + (4 + 2/\sqrt{2})\}^2 + (2 + 4)^2]$$

$$= \tfrac{1}{2}\sqrt{\pi}(74 + 12\sqrt{2})$$

$$F_1(4 - 1/\sqrt{2}) = \tfrac{1}{2}\sqrt{\pi}[\{2 + (4 - 2/\sqrt{2})\}^2 + (2 + 4)^2]$$

$$= \tfrac{1}{2}\sqrt{\pi}(74 - 12\sqrt{2})$$

$$F_2(4) = \tfrac{1}{2}\sqrt{\pi}\{\tfrac{1}{2}\sqrt{\pi}(74 + 12\sqrt{2}) + \tfrac{1}{2}\sqrt{\pi}(74 - 12\sqrt{2})\}$$

$$= 37\pi.$$

(In general, we can show $F_n(x) = \pi^{n/2}\{(2 + x)^2 + \tfrac{1}{2}n\}$.)

A.M.C.

REFERENCES

Anderssen, A. S. and White, E. T. (1971). Improved Numerical Methods for Volterra Integral Equations of the First Kind, *The Computer J.*, **14**, 442–443.

Baker, C. T. H. (1977). *The Numerical Treatment of Integral Equations*, Oxford University Press, London.

Baker, C. T. H., Fox, L., Mayers, D. F. and Wright, K. (1964). Numerical Solution of Fredholm Integral Equations of the First Kind, *The Computer J.*, **7**, 141–147.

Bakushinskii, A. B. (1965). A Numerical Method for Solving Fredholm Integral Equations of the First Kind, *USSR Comput. Maths. and Math. Phys.*, **5**, 226–233.

Chambers, Ll. G. (1976). *Integral Equations—A Short Course*, International Text Book Co. Ltd., London.

Chandrasekhar, S. (1960). *Radiative Transfer*, Dover Publications Inc., N.Y.

Cochran, J. A. (1972). *Analysis of Linear Integral Equations*, McGraw-Hill Book Co. Ltd., N.Y.

Day, J. T. (1968). On the Numerical Solution of Volterra Integral Equations, *BIT*, **8**, 134–137.

Delves, L. M. and Walsh, J. (1974). *Numerical Solution of Integral Equations*, Oxford University Press, London.

Downton, F. (1969). An Integral Equation Approach to Equipment Failure, *J. Roy. Stat. Soc.*, **31**, 335–349.

Elliot, D. (1963). A Chebyshev Series Method for the Numerical Solution of Fredholm Integral Equations, *The Computer J.*, **6**, 102–111.

Fox, L. (Ed.) (1962). *The Numerical Solution of Ordinary and Partial Differential Equations*, Pergamon Press, Oxford.

Fox, L. and Goodwin, E. T. (1953). The Numerical Solution of Non-Singular Linear Integral Equations, *Phil. Trans. Roy. Soc.*, *A*, **245**, 501–534.

Jones, J. G. (1961). On the Numerical Solution of Convolution Integral Equations and Systems of such Equations, *Math. Comp.*, **18**, 491–496.

Lewis, B. A. (1975). On the Numerical Solution of Fredholm Integral Equations of the First Kind, *J. Inst. Maths. Applics.*, **16**, 207–220.

Linz, P. (1967). *The Numerical Solution of Volterra Integral Equations by Finite Difference Methods*, MRC Tech. Report 825, Madison, Wisconsin.

Linz, P. (1969). Linear Multistep Methods for Volterra Integro-Differential Equations, *J. Assoc. Comp. Mach.*, **16**, 295–301.

Pouzet, P. (1960). In *Symposium on the Numerical Treatment of Ordinary Differential Equations, Integral and Integro-Differential Equations*, Birkauser Verlag, Basel, 362–368.

Stroud, A. H. and Secrest, D. (1966). *Gaussian Quadrature Formulas*, Prentice-Hall Inc., Englewood Cliffs, N.J.

Symm, G. T. (1963). Integral Equation Methods in Potential Theory II, *Proc. Roy. Soc.*, *A*, **275**, 33–46.

Titchmarsh, E. C. (1948). *Introduction to the Theory of Fourier Integrals*, Oxford University Press, London.

Young, A. (1954). The Application of Product Integration to the Numerical Solution of Integral Equations, *Proc. Roy. Soc. A*, **224**, 561–573.

APPENDIX 10.1.

Table of the Coefficients in Gregory's Formula of Order p

a_r \ p	0	1	2	3	4
a_0	0·5	0·416667	0·375000	0·348611	0·329861
a_1	1·0	1·083333	1·166667	1·245833	1·320833
a_2	1·0	1·0	0·958333	0·879167	0·766667
a_3	1·0	1·0	1·0	1·026389	1·101389
a_4	1·0	1·0	1·0	1·0	0·981250
a_5	1·0	1·0	1·0	1·0	1·0
a_{n-5}	1·0	1·0	1·0	1·0	1·0
a_{n-4}	1·0	1·0	1·0	1·0	0·981250
a_{n-3}	1·0	1·0	1·0	1·026389	1·101389
a_{n-2}	1·0	1·0	0·958333	0·879167	0·766667
a_{n-1}	1·0	1·083333	1·166667	1·245833	1·320833
a_n	0·5	0·416667	0·375000	0·348611	0·329861

APPENDIX 10.2.

Tables

Multiplier	Coefficients			RHS 1	RHS 2
	0·833333	−0·666667	−0·166667	1·000000	−0·000583
0·2	−0·166667	0·166667	−0·250000	1·648721	−0·001892
0·2	−0·166667	−1·000000	0·666667	2·718282	−0·003200
0·029412		0·033334	−0·283333	1·848721	−0·002009
		−1·133333	0·633334	2·918282	−0·003317
			−0·264705	1·934554	−0·002107

The solution of the equations with right-hand side 1 is

$$\phi^*(0) = -5\cdot588903, \qquad \phi^*(\tfrac{1}{2}) = -6\cdot659033, \qquad \phi^*(1) = -7\cdot308339$$

The solution of the equations with right-hand side 2 is

$$\phi_1^*(0) = 0\cdot006793, \qquad \phi_1^*(\tfrac{1}{2}) = 0\cdot007375, \qquad \phi_1^*(1) = 0\cdot007960$$

Table 10.2.1: Solution of the linear equations which approximate to the integral equation $\phi(x) = e^x + \int_0^1 (1 + xy)\phi(y)\, dy$.

y	$\phi^*(y)$	$g(\tfrac12, y)$		δ^2		δ^4
$-\tfrac12$	$-4\cdot263498$	$-3\cdot197624$				
			-1134204			
$-\tfrac14$	$-4\cdot950661$	$-4\cdot331828$		-122871		
			-1257075		34007	
0	$-5\cdot588903$	$-5\cdot588903$		-88864		11847
			-1345939		45854	
$\tfrac14$	$-6\cdot164304$	$-6\cdot934842$		-43010		15878
			-1388949		61732	
$\tfrac12$	$-6\cdot659033$	$-8\cdot323791$		18722		21282
			-1370227		83014	
$\tfrac34$	$-7\cdot050195$	$-9\cdot694018$		101736		28205
			-1268491		111219	
1	$-7\cdot308339$	$-10\cdot962509$		212955		37743
			-1055536		148962	
$\tfrac54$	$-7\cdot395722$	$-12\cdot018045$		361917		
			-693619			
$\tfrac32$	$-7\cdot263808$	$-12\cdot711664$				

Correction term $= -\left\{\dfrac{1}{2880(\frac14)^3}[(0\cdot083014-0\cdot061732)+1\cdot5(0\cdot021282+0\cdot021282)]\right\}$

$= -0\cdot001892$

Table 10.2.2: Differences of the function $g(\tfrac12, y) = K(\tfrac12, y)\phi^*(y)$, $K(\tfrac12, y) = 1 + \tfrac12 y$.

y	$g(0, y)$		δ^2		δ^4
0	$-5\cdot588903$				
		-575401			
$\tfrac14$	$-6\cdot164304$		80672		
		-494729		22895	
$\tfrac12$	$-6\cdot659033$		103567		6556
		-391162		29451	
$\tfrac34$	$-7\cdot050195$		133018		
		-258144			
1	$-7\cdot308339$				

Correction term $= -\left\{\dfrac{1}{2880(\frac14)^3}[(0\cdot029451-0\cdot022895)+1\cdot5(0\cdot006556+0\cdot006556)]\right\}$

$= -0\cdot000583$

Table 10.2.3: Difference table for $g(0, y) = K(0, y)\phi^*(y)$, $K(0, y) = 1$.

y	$g(1, y)$		δ^2		δ^4
0	$-5\cdot588903$				
		-2116477			
$\tfrac14$	$-7\cdot705380$		-166693		
		-2283170		100572	
$\tfrac12$	$-9\cdot988550$		-66121		36003
		-2349291		136575	
$\tfrac34$	$-12\cdot337841$		70454		
		-2278837			
1	$-14\cdot616678$				

Correction term $= -\left\{\dfrac{1}{2880(\frac14)^3}[(0\cdot136575-0\cdot100572)+1\cdot5(0\cdot036003+0\cdot036003)]\right\}$

$= -0\cdot003200$

Table 10.2.4: Difference table for $g(1, y) = K(1, y)\phi^*(y)$, $K(1, y) = 1 + y$.

x	$\phi(x)$ Exact	T	ST	TS	G2	P
0·0	0·00000	0·00000	0·00000	0·00000	0·00000	0·00000
0·1	0·10001	0·10001	0·10001	0·10001	0·10001	0·10001
0·2	0·20011	0·20012	0·20011	0·20011	0·20011	0·20011
0·3	0·30054	0·30057	0·30054	0·30054	0·30054	0·30054
0·4	0·40171	0·40176	0·40171	0·40171	0·40171	0·40171
0·5	0·50418	0·50427	0·50418	0·50337	0·50418	0·50418
0·6	0·60870	0·60883	0·60870	0·60870	0·60870	0·60870
0·7	0·71619	0·71637	0·71619	0·71565	0·71619	0·71619
0·8	0·82778	0·82802	0·82778	0·82775	0·82778	0·82778
0·9	0·94482	0·94514	0·94482	0·94334	0·94482	0·94482
1·0	1·06894	1·06935	1·06894	1·06887	1·06894	1·06894
1·1	1·20207	1·20260	1·20207	1·20023	1·20207	1·20207
1·2	1·34652	1·34720	1·34652	1·34637	1·34652	1·34652
1·3	1·50506	1·50593	1·50506	1·50284	1·50506	1·50506
1·4	1·68103	1·68215	1·68103	1·68074	1·68104	1·68103
1·5	1·87849	1·87993	1·87849	1·87584	1·87849	1·87848
1·6	2·10238	2·10425	2·10238	2·10187	2·10240	2·10238
1·7	2·35882	2·36124	2·35883	2·35568	2·35884	2·35882
1·8	2·65538	2·65852	2·65538	2·65453	2·65540	2·65537
1·9	3·00152	3·00564	3·00154	2·99778	3·00156	3·00152
2·0	3·40921	3·41463	3·40922	3·40783	3·40926	3·40920

Key: T—trapezium rule; ST—Simpson rule with three-eights rule at top end; TS—Simpson rule with three-eights rule at bottom end; G2—Gregory second order formula; P—Pouzet method.

Table 10.3.1: Solution of the Volterra equation $\phi(x) = x + \frac{1}{5}\int_0^x xy\phi(y)\,dy$.

x $\phi(x)$	T	M*	S	G2	SK
0·0	1·00000	0·99958	1·00000	1·00000	1·00000
0·1	1·00166	0·99958	1·00000	1·00000	1·00000
0·2	1·00000	0·99959	0·99999	1·00000	1·00000
0·3	1·00166	0·99958	1·00001	1·00000	1·00000
0·4	1·00000	0·99958	0·99998	1·00000	1·00000
0·5	1·00166	0·99959	1·00004	0·99998	1·00000
0·6	1·00000	0·99958	0·99983	1·00004	1·00000
0·7	1·00166	0·99959	1·00039	0·99991	1·00000
0·8	1·00001	0·99958	0·99864	1·00021	1·00000
0·9	1·00166	0·99958	1·00308	0·99950	1·00000
1·0	0·99998	0·99958	0·98907	1·00116	1·00000
1·1	1·00167	0·99959	1·02488	0·99724	1·00000
1·2	0·99999	0·99958	0·91188	1·00651	1·00000
1·3	1·00168	0·99959	1·20061	0·98466	1·00000
1·4	0·99996	0·99958	0·28972	1·03617	1·00000
1·5	1·00171	0·99959	2·61690	0·91470	1·00000
1·6	0·99994	0·99959	−4·72502	1·20117	1·00000
1·7	1·00172	0·99958	14·0325	0·52557	1·00000
1·8	0·99992	0·99959	−45·1447	2·11889	1·00000
1·9	1·00177	0·99957	106·044	−1·63875	1·00000
2·0	0·99985		−360·934	7·22313	1·00000

Key: T—trapezium rule; M—mid-point rule (* the values of $\phi(x)$ are estimated at $x+0\cdot5$); S—Simpson rule with three-eighths rule at top end, G2—Gregory second order formula; SK—evaluation by transforming into a Volterra equation of the second kind.

Table 10.4.1: Solution of the Volterra equation $\int_0^x \cos(x-y)\phi(y)\,dy = \sin x$ (exact solution is $\phi(x)=1$).

		Estimated values of		
k	$\alpha = 2^{-k}$	$\phi(0)$	$\phi(\tfrac{1}{2})$	$\phi(1)$
4	$0 \cdot 62500 \cdot 10^{-1}$	18·01677	2·89075	−12·23528
5	$0 \cdot 31250 \cdot 10^{-1}$	6·34789	1·31458	−3·71873
6	$0 \cdot 15625 \cdot 10^{-1}$	4·84583	1·11655	−2·61280
7	$0 \cdot 78125 \cdot 10^{-2}$	4·34350	1·05145	−2·24064
8	$0 \cdot 39063 \cdot 10^{-2}$	4·13175	1·02428	−2·08325
9	$0 \cdot 19531 \cdot 10^{-2}$	4·03401	1·01181	−2·01045
10	$0 \cdot 97656 \cdot 10^{-3}$	3·98699	1·00583	−1·97540
. . .	. . .	. . .	. . .	. . .
19	$0 \cdot 19074 \cdot 10^{-5}$	3·97558	1·02901	−1·94369
20	$0 \cdot 95367 \cdot 10^{-6}$	4·00728	1·05653	−1·94369
21	$0 \cdot 47684 \cdot 10^{-6}$	4·09406	1·14221	−2·05832
22	$0 \cdot 23842 \cdot 10^{-6}$	4·12083	1·33006	−1·90723
23	$0 \cdot 11921 \cdot 10^{-6}$	4·38012	1·73238	−0·58201
Value of $4 - 6x$		4	1	−2

Table 10.4.2: The method of regularization for the solution of
$\int_0^1 (x + y)\phi(y)\, dy = x$.

N	Estimate of S
8	−5·3127
16	−6·0228
32	−6·4538
64	−6·6933
Correctly rounded value of	
$S = -6 \cdot 9521$	

Table 10.4.3: Estimate of the total
charge, S, by Symm's method.

r	0	1	2	3	Exact
$\phi_0^{(r)}$	1	1	1	1	1
$\phi_{1/2}^{(r)}$	1	1·002314	1·002414	1·002397	1·00250
$\phi_1^{(r)}$	1	1·010332	1·010000	1·010007	1·01005
$\phi_2^{(r)}$	1	1·043482	1·040687	1·040824	1·04081
$V_0^{(r)}$	0	0	0	0	
$V_{1/2}^{(r)}$	0·001968	0·001898	0·001903	0·001903	
$V_1^{(r)}$	0·012020	0·012059	0·012059	0·012059	
$V_2^{(r)}$	0·056374	0·057169	0·057131	0·057132	

Table 10.6.1: Numerical results in the solution of the integro-differential equation (10.6.14).

PROGRAMS

In the Appendix there are programs (1) for the solution of the Volterra integral equation of the second kind, (2) for the solution of the Fredholm integral equation.

CHAPTER 11

Numerical Optimization

11.1. SINGLE VARIABLE MINIMIZATION

Most of the methods for minimization of a function of one variable work on the principle of reducing an interval within which the function is known, or assumed, to be *unimodal*, that is it has only one turning point within this interval.

To obtain a bracket for the minimum of a function f without using any derivative information [see IV, § 3.5] we require three points $x < y < z$ such that

$$f(x) > f(y), \qquad f(z) > f(y)$$

or, if the derivative can be used, two points $x < y$ with $f'(x) < 0 < f'(y)$.

In general we start from a point x_0 and select a steplength h. Subsequent points x_i $(i = 1, 2, \ldots)$ are generated by the relation

$$x_i = x_{i-1} + 2^{i-1} h \qquad (11.1.1)$$

until one of the conditions above is satisfied. In (11.1.1) the step taken is doubled each time so that if x_0 is well removed from the minimum we do not require a large number of small steps. This is because in general the refining process is much more efficient and so we make an overall saving as the following example shows.

EXAMPLE 11.1.1. Suppose we wish to obtain a bracket of length less than $0 \cdot 1$ for the minimum of $(x - 10 \cdot 01)^2$ starting with $x_0 = 0$ and $h = 0 \cdot 1$.

If the step was kept fixed at $0 \cdot 1$ then after 102 evaluations we would have the bracket $[9 \cdot 9, 10 \cdot 1]$ for the minimum whereas doubling the step we evaluate the function at the points $0, 0 \cdot 1, 0 \cdot 3, 0 \cdot 7, 1 \cdot 5, 3 \cdot 1, 6 \cdot 3, 12 \cdot 7$ and $25 \cdot 5$ to obtain the much wider bracket $[6 \cdot 3, 25 \cdot 5]$ after only 9 function evaluations. Using a Fibonacci search [see IV, § 15.9.5] in the first case, since $F_3 = 3 > 0 \cdot 2 / 0 \cdot 1$ we would require a further four evaluations making 106. The second bracket is of length $19 \cdot 2$ and so we require the Fibonacci number $F_n > 192$ (= $19 \cdot 2 / 0 \cdot 1$). Now $F_{11} = 144$ and $F_{12} = 233$ and so a total of 13 more (making 22 in all) evaluations will satisfy our requirements.

For the theoretical explanation of these numbers for the Fibonacci search see the Nonlinear Programming chapter (Volume IV, Chapter 15). The method itself will be discussed shortly but the purpose of the example is simply to show the advantage of the somewhat cruder initial bracket.

This strategy has its drawbacks however. In particular we could be searching in the wrong direction; or if the steplength becomes large and the function is not unimodal we could miss the minimum.

If derivatives are available the first of these drawbacks is overcome by the simple expedient of searching in the downhill direction; that is if $f'(x_0)>0$ we choose $h<0$ and vice versa. Alternatively if the function is known to be unimodal and $f(x_1)>f(x_0)$ we would then reverse the direction of search.

More generally if, say, ten steps are taken in one direction without obtaining a bracket then we would change direction. This is probably best achieved by resetting x_0 to x_{10} and h to the most recent steplength but with its sense reversed. This means the initial point is passed quickly and the possibility of a bracket around that point is not lost.

It is easy to see that many different forms of this procedure could be used for specific cases if, for example, we have greater knowledge of the function or strict limits on the range of values in which to search. The particular form described here can be formalized as follows.

ALGORITHM 11.1.1.   *Select x_0, h.*

1. $f_0 = f(x_0)$; $x_1 = x_0 + h$; $f_1 = f(x_1)$ and $n = 2$.
2. $h = 2h$; $x_n = x_{n-1} + h$; $f_n = f(x_n)$.
3. If $f_{n-2} > f_{n-1}$ and $f_{n-1} < f_n$ go to step 7.
4. If $n = 10$ go to step 6.
5. $n = n + 1$; go to step 2.
6. $x_0 = x_{10}$; $h = -h$; go to step 1.
7. Bracket is $[x_{n-2}, x_n]$ (or $[x_n, x_{n-2}]$ if $h < 0$). Stop.

EXAMPLE 11.1.2.   Bracket the minimum of

$$f(x) = [(x+1)^2 + 1](x-2)^2$$

starting from $x_0 = -0.2$ with $h = 0.1$.

Following the algorithm we get the following table;

n	x_n	$f(x_n)$	h
0	−0·2	7·9376	0·1
1	−0·1	7·9821	0·2
2	0·1	7·9781	0·4
3	0·5	7·3125	0·8
4	1·3	3·0821	1·6
5	2·9	13·1301	

which gives us the bracket $[0·5, 2·9]$. Since $f(x_1) > f(x_0)$ if the (false) assumption of unimodality were used the search would have been reversed and the local minimum at $-0·5$ would have been bracketed.

The other point to make here is that if the derivative is available it is usually advantageous to use it. Since only two consecutive points are necessary to identify a bracket and the steplength doubles each time, the bracket obtained with derivatives is at most two-thirds of the length of that obtained otherwise.

EXAMPLE 11.1.3.   Bracket the minimum of

$$f(x) = x + 1/x^2$$

starting with $x_0 = 1$ and $h = 0·1$ both with and without derivatives.

Using derivatives we have

n	x_n	$f'(x_n)$	h
0	1·0	−1·0000	0·1
1	1·1	−0·5026	0·2
2	1·3	0·0897	

giving the bracket $[1·1, 1·3]$.

If derivatives were not used the function values for these points are 2, 1·9264 and 1·8917 and so we try $x_3 = 1·7$ which gives $f_3 = 2·0460$ and, therefore, the bracket $[1·1, 1·7]$.

Once a bracket has been obtained there are many ways of either reducing this to a smaller bracket or estimating the actual position of the minimum to any required accuracy.

The Fibonacci search is the most efficient for reducing the interval length to any required tolerance [cf. Example 11.1.1]. If the initial bracket is of length L and the required length is ε then we first find the smallest Fibonacci number such that $L/F_r < \varepsilon$ where

$$F_r = F_{r-1} + F_{r-2}; \qquad F_0 = F_1 = 1.$$

Now $r + 1$ further evaluations will lead to a bracket of the required length. (The number $r + 1$ takes account of the need to evaluate either side of the mid-point of the final interval—this will usually occur automatically because of rounding errors.)

The strategy at any stage is that two points a_3, a_4 are placed symmetrically in the bracket $[a_1, a_2]$ in the sense that $a_2 - a_4 = a_3 - a_1$. [Initially they are such that $a_3 - a_1 = (a_2 - a_1)F_{r-2}/F_r$.] Either $[a_1, a_4]$ or $[a_3, a_2]$ now gives a new bracket. The points are relabelled and *one* further point placed in the new interval. In the situation in Figure 11.1.1, $[a_1, a_4]$ gives a new bracket and so we relabel a_4 as a_2, a_3 as a_4 and put a new a_3 such that $a_3 - a_1 = a_2 - a_4$.

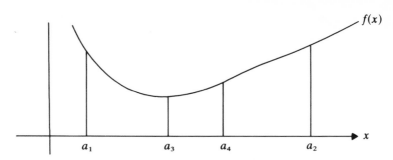

Figure 11.1.1: Fibonacci search.

ALGORITHM 11.1.2.   (Fibonacci search).   *Given a bracket* $[a_1, a_2]$ *for the minimum of* f *and tolerance* ε.

1. Find   smallest r such   that $F_r > (a_2 - a_1)/\varepsilon$;   put $\alpha = F_{r-2}/F_r$; $a_3 = a_1 + \alpha(a_2 - a_1)$; $a_4 = a_1 + a_2 - a_3$; evaluate f_1, f_3, f_4, f_2.
2. For $i = 1$ to $r - 2$:

   If $f_3 < f_4$ then $a_2 := a_4$, $a_4 := a_3$, $f_2 := f_4$, $f_4 := f_3$

   $\qquad\qquad\qquad a_3 := a_1 + a_2 - a_4$, $f_3 := f(a_3)$

   If $f_4 \leq f_3$ then $a_1 := a_3$, $a_3 := a_4$, $f_1 := f_3$, $f_3 := f_4$

   $\qquad\qquad\qquad a_4 := a_1 + a_2 - a_3$, $f_4 := f(a_4)$

   Continue.
3. Evaluate f a small distance either side of a_3 to determine which half-interval contains the minimum. Stop.

EXAMPLE 11.1.4.   Use Fibonacci search to reduce the bracket $[1 \cdot 1, \; 1 \cdot 7]$ for the minimum of

$$f(x) = x + 1/x^2$$

to a length less than $0 \cdot 1$.

   Here $a_2 - a_1 = 0 \cdot 6$ and so we want $F_r > 0 \cdot 6/0 \cdot 1 = 6$. $F_0 = F_1 = 1$ so $F_2 = 2$, $F_3 = 3$, $F_4 = 5$ and $F_5 = 8 > 6$ so $r = 5$.
   Thus $\alpha = F_3/F_5 = 3/8$ so that

$\qquad\qquad a_1 = 1 \cdot 1 \qquad a_3 = 1 \cdot 325 \qquad a_4 = 1 \cdot 475 \qquad a_2 = 1 \cdot 7$

$\qquad\qquad f_1 = 1 \cdot 9264 \quad f_3 = 1 \cdot 8946 \quad f_4 = 1 \cdot 9346 \quad f_2 = 2 \cdot 0460$

$i = 1 \quad f_3 < f_4$ so

$\qquad\qquad a_1 = 1 \cdot 1 \qquad a_3 = 1 \cdot 25 \qquad a_4 = 1 \cdot 325 \qquad a_2 = 1 \cdot 475$

$\qquad\qquad f_1 = 1 \cdot 9264 \quad f_3 = 1 \cdot 8900 \quad f_4 = 1 \cdot 8946 \quad f_2 = 1 \cdot 9346$

$i = 2$ $f_3 < f_4$ so

$a_1 = 1 \cdot 1$ $a_3 = 1 \cdot 175$ $a_4 = 1 \cdot 25$ $a_2 = 1 \cdot 325$

$f_1 = 1 \cdot 9264$ $f_3 = 1 \cdot 8993$ $f_4 = 1 \cdot 8900$ $f_2 = 1 \cdot 8946$

$i = 3$ $f_4 < f_3$ so

$a_1 = 1 \cdot 175$ $a_3 = 1 \cdot 25$ $a_4 = 1 \cdot 25$ $a_2 = 1 \cdot 325$

$f_1 = 1 \cdot 8993$ $f_3 = 1 \cdot 8900$ $f_4 = 1 \cdot 8900$ $f_2 = 1 \cdot 8946$.

We now evaluate to left and right of a_3 to check which half-interval is our final bracket. Since

$$f(1 \cdot 24) = 1 \cdot 8904 \quad \text{and} \quad f(1 \cdot 26) = 1 \cdot 8899$$

we obtain the final bracket $[1 \cdot 25, 1 \cdot 325]$.

Most of the methods for locating the actual position of the minimum are based on finding the minimum point of an interpolating polynomial [see § 2.3] and then adjusting the points and information to be used. The simplest is to use quadratic interpolation [see (2.3.3)]. Any three points can be used and a formula obtained for the minimum point of the quadratic function but the process is greatly simplified if the points are equally spaced [see § 2.3.1].

If x_0, x_1 and x_2 are equally spaced points bracketing the minimum of a function f, so that

$$x_2 = x_1 + h = x_0 + 2h$$

and $f_1 < f_0, f_2$, then the quadratic interpolation polynomial which agrees with f at these points has its minimum at

$$x^* = x_1 + \frac{(f_0 - f_2)}{(f_0 - 2f_1 + f_2)} \frac{h}{2} \tag{11.1.2}$$

which lies in $[x_1 - h/2, x_1 + h/2]$.

Now if $f(x^*) < f_1$ we can use x^* as the mid-point of the next interval which we take to be half the length of the previous one. Otherwise we keep x_1 as mid-point but still halve the interval length. The other necessary check in this routine is to ensure that the three points of the new interval provide a bracket. Provided this is done then this method can be proved to converge to the minimum of a unimodal function.

EXAMPLE 11.1.5. Use a quadratic search to minimize $x + 1/x^2$ in $[1 \cdot 1, 1 \cdot 7]$.

Initially $h = 0 \cdot 3$ and we have

$x_0 = 1 \cdot 1$ $x_1 = 1 \cdot 4$ $x_2 = 1 \cdot 7$

$f_0 = 1 \cdot 9264$ $f_1 = 1 \cdot 9102$ $f_2 = 2 \cdot 0460$.

Hence

$$x^* = 1 \cdot 4 + \frac{(1 \cdot 9264 - 2 \cdot 0460) \times 0 \cdot 3}{2(1 \cdot 9264 - 2 \times 1 \cdot 9102 + 2 \cdot 0460)}$$

$$= 1 \cdot 282$$

and

$$f(x^*) = 1 \cdot 8904 < f_1.$$

The new value for h is $0 \cdot 3/2 = 0 \cdot 15$ and x_1 is replaced by $1 \cdot 282$ so we obtain the following results.

x_0	1·132	1·1963	1·2251
x_1	1·282	1·2713	1·2626
x_2	1·432	1·3463	1·3001
f_0	1·9124	1·8950	1·8914
f_1	1·8904	1·8900	1·8899
f_2	1·9197	1·8980	1·8917
x^*	1·2713	1·2626	1·2609
$f(x^*)$	1·8900	1·8899	1·88988
New h	0·075	0·0375	
New x_1	1·2713	1·2626	

So after 11 function evaluations in addition to the original bracket we have a best point $1 \cdot 2609$ with the value of $f = 1 \cdot 88988$ to 5 d.p.

If the derivative of f is used then we can find a quadratic which agrees with f at two points and its derivative at one of them. However, more efficient use of the derivative can be obtained by using cubic interpolation. Here we use the cubic which agrees with f and f' at two points which bracket the minimum. This cubic must have a minimum in the interval, at x^* say, which can then be used with one of the original points to provide a smaller bracket and therefore a more accurate approximation.

If the cubic which agrees with f and f' at two points x_0, x_1 where

$$f'(x_0) < 0 < f'(x_1)$$

is $ax^3 + bx^2 + cx + d = p(x)$ say, then p has turning points at the points

$$\frac{1}{3a}\{-b \pm (b^2 - 3ac)^{1/2}\} \tag{11.1.3}$$

[see IV, § 3.5]. Since $p'(x_0) < 0 < p'(x_1)$ one of these lies in (x_0, x_1). We choose this to be x^*. If $f'(x^*) < 0$ then we replace x_0 with x^* whereas if $f'(x^*) > 0$ we replace x_1 with x^*. The coefficients a, b, c and d are the solutions of

$$\begin{pmatrix} x_0^3 & x_0^2 & x_0 & 1 \\ x_1^3 & x_1^2 & x_1 & 1 \\ 3x_0^2 & 2x_0 & 1 & 0 \\ 3x_1^2 & 2x_1 & 1 & 0 \end{pmatrix} \begin{pmatrix} a \\ b \\ c \\ d \end{pmatrix} = \begin{pmatrix} f_0 \\ f_1 \\ f_0' \\ f_1' \end{pmatrix}$$

which [see I, § 5.10] are given by

$$a = X(G - 2H)$$
$$b = H - (x_1 + 2x_0)a \qquad\qquad (11.1.4)$$
$$c = F - (x_0^2 + x_0 x_1 + x_1^2)a - (x_0 + x_1)b$$

where

$$X = 1/(x_1 - x_0), \qquad F = X(f_1 - f_0), \qquad G = X(f_1' - f_0')$$

and $H = X(F - f_0')$. (Note that we do not require the value for d since it does not affect the position of the stationary points.)

EXAMPLE 11.1.6.   Minimize $x + 1/x^2$ by cubic search.

The bracket obtained in Example 11.1.3 using derivatives is $[1 \cdot 1, 1 \cdot 3]$, so for the first iteration we have

$$x_0 = 1 \cdot 1 \qquad\qquad x_1 = 1 \cdot 3$$
$$f_0 = 1 \cdot 9264 \qquad f_1 = 1 \cdot 8917$$
$$f_0' = -0 \cdot 5026 \quad f_1' = 0 \cdot 0897.$$

Thus

$$X = 1/(1 \cdot 3 - 1 \cdot 1) = 5,$$
$$F = 5(1 \cdot 8917 - 1 \cdot 9264) = -0 \cdot 1735,$$
$$G = 5(0 \cdot 0897 + 0 \cdot 5026) = 2 \cdot 9615,$$
$$H = 5(-0 \cdot 1735 + 0 \cdot 5026) = 1 \cdot 6455$$

which gives

$$a = -1 \cdot 6475, \qquad b = 7 \cdot 4118, \qquad c = -10 \cdot 8281.$$

The turning points are at

$$1 \cdot 2588 \quad \text{and} \quad 1 \cdot 7404.$$

Hence we have $x^* = 1 \cdot 2588$ and

$$f'(x^*) = 1 - 2/x^{*3} = -0 \cdot 0027 < 0$$

and so for the next iteration

$$x_0 = 1\cdot2588 \qquad x_1 = 1\cdot3$$
$$f_0 = 1\cdot8899 \qquad f_1 = 1\cdot8917$$
$$f_0' = -0\cdot0027 \qquad f_1' = 0\cdot0897.$$

These give us the turning points

$$1\cdot2600 \quad \text{and} \quad 4\cdot4746.$$

Hence

$$x^* = 1\cdot2600, \qquad f'(x^*) = 0\cdot0002$$

to 4 d.p. Thus after four evaluations of either the function or its derivative in addition to the original bracket the next iteration would use the information

$$x_0 = 1\cdot2588 \qquad x_1 = 1\cdot2600$$
$$f_0 = 1\cdot8898831 \qquad f_1 = 1\cdot8898816$$
$$f_0' = -0\cdot0027 \qquad f_1' = 0\cdot0002.$$

This can be seen to be approaching the actual solution $2^{1/3} = 1\cdot259921$ very rapidly.

Of course if the derivative is to be used one other possibility is to use one of the methods of solving a nonlinear equation to find a root of $f'(x) = 0$. (See Chapter 5 for a description of suitable methods.) The drawback to such an approach is that the solution found could be a maximum or a point of inflexion so that great care must be taken to ensure that a minimum is approached.

11.2. FUNCTIONS OF SEVERAL VARIABLES

In this section we consider methods for locating the minimum of a function of several variables without using any derivatives—these are often called *direct* search methods. There is a large class of methods from which to choose; for a much wider selection than is covered here the reader is referred to any standard textbook on optimization such as Walsh (1975) or Wolfe (1978).

We deal here with just two quite different methods which give some idea of the wide diversity of approach. They are also regarded as two of the more efficient techniques for this type of problem.

Suppose then that we wish to minimize a function $f: \mathbb{R}^n \to \mathbb{R}$.

We begin with the simplex method of Nelder and Mead (1965). Suppose we have a simplex in $\mathbb{R}^n$, that is [see V, § 5.2] a set of $n + 1$ points $\mathbf{x}_0, \mathbf{x}_1, \ldots, \mathbf{x}_n$ which form a non-degenerate polyhedron, or, in other words, the set $\{\mathbf{x}_i - \mathbf{x}_0 : i = 1, 2, \ldots, n\}$ is linearly independent [see I, Definition 5.3.2]. In a fairly sophisticated though essentially intuitive manner one of these vertices is replaced by a new point which then gives the next simplex for the process.

The method employs a variety of alternative moves which we illustrate in two dimensions but define for the general case.

First of all let $\mathbf{x}_g$, $\mathbf{x}_h$ and $\mathbf{x}_s$ denote the vertices with respectively the greatest, second largest and smallest values of f. We also use $\bar{\mathbf{x}}$ to denote the centroid of all the points except $\mathbf{x}_g$, so that

$$\bar{\mathbf{x}} = \frac{1}{n}\left(\sum_{i=0}^{n} \mathbf{x}_i - \mathbf{x}_g\right).$$

The first step of an iteration is to reflect $\mathbf{x}_g$ in $\bar{\mathbf{x}}$ to yield the point

$$\mathbf{x}_r = \bar{\mathbf{x}} + \alpha(\bar{\mathbf{x}} - \mathbf{x}_g)$$

where $0 < \alpha \le 1$ is the *reflection coefficient* which is usually taken to be unity.

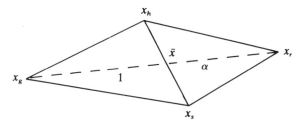

Figure 11.2.1: Reflection step.

Now suppose that $f_r = f(\mathbf{x}_r) < f(\mathbf{x}_s) = f_s$ so that the reflection has produced a new best point. This suggests that further exploration in this direction is worthwhile and so an expansion step is tried. For this we put

$$\mathbf{x}_e = \bar{\mathbf{x}}_1 + \gamma(\mathbf{x}_r - \bar{\mathbf{x}})$$

where $\gamma > 1$ is the *coefficient of expansion*. We then evaluate $f_e = f(\mathbf{x}_e)$ and if $f_e < f_r$ the expansion is successful. In this case $\mathbf{x}_g$ is replaced in the simplex by $\mathbf{x}_e$ and the process repeated (beginning, of course, with relabelling the points $\mathbf{x}_g, \mathbf{x}_h, \mathbf{x}_s$). If on the other hand $f_e > f_r$ then we replace $\mathbf{x}_g$ with $\mathbf{x}_r$.

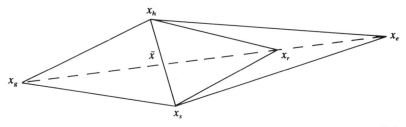

Figure 11.2.2: Expansion step: $f_r < f_s$ and if $f_e < f_r$ next simplex is $\mathbf{x}_e, \mathbf{x}_s, \mathbf{x}_h$ or if $f_e > f_r$ next simplex is $\mathbf{x}_r, \mathbf{x}_s, \mathbf{x}_h$.

The next possibility is that f_r lies somewhere in the range of the other values, that is $f_s < f_r < f_h = f(\mathbf{x}_h)$. In this case the reflection is accepted and $\mathbf{x}_g$ replaced by $\mathbf{x}_r$ to give the new simplex.

The remaining possibilities both lead to a contraction step. Whichever is the better of $\mathbf{x}_r$ and $\mathbf{x}_g$ is used to generate a new point $\mathbf{x}_c$ on the line segment between that point and $\bar{\mathbf{x}}$. Thus if $f_h < f_r < f_g$ then $\mathbf{x}_g$ is replaced by $\mathbf{x}_r$, while if $f_r > f_g$, $\mathbf{x}_g$ is left unaltered; we then put $\mathbf{x}_c = \bar{\mathbf{x}} + \beta(\mathbf{x}_g - \bar{\mathbf{x}})$ where the *contraction coefficient* satisfies $0 < \beta < 1$.

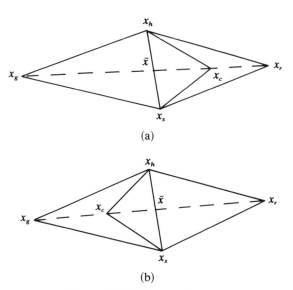

(a)

(b)

Figure 11.2.3: Contraction step.

(a) $f_h < f_r < f_g$ so $\mathbf{x}_g$ replaced by $\mathbf{x}_r$ to obtain $\mathbf{x}_c$.
(b) $f_r > f_g$ so $\mathbf{x}_c$ is obtained from $\mathbf{x}_g$.

Now, if $\mathbf{x}_c$ is an improvement, that is if $f_c < f_h$ then $\mathbf{x}_g$ is replaced by $\mathbf{x}_c$ to give the new simplex.

Finally if $\mathbf{x}_c$ is still unsatisfactory ($f_c > f_h$) then the whole simplex is reduced to half its size retaining the point $\mathbf{x}_s$, so that we redefine

$$\mathbf{x}_i := \tfrac{1}{2}(\mathbf{x}_i + \mathbf{x}_s) \qquad (i = 0, 1, \ldots, n)$$

(with, of course $\mathbf{x}_g$ replaced by $\mathbf{x}_r$ as above if $f_r < f_g$).

The stopping condition for this method is usually that every edge of the simplex is less than some prescribed length.

The usual choices for the various coefficients are:

$$\text{reflection} \quad \alpha = 1,$$

$$\text{expansion} \quad \gamma = 2,$$

and

$$\text{contraction} \quad \beta = \tfrac{1}{2}.$$

EXAMPLE 11.2.1. Minimize the function

$$f(x, y) = (x - y)^2 + y^2$$

by the method of Nelder and Mead starting with the simplex whose vertices are at the points $(5, 0)$, $(0, 1)$ and $(6, 1)$.

For the first iteration we have

	$\mathbf{x}_g$	$\mathbf{x}_h$	$\mathbf{x}_s$	$\bar{\mathbf{x}}$	$\mathbf{x}_r$	$\mathbf{x}_e$
x	6	5	0	2·5	−1	−4·5
y	1	0	1	0·5	0	−0·5
f	26	25	2		1	16·25

So the expansion fails and we now have

	$\mathbf{x}_g$	$\mathbf{x}_h$	$\mathbf{x}_s$	$\bar{\mathbf{x}}$	$\bar{\mathbf{x}}_r$	$\mathbf{x}_c = \frac{1}{2}(\mathbf{x}_g + \bar{\mathbf{x}})$
x	5	0	−1	−0·5	−6	2·25
y	0	1	0	0·5	1	0·25
f	25	2	1		50	4·0625

Here $f_c > f_h$ and so we reduce the simplex keeping the point $(-1, 0)$. Thus we obtain for the next two iterations

	$\mathbf{x}_g$	$\mathbf{x}_h$	$\mathbf{x}_s$	$\bar{\mathbf{x}}$	$\mathbf{x}_r$	$\mathbf{x}_c$
x	2	−0·5	−1	−0·75	−3·5	0·625
y	0	0·5	0	0·25	0·5	0·125
f	4	1·25	1		16·25	0·2656
x	−0·5	−1	0·625	−0·1875	0·125	
y	0·5	0	0·125	0·0625	−0·375	
f	1·25	1	0·2656		0·3906	

Here the reflection is accepted and the next iteration begins with $\mathbf{x}_g = (-1, 0)$, $\mathbf{x}_h = (0·125, -0·375)$ and $\mathbf{x}_s = (0·625, 0·125)$.

The other method considered here is due to Powell (1964). In its nature Powell's method is much like some of the gradient methods to be discussed in the next section. The strategy of this method is that on each iteration a line search is carried out for the minimum of f in each of n linearly independent directions. One further search is then made in the direction of the total step taken by these n line searches. Then this direction can be used to replace one of the original set to set up the next iteration.

It can be proved (see, for example Wolfe (1978)) that if f is a quadratic function then this process yields a system of conjugate directions [see IV, § 15.2.8] so the process terminates, for such a function, after at most n iterations. (A similar result is quoted for the *conjugate gradient* and *variable metric* methods in Volume IV, §§ 15.9.7 and 15.9.8. For a full treatment of these results the reader is again referred to a textbook on Optimization such as Wolfe (1978).)

Before describing the method in full we summarize the fundamental approach.

ALGORITHM 11.2.1   (Powell's basic method).   *Choose* $\mathbf{x}_0$ *and set* $\mathbf{s}_i = \mathbf{e}_i$, *the ith unit coordinate vector, for* $i = 1, 2, \ldots, n$.

1. For $i = 1$ to n compute α_i so that

$$f(\mathbf{x}_{i-1} + \alpha_i \mathbf{s}_i) = \min_{\alpha} f(\mathbf{x}_{i-1} + \alpha \mathbf{s}_i)$$

   and put $\mathbf{x}_i = \mathbf{x}_{i-1} + \alpha_i \mathbf{s}_i$.
2. For $i = 1$ to $n-1$ replace $\mathbf{s}_i$ by $\mathbf{s}_{i+1}$ and set

$$\mathbf{s}_n = \mathbf{x}_n - \mathbf{x}_0.$$

3. Find α_n such that

$$f(\mathbf{x}_n + \alpha_n \mathbf{s}_n) = \min_{\alpha} f(\mathbf{x}_n + \alpha \mathbf{s}_n),$$

   put $\mathbf{x}_0 = \mathbf{x}_n + \alpha_n \mathbf{s}_n$ and return to 1.

There is no mention here of any stopping condition. One simple condition that can be used is that [see Definition 6.2.1]

$$\|\mathbf{x}_n + \alpha_n \mathbf{s}_n - \mathbf{x}_0\| < \varepsilon.$$

This condition can result in premature convergence and Powell (1964) proposes a more restrictive condition to overcome this. Details can again be found in Wolfe (1978).

EXAMPLE 11.2.2.   Use Powell's basic method to minimize

$$f(x, y) = (x - y)^2 + y^2$$

from (a) $\mathbf{x}_0 = (3, 1)$ and (b) $\mathbf{x}_0 = (1, 1)$.

(a)   *First Iteration*

$$\mathbf{x}_0 = (3, 1), \qquad \mathbf{s}_1 = (1, 0) \quad \text{and} \quad \mathbf{s}_2 = (0, 1).$$

This gives $\alpha_1 = -2$ so that $x_1 = (1, 1)$. Then $\alpha_2 = -\frac{1}{2}$ and $x_2 = (1, \frac{1}{2})$ which completes step 1. Now we put $s_1 = (0, 1)$, $s_2 = (1, \frac{1}{2}) - (3, 1) = (-2, -\frac{1}{2})$. Thus the α_2 of step 3 is the value of α which minimizes $f(1-2\alpha, \frac{1}{2}(1-\alpha)) = \frac{1}{4}(1-3\alpha)^2 + \frac{1}{4}(1-\alpha)^2$. Hence $\alpha_2 = \frac{2}{5}$ and we put

$$x_0 = (1, \tfrac{1}{2}) - \tfrac{2}{5}(2, \tfrac{1}{2}) = (\tfrac{1}{5}, \tfrac{3}{10}).$$

Second Iteration

$$x_0 = (\tfrac{1}{5}, \tfrac{3}{10}), \qquad s_1 = (0, 1), \qquad s_2 = (-2, -\tfrac{1}{2}).$$

This gives $\alpha_1 = -\frac{1}{5}$, $x_1 = (\frac{1}{5}, \frac{1}{10})$; $\alpha_2 = \frac{2}{25}$ and $x_2 = (\frac{1}{25}, \frac{3}{50})$. Hence we put $s_1 = (-2, -\frac{1}{2})$, $s_2 = (-\frac{4}{25}, -\frac{6}{25})$ from which we obtain $\alpha_2 = \frac{1}{4}$ and $x_0 = (0, 0)$, the minimum of f.

(b)   Here we have $x_0 = (1, 1)$, $s_1 = (1, 0)$, $s_2 = (0, 1)$ and $\alpha_1 = 0$, $\alpha_2 = -\frac{1}{2}$ and $x_2 = (1, \frac{1}{2})$.

The new s_1 and s_2 are thus $s_1 = (0, 1)$ and $s_2 = (0, -\frac{1}{2})$ which are linearly dependent [see I, Definition 5.3.1]. Since neither has an x component and we are already at the line minimum in the y direction no further progress can be made.

In the example we see that Powell's basic method can fail if the system of search directions becomes linearly dependent and the method must be modified to overcome this difficulty. Powell's modification of the basic method is based on the fact that if f is a positive definite quadratic function [see I, § 9.2] with Hessian matrix G [see IV (5.7.1)] and the search directions $s_1, \ldots, s_n$ are scaled so that $\frac{1}{2}s_i'Gs_i = 1$ $(i = 1, 2, \ldots, n)$ then the determinant of the matrix whose columns are these s_i is maximized if and only if they are conjugate with respect to G.

The policy is then to replace one of the old directions with the new one in such a way as to increase this determinant as much as possible. It can happen that this results in using the same directions again or in discarding one of the conjugate directions obtained previously. This of course increases the number of iterations required.

It can be proved, though not easily (see Walsh (1975), § 4.9), that these considerations lead to the following strategy.

ALGORITHM 11.2.2   (Powell's method).   *Choose* x_0 *and set* $s_i = e_i (i = 1, 2, \ldots, n)$.

1. For $i = 1$ to n compute α_i so that

$$f(x_{i-1} + \alpha_i s_i) = \min_\alpha f(x_{i-1} + \alpha s_i)$$

and put $x_i = x_{i-1} + \alpha_i s_i$.

2. Find q so that

$$f(\mathbf{x}_{q-1}) - f(\mathbf{x}_q) = \max_{1 \le i \le n} [f(\mathbf{x}_{i-1}) - f(\mathbf{x}_i)]$$

and put $\Delta = f(\mathbf{x}_{q-1}) - f(\mathbf{x}_q)$.

3. Define $\mathbf{s}_{n+1} = \mathbf{x}_n - \mathbf{x}_0$, $f_1 = f(\mathbf{x}_0)$, $f_2 = f(\mathbf{x}_n)$ and evaluate $f_3 = f(\mathbf{x}_n + \mathbf{s}_{n+1})$.

4. If *either* $f_3 \ge f_1$ or

$$(f_1 - 2f_2 + f_3)(f_1 - f_2 - \Delta)^2 \ge \tfrac{1}{2}\Delta(f_1 - f_3)^2$$

then put $\mathbf{x}_0 = \mathbf{x}_n$ and return to step 1. Otherwise compute α_{n+1} so that $f(\mathbf{x}_n + \alpha_{n+1}\mathbf{s}_{n+1}) = \min_\alpha f(\mathbf{x}_n + \alpha\mathbf{s}_{n+1})$, put $\mathbf{x}_0 = \mathbf{x}_n + \alpha_{n+1}\mathbf{s}_{n+1}$ and for $i = q$ to n set $\mathbf{s}_i = \mathbf{s}_{i+1}$ and return to step 1.

EXAMPLE 11.2.3. Use Powell's method with $\mathbf{x}_0 = (1, 1)$ to minimize $(x - y)^2 + y^2$.

First Iteration

$\mathbf{x}_0 = (1, 1)$, $\mathbf{s}_1 = (1, 0)$ and $\mathbf{s}_2 = (0, 1)$ so that $\alpha_1 = 0$, $\alpha_2 = -\tfrac{1}{2}$ and $\mathbf{x}_2 = (1, \tfrac{1}{2})$ as before. Here $f(\mathbf{x}_0) = f(\mathbf{x}_1) = 1$ and $f(\mathbf{x}_2) = \tfrac{1}{2}$ so $q = 2$ and $\Delta = \tfrac{1}{2}$. Then $\mathbf{s}_3 = (0, -\tfrac{1}{2})$, $f_1 = 1$, $f_2 = \tfrac{1}{2}$ and $f_3 = f(1, 0) = 1 \ge f_1$. Hence we put $\mathbf{x}_0 = (1, \tfrac{1}{2})$ and leave the search directions unaltered.

Second Iteration

$\mathbf{x}_0 = (1, \tfrac{1}{2})$, $\mathbf{s}_1 = (1, 0)$ and $\mathbf{s}_2 = (0, 1)$. This gives $\alpha_1 = -\tfrac{1}{2}$, $\mathbf{x}_1 = (\tfrac{1}{2}, \tfrac{1}{2})$ and $f(\mathbf{x}_1) = \tfrac{1}{4}$. Then $\alpha_2 = -\tfrac{1}{4}$, $\mathbf{x}_2 = (\tfrac{1}{2}, \tfrac{1}{4})$ and $f(\mathbf{x}_2) = \tfrac{1}{8} = f_2$. Thus $q = 1$, $\Delta = \tfrac{1}{4}$, $\mathbf{s}_3 = (-\tfrac{1}{2}, -\tfrac{1}{4})$ and

$$f_3 = f(0, 0) = 0.$$

Then $f_3 < f_1$ and

$$(f_1 - 2f_2 + f_3)(f_1 - f_2 - \Delta)^2 = (\tfrac{1}{2} - \tfrac{1}{4})(\tfrac{1}{2} - \tfrac{1}{8} - \tfrac{1}{4})^2$$

$$= \tfrac{1}{64} < \tfrac{1}{32} = \tfrac{1}{2} \cdot \tfrac{1}{4}(\tfrac{1}{2} - 0)^2$$

$$= \tfrac{1}{2}\Delta(f_1 - f_3)^2.$$

Hence we search for the minimum of $f(\tfrac{1}{2} - \tfrac{1}{2}\alpha, \tfrac{1}{4} - \tfrac{1}{4}\alpha)$ which of course yields the minimum point $(0, 0)$. (This point had in fact already been tried as f_3 but would not be recognized as the minimum at that stage of the process.)

11.3. GRADIENT METHODS

As in the case of functions of a single variable so with those of several variables considerable benefits can be obtained by making use of derivative information if this is fairly readily available.

Most of the gradient methods are of the following basic form.
Given an initial estimate $\mathbf{x}_0$ and $i = 0$

1. Define a search direction $\mathbf{s}_i$.
2. Find α_i to minimize $f(\mathbf{x}_i + \alpha \mathbf{s}_i)$.
3. Put $\mathbf{x}_{i+1} = \mathbf{x}_i + \alpha_i \mathbf{s}_i$.
4. If the minimum has not yet been found to sufficient accuracy, set $i = i + 1$ and return to 1.

We shall see later that the line search for the minimum of $f(\mathbf{x}_i + \alpha \mathbf{s}_i)$ is sometimes dropped in favour of finding a value α_i which reduces the value of the objective function.

The convergence criteria adopted for these methods [see Definition 6.2.1] are usually a combination of $\|g_i\| < \varepsilon$, $\|\mathbf{x}_{i+1} - x_i\| < \varepsilon$ and $f(\mathbf{x}_i) - f(\mathbf{x}_{i+1}) < \varepsilon$ where we use $\mathbf{g}_i$ to denote the gradient vector of f at $\mathbf{x}_i$ so

$$\mathbf{g}_i = \mathbf{g}(\mathbf{x}_i) = \nabla f(\mathbf{x}_i) = \left(\frac{\partial f}{\partial x_1}, \frac{\partial f}{\partial x_2}, \ldots, \frac{\partial f}{\partial x_n} \right)'.$$

Methods differ then essentially in the manner of obtaining the search direction $\mathbf{s}_i$ (from $\mathbf{g}_i$ and other information based on past performance or second derivatives).

The simplest of the gradient techniques is the *method of steepest descent* in which we simply take

$$\mathbf{s}_i = -\mathbf{g}_i.$$

However, even on quite well-behaved functions this can be a very slow process.

EXAMPLE 11.3.1.   Use the method of steepest descent to minimize

$$f(x, y) = (x - y)^2 + y^2$$

starting with $\mathbf{x}_0 = (1, 1)'$.
   Here $\mathbf{g}_0 = (0, 2)'$, $\mathbf{s}_0 = (0, -2)'$ and $\alpha_0 = \frac{1}{4}$. Thus $\mathbf{x}_1 = (1, \frac{1}{2})'$ with $\mathbf{g}_1 = (1, 0)'$ giving $\alpha_1 = \frac{1}{2}$. In this case the search directions simply alternate between the (negative) x- and y-directions and the sequence of $\mathbf{x}_i$'s continues

$$(\tfrac{1}{2}, \tfrac{1}{2})', (\tfrac{1}{2}, \tfrac{1}{4})', (\tfrac{1}{4}, \tfrac{1}{4})', \ldots, \left(\frac{1}{2^{r-1}}, \frac{1}{2^r} \right)', \left(\frac{1}{2^r}, \frac{1}{2^r} \right)', \ldots.$$

It seems natural given the poor performance of the steepest descent technique and the usefulness of searching in conjugate directions to try to marry the use of the gradient to the generation of conjugate directions of search.

This is achieved in the conjugate gradient method of Fletcher and Reeves (1964) and by other conjugate gradient methods which we shall not discuss here. (See Wolfe (1978) for a detailed treatment of these methods.)

If we assume all line searches are performed exactly then it follows that

$$g_i's_{i-1} = 0 \qquad (i = 1, 2, \ldots) \tag{11.3.1}$$

since the point x_i is a minimum in the direction s_{i-1} so that the gradient g_i has no component in that direction. If further it is assumed that the objective function is a positive definite quadratic [see I, § 9.2], so that f is of the form

$$f(x) = a + x'w + \tfrac{1}{2}x'Gx \tag{11.3.2}$$

where G is a positive definite symmetric matrix and $a \in \mathbb{R}$ and $w \in \mathbb{R}^n$ are constant, then defining $s_0 = -g_0$ and

$$s_i = -g_i + \beta_{i-1}s_{i-1} \qquad (i = 1, 2, \ldots) \tag{11.3.3}$$

where $\beta_{i-1} = g_i'g_i/g_{i-1}'g_{i-1}$ we obtain a system of conjugate directions. That is

$$s_i'Gs_j = 0 \qquad (i \neq j)$$

and hence the minimum of the quadratic function given by (11.3.2) is reached in at most n iterations.

EXAMPLE 11.3.2.   Use the conjugate gradient method to minimize $(x - y)^2 + y^2$ with $x_0 = (1, 1)'$.

As before $g_0 = -s_0 = (0, 2)'$, $\alpha_0 = \tfrac{1}{4}$, $x_1 = (1, \tfrac{1}{2})'$ and $g_1 = (1, 0)'$. Then

$$\beta_0 = \frac{g_1'g_1}{g_0'g_0} = \tfrac{1}{4}$$

so that

$$s_1 = -g_1 + \beta_0 s_0 = (-1, 0)' + \tfrac{1}{4}(0, -2)'$$
$$= (-1, -\tfrac{1}{2})'.$$

This yields $\alpha_1 = 1$ and $x_2 = (0, 0)'$ as required.

This example demonstrates the efficiency of the method; it has been found able to cope well with non-quadratic functions of quite large numbers of variables.

A further example of a quadratic function illustrates the conjugacy of the directions.

EXAMPLE 11.3.3.   Minimize $5x^2 + 2y^2 + 2z^2 + 2xy + 2yz - 2zx - 6z$ starting with $x_0 = (0, 0, 0)'$.

Here

$$g(x) = \begin{pmatrix} 10x + 2y - 2z \\ 2x + 4y + 2z \\ -2x + 2y + 4z - 6 \end{pmatrix}$$

and the Hessian is

$$\mathbf{G} = \begin{pmatrix} 10 & 2 & -2 \\ 2 & 4 & 2 \\ -2 & 2 & 4 \end{pmatrix}.$$

Hence $\mathbf{g}_0 = -\mathbf{s}_0 = (0, 0, -6)'$ which gives $\alpha_0 = \frac{1}{4}$ and $\mathbf{x}_1 = (0, 0, \frac{3}{2})'$. Then $\mathbf{g}_1 = (-3, 3, 0)'$ so that $\beta_0 = \frac{18}{36} = \frac{1}{2}$ and so

$$\mathbf{s}_1 = -(-3, 3, 0)' + \tfrac{1}{2}(0, 0, 6)' = (3, -3, 3)'.$$

Note that $\mathbf{G}\mathbf{s}_0 = (-12, 12, 24)'$ and so $\mathbf{s}_1'\mathbf{G}\mathbf{s}_0 = 0$ showing that $\mathbf{s}_0$ and $\mathbf{s}_1$ are conjugate.

Now to obtain α_1 we minimize $f(\mathbf{x}_1 + \alpha\mathbf{s}_1)$ and $\mathbf{x}_1 + \alpha\mathbf{s}_1 = (3\alpha, -3\alpha, \frac{3}{2}+3\alpha)'$ giving $\alpha_1 = \frac{1}{3}$ and $\mathbf{x}_2 = (1, -1, \frac{5}{2})'$. Then $\mathbf{g}_2 = (3, 3, 0)'$ and $\beta_1 = \frac{18}{18} = 1$. Hence

$$\mathbf{s}_2 = -(3, 3, 0)' + (3, -3, 3)' = (0, -6, 3)'.$$

Therefore

$$\mathbf{s}_2'\mathbf{G}\mathbf{s}_0 = (0, -6, 3)(-12, 12, 24)' = 0$$

and

$$\mathbf{s}_2'\mathbf{G}\mathbf{s}_1 = (0, -6, 3)(18, 0, 0)' = 0$$

which establishes the conjugacy.

Finally we obtain $\alpha_2 = \frac{1}{6}$ which gives $\mathbf{x}_3 = (1, -1, \frac{5}{2})' + \frac{1}{6}(0, -6, 3)' = (1, -2, 3)'$ and $\mathbf{g}_3 = (0, 0, 0)'$ showing that $\mathbf{x}_3$ is indeed the minimum required.

There are many other gradient methods which generate systems of conjugate directions for quadratic functions. Most notable amongst these are the *quasi-Newton methods*. These owe their derivation to the twin objectives of obtaining conjugate directions (and the resulting finite termination property) for quadratic functions and of trying to imitate the exceptional power of Newton's method in the vicinity of the minimum where a quadratic approximation to the function is expected to be very accurate.

Newton's method itself uses the iteration

$$\mathbf{x}_{i+1} = \mathbf{x}_i - \mathbf{J}(\mathbf{x}_i)^{-1}\mathbf{g}_i \tag{11.3.4}$$

where $\mathbf{J}(\mathbf{x}_i)$ is the Hessian matrix of second derivatives of f [see IV, § 5.7], so that it is just the Newton–Raphson technique for solving the system of equations

$$\mathbf{g}(\mathbf{x}) = 0$$

[see § 5.4.1]. This method is unreliable; it will fail if for example $\mathbf{J}(\mathbf{x}_i)$ is singular [see I, Theorem 6.4.2] or if $\mathbf{J}(\mathbf{x}_i)^{-1}\mathbf{g}_i$ and $\mathbf{g}_i$ are orthogonal [see I, Definition 10.2.1]. Many safeguards including the use of a line search along $\mathbf{x}_i + \alpha\mathbf{s}_i$ where $\mathbf{s}_i = -\mathbf{J}(\mathbf{x}_i)^{-1}\mathbf{g}_i$ and the use of the steepest descent direction if $\mathbf{J}(\mathbf{x}_i)$ is singular can be incorporated into Newton's method. Some of these modified Newton methods are described in some detail by Wolfe (1978).

The strategy of the quasi-Newton methods is that a sequence of positive definite symmetric matrices $\mathbf{H}_i$ [see I, § 6.7(v)] is generated and these are used to define the search directions by

$$\mathbf{s}_i = -\mathbf{H}_i\mathbf{g}_i. \tag{11.3.5}$$

The idea behind this is that the $\mathbf{H}_i$'s are generated so that the search directions are similar to steepest descent in the early stages and to the Newton direction in the final stages. Thus the $\mathbf{H}_i$ can be regarded as approximations to $\mathbf{J}(\mathbf{x}_i)^{-1}$ which we hope will improve as i increases.

Now if we use

$$\boldsymbol{\delta}_i = \mathbf{x}_{i+1} - \mathbf{x}_i \quad \text{and} \quad \boldsymbol{\gamma}_i = \mathbf{g}_{i+1} - \mathbf{g}_i \tag{11.3.6}$$

to denote the changes in position and gradient on the ith iteration then if the objective function is the quadratic given by (11.3.2) we have

$$\boldsymbol{\gamma}_i = \mathbf{G}\boldsymbol{\delta}_i.$$

In this case the Hessian matrix is $\mathbf{G}$ and so if $\mathbf{H}_{i+1}$ is to be an approximation to $\mathbf{G}^{-1}$ it is natural to ask that $\mathbf{H}_{i+1}$ satisfy the quasi-Newton equation

$$\mathbf{H}_{i+1}\boldsymbol{\gamma}_i = \boldsymbol{\delta}_i. \tag{11.3.7}$$

We also want $\mathbf{H}_{i+1}$ to be positive definite symmetric and to be obtained in a simple way from the information already available. The other objective of the quasi-Newton methods mentioned above is that they should generate conjugate directions when applied to a quadratic objective function.

If the line searches are all performed exactly then all these requirements are satisfied, for a quadratic function, if the $\mathbf{H}_i$'s are generated by the famous *Davidon–Fletcher–Powell* (DFP) *formula* (Davidon (1959), Fletcher and Powell (1963)):

$$\mathbf{H}_{i+1} = \mathbf{H}_i - \frac{\mathbf{H}_i\boldsymbol{\gamma}_i(\mathbf{H}_i\boldsymbol{\gamma}_i)'}{\boldsymbol{\gamma}_i'\mathbf{H}_i\boldsymbol{\gamma}_i} + \frac{\boldsymbol{\delta}_i\boldsymbol{\delta}_i'}{\boldsymbol{\delta}_i'\boldsymbol{\gamma}_i}. \tag{11.3.8}$$

We summarise this in the following theorem.

THEOREM 11.3.1.   *Let the objective function f be given by (11.3.2) and let $\mathbf{H}_0$ be a positive definite symmetric matrix [see I, § 6.7(v)]. Let $\mathbf{s}_i = -\mathbf{H}_i\mathbf{g}_i$, and the $\mathbf{H}_i$'s be generated by (11.3.8). If the line searches are performed exactly then*

(i)   *$\mathbf{H}_i$ is positive definite symmetric for each i,*
(ii) *the search directions are $\mathbf{G}$-conjugate:*

$$\mathbf{s}_i'\mathbf{G}\mathbf{s}_j = 0 \qquad (i \neq j),$$

and

(iii) *if $\mathbf{g}_i \neq \mathbf{0}$ $(i = 0, 1, \ldots, n-1)$ then $\mathbf{g}_n = \mathbf{0}$, $\mathbf{x}_n$ is the minimum point of f and $\mathbf{H}_n = \mathbf{G}^{-1}$. (If $\mathbf{g}_i = \mathbf{0}$ for some $i < n$ then of course $\mathbf{x}_i$ is the minimum point.)*

The DFP formula is just one from a large class of quasi-Newton formula discovered by Broyden (1967). The general formula is given in Fletcher's (1970) parameterization by

$$H_{i+1} = H_i - \frac{H_i\gamma_i(H_i\gamma_i)'}{\gamma_i'H_i\gamma_i} + \frac{\delta_i\delta_i'}{\delta_i'\gamma_i} + \phi v_i v_i' \tag{11.3.9}$$

where

$$v_i = (\gamma_i'H_i\gamma_i)^{1/2}\left(\frac{\delta_i}{\delta_i'\gamma_i} - \frac{H_i\gamma_i}{\gamma_i'H_i\gamma_i}\right)$$

and $\phi \geq 0$ is the free parameter.

Provided that $\delta_i'\gamma_i > 0$ it can be shown that if H_i is positive definite symmetric then so is any H_{i+1} given by (11.3.9) even if the objective function is not quadratic and that Theorem 11.3.1 holds for all such formulae.

A great deal of research, both analytic and computational, has been devoted to the attempt to find the 'best' formula in this class and indeed in a much larger three parameter family of variable metric formulae developed by Huang (1970). The particular formula which has the widest support so far is the BFGS formula due to Broyden (1970), Fletcher (1970), Goldfarb (1970) and Shanno (1970). This is obtained by putting $\phi = 1$ in (11.3.9) which then reduces to

$$H_{i+1} = H_i - \frac{H_i\gamma_i\delta_i'}{\delta_i'\gamma_i} - \frac{\delta_i(H_i\gamma_i)'}{\delta_i'\gamma_i} + \left\{1 + \frac{\gamma_i'H_i\gamma_i}{\delta_i'\gamma_i}\right\}\frac{\delta_i\delta_i'}{\delta_i'\gamma_i}. \tag{11.3.10}$$

If all the line searches are performed exactly then from the same starting point and same H_0 all the quasi-Newton methods using formulae from (11.3.9) with $\delta_i'\gamma_i > 0$ generate the same sequence of points. This remarkable result was proved by Dixon (1972). It does however depend critically on the line searches being exact—something which cannot of course be achieved in practice. We demonstrate the fact for a quadratic function where we can minimize exactly.

EXAMPLE 11.3.4. Minimize the function of Example 11.3.3 with $x_0 = (0, 0, 0)'$ and $H_0 = I$ by (a) the DFP method and (b) the BFGS method.

(a) DFP	(b) BFGS
As before: $g_0 = (0, 0, -6)'$	$g_0 = (0, 0, -6)'$
$s_0 = (0, 0, 6)', \alpha_0 = \frac{1}{4}$	$s_0 = (0, 0, 6)', \qquad \alpha_0 = \frac{1}{4}$
$x_1 = (0, 0, \frac{3}{2})', \qquad g_1 = (-3, 3, 0)'$	$x_1 = (0, 0, \frac{3}{2})', \qquad g_1 = (-3, 3, 0)'$
$\delta_0 = (0, 0, \frac{3}{2})', \qquad \gamma_0 = (-3, 3, 6)'$	$\delta_0 = (0, 0, \frac{3}{2})', \qquad \gamma_0 = (-3, 3, 6)'$
$H_0 = I$ so $H_0\gamma_0 = \gamma_0$	$H_0 = I$ so $H_0\gamma_0 = \gamma_0$
$\gamma_0'H_0\gamma_0 = 54, \qquad \delta_0'\gamma_0 = 9$	$\gamma_0'H_0\gamma_0' = 54, \qquad \delta_0'\gamma_0 = 9$

DFP (*cont.*)

$$\boldsymbol{\delta}_0\boldsymbol{\delta}_0' = \begin{pmatrix} 0 & 0 & 0 \\ 0 & 0 & 0 \\ 0 & 0 & \frac{9}{4} \end{pmatrix}$$

$$\mathbf{H}_0\boldsymbol{\gamma}_0(\mathbf{H}_0\boldsymbol{\gamma}_0)' = \begin{pmatrix} 9 & -9 & -18 \\ -9 & 9 & 18 \\ -18 & 18 & 36 \end{pmatrix}$$

$$\mathbf{H}_1 = \begin{pmatrix} \frac{5}{6} & \frac{1}{6} & \frac{1}{3} \\ \frac{1}{6} & \frac{5}{6} & -\frac{1}{3} \\ \frac{1}{3} & -\frac{1}{3} & \frac{7}{12} \end{pmatrix}$$

$$\mathbf{s}_1 = -\mathbf{H}_1\mathbf{g}_1 = (2, -2, 2)'$$

$$\alpha_1 = \tfrac{1}{2}$$

$$\mathbf{x}_2 = (1, -1, \tfrac{5}{2})', \qquad \mathbf{g}_2 = (3, 3, 0)'$$

$$\boldsymbol{\delta}_1 = (1, -1, 1)', \qquad \boldsymbol{\gamma}_1 = (6, 0, 0)'$$

Hence we get

$$\mathbf{H}_2 = \begin{pmatrix} \frac{1}{6} & -\frac{1}{6} & \frac{1}{6} \\ -\frac{1}{6} & \frac{29}{30} & -\frac{17}{30} \\ \frac{1}{6} & -\frac{17}{30} & \frac{37}{60} \end{pmatrix}$$

giving $\mathbf{s}_2 = (0, -\frac{12}{5}, \frac{6}{5})'$,

$$\alpha_2 = \tfrac{5}{12} \text{ and } \mathbf{x}_3 = (1, -2, 3)'$$

BFGS (*cont.*)

$$\boldsymbol{\delta}_0\boldsymbol{\delta}_0' = \begin{pmatrix} 0 & 0 & 0 \\ 0 & 0 & 0 \\ 0 & 0 & \frac{9}{4} \end{pmatrix}$$

$$\boldsymbol{\delta}_0(\mathbf{H}_0\boldsymbol{\gamma}_0)' = \begin{pmatrix} 0 & 0 & 0 \\ 0 & 0 & 0 \\ -\frac{9}{2} & \frac{9}{2} & 9 \end{pmatrix}$$

$$\mathbf{H}_0\boldsymbol{\gamma}_0\boldsymbol{\delta}_0' = \begin{pmatrix} 0 & 0 & -\frac{9}{2} \\ 0 & 0 & \frac{9}{2} \\ 0 & 0 & 9 \end{pmatrix}$$

$$\mathbf{H}_1 = \begin{pmatrix} 1 & 0 & \frac{1}{2} \\ 0 & 1 & -\frac{1}{2} \\ \frac{1}{2} & -\frac{1}{2} & \frac{3}{4} \end{pmatrix}$$

$$\mathbf{s}_1 = -\mathbf{H}_1\mathbf{g}_1 = (3, -3, 3)'$$

$$\alpha_1 = \tfrac{1}{3}$$

$$\mathbf{x}_2 = (1, -1, \tfrac{5}{2})', \qquad \mathbf{g}_2 = (3, 3, 0)'$$

$$\boldsymbol{\delta}_1 = (1, -1, 1)', \qquad \boldsymbol{\gamma}_1 = (6, 0, 0)'$$

Hence we get

$$\mathbf{H}_2 = \begin{pmatrix} \frac{1}{6} & -\frac{1}{6} & \frac{1}{6} \\ -\frac{1}{6} & \frac{13}{6} & -\frac{7}{6} \\ \frac{1}{6} & -\frac{7}{6} & \frac{11}{12} \end{pmatrix}$$

giving $\mathbf{s}_2 = (0, -6, 3)'$

$$\alpha_2 = \tfrac{1}{6} \quad \text{and} \quad \mathbf{x}_3 = (1, -2, 3)$$

Thus we see that although the sequences of $\mathbf{H}_i$'s are not the same, the $\mathbf{s}_i$'s are parallel and the same points $\mathbf{x}_1, \mathbf{x}_2, \mathbf{x}_3$ are generated by the two methods.

We have already observed that the quasi-Newton algorithms with exact line searches generate the exact minimum of a positive definite quadratic function in a finite number of iterations. This result is useful in a more general context since we expect that in the vicinity of its minimum a function will usually behave much like a quadratic. Much more important in the context of general functions, however, are the results establishing the convergence of quasi-Newton methods.

Powell (1971) establishes the superlinear convergence of quasi-Newton methods with exact line searches for a function whose Hessian matrix is positive definite everywhere. Much more work recently has been devoted to the question of convergence of the methods when the line searches are dropped in favour of obtaining a significant decrease in the value of the objective function. Under certain conditions on the second derivatives including the positive definiteness of the Hessian at the solution Broyden, Dennis and Moré (1973) established superlinear convergence when each α_i is taken as unity. More recently Powell (1976) established convergence of the BFGS algorithm without line searches for arbitrary convex functions [see IV, § 15.2.6] and that this convergence is superlinear if the Hessian matrix at the solution is positive definite.

Many different suggestions have been made for the method of choosing the steplength parameter α_i. The requirement that $f(\mathbf{x}_{i+1}) < f(\mathbf{x}_i)$ is not sufficient on its own. Most use the fact that

$$\frac{f(\mathbf{x}_{i+1}) - f(\mathbf{x}_i)}{\delta_i' \mathbf{g}_i} \to 1 \quad \text{as } \alpha_i \to 0$$

to ensure that the steplength is not too small and so prevent premature convergence. Similarly the decrease in the value of the function should be sufficient to ensure that the line minimum has not been too wildly over-estimated. Fletcher's (1970) original suggestion is that for some small positive μ (10^{-4} is the value he used) α_i is chosen so that

$$\mu \le \frac{f(\mathbf{x}_{i+1}) - f(\mathbf{x}_i)}{\delta_i' \mathbf{g}_i} \le 1 - \mu.$$

It is also desirable to insist that

$$\delta_i' \boldsymbol{\gamma}_i > 0$$

since this guarantees the positive definiteness of $\mathbf{H}_{i+1}$.

Powell imposes a similar left-hand condition

$$f(\mathbf{x}_{i+1}) \le f(\mathbf{x}_i) + c_1 \delta_i' \mathbf{g}_i$$

but replaces the right-hand half with the condition

$$\delta_i' \mathbf{g}_{i+1} \ge c_2 \delta_i' \mathbf{g}_i$$

where $0 < c_1 < c_2 < 1$ and $c_1 < \frac{1}{2}$. Powell suggests that $c_1 = 10^{-4}$ and $c_2 = \frac{1}{2}$ are suitable values.

These conditions both satisfy the requirements of his convergence results and permit a suitable value of α_i to be found by a simple bracketing procedure. If a particular trial value of α_i does not satisfy the first condition then it can be used as the upper end of the bracket whereas if it satisfies the first but not the second condition it is a suitable lower bound.

In all the various strategies for determining a suitable value it is suggested that $\alpha_i = 1$ is a sensible first trial value. Indeed in the later stages of the process it is anticipated that this value will satisfy the conditions imposed.

The final point in this section is the selection of the initial matrix $\mathbf{H}_0$. The only requirement imposed by the theory is that $\mathbf{H}_0$ should be positive definite [see I, § 6.7(v)]. Consequently the most common choice is to set $\mathbf{H}_0 = \mathbf{I}$ and this can usually be expected to perform well.

P.R.T.

REFERENCES

Broyden, C. G. (1967). Quasi-Newton Methods and their Application to Function Minimization, *Maths. Comp.* **21**, 368–381.

Broyden, C. G. (1970). The Convergence of a Class of Double-Rank Minimization Algorithms, *J.I.M.A.* **6**, 222–231.

Broyden, C. G., Dennis, J. E. and Moré, J. J. (1973). On the Local and Superlinear Convergence of Quasi-Newton Methods, *J.I.M.A.* **12**, 223–245.

Davidon, W. C. (1959). Variable Metric Methods of Minimization, *Argonne Nat. Lab. Report* ANL-5990.

Dixon, L. C. W. (1972). Variable Metric Algorithms: Necessary and Sufficient Conditions for Identical Behaviour on Non-Quadratic Functions, *J. Opt. Theory and Applic.* **10**, 34–40.

Fletcher, R. (1970). A New Approach to Variable Metric Algorithms, *Comp. J.* **13**, 317–322.

Fletcher, R. and Powell, M. J. D. (1963). A Rapidly Convergent Descent Method for Minimization. *Comp. J.* **6**, 163–168.

Fletcher, R. and Reeves, C. M. (1964). Function Minimization by Conjugate Gradients, *Comp. J.* **7**, 149–154.

Goldfarb, D. (1970). A Family of Variable Metric Methods Derived by Variational Means, *Maths. Comp.* **24**, 23–26.

Huang, H. Y. (1970). Unified Approach to Quadratically Convergent Algorithms for Function Minimization, *J.O.T.A.* **5**, 405–423.

Nelder, J. A. and Mead, R. (1965). A Simplex Method for Function Minimization, *Comp. J.* **7**, 308–313.

Powell, M. J. D. (1964). An Efficient Method for Finding the Minimum of a Function of Several Variables Without Calculating Derivatives, *Comp. J.* **7**, 155–162.

Powell, M. J. D. (1971). On the Convergence of the Variable Metric Algorithm, *J.I.M.A.* **7**, 21–36.

Powell, M. J. D. (1976). Some Glogal Convergence Properties of a Variable Metric Algorithm for Minimization without Exact Line Searches, *SIAM-AMS Proc.* **9**, 53–72.

Shanno, D. F. (1970). Conditioning of Quasi-Newton Methods for Function Minimization, *Maths. Comp.* **24**, 647–656.

Walsh, G. R. (1975). *Methods of Optimization*, John Wiley and Sons Ltd. (London).

Wolfe, M. A. (1978). *Numerical methods for unconstrained optimization*, Van Nostrand Reinhold Company (London).

Appendix of Fortran Programs

LIST OF PROGRAMS

1. Determination of interpolated values of $f(x)$ using a table of Newton's divided differences.
2. Determination of interpolated values of $f(x)$ using Everett's method with a table of central differences.
3. Solution of linear equations using the Gauss–Seidel method.
4. Solution of linear equations using the Gauss–Jordan method.
5. To find the largest eigenvalue and associated eigenvector of a real matrix using the power method.
6. Determination of an accurate root for an equation using the Newton–Raphson method.
7. Determination of an accurate root for an equation using either regula–falsi or the secant method.
8. Determination of an accurate root for an equation using the method of bisection.
9. To find the best straight line through a series of points using the technique of least squares.
10. To find the best parabola through a series of points using the technique of least squares.
11. Integration using the trapezium rule followed by Romberg integration (double precision).
12. Integration using Gaussian quadrature and Robinson's method.
13. Evaluation of a multiple integral over an n-dimensional sphere.
14. To solve a first-order differential equation using the Runge–Kutta method and a fourth-order formula.
15. To solve a second-order differential equation using the Runge–Kutta method and a fourth-order formula and a substitution.
16. To solve the Volterra integral equation of the second kind using a multi-step method.
17. To solve the Fredholm integral equation.

NUMERICAL PROGRAMS—INTRODUCTION

Some of the calculations and computations discussed in this book have been programmed in the Fortran IV programming language. The following programs refer to the methods of computation described in the preceding text. Each of the programs adheres to the method of solution indicated in the text as far as is possible.

All the programs have been run on an ICL System 4-70 computer and the data and results shown refer to these runs. The System 4-70 computer usually stores real numbers as 32-bit binary numbers leading to an accuracy of about seven decimal figures. It is possible to obtain greater accuracy and one of the programs shows how this can be achieved.

The detailed description of each program in the suite has a common structure. There is a brief introduction to the program followed first by a listing of the program and then by sample data and results. Each program is well annotated so that if any of the arrays need to be extended or any of the routines need to be modified the changes can be easily incorporated.

The programs are written in a structured form using subprograms. There are occasions when the algorithms can be programmed as one subroutine but generally more than one subprogram is needed.

All the programs contain a small main program whose purpose is to read the data, print the results and call the various routines. The main program binds the routines together making a complete program. It also serves as a simple example of the main program which the reader could use when running the program himself.

The way that the major subroutines are to be called has been demonstrated in the programs and also described in the documentation.

Input parameters must have a value assigned to them when the subroutine is called. Output parameters are given a value in the subroutine. Sometimes a parameter is an input–output parameter and this means that it must have a value when the subroutine is called but the value is changed in the subroutine.

The layout of the data and the results can be varied to suit the user by changing the format statements. Care should be taken in preparing the data to ensure that it matches the corresponding format. To avoid unnecessary duplication in the detailed description of the data for each program the actual formats will be given. As a guide to the format conventions in Fortran IV.

In   e.g.   *I*5,

indicates an integer occupying *n* columns, in the example *n*, an integer itself is 5.

Fn.0   e.g.   *F*10.0,

indicates a real number being read and occupying *n* columns, in the example *n*, an integer, is 10.

Fn.d   e.g.   *F*10.4,

indicates a real number being printed and occupying n columns of which d are decimal places. In the example n is 10 and d is 4.

mFn.0, mIn   e.g. $3F10.0, 4I5,$

indicates that there are m real numbers or m integers to be read. m is itself an integer and in the example there are 3 real numbers and 4 integers, on the line of the data. When being read real numbers must have a decimal point. All numerical information must be right-aligned, it must be positioned so that the least significant digit lies in the extreme right hand column, i.e. column n.

PROGRAM 1

Determination of Interpolated Values of $f(x)$ using a table of Newton's divided differences.

Program

The program consists of a main program and three subroutines. The first subroutine NDD is the main controlling subroutine and it calls the other two routines. The second routine NEWTON creates the table of differences and the third routine INTERP finds the values of $f(x)$ for a series of values of x by interpolation. There is no need to express the polynomial approximation to the function in its simplest form.

Subroutine Call

CALL NDD $(X, TABLE, N, NUM, M, A, B)$,

X is a one-dimensional array with N entries of X to form the table,
$TABLE$ is a two-dimensional array with N entries and NUM columns,
The N values of $f(x)$ are placed in column 1 of the $TABLE$,
N is the number of values of X and $f(x)$,
NUM is the number of differences required $+ 1$,
M is the number of values of x requiring interpolated values of $f(x)$ to be determined,
A is a one-dimensional array holding the x values,
B is a one-dimensional array holding the values of $f(x)$,
X, N, NUM, M and A are input parameters, B is an output parameter and $TABLE$ is an input and output parameter.

Data

N, M, NUM with a format $3I5$,
$X(I)$ with a format $6F10.0$ the N values of X written across the line, 6 to a line,
$TABLE(I, 1)$ with a format $6F10.0$,
The N values of $f(x)$ are written across the line as above,
$A(I)$ with a format $6F10.0$,
The M values of X for which interpolated values are required.

Results

The annotated results give x and the table of differences. The values of x and the interpolated values are printed below.

Program 1. Determination of interpolated values of $f(x)$ using a table of Newton's divided differences.

```
00000100 C    THIS PROGRAM FINDS THE VALUES OF F(X) FOR A
00000200 C    SERIES OF VALUES OF X USING INTERPOLATION
00000300 C    BASED UPON NEWTON'S DIVIDED DIFFERENCES .
00000400 C    THE FIRST STEP IN THE EVALUATION IS TO PRODUCE
00000500 C    A TABLE OF DIVIDED DIFFERENCES . FROM THE
00000600 C    TABLE THE VALUE OF F(X) FOR A GIVEN X CAN BE
00000700 C    EVALUATED USING THE STANDARD FORMULA FOR F(X)
00000800 C    IN TERMS OF THE DIFFERENCES . IT IS NOT
00000900 C    NECESSARY TO EXPRESS THE POLYNOMIAL F(X) IN
00001000 C    ITS SIMPLEST FORM .
00001100 C    THE DATA CONSISTS OF N, THE NUMBER OF ENTRIES
00001200 C    IN THE TABLE ,M,THE NUMBER OF VALUES OF X FOR
00001300 C    WHICH INTERPOLATED VALUES OF X ARE REQUIRED
00001400 C    AND NUM ,THE NUMBER OF DIFFERENCES REQUIRED
00001500 C    PLUS ONE ,NUM IS THE NUMBER OF COLUMNS IN THE
00001600 C    DIFFERENCE TABLE INCLUDING THE VALUES OF F(X).
00001700 C    THE N VALUES OF X ARE INPUT ACROSS THE LINE
00001800 C    AND THEN THE N VALUES OF F(X) ARE INPUT IN
00001900 C    LIKEWISE FASHION .THE OUTPUT CONSISTS OF THE
00002000 C    FULL TABLE OF DIFFERENCES FOLLOWED BY THE M
00002100 C    VALUES OF X AND THE INTERPOLATED VALUES OF
00002200 C    F(X). THERE IS A MAIN CALLING PROGRAM WHICH
00002300 C    READS THE DATA,PRINTS THE RESULTS   AND CALLS
00002400 C    THE SUBROUTINES .THERE ARE THREE SUBROUTINES .
00002500 C    THE FIRST CALLED NDD IS THE CONTROL SUBROUTINE
00002600 C    AND ITS PURPOSE IS TO CALL THE OTHER TWO
00002700 C    SUBROUTINES NEWTON AND INTERP . THE SECOND
00002800 C    CALLED NEWTON PRODUCES THE TABLE OF DIFFERENCES
00002900 C    AND THE THIRD CALLED INTERP FINDS THE
00003000 C    INTERPOLATED VALUES OF F(X) FROM THE M VALUES
00003100 C    OF X AND THE TABLE OF DIFFERENCES .
00003200 C    IT IS ASSUMED THAT M AND N ARE NOT GREATER
00003300 C    THAN 20 ,IF THEY ARE REQUIRED TO BE BIGGER
00003400 C    THE ARRAY DECLARATIONS SHOULD BE ALTERED .
00003500        REAL X(20),TABLE(20,20),A(20),B(20)
00003600 C    INPUT THE VARIOUS COUNTS ,
00003700 C     N-THE NUMBER OF ENTRIES IN THE TABLE
00003800 C     M-THE NUMBER OF INTERPOLATED VALUES
00003900 C    NUM-THE NUMBER OF COLUMNS IN THE TABLE
00004000        PRINT 1
00004100 1    FORMAT(' A PROGRAM TO PERFORM INTERPOLATION')
00004200        PRINT 2
00004300 2    FORMAT(' USING NEWTONS METHOD OF DIVIDED',
00004400       1' DIFFERENCES',/)
00004500        READ 10,N,M,NUM
00004600 10   FORMAT(3I5)
00004700 C    INPUT THE X VALUES
00004800        READ 20,(X(I),I=1,N)
00004900 20   FORMAT(6F10.0)
00005000 C    INPUT THE VALUES OF F(X)
00005100        READ 20,(TABLE(I,1),I=1,N)
00005200 C    READ THE M VALUES OF X STORED AS A(I)
00005300        READ 20,(A(I),I=1,M)
00005400 C    CALL THE CONTROL ROUTINE.
00005500        CALL NDD(X,TABLE,N,NUM,M,A,B)
00005600        STOP
00005700        END
00005800 C    THE CONTROL SUBROUTINE WHICH CALLS
00005900 C    NEWTON AND INTERP
00006000        SUBROUTINE NDD(X,TABLE,N,NUM,M,A,B)
```

```
00006100          REAL X(20),TABLE(20,20),A(20),B(20)
00006200 C        CALL THE NEWTON ROUTINE TO PRODUCE
00006300 C        THE TABLE OF DIFFERENCES .
00006400          CALL NEWTON(X,TABLE,N,NUM)
00006500          PRINT 22
00006600 22       FORMAT(1H ,9X,'X',6X,'F(X)',5X,'DIFFERENCES',//)
00006700          DO 30 I=1,N
00006800          N1=N+1-I
00006900          IF(N1.GT.NUM) N1=NUM
00007000 30       PRINT 35,X(I),(TABLE(I,J),J=1,N1)
00007100 C        FORMAT CAN BE CHANGED TO ALLOW GREATER
00007200 C        ACCURACY OR MORE ENTRIES PER LINE .
00007300 35       FORMAT(1H ,8F10.3)
00007400 C        FIND THE INTERPOLATED VALUES USING INTERP
00007500          CALL INTERP(A,B,M,X,TABLE,N,NUM)
00007600 C        PRINT THE RESULTS
00007700          PRINT 32
00007800 32       FORMAT(//,' THE INTERPOLATED VALUES',//)
00007900          PRINT 33
00008000 33       FORMAT(1H ,9X,'X',6X,'F(X)',//)
00008100          DO 40 I=1,M
00008200 40       PRINT 35,A(I),B(I)
00008300          RETURN
00008400          END
00008500 C        THIS IS THE MAIN SUBROUTINE TO FORM THE TABLE
00008600 C        AND VALUES .
00008700          SUBROUTINE NEWTON(X,TABLE,N,NUM)
00008800          REAL X(20),TABLE(20,20)
00008900 C        THE TABLE OF DIFFERENCES IS STORED AS A
00009000 C        2-DIMENSIONAL ARRAY AND THE NUMBER OF ENTRIES
00009100 C        IN EACH ROW IS CONTROLLED BY N1 WHICH IS A
00009200 C        FUNCTION OF THE COLUMN J
00009300          DO 10 J=2,NUM
00009400          N1=N+1-J
00009500          DO 10 I=1,N1
00009600          IJ=I+J-1
00009700 10       TABLE(I,J)=(TABLE(I+1,J-1)-TABLE(I,J-1))
00009800         1/(X(IJ)-X(I))
00009900          RETURN
00010000          END
00010100 C        THE INTERPOLATION ROUTINE
00010200          SUBROUTINE INTERP(A,B,M,X,TABLE,N,NUM)
00010300          REAL A(20),B(20),X(20),TABLE(20,20)
00010400          DO 10 I=1,M
00010500 C        THE INTERPOLATED VALUES ARE STORED IN B(I)
00010600          B(I)=TABLE(1,1)
00010700          PROD =1.0
00010800          DO 20 J=2,NUM
00010900          PROD =PROD*(A(I)-X(J-1))
00011000 20       B(I)=B(I)+PROD*TABLE(1,J)
00011100 10       CONTINUE
00011200          RETURN
00011300          END
```

INPUT FOR PROGRAM 1

```
7    4    5
   -2.0      -1.0       1.0       2.0       4.0       5.0
    6.0
   12.0       1.0       3.0      16.0     246.0     607.0
 1268.0
    0.0       3.0       0.5      -0.5
```

OUTPUT FROM PROGRAM 1

A PROGRAM TO PERFORM INTERPOLATION
USING NEWTONS METHOD OF DIVIDED DIFFERENCES

```
        X       F(X)        DIFFERENCES

   -2.000      12.000     -11.000      4.000      0.000      1.000
   -1.000       1.000       1.000      4.000      6.000      1.000
    1.000       3.000      13.000     34.000     12.000      1.000
    2.000      16.000     115.000     82.000     17.000
    4.000     246.000     361.000    150.000
    5.000     607.000     661.000
    6.000    1268.000
```

THE INTERPOLATED VALUES

```
        X       F(X)

    0.000       2.000
    3.000      77.000
    0.500       2.313
   -0.500       1.313
```

**FORTRAN ** STOP

PROGRAM 2

Determination of Interpolated values of $f(x)$ using Everett's method with a table of central differences.

Program

The program consists of a main program and three subroutines. The first subroutine EVAL is the main controlling subroutine and it calls the other two routines. The third routine EVERET creates the table of central differences and the second routine INTERP forms the values of $f(x)$ for a series of values of x by interpolation using the standard formula

$$f(x_n + ph) = qf_n + \frac{(q+1)(q-1)\delta^2}{3!} f_n + \frac{(q+2)(q+1)q(q-1)(q-2)\delta^4}{5!} f_n + \ldots$$

$$+ pf_{n+1} + \frac{(p+1)p(p-1)\delta^2}{3!} f_{n+1} + \frac{(p+2)(p+1)p(p-1)(p-2)\delta^4}{5!} f_{n+1} + \ldots$$

where $0 < p < 1$, $q = 1 - p$ and $f_n = f(x_n)$, $\delta^2 f_n = f_{n+1} - 2f_n + f_{n-1}$ etc.

Subroutine Call

CALL EVAL $(A, X, Y, TABLE, N, M, NUM, H, VALUE)$,

A is a one-dimensional array containing M values of X for which interpolated values are required,
X is a one-dimensional array containing the N values of X to form the table,
Y is a one-dimensional array containing the N values of $f(x)$ to form the table,
$TABLE$ is a two-dimensional array containing the table of central differences N rows NUM columns,
N is number of entries,
M is number of values of X requiring interpolation,
NUM is number of differences $+ 1$,
H is the interval in values of X,
$VALUE$ is a one-dimensional array holding the interpolated values,
A, X, Y, N, M, NUM, H are input parameters and $TABLE$ and $VALUE$ are output parameters.

Data

N, M, NUM, H with format $3I5, F10.0$,
$X(I)$ with a format $6F10.0$, N entries across the line,
$Y(I)$ with a format $6F10.0$, N entries across the line,
$A(I)$ with a format $6F10.0$, M entries across the line.

Results

The annotated results give x and the table of central differences followed by the values of x and the interpolated values of $f(x)$.

Program 2. Determination of interpolated values of $f(x)$ using Everett's method with a
table of central differences.

PROGRAM 2

```
00000100 C    THIS PROGRAM FINDS THE VALUE OF F(X) FOR A
00000200 C    SERIES OF VALUES OF X USING INTERPOLATION
00000300 C    BASED UPON EVERETT'S METHOD .
00000400 C    THE FIRST STEP IN THE EVALUATION IS TO PRODUCE
00000500 C    A TABLE OF CENTRAL DIFFERENCES . FROM THE
00000600 C    TABLE THE VALUE OF F(X) FOR A GIVEN X CAN BE
00000700 C    EVALUATED USING THE STANDARD FORMULA FOR F(X)
00000800 C    IN TERMS OF THE DIFFERENCES.IT IS NOT NECESSARY
00000900 C    TO EXPRESS THE POLYNOMIAL IN ITS SIMPLEST FORM .
00001000 C    THE DATA CONSISTS OF N,THE NUMBER OF ENTRIES
00001100 C    IN THE TABLE,M, THE NUMBER OF VALUES OF X FOR
00001200 C    WHICH INTERPOLATED VALUES OF X ARE REQUIRED
00001300 C    AND NUM, THE NUMBER OF DIFFERENCES REQUIRED
00001400 C    PLUS 1. NUM IS THE NUMBER OF COLUMNS IN THE
00001500 C    DIFFERENCE TABLE INCLUDING THE VALUES OF F(X)
00001600 C    THE N VALUES OF X ARE INPUT ACROSS THE LINE
00001700 C    AND THEN THE N VALUES OF F(X) ARE INPUT IN
00001800 C    LIKEWISE FASHION .
00001900 C    THE OUTPUT CONSISTS OF THE FULL TABLE OF
00002000 C    DIFFERENCES FOLLOWED BY THE M VALUES OF X AND
00002100 C    THE INTERPOLATED VALUES OF F(X).
00002200 C    THERE IS A MAIN CALLING PROGRAM WHICH READS
00002300 C    THE DATA,PRINTS THE RESULTS AND CALLS THE
00002400 C    SUBROUTINES. THERE ARE THREE SUBROUTINES .
00002500 C    THE FIRST CALLED EVAL IS THE CONTROL ROUTINE
00002600 C    AND ITS PURPOSE IS TO CALL THE OTHER TWO
00002700 C    SUBROUTINES EVERET AND INTERP. THE SECOND
00002800 C    CALLED EVERET PRODUCES THE TABLE OF DIFFERENCES
00002900 C    AND THE THIRD CALLED INTERP FINDS THE
00003000 C    INTERPOLATED VALUES OF F(X) FROM THE M VALUES
00003100 C    OF X AND THE TABLE OF DIFFERENCES .
00003200 C    IT IS ASSUMED THAT M AND N ARE NOT GREATER
00003300 C    THAN 20 , IF THEY ARE REQUIRED TO BE BIGGER
00003400 C    THE ARRAY DECLARATIONS SHOULD BE ALTERED .
00003500      REAL A(20),X(20),Y(20),TABLE(20,6),VALUE(20)
00003600      PRINT 1
00003700 1    FORMAT(' A PROGRAM TO PERFORM INTERPOLATION')
00003800      PRINT 2
00003900 2    FORMAT(' USING EVERETTS METHOD OF CENTRAL',
00004000     1' DIFFERENCES')
00004100 C    INPUT THE VARIOUS COUNTS
00004200 C    N   THE NUMBER OF ENTRIES IN TH TABLE
00004300 C    M   THE NUMBER OF INTERPOLATED VALUES
00004400 C    NUM   THE NUMBER OF DIFFERENCES +1
00004500 C    AND H THE STEP BETWEEN THE VALUES OF X
00004600      READ 10,N,M,NUM,H
00004700 10   FORMAT(3I5,F10.0)
00004800 C    INPUT THE VALUES OF X
00004900      READ 20,(X(I),I=1,N)
00005000 C    INPUT THE VALUES OF THE FUNCTION
00005100      READ 20,(Y(I),I=1,N)
00005200 20   FORMAT(6F10.0)
00005300 C    READ THE M VALUES OF X STORED AS A(I)
00005400      READ 20,(A(I),I=1,M)
00005500 C    CALL THE CONTROL SUBROUTINE
00005600      CALL EVAL(A,X,Y,TABLE,N,M,NUM,H,VALUE)
00005700      STOP
00005800      END
00005900 C    THE CONTROL SUBROUTINE FOR EVERETT'S METHOD
00006000      SUBROUTINE EVAL(A,X,Y,TABLE,N,M,NUM,H,VALUE)
```

```
00006100          REAL A(20),X(20),Y(20),TABLE(20,6),VALUE(20)
00006200 C        CALL THE ROUTINE TO FORM THE CENTRAL DIFFERENCES
00006300          CALL EVERET(Y,TABLE,N,NUM)
00006400          NJ=0
00006500          INC=1
00006600 C        THE PRINT SECTION
00006700 C        INC CONTROLS THE LAYOUT OF THE TABLE
00006800          PRINT 20
00006900 20       FORMAT(1H ,9X,'X',6X,'F(X)',5X,'DIFFERENCES',/)
00007000          DO 30 I=1,N
00007100          NJ=NJ+INC
00007200          IF(NJ.GT.NUM) NJ=NUM
00007300          IF(I.EQ.N-3) INC=-1
00007400 30       PRINT 35,X(I),(TABLE(I,J),J=1,NJ)
00007500 35       FORMAT(1H ,6F10.5)
00007600 C        CALL THE INTERPOLATION ROUTINE
00007700          CALL INTERP(A,X,TABLE,N,M,NUM,H,VALUE)
00007800 C        PRINT THE INTERPOLATED VALUES
00007900          PRINT 37
00008000 37       FORMAT(//,'   THE INTERPOLATED VALUES ',//)
00008100          PRINT 38
00008200 38       FORMAT(1H ,9X,'X',6X,'F(X)',//)
00008300          DO 40 I=1,M
00008400 40       PRINT 45,A(I),VALUE(I)
00008500 45       FORMAT(1H ,2F12.5)
00008600          RETURN
00008700          END
00008800 C        THE INTERPOLATION ROUTINE
00008900          SUBROUTINE INTERP(A,X,TABLE,N,M,NUM,H,VALUE)
00009000          REAL X(20),A(20),TABLE(20,6),VALUE(20)
00009100 C        THE INTERPOLATED VALUES ARE STORED AS VALUE(I)
00009200          DO 10 I=1,M
00009300          K=(A(I)-X(1))/H+1
00009400          P=(A(I)-X(K))/H
00009500          Q=1-P
00009600          PROD1=P
00009700          PROD2=Q
00009800          V=0
00009900          DO 20 J=1,NUM
00010000          V=V+PROD1*TABLE(K+1,J)+PROD2*TABLE(K,J)
00010100          PROD1=PROD1*(P*P-J*J)/((2*J+1)*2*J)
00010200 20       PROD2=PROD2*(Q*Q-J*J)/((2*J+1)*2*J)
00010300 10       VALUE(I)=V
00010400          RETURN
00010500          END
00010600 C        THE ROUTINE TO FIND THE TABLE OF
00010700 C        CENTRAL DIFFERENCES
00010800          SUBROUTINE EVERET(Y,TABLE,N,NUM)
00010900          REAL Y(20),TABLE(20,6)
00011000          DO 10 I=1,N
00011100 10       TABLE(I,1)=Y(I)
00011200          N1=N
00011300          I1=1
00011400 C        THE NUMBER OF ENTRIES IN THE TABLE
00011500 C        CONTROLLED BY N1,I1
00011600          DO 20 J=2,NUM
00011700          N1=N1-1
00011800          I1=I1+1
00011900          DO 20 I=I1,N1
00012000 20       TABLE(I,J)=TABLE(I+1,J-1)+TABLE(I-1,J-1)
00012100          1-2*TABLE(I,J-1)
00012200          RETURN
00012300          END
```

INPUT FOR PROGRAM 2

```
10    6    3        10.0
      0.0       10.0       20.0        30.0       40.0       50.0
     60.0       70.0       80.0        90.0
      0.0     0.17365    0.34202     0.50000    0.64279    0.76604
  0.86603     0.93969    0.98481         1.0
     36.0       45.0       54.0        32.0       48.0       63.0
```

OUTPUT FROM PROGRAM 2

A PROGRAM TO PERFORM INTERPOLATION
USING EVERETTS METHOD OF CENTRAL DIFFERENCES

X	F(X)	DIFFERENCES	
0.00000	0.00000		
10.00000	0.17365	-0.00528	
20.00000	0.34202	-0.01039	0.00031
30.00000	0.50000	-0.01519	0.00045
40.00000	0.64279	-0.01954	0.00063
50.00000	0.76604	-0.02326	0.00065
60.00000	0.86603	-0.02633	0.00086
70.00000	0.93969	-0.02854	
80.00000	0.98481		
90.00000	1.00000		

THE INTERPOLATED VALUES

X	F(X)
36.00000	0.58779
45.00000	0.70710
54.00000	0.80902
32.00000	0.52992
48.00000	0.74314
63.00000	0.89101

**FORTRAN ** STOP

PROGRAM 3

Solution of Linear Equations using the Gauss–Seidel method.

A set of N linear equations is solved given initial approximations to the roots.

Program

The program consists of a main program and a subroutine which carries out the solution process. The process is terminated if the sum of the residues is less than a specified accuracy and the logical variable *SOLVED* is made *TRUE*. Up to 50 iterations allowed. If the process is incomplete a message is printed and a logical variable *SOLVED* is made *FALSE*. The main program has three linear equations but this should be modified for any specific case.

Subroutine Call

$$\text{CALL SEIDEL } (A, N, NP1, X, EPS, SOLVED),$$

A is a two dimensional array, to store the coefficients,

N is the number of rows,

$NP1$ is $N+1$ the number of columns,

X is a one-dimensional array to store the solution,

EPS is the required accuracy and *SOLVED* is a logical variable,

A, N, $NP1$ and *EPS* are input parameters and X and *SOLVED* are output parameters.

Data

N, *EPS* with a format $I5$, $F10.0$,

N lines of $A(I, J)$ with a format $5F10.0$,

The coefficients of each equation are written across the line,

The initial estimates $X(I)$, format $5F10.0$, are written across the line.

Results

The results consist of the coefficients of the N linear equations, followed by the initial estimates. The required accuracy is then printed and finally the solution. The results are annotated.

Program 3. Solution of linear equations using the Gauss–Seidel method.

```
00000100 C     THIS IS A PROGRAM TO SOLVE A SET OF N LINEAR
00000200 C     EQUATIONS USING THE METHOD OF GAUSS-SEIDEL .
00000300 C     THE LISTING CONSISTS OF A SHORT MAIN PROGRAM
00000400 C     FOLLOWED BY THE ROUTINE TO SOLVE THE EQUATIONS
00000500 C     IF THERE IS NO SOLUTION AFTER 50 ITERATIONS AN
00000600 C     APPROPRIATE MESSAGE IS PRINTED .
00000700 C     THE DATA CONSISTS OF N, THE NUMBER OF EQUATIONS
00000800 C     EPS, THE REQURED ACCURACY FOLLOWED BY THE
00000900 C     COEFFICIENTS OF THE EQUATIONS . THE LAST ITEM
00001000 C     OF DATA IS THE INITIAL GUESSES FOR THE VALUES
00001100 C     OF X.
00001200       LOGICAL SOLVED
00001300 C     THE PROGRAM ASSUMES 3 LINEAR EQUATIONS,
00001400 C     WHICH ARE INPUT LINE BY LINE . THE ESTIMATES
00001500 C     FOR THE SOLUTION ARE READ AND STORED AS X(3).
00001600       REAL X(3),A(3,4)
00001700       PRINT 8
00001800 8     FORMAT(' A PROGRAM TO SOLVE A SET OF LINEAR')
00001900       PRINT 9
00002000 9     FORMAT(' EQUATIONS USING GAUSS-SEIDEL METHOD')
00002100 C     INPUT THE NUMBER OF EQUATIONS AND THE ACCURACY
00002200 C     REQUIRED FOR THE SOLUTIONS .
00002300       READ 1,N,EPS
00002400 1     FORMAT(I5,F10.0)
00002500       NP1=N+1
00002600 C     INPUT THE EQUATIONS LINE BY LINE
00002700       PRINT 5,N
00002800 5     FORMAT(' THERE ARE ',I5,' EQUATIONS VIZ')
00002900       DO 3 I=1,N
00003000       READ 2,(A(I,J),J=1,NP1)
00003100       PRINT 4,(A(I,J),J=1,NP1)
00003200 4     FORMAT(1H ,5F10.5)
00003300 2     FORMAT(5F10.0)
00003400 3     CONTINUE
00003500 C     INPUT THE ESTIMATES
00003600       READ 2,(X(I),I=1,N)
00003700       PRINT 6,(X(I),I=1,N)
00003800 6     FORMAT('  THE ESTIMATES ARE ',8F10.4)
00003900       PRINT 7,EPS
00004000 7     FORMAT('  REQUIRED ACCURACY IS ',E15.4)
00004100 C     CALL THE SUBROUTINE TO SOLVE THE EQUATIONS
00004200 C     USING THE GAUSS-SEIDEL METHOD .
00004300       CALL SEIDEL(A,N,NP1,X,EPS,SOLVED)
00004400 C     SOLVED IS TRUE IF A SOLUTION EXISTS
00004500 C     OTHERWISE SOLVED IS FALSE .
00004600       IF(SOLVED) PRINT 10,(X(I),I=1,N)
00004700 10    FORMAT('  SOLUTION IS ',8F12.4)
00004800       STOP
00004900       END
00005000       SUBROUTINE SEIDEL(A,N,M,X,EPS,SOLVED)
00005100       LOGICAL SOLVED
00005200       REAL X(N),A(N,M)
00005300 C     ALLOW 50 ITERATIONS
00005400       DO 10 K=1,50
00005500       RES=0.0
00005600       DO 20 I=1,N
00005700       R=-A(I,M)
00005800       DO 30 J=1,N
00005900 30    R=R+A(I,J)*X(J)
00006000       X(I)=X(I)-R/A(I,I)
```

```
PROGRAM 3                                          PAGE 2
00006100 20     RES=RES+ABS(R)
00006200 C      IS THE SUM OF THE RESIDUES VERY SMALL?
00006300        IF(RES.LT.EPS)GOTO 35
00006400 10     CONTINUE
00006500 C      SUM OF RESIDUES NOT SMALL ENOUGH
00006600 C      AFTER FIFTY ITERATIONS .
00006700 C      THEREFORE SOLVED = FALSE.
00006800        PRINT 38,(X(I),I=1,N)
00006900 38     FORMAT(' NO SOLUTION AFTER 50 ITERATIONS',
00007000        1//,1X,4F10.4)
00007100        SOLVED =.FALSE.
00007200        RETURN
00007300 C      SOLUTION EXISTS, THEREFORE SOLVED = TRUE.
00007400 35     SOLVED = .TRUE.
00007500        RETURN
00007600        END
```

INPUT FOR PROGRAM 3

```
3    0.0001
   3.0         1.0        -1.0        1.0
   1.0        -3.0         1.0        2.0
   1.0         1.0         3.0        3.0
   0.0         0.0         0.0
```

OUTPUT FROM PROGRAM 3

```
A PROGRAM TO SOLVE A SET OF LINEAR
EQUATIONS USING GAUSS-SEIDEL METHOD
THERE ARE      3 EQUATIONS VIZ
   3.00000    1.00000   -1.00000    1.00000
   1.00000   -3.00000    1.00000    2.00000
   1.00000    1.00000    3.00000    3.00000
THE ESTIMATES ARE      0.0000      0.0000      0.0000
REQUIRED ACCURACY IS        0.1000E-03
SOLUTION IS            0.6667     -0.1667      0.8333


**FORTRAN ** STOP
```

PROGRAM 4

Solution of Linear Equations using the Gauss–Jordan method. A set of N linear equations is solved, provided there is a solution.

Program

The program consists of a main program, and a subroutine which obtains the solution. It is assumed in the main program that there are four linear equations but this can easily be altered in any specific case. The subroutine checks that the pivotal elements do not become too small. If the element is less than 10^{-4} the process is terminated with a suitable warning.

Subroutine Call

CALL JORDAN $(A, N, N1)$,

A is the array of coefficients,
N is the number of rows,
$N1$ is $N+1$, the number of columns,
$A(I, N1)$ $I = 1, N$ holds the solution,
A is an input and output parameter and N and $N1$ are input parameters.

Data

N with a format $I5$,
N lines of $A(I, J)$ with a format $5F10.0$, the coefficients of the equations are written across the line.

Results

The annotated results list the coefficients of the equations to be solved followed by the solution.

Program 4. Solution of linear equations using the Gauss–Jordan method.

```
PROGRAM 4                                                      PAGE 1

00000100 C      THIS PROGRAM IS TO SOLVE A SET OF LINEAR
00000200 C      EQUATIONS USING THE GAUSS-JORDAN METHOD.
00000300 C      THE DATA CONSISTS OF N THE NUMBER OF
00000400 C      EQUATIONS FOLLOWED BY THE N COEFFICIENTS
00000500 C      OF THE N UNKNOWNS AND THE RIGHT-HAND-SIDE
00000600 C      OF EACH EQUATION WHICH IS TREATED AS THE
00000700 C      N+1TH COEFFICIENT .EACH EQUATION IS INPUT
00000800 C      FROM A SEPARATE LINE .
00000900 C      THE LISTING CONSISTS OF A SMALL MAIN
00001000 C      PROGRAM FOLLOWED BY THE MAIN SUBPROGRAM
00001100 C      WHICH SOLVES THE EQUATIONS IF POSSIBLE .
00001200 C      THE ARRAY IS SET TO 4*5 TO ALLOW FOR
00001300 C      FOUR EQUATIONS . GENERALLY N BY N+1.
00001400        REAL A(4,5)
00001500        PRINT 5
00001600 5      FORMAT(' A PROGRAM TO SOLVE A SET OF LINEAR')
00001700        PRINT 6
00001800 6      FORMAT(' EQUATIONS USING GAUSS-JORDAN METHOD')
00001900 C      N. IS THE NUMBER OF LINEAR EQUATIONS
00002000        READ 3,N
00002100 3      FORMAT(I5)
00002200        PRINT 7,N
00002300 7      FORMAT(' THERE ARE',I5,'EQUATIONS VIZ')
00002400        N1=N+1
00002500 C      INPUT THE COEFFICIENTS AND RIGHT-HAND-SIDES
00002600 C      LINE BY LINE
00002700        DO 1 I=1,N
00002800        READ 2,(A(I,J),J=1,N1)
00002900 1      PRINT 11,(A(I,J),J=1,N1)
00003000 2      FORMAT(5F5.0)
00003100 11     FORMAT(1H ,5F10.4)
00003200 C      CALL THE SUBROUTINE TO SOLVE THE EQUATIONS
00003300 C      USING THE GAUSS-JORDAN  METHOD
00003400        CALL JORDAN(A,N,N1)
00003500        PRINT 12,(A(I,N1),I=1,N)
00003600 12     FORMAT(1H ,' SOLUTION IS',8F10.4)
00003700        STOP
00003800        END
00003900        SUBROUTINE JORDAN(A,N,M)
00004000        REAL A(N,M)
00004100        DO 10 K=1,N
00004200        K1=K+1
00004300        IF(K.EQ.N) GO TO 15
00004400        L=K
00004500        DO 20 J=K1,N
00004600 C      FIND THE LARGEST ELEMENT IN THE K-TH COLUMN
00004700 C      AND STORE IT AS ELEMENT L,K. .
00004800        IF(ABS(A(J,K)).GT.ABS(A(L,K))) L=J
00004900 20     CONTINUE
00005000        IF(L.EQ.K) GO TO 15
00005100 C      INTERCHANGE ROWS K AND L
00005200        DO 16 J=K,M
00005300        T=A(K,J)
00005400        A(K,J)=A(L,J)
00005500 16     A(L,J)=T
00005600 C      IF ANY DIAGONAL ELEMENTS ARE LESS THAN 1.0E-4
00005700 C      STOP THE PROCESS
00005800 15     DO 21 J=K1,M
00005900        IF(ABS(A(K,K)).LE.1.0E-4) GO TO 30
00006000 C      REDUCE THE K-TH ROW BY THE DIAGONAL ELEMENT
```

```
00006100 C     TO AVOID BIG NUMBERS.
00006200 21    A(K,J)=A(K,J)/A(K,K)
00006300       IF(K.EQ.N) GO TO 10
00006400 C     PERFORM THE PIVOTAL REDUCTION
00006500       DO 12 I=K1,N
00006600       DO 12 J=K1,M
00006700 12    A(I,J)=A(I,J)-A(I,K)*A(K,J)
00006800 10    CONTINUE
00006900 C     PERFORM THE BACK SUBSTITUTION
00007000       DO 14 J=2,N
00007100       K=N+2-J
00007200       K1=K-1
00007300       DO 14 I=1,K1
00007400 14    A(I,M)=A(I,M)-A(I,K)*A(K,M)
00007500 C     SOLUTIONS ARE NOW IN A(I,M),M=1,N,
00007600 C     WHERE M=N+1-THE RIGHT-HAND-SIDE-.
00007700       RETURN
00007800 30    PRINT 31
00007900 31    FORMAT('   NO SOLUTION EXISTS DETERMINANT =0')
00008000       RETURN
00008100       END
```

INPUT FOR PROGRAM 4

```
    4
 5.0   5.0   7.0   9.0   7.3
 3.0   4.0   10.   5.0   5.5
 4.0   2.0   1.0   1.0   5.0
-4.0  -5.0   1.0  -1.0  -6.3
```

OUTPUT FROM PROGRAM 4

```
A PROGRAM TO SOLVE A SET OF LINEAR
EQUATIONS USING GAUSS-JORDAN METHOD
THERE ARE      4EQUATIONS VIZ
      5.0000      5.0000     7.0000      9.0000    7.3000
      3.0000      4.0000    10.0000      5.0000    5.5000
      4.0000      2.0000     1.0000      1.0000    5.0000
     -4.0000     -5.0000     1.0000     -1.0000   -6.3000
   SOLUTION IS   1.0000      0.5000      0.1000    -0.1000
```

**FORTRAN ** STOP

PROGRAM 5

To find the largest eigenvalue and the associated eigenvector of a real matrix using the power method.

The multiplication process

$$\mathbf{A}\mathbf{x} = \lambda \mathbf{x}$$

is applied iteratively until λ converges to a value. $\mathbf{A}$ is the given matrix.

Program

The program is made up of a small main program and a subroutine to carry out the iterative process. Up to 100 iterations are allowed. The initial eigenvector can be read from cards or assumed to be **1** of order n. If convergence is not achieved then a suitable message is printed.

Subroutine Call

CALL POWER $(A, X, V, N, EPS, SOLVED)$,

A is the given matrix, a two-dimensional array,
X is the initial eigenvector, a one-dimensional array,
V is the eigenvalue,
N is the order of the matrix,
EPS is the accuracy required,
$SOLVED$ is a logical flag, true if the process has converged, false otherwise,
A, N and EPS are input parameters,
X is an input and output parameter and V and $SOLVED$ are output parameters.

Data

$FLAG$, N, EPS with a format $L1$, $4X$, $I5$, $F10.0$,
$FLAG$ is true if the initial eigenvector is to be input, false if it is assumed to be the vector 1 of order n,
A, the matrix, read line by line with a format of $6F10.0$. The rows are written 6 elements to a line,
X, the initial eigenvector, if required with a format of $6F10.0$.

Results

The annotated results give the matrix, the initial vector, the estimate of the largest eigenvalue and the corresponding eigenvector.

Program 5. To find the largest eigenvalue and associated eigenvector of a real matrix
using the power method.

PROGRAM 5   PAGE 1

```
00000100 C    THIS PROGRAM FINDS THE LARGEST EIGENVALUE OF
00000200 C    A MATRIX USING THE POWER METHOD .
00000300 C    THE MULTIPLICATION PROCESS
00000400 C       A X = LAMBDA X
00000500 C    IS CONTINUOUSLY APPLIED TO THE VECTOR X,
00000600 C    WHERE A IS THE MATRIX AND LAMBDA IS THE
00000700 C    LARGEST EIGENVALUE, UNTIL 2 ESTIMATES OF
00000800 C    LAMBDA AGREE TO THE REQUIRED ACCURACY .THE
00000900 C    CORRESPONDING EIGENVECTOR IS ALSO FOUND .
00001000 C    THE STARTING VECTOR CAN BE INPUT BY MAKING
00001100 C    FLAG =T-TRUE- OR IT WILL BE TAKEN TO BE
00001200 C    1,1,1 IF FLAG=F-FALSE-. UP TO 100 ITERATIONS
00001300 C    ARE ALLOWED AND IF THE PROCESS HAS NOT
00001400 C    CONVERGED BY THEN A MESSAGE IS PRINTED .
00001500 C    THE DATA CONSISTS OF THE FLAG,N THE ORDER OF
00001600 C    THE MATRIX,EPS THE REQUIRED ACCURACY,THE MATRIX
00001700 C    ROW BY ROW AND THE STARTING VECTOR IF NEEDED .
00001800 C    THE PROGRAM CONSISTS OF A MAIN PROGRAM TO READ
00001900 C    THE DATA AND A SUBROUTINE POWER TO CARRY OUT
00002000 C    THE ITERATIVE PROCESS
00002100      REAL A(10,10),X(10)
00002200      LOGICAL FLAG,SOLVED
00002300      PRINT 1
00002400 1    FORMAT(' TO FIND THE LARGEST EIGEN-VALUE OF ')
00002500      PRINT 2
00002600 2    FORMAT(' A MATRIX USING THE POWER METHOD')
00002700      PRINT 3
00002800 3    FORMAT(' THE GIVEN MATRIX IS ',/)
00002900 C    INPUT THE DATA
00003000      READ 5,FLAG,N,EPS
00003100 5    FORMAT(L1,4X,I5,F10.0)
00003200 C    INPUT THE MATRIX
00003300      DO 10 I=1,N
00003400      READ 20,(A(I,J),J=1,N)
00003500 20   FORMAT(6F10.0)
00003600 10   PRINT 30,(A(I,J),J=1,N)
00003700 30   FORMAT(1H ,6F10.3)
00003800      DO 32 I=1,N
00003900 32   X(I)=1.0
00004000 C    IS THE INITIAL VALUE OF X SUPPLIED ?
00004100      IF(FLAG) READ 20,(X(I),I=1,N)
00004200      PRINT 4
00004300 4    FORMAT(/,' THE INITIAL EIGEN VECTOR IS ',/)
00004400      PRINT 6,(X(I),I=1,N)
00004500 6    FORMAT(1H ,6F10.2)
00004600 C    CALL THE POWER ROUTINE
00004700 C    TO FIND THE LARGEST EIGENVALUE
00004800      CALL POWER(A,X,V,N,EPS,SOLVED)
00004900      IF(.NOT.SOLVED) PRINT 35
00005000 35   FORMAT('  PROCESS REQUIRES MORE ITERATIONS?')
00005100      IF(.NOT.SOLVED) GO TO 60
00005200      PRINT 40,V
00005300 40   FORMAT('  LARGEST EIGENVALUE =',F10.4)
00005400      PRINT 50,(X(I),I=1,N)
00005500 50   FORMAT('  EIGENVECTOR IS ',6F10.4)
00005600 60   STOP
00005700      END
00005800 C    THE ROUTINE TO FIND THE LARGEST EIGENVALUE
00005900      SUBROUTINE POWER(A,X,V,N,EPS,SOLVED)
00006000      LOGICAL SOLVED
```

```
00006100          REAL A(10,10),X(10),Y(10)
00006200          SOLVED =.TRUE.
00006300          V=1.0E4
00006400          DO 20 K=1,100
00006500          DO 30 I=1,N
00006600          Y(I)=0.0
00006700          DO 30 J=1,N
00006800 30       Y(I)=Y(I)+A(I,J)*X(J)
00006900          V1=Y(1)
00007000 C        NORMALISE THE VECTOR
00007100          DO 40 I=1,N
00007200 40       X(I)=Y(I)/V1
00007300 C        ANY MORE ITERATIONS REQUIRED?
00007400          IF(ABS(V-V1).LE.EPS) GO TO 50
00007500 20       V=V1
00007600 C        SET SOLVED IF NEEDED
00007700          SOLVED = .FALSE.
00007800 50       RETURN
00007900          END
```

INPUT FOR PROGRAM 5

```
F         3        0.001
        2.0      2.0              2.0
  0.66667    1.66667        1.66667
        1.0      2.5              5.5
```

OUTPUT FROM PROGRAM 5

TO FIND THE LARGEST EIGEN-VALUE OF
A MATRIX USING THE POWER METHOD
THE GIVEN MATRIX IS

```
    2.000      2.000      2.000
    0.667      1.667      1.667
    1.000      2.500      5.500
```

THE INITIAL EIGEN VECTOR IS

```
    1.00       1.00       1.00
LARGEST EIGENVALUE =      7.0041
EIGENVECTOR IS      1.0000     0.6906     1.8118
```

**FORTRAN ** STOP

PROGRAM 6

Determination of an accurate root for an equation using the Newton–Raphson method.

A root of the equation

$$f(x) = 0$$

is determined given an approximate root.

Program

The program consists of a main program, a subroutine which performs the root determination and two function subprograms. The first function defines $f(x)$, in this case

$$f(x) = x^3 - 8x - 10$$

and the second function *FPRIME* defines $f'(x)$ the derivative of $f(x)$, in this case

$$f'(x) = 3x^2 - 8.$$

The subroutine allows up to 30 iterations before terminating the procedure. If the root lies within the required accuracy before 30 iterations the procedure is terminated.

Subroutine Call

CALL NR (X, EPS),

X is the initial guess and the solution,
EPS is the required accuracy,
X is an input and output parameter, and *EPS* is an input parameter.

Data

X, *EPS* with a format 2F10.0.

Results

The results give the initial root, the required accuracy and the solution determined.

Program 6. Determination of an accurate root for an equation using the Newton–Raphson method.

```
00000100 C    THIS IS A PROGRAM TO FIND AN ACCURATE ROOT OF
00000200 C    AN EQUATION , GIVEN AN APPROXIMATE ONE ,USING
00000300 C    THE NEWTON-RAPHSON METHOD.
00000400 C    THE LISTING CONSISTS OF A SMALL MAIN PROGRAM
00000500 C    THE MAIN ROUTINE AND TWO SMALL FUNCTION
00000600 C    SUBPROGRAMS . ONE FOR THE FUNCTION F(X) THE
00000700 C    OTHER FOR THE DERIVATIVE OF F(X).
00000800 C    THE DATA IS MADE UP OF THE APPROXIMATE ROOT
00000900 C    AND THE REQUIRED ACCURACY.
00001000 C    INPUT THE APPROXIMATE ROOT AND THE ACCURACY .
00001100      PRINT 1
00001200 1    FORMAT(' A PROGRAM TO DETERMINE AN ACCURATE')
00001300      PRINT 2
00001400 2    FORMAT(' ROOT FOR A NON-LINEAR EQUATION ')
00001500      PRINT 3
00001600 3    FORMAT(' USING THE NEWTON-RAPHSON METHOD',//)
00001700      READ 10,X,EPS
00001800 C    CALL THE NEWTON RAPHSON ROUTINE
00001900 10   FORMAT(2F10.0)
00002000      PRINT 5,X,EPS
00002100 5    FORMAT(' INITIAL GUESS = ',F10.4,
00002200     1' REQUIRED ACCURACY ',E10.4)
00002300      CALL NR(X,EPS)
00002400      PRINT 20,X
00002500 20   FORMAT('  SOLUTION = ',F10.5)
00002600      STOP
00002700      END
00002800 C    THE SUBROUTINE TO FIND THE ROOT
00002900      SUBROUTINE NR(X,EPS)
00003000 C    ALLOW UP TO 30 ITERATIONS
00003100      DO 20 I=1,30
00003200      X=X-F(X)/FPRIME(X)
00003300      IF(ABS(F(X)).LT.EPS) GO TO 25
00003400 20   CONTINUE
00003500 25   RETURN
00003600      END
00003700 C    THE FUNCTION F(X)
00003800      FUNCTION F(X)
00003900      F=(X*X-8)*X-10
00004000      RETURN
00004100      END
00004200 C    THE DERIVATIVE OF F(X)
00004300      FUNCTION FPRIME(X)
00004400      FPRIME =3*X*X-8
00004500      RETURN
00004600      END
```

INPUT FOR PROGRAM 6

  1.0   0.00001

OUTPUT FROM PROGRAM 6

A PROGRAM TO DETERMINE AN ACCURATE
ROOT FOR A NON-LINEAR EQUATION
USING THE NEWTON-RAPHSON METHOD

INITIAL GUESS =   1.0000 REQUIRED ACCURACY 0.1000E-04
  SOLUTION =   3.31863

**FORTRAN ** STOP

PROGRAM 7

Determination of an accurate root for

$$f(x) = 0$$

using either regula–falsi or the secant method. The method is chosen by setting a flag.

Program

The program consists of a main program a subroutine which determines the root and a function subprogram to define $f(x)$. Here

$$f(x) = 3x^5 - 30x^4 + 95x^3 - 90x^2 - 16.$$

The two methods are fairly similar so that the subroutine covers both methods. A logical variable controls which method is chosen. In each method 30 iterations are allowed.

Subroutine Call

CALL REFINE $(X1, X2, EPS, REGFAL)$,

$X1$ and $X2$ are the bounds of the range in which the solution lies,
$X1$ holds the solution on output,
EPS is the required accuracy,
$REGFAL$ is a logical variable denoting the method chosen,
$X1$ is an input and output parameter and $X2$, EPS and $REGFAL$ are input parameters.

Data

$REGFAL$, $X1$, $X2$, EPS format $L1, 9X, 3F10.0$,
$REGFAL$ is a logical variable and is written in column 1. If the regula–falsi method is required place a 'T' in column 1, if the secant method then write an 'F' in column 1. In either case the letters should be followed by 9 spaces. The limits $X1$ and $X2$ follow, each occupying 10 columns and finally the desired accuracy is written. The data in this case shows the secant method being called.

Results

The results in this case show one method being called. The method used is indicated in the first line of printout. This is followed by the accuracy requested. A table is then printed showing the successive determination of the root. The method used in this example is the secant method.

Program 7. Determination of an accurate root for an equation using either regula–falsi or the secant method.

PROGRAM 7   PAGE 1

```
00000100 C    THIS PROGRAM OBTAINS AN ACCURATE ROOT FOR THE
00000200 C    FUNCTION F(X) GIVEN THAT IT LIES WITHIN TWO
00000300 C    LIMITS A AND B. THE ACCURACY OF DETERMINATION
00000400 C    IS GIVEN BY EPS .THERE IS A CHOICE OF METHOD
00000500 C    IF THE LOGICAL VARIABLE REGFAL IS TRUE THEN THE
00000600 C    REGULA-FALSI METHOD IS USED OTHERWISE THE
00000700 C    SECANT METHOD IS ADOPTED .
00000800 C    THE DATA CONSISTS OF A FLAG REGFAL THE TWO
00000900 C    LIMITS A AND B AND THE ACCURACY REQUIRED .
00001000 C    THE LISTING IS MADE UP OF A SMALL MAIN PROGRAM
00001100 C    FOLLOWED BY THE MAIN SUBROUTINE AND  FINALLY
00001200 C    THE FUNCTION F(X) SUBPROGRAM .
00001300      LOGICAL REGFAL
00001400 C    INPUT THE FLAG REGFAL=TRUE REGULA-FALSI
00001500 C                        =FALSE SECANT,
00001600 C    THE TWO LIMITS AND THE ACCURACY
00001700      PRINT 1
00001800 1    FORMAT(' A PROGRAM TO DETERMINE AN ACCURATE')
00001900      PRINT 3
00002000 3    FORMAT(' ROOT FOR A NON-LINEAR EQUATION')
00002100      PRINT 2
00002200 2    FORMAT(' USING THE REGOLA-FALSI METHOD')
00002300      PRINT 4
00002400 4    FORMAT(' OR THE SECANT METHOD',////)
00002500      READ 10,REGFAL,X1,X2,EPS
00002600 10   FORMAT(L1,9X,3F10.0)
00002700      PRINT 5,EPS
00002800 5    FORMAT('   REQUIRED ACCURACY IS',E10.4)
00002900 C    CALL THE MAIN ROUTINE
00003000      CALL REFINE(X1,X2,EPS,REGFAL)
00003100      PRINT 20,X1
00003200 20   FORMAT('   SOLUTION X = ',F10.4)
00003300      STOP
00003400      END
00003500 C    THE ROUTINE TO DETERMINE THE ROOT
00003600      SUBROUTINE REFINE(X1,X2,EPS,REGFAL)
00003700      LOGICAL REGFAL
00003800      IF(REGFAL) PRINT 1
00003900 1    FORMAT('   USING THE REGULA-FALSI METHOD',//)
00004000      IF(.NOT.REGFAL) PRINT 2
00004100 2    FORMAT('   USING THE SECANT METHOD ',//)
00004200      Y1=F(X1)
00004300      Y2=F(X2)
00004400      N=1
00004500      PRINT 15
00004600 15   FORMAT('   ESTIMATE          X        F(X)',/)
00004700 C    PRINT THE VALUES OF THE FUNCTION AT THE LIMITS
00004800      PRINT 20,N,X1,Y1
00004900      N=2
00005000      PRINT 20,N,X2,Y2
00005100 20   FORMAT(1H ,I10,2F10.5)
00005200 C    ALLOW UP TO 30 ITERATIONS
00005300      DO 30 N=3,30
00005400      X3=(X1*Y2-X2*Y1)/(Y2-Y1)
00005500      Y3=F(X3)
00005600      PRINT 20,N,X3,Y3
00005700 C    IS THE ROOT ACCURATE ENOUGH?
00005800      IF(ABS(Y3*Y2).LE.EPS.AND.REGFAL) GO TO 40
00005900 C    THIS TEST ONLY APPLIES TO THE
00006000 C    REGULA-FALSI METHOD
```

```
00006100          IF(Y3*Y2.GT.0.AND.REGFAL) GO TO 35
00006200          X1=X2
00006300          Y1=Y2
00006400 35       X2=X3
00006500          Y2=Y3
00006600 C        IS THE ROOT ACCURATE ENOUGH?
00006700          IF(ABS(X1-X2).LT.EPS) GO TO 40
00006800 30       CONTINUE
00006900 40       X1=X3
00007000          RETURN
00007100          END
00007200 C        THE FUNCTION F(X)
00007300          FUNCTION F(X)
00007400          F=(((3*X-30)*X+95)*X-90)*X*X-16
00007500          RETURN
00007600          END
```

INPUT FOR PROGRAM 7

```
F                    1.0        4.0      0.0001
```

OUTPUT FROM PROGRAM 7

```
A PROGRAM TO DETERMINE AN ACCURATE
ROOT FOR A NON-LINEAR EQUATION
USING THE REGOLA-FALSI METHOD
OR THE SECANT METHOD


REQUIRED ACCURACY IS0.1000E-03
USING THE SECANT METHOD

ESTIMATE         X        F(X)

       1    1.00000  -38.00000
       2    4.00000   16.00000
       3    3.11111   37.45488
       4    4.66289   89.40295
       5    1.99227   -0.46388
       6    2.00605    0.36314
       7    2.00000    0.00000
       8    2.00000   -0.00006
SOLUTION X =      2.0000


**FORTRAN ** STOP
```

PROGRAM 8

Determination of an accurate root for

$$f(x) = 0$$

using the method of bisection.

Program

The program consists of a main program, a subroutine to obtain the root and a function defining $f(x)$. Here

$$f(x) = 3x^5 - 30x^4 + 95x^3 - 90x^2 - 16.$$

The subroutine allows up to 30 iterations and if the required accuracy has not been achieved before 30 iterations the process is terminated.

Subroutine Call

CALL BISECT $(X, X1, X2, EPS)$,

X is the solution, an output parameter,
$X1$ is the lower limit,
$X2$ is the higher limit,
EPS is the accuracy required,
X is an output parameter, and $X1$, $X2$ and EPS are input parameters.

Data

$X1$, $X2$, EPS with a format $3F10.0$.

Results

The annotated results give first the limits between which the root lies, followed by the required accuracy. A table is then printed showing the successive determinations of the root and the residue. The final solution is indicated.

Program 8. Determination of an accurate root for an equation using the method of bisection.

PROGRAM 8   PAGE 1

```
00000100 C      THIS PROGRAM FINDS AN ACCURATE ROOT FOR THE
00000200 C      FUNCTION F(X) GIVEN THAT IT LIES WITHIN TWO
00000300 C      LIMITS A AND B. THE ROOT IS FOUND TO AN
00000400 C      ACCURACY EPS USING  THE METHOD OF BISECTION .
00000500 C      THE DATA CONSISTS OF THE TWO LIMITS A AND B
00000600 C      AND THE ACCURACY EPS .
00000700 C      THE LISTING CONSISTS OF A SHORT MAIN PROGRAM
00000800 C      FOLLOWED BY THE MAIN ROUTINE AND THEN THE
00000900 C      FUNCTION SUBPROGRAM .
00001000 C      INPUT THE LIMITS AND THE ACCURACY
00001100        PRINT 5
00001200 5      FORMAT(' A PROGRAM TO FIND AN ACCURATE ROOT')
00001300        PRINT 6
00001400 6      FORMAT(' FOR A NON-LINEAR EQUATION ')
00001500        PRINT 7
00001600 7      FORMAT(' USING THE METHOD OF BISECTION',//)
00001700        READ 10,X1,X2,EPS
00001800 10     FORMAT(3F10.0)
00001900        PRINT 2,X1,X2
00002000 2      FORMAT(' SOLUTION LIES BETWEEN THE LIMITS ',
00002100        12F10.4)
00002200        PRINT 3,EPS
00002300 3      FORMAT('  THE REQUIRED ACCURACY IS ',E10.4)
00002400 C      CALL THE ROUTINE
00002500        CALL BISECT(X,X1,X2,EPS)
00002600        PRINT 20,X
00002700 20     FORMAT('  SOLUTION IS X =',F10.4)
00002800        STOP
00002900        END
00003000 C      THE ROUTINE TO FIND THE ROOT
00003100 C      USING THE METHOD OF BISECTION
00003200        SUBROUTINE BISECT(X,X1,X2,EPS)
00003300        Y1=F(X1)
00003400        Y2=F(X2)
00003500        N=1
00003600        PRINT 15
00003700 15     FORMAT('    ESTIMATE          X       F(X)',//)
00003800 C      PRINT THE VALUES OF THE FUNCTION AT THE LIMITS
00003900        PRINT 20,N,X1,Y1
00004000        N=2
00004100        PRINT 20,N,X2,Y2
00004200 20     FORMAT(1H ,I10,2F10.5)
00004300 C      ALLOW UP TO 30 ITERATIONS
00004400        DO 30 N=3,30
00004500        X=(X1+X2)/2
00004600        Y=F(X)
00004700        PRINT 20,N,X,Y
00004800 C      IS THE ROOT ACCURATE ENOUGH?
00004900        IF(ABS(Y).LT.EPS) GO TO 40
00005000        IF(Y*Y1.GT.0) X1=X
00005100        IF(Y*Y2.GT.0) X2=X
00005200 C      IS THE ROOT ACCURATE ENOUGH?
00005300        IF(ABS(X1-X2).LT.EPS) GO TO 40
00005400 30     CONTINUE
00005500 40     RETURN
00005600        END
00005700 C      THE FUNCTION F(X)
00005800        FUNCTION F(X)
00005900        F=(((3*X-30)*X+95)*X-90)*X*X-16
00006000        RETURN
00006100        END
```

INPUT FOR PROGRAM 8

    0.0   3.0   0.0001

OUTPUT FROM PROGRAM 8

A PROGRAM TO FIND AN ACCURATE ROOT
FOR A NON-LINEAR EQUATION
USING THE METHOD OF BISECTION

SOLUTION LIES BETWEEN THE LIMITS   0.0000   3.0000
   THE REQUIRED ACCURACY IS 0.1000E-03
     ESTIMATE   X   F(X)

           1   0.00000 -16.00000
           2   3.00000   38.00000
           3   1.50000 -26.96875
           4   2.25000   14.61230
           5   1.87500   -7.45126
           6   2.06250   3.74390
           7   1.96875   -1.87424
           8   2.01563   0.93735
           9   1.99219   -0.46879
          10   2.00391   0.23436
          11   1.99805   -0.11719
          12   2.00098   0.05864
          13   1.99951   -0.02923
          14   2.00024   0.01471
          15   1.99988   -0.00726
          16   2.00006   0.00372
          17   1.99997   -0.00177
SOLUTION IS X =   2.0000

**FORTRAN ** STOP

PROGRAM 9

To find the best straight line through a series of points using the technique of least squares. The parameters a and b in the equation

$$y = ax + b$$

are determined.

Program

The program consists of a main program and a subroutine which determines the parameters a and b. The subroutine determines the parameters by solving the usual two normal equations obtained when using the least squares method. If the logical variable FLAG is true the deviation of the points from the best straight line are found.

Subroutine Call

$$\text{CALL LSTSQ } (X, Y, N, A, B),$$

X is the array of N x-coordinates,
Y is the array of N y-coordinates,
N is the number of points,
A is one parameter of the line,
B is the other parameter of the line,
X, Y and N are input parameters, while A and B are output parameters.

Data

N, *FLAG* with a format $I5, 4X, L1$,
FLAG is logical, if true a T is placed in column 10, if false F is placed there,
$X(I)$ with a format $6F10.0$,
$Y(I)$ with a format $6F10.0$,
where I runs from 1 to N in each case,
the N values of the x coordinates and the N values of the y coordinates are written separately 6 to a line across the line.

Results

If FLAG is true then for each point the coordinates x and y are printed together with the deviation of the point from the best line. The parameters A and B of the best straight line are then printed.

Program 9. To find the best straight line through a series of points using the technique of least squares.

PROGRAM 9

```
00000100 C    THIS PROGRAM FINDS THE PARAMETERS A AND B OF
00000200 C    THE BEST STRAIGHT LINE
00000300 C       Y = A*X +B
00000400 C    THROUGH A SERIES OF N POINTS X,Y USING LEAST
00000500 C    SQUARES. THE PARAMETERS ARE FOUND BY MINIMISING
00000600 C    THE SUM OF THE SQUARES OF THE Y-DEVIATIONS FROM
00000700 C    THE BEST LINE I.E. THE SUM OF
00000800 C    (Y(I)-A*X(I)-B)**2
00000900 C    IS MINIMISED I RUNNING FROM 1 TO N.
00001000 C    THE TWO EQUATIONS FORMED BY DIFFERENTIATING
00001100 C    PARTIALLY WITH RESPECT TO A AND B ARE SOLVED
00001200 C    TO GIVE VALUES FOR A AND B.
00001300 C    THE DATA CONSISTS OF N,THE NUMBER OF POINTS
00001400 C    A FLAG ,SEE LATER,THE VALUES OF X READ ACROSS
00001500 C    THE LINE AND THE VALUES OF  Y READ LIKEWISE.
00001600 C    THE OUTPUT CONSISTS OF THE VALUES OF X AND Y
00001700 C    AND THE VALUES OF THE PARAMETERS A AND B.
00001800 C    IF THE FLAG IS SET TO TRUE THEN THE DEVIATIONS
00001900 C    OF THE POINTS FROM THE BEST LINE ARE OUTPUT.
00002000 C    THERE IS A SHORT MAIN PROGRAM TO READ THE DATA
00002100 C    PRINT THE RESULTS AND CALL THE SUBROUTINE.
00002200 C    THE LEAST SQUARES SUBROUTINE IS CALLED LSTSQ
00002300 C    AND IT FORMS THE SUMS AS THE COEFFICIENTS
00002400 C    OF THE NORMAL EQUATIONS AND THEN FINDS A&B.
00002500      COMMON FLAG
00002600      LOGICAL FLAG
00002700      REAL X(50),Y(50)
00002800      PRINT 1
00002900 1    FORMAT(' TO FIND THE PARAMETERS OF THE BEST')
00003000      PRINT 2
00003100 2    FORMAT(' LINE THROUGH A SERIES OF POINTS',/)
00003200 C    INPUT N THE NUMBER OF POINTS AND FLAG
00003300 C     IF FLAG =TRUE THEN THE DEVIATIONS ARE PRINTED
00003400      READ 10,N,FLAG
00003500 10   FORMAT(I5,4X,L1)
00003600 C    INPUT THE X COORDINATES
00003700      READ 20,(X(I),I=1,N)
00003800 20   FORMAT(6F10.0)
00003900 C    INPUT THE Y COORDINATES
00004000      READ 20,(Y(I),I=1,N)
00004100 C    CALL THE LEAST SQUARES SUBROUTINE
00004200      CALL LSTSQ(X,Y,N,A,B)
00004300 C    PRINT THE PARAMETERS A AND B
00004400      PRINT 50,A,B
00004500 50   FORMAT(' PARAMETERS OF THE BEST LINE ARE',
00004600      11H ,2F10.4)
00004700      STOP
00004800      END
00004900 C    THE LEAST SQUARES ROUTINE
00005000      SUBROUTINE LSTSQ(X,Y,N,A,B)
00005100      COMMON FLAG
00005200      LOGICAL FLAG
00005300      REAL X(N),Y(N)
00005400      SUMX2=0
00005500      SUMX=0
00005600      SUMXY=0
00005700      SUMY=0
00005800 C    FORM THE VARIOUS SUMS
00005900      DO 10 I=1,N
00006000      SUMX2=SUMX2+X(I)*X(I)
```

PROGRAM 9

```
00006100           SUMX=SUMX+X(I)
00006200           SUMXY=SUMXY+Y(I)*X(I)
00006300 10        SUMY=SUMY+Y(I)
00006400           DENOM = N*SUMX2-SUMX*SUMX
00006500           A=(N*SUMXY-SUMX*SUMY)/DENOM
00006600           B=(SUMX2*SUMY-SUMXY*SUMX)/DENOM
00006700 C         ARE THE DEVIATIONS REQUIRED?
00006800           IF(.NOT.FLAG) RETURN
00006900 C         YES
00007000           PRINT 18
00007100 18        FORMAT(1H ,9X,'X',9X,'Y',' DEVIATION',//)
00007200           DO 20 I=1,N
00007300           D=Y(I)-A*X(I)-B
00007400 20        PRINT 25,X(I),Y(I),D
00007500 25        FORMAT(1H ,3F10.4)
00007600           RETURN
00007700           END
```

INPUT FOR PROGRAM 9

```
6    T
   1.0        2.0        3.0        4.0        5.0        6.0
   2.09       4.89       8.13      11.22      13.78      16.99
```

OUTPUT FROM PROGRAM 9

```
TO FIND THE PARAMETERS OF THE BEST
LINE THROUGH A SERIES OF POINTS

           X           Y DEVIATION


   1.0000     2.0900      0.0205
   2.0000     4.8900     -0.1584
   3.0000     8.1300      0.1028
   4.0000    11.2200      0.2139
   5.0000    13.7800     -0.2050
   6.0000    16.9900      0.0262
PARAMETERS OF THE BEST LINE ARE      2.9789    -0.9093
```

```
**FORTRAN ** STOP
```

PROGRAM 10

To find the best parabola through a series of points using the technique of least squares.

The parameters a, b and c in the equation

$$y = ax^2 + bx + c$$

are determined.

Program

The program consists of a main program and a subroutine which forms the coefficients of the normal equations. The subroutine, previously described, to solve a set of linear equations is used to solve the normal equations. This routine is called from the least squares routine and so is of no concern in this program to the reader. If the logical FLAG is true then the deviations of the points from the parabola are found.

Subroutine Call

CALL PARLSQ (X, Y, N, A, B, C),

X is an array of N x-coordinates,
Y is an array of N y-coordinates,
N is the number of points,
A, B, C are parameters of the equation to the best parabola,
X, Y and N are input parameters, while A, B and C are output parameters.

Data

N, $FLAG$ with a format $I5$, $4X$, $L1$,
$FLAG$ is logical if true T is placed in column 10 and if false F is placed there,
$X(I)$ with a format $6F10.0$,
$Y(I)$ with a format $6F10.0$,
where I runs from 1 to N in each case,
The N values of the x-coordinates and the N values of the y-coordinates are written separately six to a line across the line.

Results

If FLAG is true then the coordinates of each point and its deviation from the parabola are printed on a separate line. The parameters A, B and C of the best parabola are then printed.

Program 10. To find the best parabola through a series of points using the technique of
least squares.

```
00000100 C    THIS PROGRAM FINDS THE PARAMETERS A,B AND C
00000200 C    OF THE BEST PARABOLA THROUGH A SERIES OF N
00000300 C    POINTS X(I),Y(I),I=1,N, USING LEAST SQUARES.
00000400 C    THE EQUATION OF THE PAROBOLA IS
00000500 C        Y=A*X**2 +B*X +C
00000600 C    THE PARAMETERS ARE FOUND BY MINIMISING THE SUM
00000700 C    OF THE Y DEVIATIONS OF THE POINTS FROM THE
00000800 C    PARABOLA I.E. BY MINIMISING THE SUM OF
00000900 C    (Y(I)-A*X(I)**2-B*X(I)-C)**2
00001000 C    FOR ALL THE POINTS X(I),Y(I),I=1 TO N.
00001100 C    THE THREE EQUATIONS  ARE FORMED BY PARTIALLY
00001200 C    DIFFERENTIATING WITH RESPECT TO A,B AND C. THE
00001300 C    EQUATIONS ARE THEN SOLVED TO GIVE A,B AND C.
00001400 C    THE DATA CONSISTS OF N,THE NUMBER OF POINTS
00001500 C    A FLAG ,SEE LATER, THE VALUES OF X ACROSS
00001600 C    THE LINE AND THEN OF Y LIKEWISE.
00001700 C    THE OUTPUT CONSISTS OF THE VALUES OF THE
00001800 C    COORDINATES OF EACH POINT ON A SEPARATE
00001900 C    LINE AND THE PARAMETERS A,B AND C.
00002000 C    IF THE FLAG IS TRUE THEN THE Y-DEVIATIONS ARE
00002100 C    ALSO PRINTED.
00002200 C    THERE IS A SHORT MAIN PROGRAM TO READ THE DATA
00002300 C    AND PRINT THE RESULTS AND ALSO TO CALL THE
00002400 C    LEAST SQUARES ROUTINE. THERE ARE TWO ROUTINES
00002500 C    PARLSQ WHICH PERFORMS THE CALCULATIONS TO
00002600 C    SET UP THE NORMAL EQUATIONS AND JORDAN WHICH
00002700 C    SOLVES THEM .
00002800      COMMON FLAG
00002900      LOGICAL FLAG
00003000      REAL X(50),Y(50)
00003100      PRINT 1
00003200 1    FORMAT(' TO FIND THE PARAMETERS OF THE BEST')
00003300      PRINT 2
00003400 2    FORMAT(' PARABOLA THROUGH A SERIES OF POINTS')
00003500 C    INPUT N,THE NUMBER OF POINTS AND THE FLAG
00003600      READ 10,N,FLAG
00003700 10   FORMAT(I5,4X,L1)
00003800 C    INPUT THE X-COORDINATES ACROSS THE LINE
00003900      READ 20,(X(I),I=1,N)
00004000 20   FORMAT(6F10.0)
00004100 C    INPUT THE Y-COORDINATES ACROSSTHE LINE
00004200      READ 20,(Y(I),I=1,N)
00004300 C    CALL THE LEAST SQUARES ROUTINE
00004400 C    TO FIT A PARABOLA TO THE POINTS
00004500      CALL PARLSQ(X,Y,N,A,B,C)
00004600 C    PRINT THE PARAMETERS OF THE PARABOLA
00004700      PRINT 50,A,B,C
00004800 50   FORMAT(' PARAMETERS OF THE BEST PARABOLA ARE',
00004900      1//,1H ,3F10.4)
00005000      STOP
00005100      END
00005200 C    THE LEAST SQUARES ROUTINE TO FIT THE PARABOLA
00005300      SUBROUTINE PARLSQ(X,Y,N,A,B,C)
00005400      COMMON FLAG
00005500      LOGICAL FLAG
00005600      REAL X(N),Y(N),T(3,4)
00005700      SUMX4=0
00005800      SUMX3=0
00005900      SUMX2=0
00006000      SUMX=0
```

```
00006100        SUMX2Y=0
00006200        SUMXY=0
00006300        SUMY=0
00006400 C      FIND THE VARIOUS SUMS I.E.
00006500 C      THE COEFFICIENTS OF THE EQUATIONS
00006600        DO 10 I=1,N
00006700        X2=X(I)*X(I)
00006800        SUMY=SUMY+Y(I)
00006900        SUMXY=SUMXY+X(I)*Y(I)
00007000        SUMX2Y=SUMX2Y+X2*Y(I)
00007100        SUMX=SUMX+X(I)
00007200        SUMX2=SUMX2+X2
00007300        SUMX3=SUMX3+X(I)*X2
00007400 10     SUMX4=SUMX4+X2*X2
00007500        T(1,1)=SUMX4
00007600        T(1,2)=SUMX3
00007700        T(1,3)=SUMX2
00007800        T(1,4)=SUMX2Y
00007900        T(2,1)=SUMX3
00008000        T(2,2)=SUMX2
00008100        T(2,3)=SUMX
00008200        T(2,4)=SUMXY
00008300        T(3,1)=SUMX2
00008400        T(3,2)=SUMX
00008500        T(3,3)=N
00008600        T(3,4)=SUMY
00008700        CALL JORDAN(T,3,4)
00008800        A=T(1,4)
00008900        B=T(2,4)
00009000        C=T(3,4)
00009100        IF(.NOT.FLAG) RETURN
00009200        PRINT 22
00009300 22     FORMAT(1H ,9X,'X',9X,'Y', ' DEVIATION',//)
00009400        DO 20 I=1,N
00009500        D=Y(I)-A*X(I)**2-B*X(I)-C
00009600 20     PRINT 25,X(I),Y(I),D
00009700 25     FORMAT(1H ,3F10.4)
00009800        RETURN
00009900        END
00010000 C      THIS IS THE SUBROUTINE TO SOLVE A SET OF LINEAR
00010100 C      EQUATIONS USING THE GAUSS-JORDAN METHOD.
00010200        SUBROUTINE JORDAN(A,N,M)
00010300        REAL A(N,M)
00010400        DO 10 K=1,N
00010500        K1=K+1
00010600        IF(K.EQ.N) GO TO 15
00010700        L=K
00010800        DO 20 J=K1,N
00010900 C      FIND THE LARGEST ELEMENT IN THE K-TH COLUMN AND
00011000 C      STORE IT AS ELEMENT L,K
00011100        IF(ABS(A(J,K)).GT.ABS(A(L,K))) L=J
00011200 20     CONTINUE
00011300        IF(L.EQ.K) GO TO 15
00011400 C      INTERCHANGE ROWS K AND L
00011500        DO 16  J=K,M
00011600        T=A(K,J)
00011700        A(K,J)=A(L,J)
00011800 16     A(L,J)=T
00011900 C      IF ANY DIAGONAL ELEMENTS ARE LESS THAN 1.0E-4
00012000 C      STOP THE PROCESS
```

PROGRAM 10   PAGE 3

```
00012100 15      DO 21 J=K1,M
00012200         IF(ABS(A(K,K)).LE.1.0E-4) GO TO 30
00012300 C       REDUCE THE K-TH ROW BY THE DIAGONAL ELEMENT
00012400 C       TO AVOID LARGE NUMBERS.
00012500 21      A(K,J)=A(K,J)/A(K,K)
00012600         IF(K.EQ.N) GO TO 10
00012700 C       PERFORM THE PIVOTAL REDUCTION
00012800         DO 12 I=K1,N
00012900         DO 12 J=K1,M
00013000 12      A(I,J)=A(I,J)-A(I,K)*A(K,J)
00013100 10      CONTINUE
00013200 C       PERFORM THE BACK SUBSTITUTION
00013300         DO 14 J=2,N
00013400         K=N+2-J
00013500         K1=K-1
00013600         DO 14 I=1,K1
00013700 14      A(I,M)=A(I,M)-A(I,K)*A(K,M)
00013800 C       SOLUTIONS ARE NOW IN A(I,M),I=1,N.
00013900 C       M=N+1,THE RIGHT-HAND-SIDE.
00014000         RETURN
00014100 30      PRINT 31
00014200 31      FORMAT('   NO SOLUTION EXISTS DETERMINANT =0')
00014300         RETURN
00014400         END
```

INPUT FOR PROGRAM 10

```
6    T
   1.0       2.0       3.0       4.0       5.0       6.0
   6.12      10.82     18.21     27.01     37.89     51.11
```

OUTPUT FROM PROGRAM 10

TO FIND THE PARAMETERS OF THE BEST
PARABOLA THROUGH A SERIES OF POINTS
            X   Y DEVIATION

    1.0000   6.1200   0.0572
    2.0000   10.8200   -0.2017
    3.0000   18.2100   0.2094
    4.0000   27.0100   0.0105
    5.0000   37.8900   -0.1283
    6.0000   51.1100   0.0530
PARAMETERS OF THE BEST PARABOLA ARE

    1.0100   1.9290   3.1237

**FORTRAN ** STOP

PROGRAM 11

Integration using the trapezium rule followed by Romberg integration. The integral

$$\int_a^b f(x)\,dx$$

is evaluated. This program demonstrates how extra accuracy can be obtained by using 'double precision' variables.

Program

The program consists of a main program, a subroutine which evaluates the integral and a function subprogram which defines the function. In this case

$$f(x) = \frac{1}{1+x^2}.$$

The subroutine continually halves the step-size until two estimates agree to the required accuracy. Up to 10 iterations are allowed. The function subprogram should be changed to define the specific function that the reader requires.

Subroutine Call

CALL TRAP $(A, B, EPS, RESULT)$,

A and B are the limits of the integral,
EPS is the required accuracy,
$RESULT$ is the value of the integral,
A, B, EPS are input parameters and $RESULT$ is an output parameter.

Data

A, B, EPS with a format $2F10.0, D10.0$.
The limits A and B each occupy ten columns and the accuracy is expressed in exponent form, e.g.

$$1 \cdot 0 \times 10^{-8} \quad \text{is written} \quad 0 \cdot 1D - 7.$$

Results

The results, which are annotated, consist of two lines the value of the integral, the accuracy and the limits.

Program 11. Integration using the trapezium rule followed by Romberg integration
(double precision).

PROGRAM 11   PAGE 1

```
00000100 C      THIS IS A DOUBLE PRECISION ROUTINE .
00000200 C      THIS IS A PROGRAM TO FIND THE INTEGRAL OF
00000300 C      F(X)DX FROM A TO B USING THE TRAPEZIUM RULE
00000400 C      FOLLOWED BY ROMBERG INTEGRATION .
00000500 C      THE DATA FOR THE PROGRAM IS MADE UP OF THE
00000600 C      VALUES OF THE LIMITS A AND B FOLLOWED BY
00000700 C      EPS THE REQUIRED ACCURACY .
00000800 C      THE LISTING CONSISTS OF A SMALL MAIN PROGRAM
00000900 C      FOLLOWED BY THE MAIN SUBPROGRAM WHICH
00001000 C      PERFORMS THE INTEGRATION AND FINALLY A
00001100 C      FUNCTION SUBPROGRAM DEFINING F(X).
00001200 C      DECLARE ALL THE REAL VARIABLES
00001300 C      TO BE DOUBLE PRECISION .
00001400        IMPLICIT REAL*8(A-H,O-Z)
00001500        PRINT 9
00001600 9      FORMAT(' THIS IS A DOUBLE PRECISION ROUTINE',//)
00001700        PRINT 10
00001800 10     FORMAT(' A PROGRAM TO FIND THE INTEGRAL OF')
00001900        PRINT 11
00002000 11     FORMAT(' F(X) USING THE TRAPEZIUM RULE ')
00002100        PRINT 12
00002200 12     FORMAT(' FOLLOWED BY ROMBERG INTEGRATION')
00002300 C      INPUT THE LIMITS OF INTEGRATION
00002400 C      AND THE DESIRED ACCURACY
00002500        READ 1,A,B,EPS
00002600 1      FORMAT(2F10.0,D10.0)
00002700 C      CALL THE SUBROUTINE TO EVALUATE THE INTEGRAL.
00002800        CALL TRAP(A,B,EPS,RESULT)
00002900        PRINT 2,RESULT,EPS
00003000 C      MODIFY F20.6,D20.6 IF NECESSARY
00003100 2      FORMAT(' VALUE OF INTEGRAL = ',F20.6,
00003200       1/,' TO AN ACCURACY',D20.6)
00003300        PRINT 3,A,B
00003400 3      FORMAT(' THE LIMITS ARE ',2F10.4)
00003500        STOP
00003600        END
00003700        SUBROUTINE TRAP(A,B,EPS,VALUE)
00003800 C       TRAPEZIUM RULE WITH ROMBERG INTEGRATION
00003900 C      DECLARE ALL THE REAL VARIABLES
00004000 C      TO BE DOUBLE PRECISION
00004100        IMPLICIT REAL*8 (A-H,O-Z)
00004200        REAL*8 VI(10,10)
00004300 C      SETUP THE INITIAL VALUES,N THE NUMBER OF STEPS,
00004400 C      NI THE NUMBER OF ITERATIONS,H THE STEP SIZE.
00004500        N=2
00004600        NI=2
00004700        H=(B-A)/N
00004800        NM1=N-1
00004900        VALUE=F(A)+F(B)
00005000 C      VI(I,J) IS THE TABLE OF VALUES.
00005100 C      VI(I,1) ARE GIVEN BY TRAPEZIUM RULE
00005200        VI(1,1)=VALUE/2
00005300        VALUE= VALUE+2*F(A+H)
00005400        VI(2,1)=VALUE/4
00005500        IDIV=4
00005600 C      TO FORM THE OTHER COLUMNS FROM THE FIRST
00005700 9      IFAC=4
00005800        DO 20 J=2,NI
00005900        I=NI+1-J
00006000 21     VI(I,J)=(IFAC*VI(I+1,J-1)-VI(I,J-1))/(IFAC-1)
```

PROGRAM 11   PAGE 2

```
00006100 20     IFAC=IFAC*4
00006200        ACC=DABS(EPS*VI(1,NI))
00006300 C      IS THE ESTIMATE ACCURATE ENOUGH?
00006400        IF(DABS(VI(1,NI)-VI(2,NI-1)).LE.ACC) GO TO 14
00006500        NI=NI+1
00006600        N=N*2
00006700 C      ARE THERE MORE ITERATIONS ALLOWED?
00006800        IF(NI.GT.10)GO TO 16
00006900        H=H/2
00007000        NM1=N-1
00007100        IDIV=IDIV*2
00007200        DO 13 K=1,N,2
00007300 13     VALUE =VALUE+2*F(A+K*H)
00007400        VI(NI,1)=VALUE/IDIV
00007500        GO TO 9
00007600 16     PRINT 17
00007700 17     FORMAT(' TERMINATED AFTER 10 ITERATIONS')
00007800 C      CALCULATE THE FINAL VALUE OF THE INTEGRAL.
00007900 14     VALUE=VI(1,NI)*(B-A)
00008000        RETURN
00008100        END
00008200 C      F(X) IS INTEGRATED FROM A TO B .
00008300 C      DECLARE THE FUNCTION TO BE DOUBLE PRECISION.
00008400        REAL FUNCTION F*8(X)
00008500 C      DECLARE THE ARGUMENT X,TO BE DOUBLE PRECISION.
00008600        REAL*8 X
00008700        F= 1.0/(1+X*X)
00008800        RETURN
00008900        END
```

INPUT FOR PROGRAM 11

     -1.0   1.0   1.0 D-6

OUTPUT FROM PROGRAM 11
THIS IS A DOUBLE PRECISION ROUTINE

A PROGRAM TO FIND THE INTEGRAL OF
F(X) USING THE TRAPEZIUM RULE
FOLLOWED BY ROMBERG INTEGRATION
VALUE OF INTEGRAL =   1.570796
TO AN ACCURACY   0.100000D-05
THE LIMITS ARE   -1.0000   1.0000

**FORTRAN ** STOP

PROGRAM 12

Integration using Gaussian quadrature and Robinson's method; the integral

$$\int_a^b f(x)\,dx$$

is evaluated.

Program

The program consists of a main program, a subroutine to evaluate the integral and a function subprogram, which defines the function. The ranges of the integrals are sub-divided in a special way in Robinson's Method, and a tree is used. There are some special variable transformations that need to be applied to store the subtotals. The process continues when the ranges are split up until two estimates agree to the required accuracy. The array allows up to four iterations but it can be changed to allow more. If the logical variable MORE is true on exit from the main subroutine it indicates more iterations are required. In this example $f(x) = \sin x$.

Subroutine Call

CALL ROBSON $(A, B, VALUE, EPS, MORE)$,

A and B are the limits of the integral,
VALUE is the value of the integral,
EPS is the required accuracy,
MORE is a logical variable,
A, B and *EPS* are input parameters and *VALUE* and *MORE* are output parameters.

Data

A, B, *EPS* with a format $3F10.0$.

Results

The results consist of several lines showing the intermediate estimates followed by the final value with the accuracy of the estimate. The limits of the integral are also printed. The results are annotated.

Program 12. Integration using Gaussian quadrature and Robinson's method.

```
00000100 C    THE PROGRAM PERFORMS GAUSSIAN QUADRATURE
00000200 C    USING THE GAUSS-LEGENDRE 3-POINT FORMULA AND
00000300 C    ROBINSON'S METHOD .THE INTEGRAL OF F(X)DX
00000400 C    IS EVALUATED IN THE RANGE A TO B, BY
00000500 C    DIVIDING THE RANGE SUCCESSIVELY INTO 3 PARTS
00000600 C    THE DATA CONSISTS OF THE LIMITS A AND B.
00000700 C    THE LISTING IS MADE UP OF A SMALL MAIN
00000800 C    PROGRAM, THE ROUTINE FOR EVALUATION OF THE
00000900 C    INTEGRAL AND A FUNCTION SUBPROGRAM TO
00001000 C    DEFINE THE FUNCTION F(X).
00001100      LOGICAL MORE
00001200      PRINT 5
00001300 5    FORMAT(' A PROGRAM TO PERFORM GAUSSIAN ')
00001400      PRINT 6
00001500 6    FORMAT(' QUADRATURE USING THE GAUSS-LEGENDRE')
00001600      PRINT 7
00001700 7    FORMAT(' 3-POINT FORMULA AND ROBINSONS METHOD')
00001800 C    INPUT THE LIMITS AND THE ACCURACY
00001900      READ 1,A,B,EPS
00002000 1    FORMAT(3F10.0)
00002100 C    CALL THE ROUTINE FOR INTEGRATION
00002200 C    USING ROBINSON'S METHOD
00002300      CALL ROBSON(A,B,VALUE,EPS,MORE)
00002400      IF(MORE) PRINT 10
00002500 10   FORMAT('  MORE ITERATIONS REQUIRED')
00002600      PRINT 2,VALUE,EPS
00002700 C    MODIFY F12.5,E20.4 IF NECESSARY
00002800 2    FORMAT(' VALUE OF INTEGRAL = ',F12.5,
00002900     1/,' TO AN ACCURACY',E20.4)
00003000      PRINT 3,A,B
00003100 3    FORMAT(' THE LIMITS ARE ',2F10.4)
00003200      STOP
00003300      END
00003400 C    THE SUBROUTINE USING ROBINSON'S METHOD
00003500      SUBROUTINE ROBSON(A,B,VALUE,EPS,MORE)
00003600      LOGICAL MORE
00003700 C    INCREASE TREE(15,3) TO TREE(127,3) IF REQUIRED
00003800      REAL TREE(15,3),ROOT(3),COEFF(3)
00003900 C    INITIALISE THE VALUES IN THE TREE
00004000      MORE =.FALSE.
00004100      TREE(1,1)=(B-A)/2
00004200      TREE(1,2)=(B+A)/2
00004300      DO 10 I=2,15
00004400 10   TREE(I,1)=0.5
00004500      DO 11 I=2,14,2
00004600      TREE(I,2)=0.5
00004700 11   TREE(I+1,2)=-0.5
00004800 C    SETUP THE COEFFICIENTS AND THE ROOTS
00004900      ROOT(1)=-SQRT(0.6)
00005000      ROOT(2) =0
00005100      ROOT(3)=-ROOT(1)
00005200      COEFF(1)=5.0/9
00005300      COEFF(2)=8.0/9
00005400      COEFF(3)=COEFF(1)
00005500      N=1
00005600      M1=1
00005700      VALUE =0.0
00005800 C    SUCCESSIVELY REDUCE THE SIZE OF THE RANGES
00005900 C    INCREASE THE CONSTANT 4 TO 7 FOR EXAMPLE
00006000 C    IF NECESSARY
```

PROGRAM 12   PAGE 2

```
00006100        DO 20 N=2,4
00006200        M1=M1*2
00006300        M2=2*M1-1
00006400        SUM=0
00006500        C=TREE(1,1)/M1
00006600        DO 30 I=M1,M2
00006700        SUM1=0
00006800        DO 40 K=1,3
00006900        J=I
00007000        ARG = ROOT(K)
00007100 41     ARG = TREE(J,1)*ARG+TREE(J,2)
00007200        J=J/2
00007300        IF(J.GT.0) GO TO 41
00007400 40     SUM1=SUM1+F(ARG)*COEFF(K)
00007500        TREE(I,3)=SUM1*C
00007600 C      FORM THE SUM OF THE INDIVIDUAL CONTRIBUTIONS
00007700 30     SUM=SUM+TREE(I,3)
00007800        N1=N-1
00007900        PRINT 2,N1,SUM
00008000 2      FORMAT(' INTERMEDIATE VALUE NO ',I5,' = ',F12.7)
00008100 C      MODIFY F12.7 IF NECESSARY
00008200        IF(ABS(VALUE-SUM).LE.EPS) GO TO 99
00008300        VALUE =SUM
00008400 20     CONTINUE
00008500        MORE =.TRUE.
00008600 99     RETURN
00008700        END
00008800 C      THE FUNCTION F(X)
00008900        FUNCTION F(X)
00009000        F=SIN(X)
00009100        RETURN
00009200        END
```

INPUT FOR PROGRAM 12

     0.0   3.14159   0.00001

OUTPUT FROM PROGRAM 12

```
A PROGRAM TO PERFORM GAUSSIAN
QUADRATURE USING THE GAUSS-LEGENDRE
3-POINT FORMULA AND ROBINSONS METHOD
INTERMEDIATE VALUE NO      1 =     2.0000143
INTERMEDIATE VALUE NO      2 =     1.9999990
INTERMEDIATE VALUE NO      3 =     1.9999971
VALUE OF INTEGRAL =        2.00000
TO AN ACCURACY            0.1000E-04
THE LIMITS ARE        0.0000     3.1416
```

**FORTRAN ** STOP

PROGRAM 13

Evaluation of a multiple integral over an n-dimensional sphere. The multiple integral

$$\int_{S_n} \cdots \int f(x_1 \ldots x_n) \, dx_1 \ldots dx_n$$

is evaluated.

The method uses a third degree formula and involves 2^n function evaluations.

Program

The program is made up of a main program, a subroutine to perform the evaluation, a function subprogram to define the function, in this case

$$f(x_1 \ldots x_n) = \exp(x_1 \times x_2 \ldots \times x_n)$$

and another function subprogram for the gamma function [see IV, § 10.2]. It is assumed that $N \leq 10$ if N is greater than ten the arrays need to be enlarged.

Subroutine Call

CALL MULTI2 (N, V),

N is the number of variables,
V is the value of the integral,
N is an input parameter and V is an output parameter.

Data

N with a format of I5.

Results

The results consist of two lines, both annotated giving the order and value of the integral.

Program 13. Evaluation of a multiple integral over an *n*-dimensional sphere.

PROGRAM 13   PAGE 1

```
00000100 C     THIS IS A PROGRAM TO FIND AN ESTIMATE OF THE
00000200 C     MULTIPLE INTEGRAL OF F(X1,X2,...XN)DX1...DXN
00000300 C     OVER AN N-DIMENSIONAL SPHERE.
00000400 C     THERE ARE 2 TO THE POWER N EVALUATIONS .
00000500 C     THE DATA CONSISTS OF N THE NUMBER OF INTEGRALS
00000600 C     THE LISTING CONSISTS OF THE SMALL MAIN PROGRAM
00000700 C     THE MAIN SUBROUTINE AND THE FUNCTION SUBPROGRAM
00000800 C     IN THIS CASE F(X1,X2,,XN)=FXP(X1*X2...*XN)
00000900 C     FINALLY THE GAMMA FUNCTION
00001000 C     INPUT N
00001100       PRINT 6
00001200 6     FORMAT(' A PROGRAM TO FIND AN ESTIMATE FOR A')
00001300       PRINT 7
00001400 7     FORMAT(' MULTIPLE INTEGRAL OVER AN N-D SPHERE')
00001500       READ 5,N
00001600 5     FORMAT(I5)
00001700       IF(N.GT.10) GO TO 10
00001800 C     CALL THE SUBROUTINE
00001900       CALL MULTI2(N,V)
00002000       PRINT 2,N
00002100 2     FORMAT(' THE ORDER OF THE INTEGRAL IS',I5)
00002200       PRINT 20,V
00002300 20    FORMAT('  VALUE OF INTEGRAL = ',F10.4)
00002400       STOP
00002500 10    PRINT 11
00002600 11    FORMAT(' N>10  AMEND PROGRAM ARRAYS')
00002700       STOP
00002800       END
00002900 C     THE MAIN SUBROUTINE
00003000       SUBROUTINE MULTI2(N,V)
00003100 C     GENERALISED TO DEAL WITH N UP TO 10
00003200       REAL A(2),X(10)
00003300       INTEGER SUB(10),P
00003400       R=1.0/SQRT(N+2.0)
00003500       A(1)=-R
00003600       A(2)=R
00003700       PI=4*ATAN(1.0)
00003800       V=0
00003900       DO 10 I=1,N
00004000 10    SUB(I)=1
00004100 15    P=1
00004200       DO 18 I=1,N
00004300       J=SUB(I)
00004400 18    X(I)=A(J)
00004500       V=V+F(X,N)
00004600 25    SUB(P)=SUB(P)+1
00004700       IF(SUB(P).LE.2) GO TO 15
00004800       SUB(P)=1
00004900       P=P+1
00005000       IF(P.LE.N) GO TO 25
00005100 C     APPLY THE FACTOR   TO THE SUM TO GIVE
00005200 C     THE VALUE OF THE INTEGRAL
00005300       V=V*2*SQRT(PI**N)/(N*GAMMA(N)*2**N)
00005400       RETURN
00005500       END
00005600 C     THE FUNCTION SUBPROGRAM
00005700       FUNCTION F(X,N)
00005800       REAL X(10)
00005900       PROD =1.0
00006000       DO 10 I=1,N
```

```
00006100 10       PROD =PROD*X(I)
00006200          F=EXP(PROD)
00006300          RETURN
00006400          END
00006500 C        THE GAMMA FUNCTION
00006600          FUNCTION GAMMA(N)
00006700          M=N
00006800          TERM=1.0
00006900          IF(N.NE.N/2*2) TERM=SQRT(4*ATAN(1.0))
00007000 5        M=M-2
00007100          X=M*0.5
00007200          IF(M.LT.1) GO TO 10
00007300          TERM=TERM*X
00007400          GO TO 5
00007500 10       GAMMA=TERM
00007600          RETURN
00007700          END
```

INPUT FOR PROGRAM 13

    3

OUTPUT FROM PROGRAM 13

```
A PROGRAM TO FIND AN ESTIMATE FOR A
MULTIPLE INTEGRAL OVER AN N-D SPHERE
THE ORDER OF THE INTEGRAL IS     3
  VALUE OF INTEGRAL =       4.2056
```


**FORTRAN ** STOP

PROGRAM 14

To solve the first order differential equation

$$\frac{dy}{dx} = f(x, y)$$

using the Runge–Kutta method and a fourth-order formula.

Program

The program consists of a small main program and a subroutine RK which evaluates $f(x, y)$ at $x = x_n$ using a certain step-size and then by halving the step-size up to five times repeats the procedure until two estimates agree to the required accuracy. The function subprogram defines $f(x, y)$ in this case $f(x, y) = \sin(x + y)$.

Subroutine Call

$$\text{CALL RK } (X1, Y1, XN, YN, H, EPS, N),$$

$X1$ is the initial value of x,
$Y1$ is the value of y at $x = x_1$,
XN is the final value of x,
YN is the value of y at $x = x_n$,
H is the initial step-size,
EPS is the required accuracy,
N is the number of steps from x_1 to x_n,
$X1$, $Y1$, XN, H, EPS and N are input parameters and YN is an output parameter.

Data

$X1$, $Y1$, XN, H, EPS, N with a format $5F10.0$, $I5$.

Results

The results consist of the intermediate values of y_n at x_n and are printed on successive lines. Each line gives the values of x_n, y_n, h and n.
The final estimate is given on a following line.

Program 14. To solve a first-order differential equation using the Runge–Kutta method and a fourth-order formula.

PROGRAM 14   PAGE 1

```
00000100 C     THIS PROGRAM SOLVES A FIRST ORDER DIFFERENTIAL
00000200 C     EQUATION OF THE FORM DY/DX = F(X,Y),
00000300 C     BY MEANS OF THE RUNGE-KUTTA METHOD AND USING
00000400 C     THE FOURTH ORDER FORMULA . IT IS ASSUMED THAT
00000500 C     THE STARTING VALUE OF Y IS KNOWN FOR X = X1,
00000600 C     THE INITIAL VALUE .THE VALUE OF Y IS DETERMINED
00000700 C     AT XN,THE METHOD IS TO DETERMINE AN ESTIMATE
00000800 C     FOR YN USING A STEP-SIZE OF H WHICH IS THEN
00000900 C     HALVED AND A NEW ESTIMATE FOUND .IF THE TWO
00001000 C     ESTIMATES AGREE WITHIN THE REQUIRED ACCURACY
00001100 C     THE PROCESS IS TERMINATED .
00001200 C     THE STEP-SIZE IS HALVED UP TO 5 TIMES .
00001300 C     THE DATA CONSISTS OF THE INITIAL VALUE OF X,X1
00001400 C     Y1 THE VALUE OF Y AT X1,XN THE FINAL VALUE OF
00001500 C     X,H THE STEP-SIZE,EPS THE REQUIRED ACCURACY AND
00001600 C     N THE NUMBER OF STEPS.
00001700 C     THE PROGRAM IS MADE UP OF A SHORT MAIN PROGRAM
00001800 C     A FUNCTION F(X,Y) AND THE MAIN SUBROUTINE RK TO
00001900 C     DETERMINE THE SOLUTION USING RUNGE-KUTTA.
00002000 C     INPUT THE DATA
00002100       PRINT 5
00002200 5     FORMAT(' TO SOLVE A FIRST ORDER DIFFERENTIAL')
00002300       PRINT 6
00002400 6     FORMAT(' EQUATION USING RUNGE-KUTTA')
00002500       READ 10,X1,Y1,XN,H,EPS,N
00002600 10    FORMAT(5F10.0,I5)
00002700 C     CALL THE RUNGE-KUTTA ROUTINE
00002800       CALL RK(X1,Y1,XN,YN,H,EPS,N)
00002900       PRINT 20,XN,YN
00003000 20    FORMAT('  VALUE OF Y AT ',F10.4,' IS ',F10.5)
00003100       STOP
00003200       END
00003300 C     THE FUNCTION F(X,Y) IN THIS CASE  SIN(X+Y)
00003400       FUNCTION F(X,Y)
00003500       F=SIN(X+Y)
00003600       RETURN
00003700       END
00003800 C     THE RUNGE-KUTTA ROUTINE
00003900       SUBROUTINE RK(X1,Y1,XN,YN,H,EPS,N)
00004000       REAL K1,K2,K3,K4,K
00004100 C     ABSURD INITIAL VALUE SET
00004200       ANS=1.0E4
00004300 C     ALLOW UP TO 5 HALVINGS OF H
00004400       PRINT 5
00004500 5     FORMAT(1H ,20X,'  NO.',9X,'X',9X,'Y',6X,
00004600      1'STEP',4X,'N')
00004700       DO 10 J=1,5
00004800       X=X1
00004900       Y=Y1
00005000       DO 20 I=1,N
00005100       K1=H*F(X,Y)
00005200       K2=H*F(X+H/2,Y+K1/2)
00005300       K3=H*F(X+H/2,Y+K2/2)
00005400       K4=H*F(X+H,Y+K3)
00005500       K=(K1+2*K2+2*K3+K4)/6
00005600       Y=Y+K
00005700 20    X=X+H
00005800       PRINT 15,J,X,Y,H,N
00005900 15    FORMAT('  INTERMEDIATE VALUES',I5,3F10.4,I5)
00006000 C     ANY MORE ITERATIONS?
```

PROGRAM 14   PAGE 2

```
00006100        IF(ABS(Y-ANS).LT.EPS) GO TO 30
00006200 C      RESET THE PARAMETERS
00006300        H=H/2
00006400        N=N*2
00006500        ANS=Y
00006600 10     CONTINUE
00006700 30     YN=Y
00006800        RETURN
00006900        END
```

INPUT FOR PROGRAM 14

   0.0   0.0   3.0   1.0   0.0001   3

OUTPUT FROM PROGRAM 14

TO SOLVE A FIRST ORDER DIFFERENTIAL
EQUATION USING RUNGE-KUTTA

	NO.	X	Y	STEP	N
INTERMEDIATE VALUES	1	3.0000	0.7755	1.0000	3
INTERMEDIATE VALUES	2	3.0000	0.7844	0.5000	6
INTERMEDIATE VALUES	3	3.0000	0.7851	0.2500	12
INTERMEDIATE VALUES	4	3.0000	0.7851	0.1250	24

VALUE OF Y AT   3.0000 IS   0.78509

**FORTRAN ** STOP-

PROGRAM 15

To solve the second-order differential equation

$$\frac{d^2y}{dx^2} = f\left(x, y, \frac{dy}{dx}\right)$$

using the Runge–Kutta method, the fourth-order formula and the substitution,

$$z = \frac{dy}{dx}$$

giving

$$\frac{dy}{dx} = g(x, y, z)$$

$$\frac{dz}{dx} = f(x, y, z).$$

Program

The listing consists of the main program which reads the data and prints the results, a subroutine RK2 which performs the solution process and two function subprograms to define f and g. The two differential equations must be solved simultaneously and the process is terminated when the required accuracy is achieved. Two successive estimates for both y and z are found by halving the step-size up to five times.

Subroutine Call

CALL RK2 $(X1, Y1, Z1, XN, YN, ZN, H, EPS, N)$,

$X1, Y1, Z1$ are the initial values of X, Y and Z respectively,
XN is the final value of x and YN and ZN are the values of y and $z = dy/dx$ at $x = XN$,
H is the initial step-size,
EPS is the required accuracy,
N is the number of steps,
$X1, Y1, Z1, XN, H, EPS$ and N are input parameters and YN and ZN are output parameters.

Data

$X1, Y1, Z1, XN, H, EPS, N$ with a format of $6F10.0, I5$.

Results

The annotated results show the intermediate values of $x, y, dy/dx$, the step-size and N before the final values are achieved.

Program 15. To solve a second-order differential equation using the Runge–Kutta
method and a fourth-order formula and a substitution.

```
00000100 C    THE PROGRAM SOLVES A SECOND ORDER DIFFERENTIAL
00000200 C    EQUATION OF THE FORM D2Y/DX2=F(X,Y,DY/DX)
00000300 C    A NEW VARIABLE Z IS CREATED WHERE Z=DY/DX AND
00000400 C    THE PROBLEM IS GENERALISED SO THAT THE TWO
00000500 C    EQUATIONS DY/DX=G(X,Y,Z)   DZ/DX=F(X,Y,Z)
00000600 C    ARE SOLVED SIMULTANEOUSLY USING THE METHOD OF
00000700 C    RUNGE-KUTTA AND THE FOURTH ORDER FORMULA AS IN
00000800 C    THE CASE OF THE FIRST ORDER EQUATION.IT IS
00000900 C    ASSUMED THAT STARTING VALUES OF Y AND Z ARE
00001000 C    KNOWN AT X=X1.THE VALUES OF Y AND Z ARE FOUND
00001100 C    AT X=XN.THE METHOD TAKES AN INITIAL VALUE OF H
00001200 C    FINDS VALUES FOR Y AND Z,HALVES H FINDS FURTHER
00001300 C    VALUES FOR Y AND Z.IF THE VALUES OF Y AND OF Z
00001400 C    AGREE TO THE REQUIRED ACCURACY THE PROCESS IS
00001500 C    TERMINATED OTHERWISE THE VALUE OF H IS HALVED
00001600 C    AGAIN (UP TO 5 TIMES).
00001700 C    THE DATA CONSISTS OF X1,Y1,Z1,INITIAL VALUES OF
00001800 C    X,YAND Z,FOLLOWED BY XN,H THE STEP-SIZE AND N
00001900 C    THE NUMBER OF STEPS.
00002000 C    THE LISTING CONSISTS OF A MAIN PROGRAM A
00002100 C    SUBROUTINE RK2 WHICH FINDS THE SOLUTION AND
00002200 C    FUNCTION SUBPROGRAMS WHICH DEFINE F AND G.
00002300      PRINT 1
00002400 1    FORMAT(' TO SOLVE A SECOND ORDER DIFFERENTIAL')
00002500      PRINT 2
00002600 2    FORMAT(' EQUATION USING RUNGE-KUTTA')
00002700 C    INPUT THE DATA
00002800      READ 10,X1,Y1,Z1,XN,H,EPS,N
00002900 10   FORMAT(6F10.0,I5)
00003000 C    CALL THE RUNGE-KUTTA ROUTINE FOR 2 EQUATIONS
00003100      CALL RK2(X1,Y1,Z1,XN,YN,ZN,H,EPS,N)
00003200      PRINT 20,XN,YN,ZN
00003300 20   FORMAT(' VALUE OF Y AT',F10.4,' =',F10.5,
00003400     1/,' VALUE OF DY/DX =',F10.5)
00003500      STOP
00003600      END
00003700 C    THE FUNCTION F(X,Y,Z)
00003800      FUNCTION F(X,Y,Z)
00003900      F=SIN(X+Y+Z)
00004000      RETURN
00004100      END
00004200 C    THE FUNCTION G(X,Y,Z)
00004300      FUNCTION G(X,Y,Z)
00004400      G=Z
00004500      RETURN
00004600      END
00004700 C    THE RUNGE-KUTTA ROUTINE
00004800      SUBROUTINE RK2(X1,Y1,Z1,XN,YN,ZN,H,EPS,N)
00004900      REAL K,K1,K2,K3,K4,L,L1,L2,L3,L4
00005000 C    ABSURD INITIAL VALUES
00005100      ANS1=1.0E4
00005200      ANS2=1.0E4
00005300      PRINT 1
00005400 1    FORMAT(' INTERMEDIATE VALUES OF X,Y,DY/DX ')
00005500      PRINT 5
00005600 5    FORMAT('    NO.',9X,'X',9X,'Y',5X,'DY/DX',6X,
00005700     1'STEP',4X,'N')
00005800 C    HALVE H UP TO 5 TIMES
00005900      DO 10 J=1,5
00006000      X=X1
```

PROGRAM 15

```
00006100         Y=Y1
00006200         Z=Z1
00006300 C       MARCH FROM X=X1 TO X=XN
00006400         DO 20 I=1,N
00006500         K1=H*G(X,Y,Z)
00006600         L1=H*F(X,Y,Z)
00006700         K2=H*G(X+H/2,Y+K1/2,Z+L1/2)
00006800         L2=H*F(X+H/2,Y+K1/2,Z+L1/2)
00006900         K3=H*G(X+H/2,Y+K2/2,Z+L2/2)
00007000         L3=H*F(X+H/2,Y+K2/2,Z+L2/2)
00007100         K4=H*G(X+H,Y+K3,Z+L3)
00007200         L4=H*F(X+H,Y+K3,Z+L3)
00007300         K=(K1+2*K2+2*K3+K4)/6
00007400         L=(L1+2*L2+2*L3+L4)/6
00007500         X=X+H
00007600         Y=Y+K
00007700 20      Z=Z+L
00007800         PRINT 15,J,X,Y,Z,H,N
00007900 15      FORMAT(1H ,I5,4F10.4,I5)
00008000 C       ANY MORE ITERATIONS REQUIRED?
00008100         IF(ABS(Y-ANS1).LE.EPS.AND.ABS(Z-ANS2).LE.EPS)
00008200 1       GO TO 30
00008300 C       YES ADJUST VARIABLES
00008400         H=H/2
00008500         N=N*2
00008600         ANS1=Y
00008700         ANS2=Z
00008800 10      CONTINUE
00008900 30      YN=Y
00009000         ZN=Z
00009100         RETURN
00009200         END
```

INPUT FOR PROGRAM 15

```
    0.0         1.0         0.0        1.0        0.5       0.001      2
    2.0         2.0         2.0
0.66667     1.66667     1.66667
    1.0         2.5         5.5
```

OUTPUT FROM PROGRAM 15

```
TO SOLVE A SECOND ORDER DIFFERENTIAL
EQUATION USING RUNGE-KUTTA
INTERMEDIATE VALUES OF X,Y,DY/DX
  NO.         X           Y         DY/DX      STEP      N
   1        1.0000      1.4302      0.7034     0.5000     2
   2        1.0000      1.4308      0.7035     0.2500     4
VALUE OF Y AT      1.0000 =      1.43078
VALUE OF DY/DX =      0.70355
```

**FORTRAN ** STOP

PROGRAM 16

To solve the Volterra integral equation of the second kind expressed in the form

$$\phi(x) = f(x) + \int_0^x k(x, y)\phi(y)\, dy \qquad x > 0$$

using a multi-step method.

Program

The program consists of a main program which reads the data and calls the main control subroutine VOLTRA. There are four subroutines and four function subprograms. The subroutine VOLTRA calls the routines DAY, for the starting procedure, SIMSON to perform quadrature using the $\frac{3}{8}$th's rule when the number of intervals are odd. The functions F, G, $DENOM$ and $EVEN$ define $f(x)$, $k(x, y)$. In this case $f(x) = x$, $k(x, y) = xy/5$ and the functions $DENOM$ and $EVEN$ form the denominator in the expression for $\phi(x)$ and check whether I is odd or even, respectively.

Subroutine Call

$$\text{CALL VOLTRA } (PHI, H, N),$$

PHI is an array which stores values of the function at desired points,
H is the step-size to give the intervals,
N is the number of steps,
PHI is an output parameter and N and H are input parameters.

Data

H and N with a format $F10.0$, $I5$.

Results

A table of $N + 1$ values showing the values of x and $\phi(x)$ at the $N + 1$ points $0, h, 2h \ldots nh$.

Program 16. To solve the Volterra integral equation of the second kind using a multi-step method.

PROGRAM 16   PAGE 1

```
00000100 C     THIS PROGRAM SOLVES A VOLTERRA INTEGRAL
00000200 C     EQUATION OF THE SECOND KIND USING A
00000300 C     MULTI-STEP METHOD.THE METHOD USES DAY'S
00000400 C     STARTING PROCEDURE IN CONJUNCTION WITH
00000500 C     SIMPSON'S RULE AND THE THREE-EIGHTH'S RULE
00000600 C     THE VALUES OF THE FUNCTION PHI ARE FOUND AT
00000700 C     THE POINTS 0,H,2H,...NH WHERE N AND H ARE
00000800 C     READ AS DATA. THE VALUES OF PHI(0),PHI(H),
00000900 C     PHI(2H),PHI(3H)ARE FOUND USING DAY'S
00001000 C     STARTING PROCEDURE.THE VALUES OF PHI(MH)
00001100 C     M=4,N ARE FOUND USING SIMPSON'S RULE AND
00001200 C     THE THREE-EIGHTH'S RULE IF M IS ODD.
00001300 C     THE DATA CONSISTS OF H THE STEP-SIZE AND N
00001400 C     THE NUMBER OF STEPS.
00001500 C     THERE IS A MAIN PROGRAM 4 SUBROUTINES AND
00001600 C     4 FUNCTION SUBPROGRAMS. THE FIRST SUBROUTINE
00001700 C     VOLTRA IS A CONTROL ROUTINE WHICH CALLS THE
00001800 C     OTHER SUBROUTINES DAY,WHICH SETS UP THE
00001900 C     STARTING PROCEDURE,SIMSON TO PERFORM
00002000 C     QUADRATURE USING SIMPSON'S RULE AND THREE
00002100 C     USING THE THREE-EIGHTH'S RULE. THE FUNCTIONS
00002200 C     ARE F(X),THE KERNEL,G(X),A LOGICAL FUNCTION
00002300 C     EVEN TO CHECK WHETHER I IS EVEN AND DENOM
00002400 C     TO EVALUATE THE DENOMINATOR IN THE
00002500 C     EXPRESSION FOR PHI(X).
00002600       REAL PHI(30)
00002700       PRINT 1
00002800 1     FORMAT(' A PROGRAM TO SOLVE THE VOLTERRA',
00002900      1' INTEGRAL EQUATION')
00003000       PRINT 2
00003100 2     FORMAT(' OF THE SECOND KIND USING DAYS',
00003200      1' STARTING PROCEDURE')
00003300 C     INPUT THE STEP-SIZE AND NUMBER OF STEPS
00003400       PRINT 5
00003500 5     FORMAT(1H ,9X,'X',6X,'PHI(X)',/)
00003600       READ 3,H,N
00003700 3     FORMAT(F10.0,I5)
00003800 C     CALL THE CONTROL ROUTINE
00003900       CALL VOLTRA(PHI,H,N)
00004000       N1=N+1
00004100 C     PRINT THE TABLE OF RESULTS
00004200       DO 10 I=1,N1
00004300       X=H*(I-1)
00004400 10    PRINT 11,X,PHI(I)
00004500 11    FORMAT(1H ,F10.2,F12.6)
00004600       STOP
00004700       END
00004800 C     THE CONTROL ROUTINE
00004900       SUBROUTINE VOLTRA(PHI,H,N)
00005000       LOGICAL EVEN
00005100       REAL PHI(30)
00005200 C     CALL THE STARTING PROCEDURE ROUTINE
00005300       CALL DAY(PHI,H)
00005400       DO 10 I=4,N
00005500       SUM= F(I*H)
00005600       A=0.0
00005700       K=0
00005800 C     IS I EVEN ? IF SO ONLY CALL SIMPSON'S RULE
00005900 C     OTHERWISE CALL SIMPSON'S RULE
00006000 C     AND THE THREE-EIGHTH'S RULE
```

```
00006100          IF(EVEN(I)) GO TO 20
00006200 C        ODD I
00006300          J=I-3
00006400 C        CALL SIMPSON'S RULE FOR THE FIRST N-3 INTERVALS
00006500 15       CALL SIMSON(A,H,I,SUM,K,PHI)
00006600          K=K+2
00006700          A=A+2*H
00006800          IF(K.LT.J) GO TO 15
00006900 C        CALL THE THREE-EIGHTH'S RULE AT THE END
00007000          CALL THREE(A,H,I,SUM,K,PHI)
00007100          V=SUM/DENOM(0.375,I,H)
00007200          GO TO 10
00007300 C        EVEN I
00007400 20       CALL SIMSON(A,H,I,SUM,K,PHI)
00007500          K=K+2
00007600          A=A+2*H
00007700          IF(K.LT.I) GO TO 20
00007800          V=SUM/DENOM(1.0/3,I,H)
00007900 10       PHI(I+1)=V
00008000          RETURN
00008100          END
00008200 C        THE STARTING PROCEDURE
00008300          SUBROUTINE DAY(PHI,H)
00008400          REAL PHI(30)
00008500 C        COMPUTE THE FIRST 4 VALUES OF PHI
00008600 C        ACCORDING TO THE FORMULAE
00008700          GHO=G(H,0.0)
00008800          F0= F(0.0)
00008900          PHI(1)=F0
00009000          FH=F(H)
00009100          PHI11=FH +H*GHO*F0
00009200          PHI12=FH+H/2*(GHO*F0+G(H,H)*PHI11)
00009300          PHI13=F(H/2)*H/4*(G(H/2,0.0)*F0+G(H/2,H/2)
00009400         1*(F0+PHI12)/2)
00009500          PHI(2)=FH+H/6*(GHO*F0+4*G(H,H/2)*PHI13+G(H,H)
00009600         1*PHI12)
00009700          PHI21=F(2*H)+2*H*G(2*H,H)*PHI(2)
00009800          PHI(3)=F(2*H)+H/3*(G(2*H,0.0)*F0+4*G(2*H,H)
00009900         1*PHI(2)+G(2*H,2*H)*PHI21)
00010000          PHI31=F(3*H)+1.5*H*(G(3*H,H)*PHI(2)+G(3*H,2*H)
00010100         1*PHI(3))
00010200          PHI(4)=F(3*H)+0.375*H*(G(3*H,0.0)*F0+3*G(3*H,H)
00010300         1*PHI(2)+3*G(3*H,2*H)*PHI(3)+G(3*H,3*H)
00010400         2*PHI31)
00010500          RETURN
00010600          END
00010700 C        THE FUNCTION F(X)
00010800          FUNCTION F(X)
00010900          F=X
00011000          RETURN
00011100          END
00011200 C        THE KERNEL K(X)IN THE TEXT HERE G(X)
00011300          FUNCTION G(X,Y)
00011400          G=X*Y/5
00011500          RETURN
00011600          END
00011700 C        THE SPECIAL ROUTINE TO PERFORM SIMPSON'S RULE
00011800 C        IT ONLY COMPUTES THE CONTRIBUTION FOR TWO
00011900 C        INTERVALS AT A TIME BECAUSE OF THE
00012000 C        VARIABILITY OF THE NUMBER  OF CALLS AND
```

PROGRAM 16   PAGE 3

```
00012100 C      THE PARITY OF I
00012200        SUBROUTINE SIMSON(A,H,I,SUM,K,PHI)
00012300        REAL PHI(30)
00012400        X=I*H
00012500        IF(K.NE.I-2) GO TO 10
00012600        SUM=SUM+(G(X,A)*PHI(K+1)+4*G(X,A+H)*PHI(K+2))
00012700       1*H/3
00012800        RETURN
00012900 10     SUM=SUM+(G(X,A)*PHI(K+1)+4*G(X,A+H)*PHI(K+2)
00013000       1+G(X,A+2*H)*PHI(K+3))*H/3
00013100        RETURN
00013200        END
00013300 C      THE THREE-EIGHTH'S RULE ROUTINE
00013400        SUBROUTINE THREE(A,H,I,SUM,K,PHI)
00013500        REAL PHI(30)
00013600        X=I*H
00013700        SUM=SUM+(G(X,A)*PHI(K+1)+3*G(X,A+H)*PHI(K+2)
00013800       1+3*G(X,A+2*H)*PHI(K+3))*H*0.375
00013900        RETURN
00014000        END
00014100 C      THE FUNCTION TO CALCULATE THE DENOMINATOR
00014200 C      IN THE EXPRESSION FOR PHI(X) ALLOWS FOR
00014300 C      THE PARITY OF I
00014400 C      FOR PHI(X) ALLOWS FOR THE PARITY OF I
00014500        FUNCTION DENOM(C,I,H)
00014600        DENOM=1-C*H*G(I*H,I*H)
00014700        RETURN
00014800        END
00014900 C      CHECKS THE PARITY OF I
00015000 C      TRUE IF EVEN - FALSE IF ODD
00015100        LOGICAL FUNCTION EVEN(I)
00015200        EVEN= I.EQ.I/2*2
00015300        RETURN
00015400        END
```

INPUT FOR PROGRAM 16

    0.1   20

OUTPUT FROM PROGRAM 16

A PROGRAM TO SOLVE THE VOLTERRA INTEGRAL EQUATION
OF THE SECOND KIND USING DAYS STARTING PROCEDURE

X	PHI(X)
0.00	0.000000
0.10	0.100003
0.20	0.200107
0.30	0.300541
0.40	0.401710
0.50	0.504184
0.60	0.608703
0.70	0.716191
0.80	0.827778
0.90	0.944821
1.00	1.068937
1.10	1.202065
1.20	1.346514
1.30	1.505055
1.40	1.681025
1.50	1.878486
1.60	2.102380
1.70	2.358824
1.80	2.655379
1.90	3.001534
2.00	3.409217

**FORTRAN ** STOP

PROGRAM 17

To solve the Fredholm integral equation expressed in the form

$$\phi(x) = f(x) + \int_0^1 k(x, y)\phi(y)\, dy \qquad 0 < x < 1.$$

The values of $\phi(x)$ are found at $x = 0, \frac{1}{4}, \frac{1}{2}, \frac{3}{4}$ and 1.

Program

The program consists of a very small main program, three subroutines and two function subprograms. The first subroutine FRED is the control subroutine which calls the second subroutine FREDHM, and the third subroutine JORDAN. FREDHM forms the coefficients of the linear equation and JORDAN as previously described solves them using Gaussian elimination. The functions F and G define $f(x)$ and $K(x, y)$, in this case $f(x) = x$ and $K(x, y) = 1 + xy$.

Subroutine Call

$$\text{CALL FRED,}$$

with no parameters.

Data

There is no data as the standard form of the equation is used.

Results

The annotated results give the values of the function at the points

$$x = 0, \tfrac{1}{2} \text{ and } 1$$

and then at

$$x = 0, \tfrac{1}{4}, \tfrac{1}{2}, \tfrac{3}{4} \text{ and } 1.$$

Program 17. To solve the Fredholm integral equation.

```
00000100 C    THIS PROGRAM SOLVES A FREDHOLM INTEGRAL
00000200 C    EQUATION USING SIMPSON'S RULE WITH TWO
00000300 C    STEP-SIZES H=0.5 AND H=0.25.THE VALUES
00000400 C    OF PHI(0),PHI(0.5) AND PHI(1) ARE FORMED BY
00000500 C    SOLVING A SET OF 3 LINEAR EQUATIONS USING
00000600 C    THE GAUSS-JORDAN METHOD AND SUBROUTINE.
00000700 C    A SIMILAR SET OF 5 VALUES OF PHI USING H=0.25
00000800 C    ARE FORMED BY SOLVING A SET OF 5 LINEAR
00000900 C    EQUATIONS. THE METHOD COULD BE EXTENDED TO
00001000 C    CATER FOR SMALLER VALUES OF H.MORE ACCURATE
00001100 C    VALUES OF PHI CAN BE DETERMINED USING THE
00001200 C    ROMBERG TECHNIQUE.
00001300 C    THERE IS NO DATA FOR THIS PROGRAM AS THE
00001400 C    STANDARD FORM FOR THE EQUATION IS CHOSEN.
00001500 C    THERE IS A VERY SMALL MAIN PROGRAM THREE
00001600 C    SUBROUTINES AND TWO FUNCTION SUBPROGRAMS.
00001700 C    THE FIRST SUBROUTINE FRED IS A CONTROL
00001800 C    ROUTINE WHICH CALLS THE SECOND SUBROUTINE
00001900 C    FREDHM WHICH FORMS THE COEFFICIENTS OF THE
00002000 C    LINEAR EQUATIONS.IT ALSO CALLS THE SUBROUTINE
00002100 C    JORDAN WHICH SOLVES THE EQUATIONS.THE
00002200 C    FUNCTIONS ARE F(X) AND THE KERNEL G(X).
00002300      PRINT 1
00002400 1    FORMAT(' A PROGRAM TO SOLVE THE FREDHOLM',
00002500     1' INTEGRAL EQUATION')
00002600      PRINT 2
00002700 2    FORMAT(' OF THE SECOND KIND USING COMPOSITE',
00002800     1' SIMPSONS RULE')
00002900      CALL FRED
00003000      STOP
00003100      END
00003200 C    THE CONTROL ROUTINE
00003300      SUBROUTINE FRED
00003400      REAL V(3),W(5),C(5),PHI(5)
00003500 C    SET UP THE CONSTANTS FOR N=3
00003600      V(1)=0.0
00003700      V(2)=0.5
00003800      V(3)=1.0
00003900      C(1)=1.0/6
00004000      C(2)=4*C(1)
00004100      C(3)=C(1)
00004200      N=3
00004300 C    CALL THE FREDHOLM ROUTINE WITH H=0.5
00004400      CALL FREDHM(V,C,PHI,N)
00004500      PRINT 15,(V(I),I=1,N)
00004600 15   FORMAT(' AT THE POINTS ',/,6F10.3)
00004700      PRINT 20,(PHI(I),I=1,N)
00004800 20   FORMAT(' SOLUTION IS GIVEN BELOW',/,6F12.6)
00004900 C    SET UP THE CONSTANTS WITH N=5
00005000      N=5
00005100      W(1)=0.0
00005200      W(2)=0.25
00005300      W(3)=0.5
00005400      W(4)=0.75
00005500      W(5)=1.0
00005600      C(1)=1.0/12
00005700      C(2)=4*C(1)
00005800      C(3)=2*C(1)
00005900      C(4)=C(2)
00006000      C(5)=C(1)
```

```
00006100 C     CALL THE FREDHOLM ROUTINE WITH H=0.25
00006200       CALL FREDHM(W,C,PHI,N)
00006300       PRINT 15,(W(I),I=1,N)
00006400       PRINT 20,(PHI(I),I=1,N)
00006500       STOP
00006600       END
00006700 C     THE FREDHOLM ROUTINE
00006800       SUBROUTINE FREDHM(V,C,PHI,N)
00006900       COMMON A(5,6)
00007000       REAL V(5),C(5),PHI(5)
00007100 C     CALCULATE THE COEFFICIENTS
00007200 C     OF THE LINEAR EQUATIONS
00007300       DO 10 I=1,N
00007400       A(I,N+1)=F(V(I))
00007500       DO 20 J=1,N
00007600 20    A(I,J)=-C(J)*G(V(I),V(J))
00007700 10    A(I,I)=A(I,I)+1
00007800       N1=N+1
00007900 C     CALL JORDAN TO SOLVE THE LINEAR EQUATIONS
00008000       CALL JORDAN(N,N+1)
00008100       DO 50 I=1,N
00008200 50    PHI(I)=A(I,N+1)
00008300       RETURN
00008400       END
00008500 C     THE FUNCTION F(X)
00008600       FUNCTION F(X)
00008700       F=EXP(X)
00008800       RETURN
00008900       END
00009000 C     THE FUNCTION K(X,Y) KNOWN AS G(X,Y)
00009100       FUNCTION G(X,Y)
00009200       G=1+X*Y
00009300       RETURN
00009400       END
00009500 C     THE GAUSS-JORDAN ROUTINE
00009600       SUBROUTINE JORDAN(N,M)
00009700       COMMON A(5,6)
00009800       DO 10 K=1,N
00009900       K1=K+1
00010000       IF(K.EQ.N) GO TO 15
00010100       L=K
00010200       DO 20 J=K1,N
00010300 C     FIND THE LARGEST ELEMENT IN THE K-TH COLUMN
00010400 C     AND STORE IT AS ELEMENT L,K
00010500       IF(ABS(A(J,K)).GT.ABS(A(L,K))) L=J
00010600 20    CONTINUE
00010700       IF(L.EQ.K) GO TO 15
00010800 C     INTERCHANGE ROWS K AND L
00010900       DO 16  J=K,M
00011000       T=A(K,J)
00011100       A(K,J)=A(L,J)
00011200 16    A(L,J)=T
00011300 C     IF ANY DIAGONAL ELEMENTS ARE < 1.0E-4
00011400 C     STOP THE PROCESS
00011500 15    DO 21 J=K1,M
00011600       IF(ABS(A(K,K)).LE.1.0E-4) GO TO 30
00011700 C     REDUCE THE K-TH ROW BY THE DIAGONAL ELEMENT
00011800 C     TO AVOID BIG NUMBERS
00011900 21    A(K,J)=A(K,J)/A(K,K)
00012000       IF(K.EQ.N) GO TO 10
```

```
00012100 C     PERFORM THE PIVOTAL REDUCTION
00012200       DO 12 I=K1,N
00012300       DO 12 J=K1,M
00012400 12    A(I,J)=A(I,J)-A(I,K)*A(K,J)
00012500 10    CONTINUE
00012600 C     PERFORM THE BACK SUBSTITUTION
00012700       DO 14 J=2,N
00012800       K=N+2-J
00012900       K1=K-1
00013000       DO 14 I=1,K1
00013100 14    A(I,M)=A(I,M)-A(I,K)*A(K,M)
00013200 C     SOLUTIONS ARE NOW IN A(I,M),I=1,N
00013300 C     M=N+1, THE RIGHT-HAND-SIDE.
00013400       RETURN
00013500 30    PRINT 31
00013600 31    FORMAT('   NO SOLUTION EXISTS DETERMINANT =0')
00013700       RETURN
00013800       END
```

```
OUTPUT FROM PROGRAM 17
A PROGRAM TO SOLVE THE FREDHOLM INTEGRAL EQUATION
OF THE SECOND KIND USING COMPOSITE SIMPSONS RULE
AT THE POINTS
     0.000      0.500      1.000
SOLUTION IS GIVEN BELOW
   -5.588865   -6.659006   -7.308304
AT THE POINTS
     0.000      0.250      0.500      0.750      1.000
SOLUTION IS GIVEN BELOW
   -5.582526   -6.157662   -6.652127   -7.043011   -7.300891


**FORTRAN ** STOP
```

Index